MARINE FOOD CHAINS

MARINE
FOOD
CHAINS

Edited by J. H. STEELE

OLIVER & BOYD
EDINBURGH

Reprint
by

 OTTO KOELTZ ANTIQUARIAT
Koenigstein-Ts./B.R.D.
1973

Printed in Western Germany
All rights of this reprint edition reserved by Otto Koeltz
ISBN 3-87429-047-6

Preface

This volume contains the proceedings of a Symposium held at the University of Aarhus, Denmark, from 23rd to 26th July 1968. The Symposium was held under the auspices of the International Council for the Exploration of the Sea (ICES), with the support of the:

Food and Agriculture Organization of the United Nations (FAO), International Commission for the Northwest Atlantic Fisheries (ICNAF), United Nations Educational, Scientific and Cultural Organization (UNESCO).

The decision to convene the Symposium originated with a recommendation of the FAO Advisory Committee on Marine Resources Research, which was subsequently endorsed by the other organizations concerned. The Symposium is regarded, additionally, as a contribution to the objectives of the International Biological Program (IBP), being conducted by the Special Committee for the IBP of the International Council of Scientific Unions, and that committee therefore co-sponsored it.

The organizing committee consisted of:

J. H. Steele, Marine Laboratory, Aberdeen, Scotland (Convener), D. H. Cushing, Fisheries Laboratory, Lowestoft, U.K., L. M. Dickie, Bedford Institute of Oceanography, Canada, G. Hempel, Institut für Meereskunde, Kiel, Germany, Mario Ruivo, Food and Agriculture Organization of the United Nations, L. B. Slobodkin, State University of New York at Stony Brook, U.S.A., J. D. H. Strickland, University of California, La Jolla, U.S.A., M. E. Vinogradov, Institute of Oceanology, Academy of Sciences of the U.S.S.R.

The purpose of the Symposium was to bring together those people interested in different aspects of marine food chains.

At present there is great interest in attempts to study the interrelation between different trophic levels, particularly in terms of the energy processes

v

involved. Our understanding of these processes in the sea is least in the intermediate levels between primary production and the commercial exploitation of fishery resources. Yet the assessment of possible new fisheries and the rational use of existing stocks must depend on our knowledge of the food webs of which these fisheries are part. Any attempt to alter the environment of selected areas for the cultivation and intensive harvesting of marine resources will require not only descriptions of the structure of the food webs but also detailed knowledge of their energetics. Similarly the control of pollution is based on insight into the changes that can take place in the pattern of food chains. In many cases there must be similarities in the definition and structure of these problems even though the species studied or the areas worked in appear very different. A comparison of research from diverse sources should be stimulating and valuable, and should lead to a better understanding of those features which aquatic food chains have in common. The assessment of the possibilities for generalizations to be made in this field may be considered as the main theme of this volume.

Because these problems are of topical interest I have tried to ensure that publication was as rapid as possible and must thank all the contributors for their help in this, particularly those members of the organizing committee who assisted with the work of editing. I would also like to thank Mr A. Eleftheriou who prepared the indexes, Mrs D. M. Crockett who helped with the proof reading, and Mrs C. M. Reid for her secretarial work.

JOHN STEELE

Contents

vii

Part Three FEEDING MECHANISMS

Part Four FOOD REQUIREMENTS FOR FISH PRODUCTION

Part Five FOOD ABUNDANCE AND AVAILABILITY IN RELATION TO PRODUCTION

Part Six THEORETICAL PROBLEMS

Part Seven SUMMARY OF THE SYMPOSIUM L. B. Slobodkin

Part One

RECYCLING OF ORGANIC MATTER

Introduction

J. D. H. STRICKLAND

The "classical" marine food web, wherein photosynthetically produced organic matter in the form of phytoplankton is consumed by herbivorous zooplankters which are, in turn, consumed by carnivores, has been postulated and documented for well over half a century. The tropho-dynamics of these processes comprise much of the subject matter of the present symposium. Nothing has been discovered in recent years to destroy our faith in the basic correctness of the classically conceived food chain, but we are becoming more aware of the fact that, when matter or energy is transferred from one trophic level to another, the procedure is not 100% efficient. Organic matter is lost in a dissolved, colloidal or particulate form and the fate of this material has begun to attract some attention.

The most obvious loss is as faeces, much of which quickly disintegrates to give finely divided particulate matter. Soluble and colloidal material is leached from faeces and is excreted by animals but the greatest amount probably originates from the soluble organic matter produced either from crushed or lysed plant cells or actively excreted by the phytoplankton during growth.

A considerable literature has accumulated in the past decade on the decretion of organic matter, but the picture is still confused and the amount excreted, as a fraction of the amount being photosynthesized, is very variable, expending both on species and growth conditions. Few authors, however, would suggest that this fraction exceeded 10% and, as the excreted material can scarcely be converted to fresh living matter by processes more efficient than 30%, we are only talking about, at most, a few per cent of the initial fixed energy being deflected from the main pathway and used elsewhere in the food chain. The subject is of little importance unless one is considering an

3

organism or community which depends on this small fraction for its sustenance. The most obvious examples of these are coprophagic animals, the bacteria and heterotrophic protozoa.

Particulate detritus, some of which may be formed from "soluble" material under the influence of bacteria and air bubbles, by mechanisms still imperfectly understood, is clearly of little use most of the time to most of the filter-feeding zooplankters. If it were, it would not persist in the water in concentrations many times greater than that of the living plant plankton. (This discrimination of zooplankters against organic detritus deserves more study.) Organic detritus, however, has been implicated as a foodstuff, in particular a source of nitrogen, for over-wintering stocks of animals or as a possible augmentation of plant food when the latter is particularly scarce. In relatively high concentrations soluble matter may be adsorbed through the integuments of some bottom dwelling animals, or absorbed from swallowed water in amounts sufficient to provide critical growth factors. Direct experimental evidence is scarce and sometimes ambiguous.

In shallow productive regions much of the primary product sinks to the bottom and reaches the sea bed largely unchanged or in the form of fresh and as yet undisintegrated faecal material. This forms a rich source of food for bottom communities, either directly or after a preliminary attack by bacteria. Just what depth of water is needed before the euphotic zone production is effectively consumed within the water column and does not accumulate on the bottom is not known. There is nearly always a drastic decline in the concentration of particulate organic material in waters deeper than a few hundred metres.

The concentration of particulate and dissolved organic matter in sea water at a given instant and place results from an equilibrium between input rates and removal rates and we can do little more than guess at these magnitudes. The ability of heterotrophs to "scavenge" dissolved organic matter is limited. The concentration of any substrate, even with a zero input rate, would be expected to approach zero only asymptotically, as micro-organisms remove substrates according to an isotherm-type relationship.

The concentration of organic matter near the surface of the sea gives an interesting insight into the evolution of heterotrophs. The concentration of simple monomers of sugars and amino acids rarely exceeds a few tens of micrograms per litre, whereas polymerized material is probably present in an order of magnitude greater, indicating the greater difficulty of depolymerization reactions with substrates at low concentrations. Finally there is organic matter (largely free from nitrogen and phosphorus) which resist attack by heterotrophs in concentrations of several tenths of milligrams per litre.

Very rapid adjustments must take place as the concentrations of dissolved organic matter and particulate organic detritus show very little increase during periods of high productivity and, in the water immediately below the euphotic zone, the concentration of particulate matter appears to be nearly independent of the level of productivity at the surface.

Probably nowhere is the loss of organic matter during a given energy transfer reclaimable by the trophic level suffering the loss. For example, it is most unlikely that the unassimilated organic matter lost by copepods will ever be reconverted to useful food for the same copepod population. The aquatic environment is rarely if ever limited by a shortage of carbon, and there seems no obvious reason why organic carbon, *per se*, should be efficiently utilized in the biosphere. The presence of fossil fuels bears witness to the fact that it sometimes has not been. However, nitrogen is at a premium, especially in the sea, and we might expect the ecosystem to have evolved some ingenious mechanisms to see that little if any nitrogen is wasted as sediments or lost by denitrification.

The means by which nature re-utilizes waste products thus makes an interesting study, separate from the mainstream of tropho-dynamics, and it seems fitting to include some papers on the subject in the present volume.

We are presenting two contributions (Marshall and Qasim) concerning the utilization of the unused pelagic production of shallow waters by bottom communities. The whole subject of the production of particulate matter and its relationship to annual productivity is discussed by Finenko. He presents an account of Russian work which makes an interesting comparison with the work of Menzel, Parsons, Strickland and co-workers in North America, Newall in Australia and Steele and co-workers in the United Kingdom. McIntyre *et al.* present evidence for the re-utilization of dissolved organic matter by a food chain involving benthic communities. Their work supports the contention of Liston and his colleagues that dissolved organic matter can be scavenged by bottom deposits by some form of adsorption. Finally, the mechanism by which the organic matter in sea water (especially high molecular weight polymers) is adsorbed onto inorganic or organic detritus, is the subject of the study by Khailov and Finenko, using carbon-14 labelled organic matter.

One is struck, on reading these contributions, by the relative importance in fertile, shallow coastal regions, of the "waste" products of the initial food chain and how important it may be for many communities to utilize these sources of energy to the fullest. Further studies of this kind in coastal ecology will be awaited with increasing interest.

Organic macromolecular compounds dissolved in sea-water and their inclusion into food chains

K. M. KHAILOV and Z. Z. FINENKO

Institute of the Biology of the Southern Seas
Academy of Sciences of the Ukraine SSR
Sevastopol, USSR

ABSTRACT. From sea water filtered through 0·45 μ pores a mixture of the surface-active organic compounds was extracted by the emulsion method. By means of gel-filtration on neutral Sephadexes the mixture of the extracted compounds was fractionated and some fractions of macromolecular compounds (MW between 200 000 and 10 000) were isolated. Macromolecular compounds were found in different samples of sea water in concentrations between 0 and 0·5 mg/l. These compounds are proteins, polysaccharides and some unknowns.

Two mechanisms of the inclusion of macromolecules dissolved in sea water in the food chains are discussed on an experimental basis: sorption of macromolecules (polysaccharides with MW $\geqslant$ 50 000) on particulate matter (natural detritus) and their hydrolysis in sea water. The rates of these two processes are different under different conditions. The role of the surface-active macromolecular and low-molecular compounds in the formation of aggregates at the air–water interface is discussed.

Modern data on molecular diversity of organic compounds dissolved in sea water suggest that mechanisms for their inclusion into food chains in the sea may differ according to the molecular weight. Organic matter with low molecular weights—simple sugars, amino acids and so on—may be utilized directly, if dissolved in sufficient concentrations. On the other hand all complex dissolved compounds—polysaccharides, long-chain peptides, proteins, etc.—cannot be utilized directly regardless of their concentration in sea water. However, such high molecular weight compounds exhibit a very specific property: high surface activity in accordance with Traube's Law.

This property offers another possibility for them to be utilized by marine organisms: through adsorption on solid water interfaces (particulate matter) populated by micro-organisms. The quantity of micro-organisms, especially bacteria, adhering to detritus is significant, about 1-5% (Jørgensen, 1966). It is therefore reasonable to suppose that the interaction of dissolved macromolecular compounds consists of physico-chemical processes (adsorption) and bio-chemical processes (a transformation of sorbed material).

The aim of the present paper is the investigation of organic macromolecular compounds dissolved in sea water, and of the interaction between dissolved macromolecules and particulate matter.

MATERIAL AND METHODS

Extraction of dissolved macromolecules of sea water

The extraction was carried out by an emulsion method used earlier (Khailov, 1966) but not described in detail. The method is based on the ability of different compounds to concentrate on the surface between two immiscible liquids. As an organic phase, chloroform was used because of its low solubility in water.

The water was taken near the rocky shore of the Barents Sea. Ten 3·0 l portions of water were filtered through a Millipore 0·45 μ filter immediately after sampling and shaken vigorously with about 200 ml of chloroform. For preparative aims a five-minute shaking is sufficient for good recovery of macromolecules. After a complete sedimentation of chloroform drops on the bottom of a glass mixer, the supernatant (extracted sea water) was decanted and a new 2·5-3·0 l portion of filtered sea water was added. Shaking and all other procedures were repeated. In this way all ten 3 l portions of sea water were extracted and as a result the stability of emulsion increased, due to accumulation of organic matter on the chloroform drops. When a stable emulsion was formed the drops of chloroform were left to settle and the water decanted. The chloroform layer was then centrifuged until there was a clear differentiation of: (*i*) chloroform, (*ii*) water and (*iii*) a white layer between them. Below we shall refer to this third layer as "intermediate layer". This intermediate layer was separated from the excess of water and chloroform, washed with distilled water, centrifuged repeatedly and transferred to a vacuum desiccator. The rest of the chloroform was easily removed in vacuum. It should be pointed out too that no reactions sensitive to $CHCl_3$ were used in the following procedures. Fractionation by gel-filtration provides further purification of macromolecular compounds from $CHCl_3$ and any other low-molecular contaminations.

Fractionation of high molecular compounds by gel-filtration

Further treatment of the intermediate layer consisted of its dissolution and molecular fractionation by means of gel-filtration (Flodin, 1961; Porath and Lindner, 1961; Khailov and Burlakova, 1965). For this aim material of the intermediate layer was dissolved in 3-5 ml of distilled water, giving a yellowish solution which foamed when shaken. Below we shall refer to this material as "water soluble". That part of organic matter which was poorly dissolved in distilled water, was then dissolved in 2-3 ml of 0·1N NaOH giving a brown solution. This basic solution was immediately neutralized to a pH of 7·0. Below we shall refer to this material as "base soluble". Water soluble and base soluble compounds were fractionated (separately) on neutral Sephadexes with a different degree of porosity: G-75, G-100 and G-200. The columns were previously stabilized with distilled water. In some instances, after elution with distilled water, 0·1N NaCl was put on the column and elution with distilled water was continued.

The extraction of "humic acids" by emulsion method and their fractionation by gel-filtration

To obtain additional information on the nature of fractionated high molecular compounds it was necessary to know whether so-called "humic compounds" which are assumed to be in sea water can be readily extracted by an emulsion method.

To check this possibility, soil humic acids, after threefold reprecipitation by acidification of base solution, were dissolved in 0·1N NaCl at pH 8·6. The solution was then extracted by the emulsion method giving a stable yellowish emulsion. The intermediate layer resulting from all procedures was vacuum dried, dissolved in 2-3 ml of 0·1N NaOH and immediately neutralized. Fractionation of this solution on Sephadex G-100 shows that this fraction is composed of two subfractions: a brown solution with properties of humic acids (MW about 100 000) and another fraction. In the same manner humic acids from peat-bog water were successfully extracted and fractionated.

The results of these experiments give us confidence that, if humic acids identical to those of fresh water or soil are also present in sea water, these ought to be extracted by the emulsion method and be found in the intermediate layer after its fractionation.

Adsorption of macromolecules on detritus particles

Physico-chemical adsorption of macromolecules from sea water was investigated in the following systems: (i) natural detritus after sterilization and dissolved ^{14}C-polysaccharides, (ii) detritus and proteins with MW $\geqslant$ 50 000 and (iii) $CaCO_3$ particles and dissolved ^{14}C-labelled protein. Both the polysaccharides and water-soluble protein for this work were obtained from ^{14}C-labelled cells of Platimonas viridis in pure culture after extraction and fractionation (macromolecules) by gel-filtration. The chemical composition of polysaccharides and water-soluble protein received by this way was not determined. Experiments were performed in glass vessels with 300 ml of sea water at 10-15°C. The concentration of particulate matter was natural detritus 40 mg/l, and $CaCO_3$ 400 mg/l. Concentrations of dissolved polysaccharide in experiments with detritus were 0·3, 0·6, 1·6, 3·2, 16·0 mg/l; those of protein were 0·007, 0·07, 0·7 mg/l. Concentrations of protein in experiments with $CaCO_3$ were 0·3, 0·75, and 2·5 mg/l. The choice of these concentrations was influenced by the data of Walsh (1965) on the quantity of dissolved carbohydrates in sea water (1-3 mg/l) and our own data on the concentration of macromolecular surfactants in sea water—about 0·5 mg/l for inshore water.

The equilibrium was established in each system and the radioactivity of detritus and sea water were counted one hour after beginning each experiment. In each case four replicate samples of sea water (1 ml each) were taken on aluminium planchettes (1·0 cm in diameter) and four samples of detritus were filtered. The self-absorption from the detritus ($\sim$0·13 mg/cm^2) was neglected; the self-absorption from salts on aluminium plates was assumed to be equal to that of a calcium sulphate precipitate of the same thickness. All samples of sea water were evaporated at 100°C before counting; all filters were water-washed (10 ml per each filter), dried and counted. The decrease of total ^{14}C in the whole system (e.g. Fig. 3 (B)) shows that ^{14}CO$_2$, released during microbial metabolism, escaped the counting in sea water and detritus samples.

The interaction between macromolecules adsorbed on to detritus and detrital micro-organisms

The following experiments were undertaken to test the supposition that interaction took place between adsorbed material and micro-organisms after adsorption of proteins and polysaccharides on detritus.

Experimental conditions: volume of sea water 300-500 ml, temperature 10-15°C, concentration of dissolved ^{14}C polysaccharides 5·0 mg/l and of dissolved ^{14}C protein 0·66 mg/l. At 1, 2, 4, 8, 24 and 45 hours after the beginning of the experiment, samples (10-15 ml) of sea water with particulate matter were filtered from each system through 0·5 μ Millipore filters and filters counted.

In the next experiment the "artificial detritus" populated by marine bacteria was used instead of natural detritus. For this aim thin-dispersed glass dust "covered" with thin layers of ^{14}C-protein was dispersed (after previous washing) in 250 ml of sea water in closed glass bottles. After 1, 3, 6 and 12 days the following measurements were performed: radioactivity of "artificial" detritus, radioactivity of solution after filtering through 0·45 μ filter and oxygen consumption (Winkler method).

RESULTS AND DISCUSSION

Macromolecular compounds dissolved in sea water

Three different samples of sea water (20-30 l each) were collected and immediately filtered. Dissolved surfactants were extracted by the emulsion method and fractionated by gel-filtration as described above.

Fig. 1 (A) shows typical results of the fractionation of the water soluble material from the intermediate layer. It divided into three main fractions. Of these, fraction 1 contains macromolecules with MW $\geqslant$ 50 000 and fraction 2, MW about 40 000. The place of elution of macromolecular compounds with MW $\geqslant$ 50 000 (solid arrow on Fig. 1) was determined by means of the polysaccharide FICOLL, which moves on Sephadex G-75 as a compound with MW $\geqslant$ 50 000. Fraction 3 consists of low-molecular compounds which move in the gel at the same rate as do inorganic salts (dotted arrow in Fig. 1 (A)).

Fig. 1 (B) shows the results of fractionation of base-soluble material from the intermediate layer under the same conditions. Fractions 2, 3 and 4 consist of compounds which cannot be measured by optical density because of their

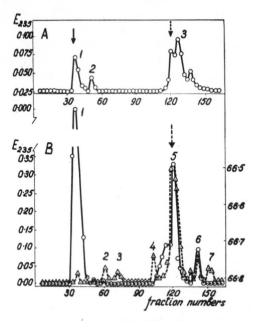

Fig. 1. Fractionation of water soluble (A) and base soluble (B) material of surfactants extracted from sea water on Sephadex G-75. The column (80·0 × 1·4 cm) was loaded with 2·0 ml of solution. Elution was with distilled water. 3·0 ml fractions were collected. Solid lines—optical density (E_{235}): dotted line—refractometer readings. 1, 2, 3, etc.—fractions—see text.

low concentration, but which exhibit definite refraction. Molecular weights of these fractions are about 40 000 to 10 000 respectively. Fraction 5 moves with the same rate as inorganic salts (dotted arrow), and may be referred to as low-molecular weight. Fractions 6 and 7 exhibit interaction with the gel matrix.

In order to determine whether or not macromolecules of the base-soluble fraction exceed MW 50 000, this fraction was concentrated and aliquots were put on the columns of Sephadex G-100 and G-200. These Sephadexes allow fractionation of compounds with MW up to 100 000 and 200 000 respectively. The results of such work on Sephadex G-100 show that the greater part of macromolecular compounds have MW $\geqslant$ 100 000. Besides this there are fractions with MW between 90 000 and 50 000, but these are insignificant in quantity. Fractionation of the second aliquot of base-soluble compounds on

Sephadex G-200, shows that at least a part of macromolecules extracted from sea water have MW ⩾ 200 000.

Considering the data on molecular fractionation of compounds extracted with chloroform emulsion one can see that at least inshore sea water contains the diverse hydrophilic macromolecules (referring to "macromolecules" all compounds with MW ⩾ 50 000). The highest defined molecular weight is about 200 000, as measured from the rates of movement in neutral Sephadex (Andrews, 1965). A large part of macromolecular material has MW between 200 000 and 100 000 and a smaller part between 100 000 and 50 000. In general these values coincide with MW of soil humic acids (to 300 000, Posner, 1963). However, such molecular weights are not only typical of humic acids, but of many other biological macromolecules—proteins, polysaccharides and their complexes with other compounds.

The total quantity of macromolecules in one sample examined (20 l) is about 9·3 mg, that is 0·46 mg/l. If the total content of organic matter in inshore water is about 5 mg/l, then macromolecules compose nearly 10% of this quantity. Taking into consideration that the extraction in our case was somewhat incomplete, the fraction of macromolecules in total dissolved organic matter may be more than 10%. This value is lower than the assumed quantity of so-called "water humus" in sea water. However, it should be noted that not all yellow-coloured compounds which are called "humus" can be extracted by the emulsion method. Good recovery was mentioned only for "humic acids" themselves, but not for so-called "fulvic'' acids. Hence a considerable part of "water humus" may be omitted by such method of extraction.

The most abundant of all the fractions, that with MW between 200 000 and 100 000, gives a clear brown colour in solution similar to that of soil humic acids. On the other hand, this high molecular fraction did not exhibit a more significant property of humic acids: the acidification of its base solution is not accompanied by the formation of brown sediment.

All these facts give an impression that macromolecules dissolved in sea water are not identical with soil and fresh-water humic acids. The same conclusion results from the investigation of dissolved macromolecular surfactants on DEAE Sephadex. But whatever the nature of macromolecular compounds dissolved in sea water from the thermo-dynamical point of view, the transformation of dissolved macromolecules into suspended matter is very probable. The reduction of solubility due to dehydration can take place in natural waters, for instance in any boundary between two phases such as air and water, or water and solid surface, especially detritus. It is well known that the bubbling of air through sea water results in the appearance of suspended organic aggregates (Riley, 1963; Sutcliffe et al., 1963). It was shown recently that the aggregation proceeds very effectively in the presence of bacteria and high molecular compounds (Barber, 1966). The essence of chloroform extraction is the same process taking place on chloroform drops instead of air bubbles. Such a mechanism can also operate on water-solid interfaces. Particulate organic and inorganic matter suspended in sea water has a great surface open to different dissolved surfactants. The quantitative relation between total particulate and total dissolved organic matter is about 1:10 in general; that is, in every part of an open ocean and especially inshore water, an interaction between the two is very probable.

Adsorption of dissolved proteins and polysaccharides on particulate matter

The isotherms of polysaccharide and protein adsorption on to detritus and $CaCO_3$ are given in Fig. 2. For comparison the isotherm of protein hydrolysate adsorption on to detritus is also given. The intensity of protein adsorption on to natural detritus is about 10 to 15 times more than on to $CaCO_3$. Such intensive adsorption may result from the nature of the surface of natural detritus in comparison with the $CaCO_3$ surface or from its chemical affinity for charged protein molecules. The difference between the relative adsorption rate of proteins and of polysaccharides (for concentrations of dissolved

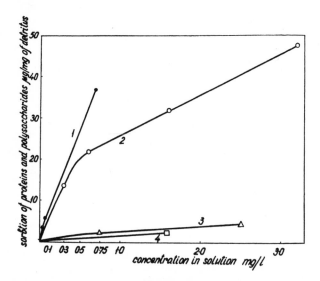

Fig. 2. Adsorption isotherms of dissolved polysaccharides and proteins on to natural detritus and $CaCO_3$. 1—protein and natural detritus, 2—polysaccharide and natural detritus, 3—protein and $CaCO_3$, 4—protein hydrolysate and detritus.

macromolecules under 0·5 mg/l) is 20%. Relative adsorption rate of low molecular weight protein hydrolysate is 20 to 30 times less than the adsorption rate of intact macromolecules.

The absolute and relative quantities of proteins and polysaccharides which may be adsorbed on to detritus from sea water cannot be very great. If, for instance, the concentration of macromolecular surfactants is about 0·5 mg/l, the quantity adsorbed on to natural detritus is about 4-5% of this value. However, this value reflects the equilibrium condition and tells us nothing of the real meaning of this process in the sea, because the equilibrium in the "dissolved macromolecular compounds ⇌ adsorbed macromolecular compounds" system may be shifted due to the activity of detrital micro-organisms. This indeed is the case in the natural systems with detritus.

The interaction between adsorbed macromolecules and detrital micro-organisms

In Fig. 3 (A) the interaction between proteins and polysaccharides adsorbed on to natural detritus and detrital micro-organisms is reflected. During the first hour physico-chemical adsorption on to particles takes place until the system is in equilibrium. When calculated as a percentage of the total quantity in solution, the quantity of adsorbed protein and polysaccharide is nearly the same for these two compounds: 64% of total protein and 54% of total polysaccharide.

From Fig. 3 (A), after the establishment of equilibrium in the physico-chemical system, the radioactivity of particulate matter decreases rapidly

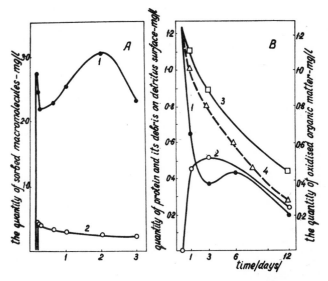

Fig. 3. The interaction of macromolecules adsorbed on to detritus with detrital micro-organisms. A—System with natural detritus: 1—the change of polysaccharide on detritus, 2—the change of protein on detritus. B—system with "artificial detritus": 1—the change of protein on glass particles, 2—the change of protein and its debris in solution, 3—the change of total protein and its debris (particulate and in solution), 4—the change of total organic matter in experimental system (calculated on O_2 consumption data).

after some hours and then slowly during about two days. Curves of this kind were repeatedly observed in different experiments. The underlying processes may be described by a theoretical model (Fig. 4). This model supposes two different processes: (*i*) hydrolysis of adsorbed macromolecules with resulting desorption of their low molecular debris in accordance with Traube's Law, and (*ii*) uptake of low-molecular debris (simple sugars or amino acids in our

case) by detrital micro-organisms and inclusion of a fraction of this radioactivity into the bacterial biomass. The first of the two of these processes
(curve 1) is very intensive and it is supported by the experiments; the second
is not so intensive (curve 2) and decreases with the mineralization of organic
matter after some increase in radioactivity. When occurring simultaneously,
two of these processes are described by curve 3, corresponding to experimental curves (Fig. 3 (A and B)). The same results were received in experiments
with protein (Fig. 3 (B)).

The rate of protein hydrolysis in this experiment may be calculated as a
function of the total protein quantity on the detritus. From such calculations

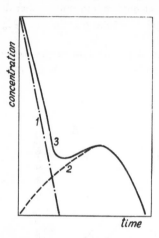

Fig. 4. The scheme of interaction between
adsorbed protein and detrital micro-organisms
1—hydrolysis of protein on detritus surface,
2—inclusion of its debris into bacterial
biomass, 3—the addition of both processes.

it may be seen that this rate is a linear function of total protein quantity
(33 % per day). The rate of such reaction may be expressed by the equation
(Skopintsev, 1949):

$$da/dt = R(A_r - a),$$

where A_r is the concentration of protein at the beginning of the experiment,
a is the quantity of protein participating in the reaction to time t and R is the
constant for reaction rate.

After the integration:

$$R = \frac{1}{t} \ln \frac{A_r}{A_r - a} \text{ or } R = \frac{2 \cdot 303}{t} \log \frac{A_r}{A_r - a}.$$

R calculated from this equation for protein hydrolysis is 0·3, the value being
twice as much as R for the oxidation of plankton organic matter (Skopintsev,
1949).

Data obtained in the experiment with "artificial detritus" populated by
micro-organisms (glass dust with absorbed protein) are given in Fig. 3 (B).
The curve reflecting filtrate radioactivity shows desorption of ^{14}C compounds
from the particles. This may be a result of protein hydrolysis taking place on
the detritus surface under the influence of bacterial enzymes. The decrease of

total radioactivity in the system reflects intensive oxidation produced by micro-organisms.

The main interest in this experiment is concerned with the comparison of total protein oxidation calculated from ^{14}C data, curve 3 in Fig. 3 (B), and

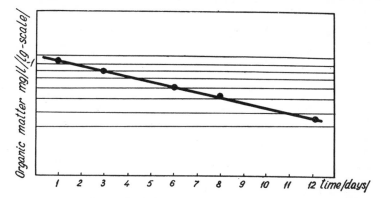

Fig. 5. Kinetics of protein oxidation on detritus.

with total organic matter oxidation calculated from O_2 consumption data, curve 4 in Fig. 3 (B). First it must be noted that the rate of both these processes decreases logarithmically. The dependence between substrate concentration and time is given in Fig. 5. The rate of reaction was calculated

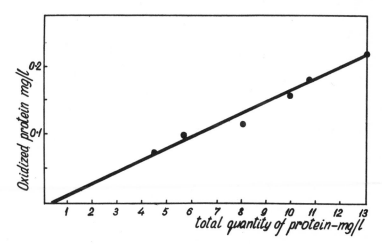

Fig. 6. The rate of protein decomposition on detritus as function of its total quantity.

from the equation mentioned above. The R-value in this reaction is 0·15 and coincides with the Skopintsev data (Skopintsev, 1949) on the decomposition of dead plankton. Hence the rate of protein oxidation is half the rate of protein hydrolysis on the detritus surface.

A question arising from this experiment with "artificial detritus" is whether or not the requirements of micro-organisms which populate detritus are satisfied solely by adsorbed organic matter (protein in our case) or whether other organic compounds in solution in the sea water are utilized as well. If we suppose that the rate of total oxidation reflected by O_2 consumption is constant during all 12 days of the experiment (Fig. 6), one may conclude that the oxidation of dissolved organic matter of sea water is insignificant. This conclusion is also supported by ^{14}C tracing of protein adsorbed on to detritus (curve 3 in Fig. 3 (B)) in comparison with O_2 consumption tracing (curve 4 in Fig. 3 (B)). The difference between two of these curves at the end of the experiment is equal to 0·2 mg/l. This value corresponds to the quantity of dissolved organic matter being oxidized in addition to protein.

The data give an impression that the traditional question: "do micro-organisms populating detritus decompose organic matter dissolved in the surrounding water or organic matter of detritus itself?" was posed inexactly. Apparently the answer is that micro-organisms populating detrital particles

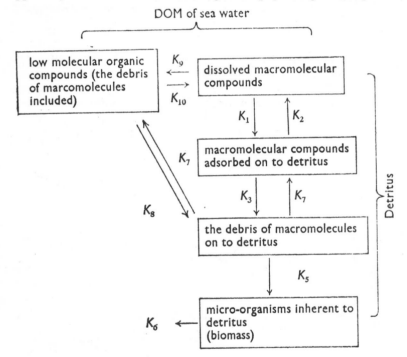

Fig. 7. The scheme of some physico-chemical and biochemical reactions in DOM $\rightleftarrows$ Detritus system. K_1 and K_2—the rates of adsorbtion and desorption of macromolecular compounds onto detritus and from it. K_3 and K_4—the rates of bacterial hydrolysis and synthesis of macromolecular compounds on detritus surface. K_5—the rate of inclusion (uptake) of low molecular debris into microbial biomass. K_6—The rate of CO_2 release by micro-organisms (the rate of mineralization of organic matter). K_7 and K_8—The rates of adsorption and desorption of low molecular debris of macromolecules. K_9 and K_{10}—the rates of bacterial hydrolysis and synthesis of macromolecular compounds in solution.

decompose dissolved surfactants of sea water, but only after their adsorption on to the detritus surface. Organic matter inherent to fresh detritus can be readily decomposed too, but the organic matter of so-called "conservative" particles apparently cannot be decomposed so successfully. Such "conservative" particles with their developed and complex (in chemical sense) surface serve as a specific catalyst. Accumulation and further biochemical transformation of dissolved surfactants take place on this surface. This function of marine detritus has not yet been described in oceanographical literature.

All known data show that dissolved macromolecular surfactants, particulate matter (detritus) and micro-organisms which populate it, compose a system integrated by different physico-chemical and bio-chemical processes. Physicochemical reactions are adsorption and desorption of compounds of different molecular weights. According to Traube's Law, low-molecular weight compounds (simple sugars, amino acids and so on) are adsorbed less than macromolecular compounds (proteins and polysaccharides in particular), that is $K_1 > K_8$ (Fig. 7). However, so far as the equilibrium in the subsystem "macromolecules adsorbed $\rightleftharpoons$ low molecular debris of macromolecules" is shifted to hydrolysis of macromolecules ($K_3 > K_4$), their concentration on detritus must constantly decrease. As a result the equilibrium in the system "dissolved macromolecules $\rightleftharpoons$ adsorbed macromolecules" can be shifted to a higher adsorption of macromolecules from sea water. Under such conditions in the chain

"Dissolved Organic Matter (DOM) → Detritus" and hence in the chain

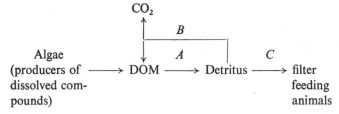

there can exist a constant flow of matter and energy. The rate of this flow is determined by the rate of all reactions which take part in the system (K_1, K_3 and K_5 especially).

To calculate the rate of this flow one needs to know what fraction of the material adsorbed on detritus at the beginning of the process is desorbed again into the water after hydrolysis to low-molecular compounds, and what part of it is mineralized by detrital micro-organisms to CO_2 (the back flow, $-B$). Obviously the total quantity of organic matter moving through the pathway, DOM → detritus → filter feeding animals is: $C = A - B$. As it follows from experimental data, the capacity of this chain may be significant. For instance, if the dissolved protein concentration has the reasonable value of 0·66 mg/l, the quantity of adsorbed protein would be 0·025-0·035 mg/(mg of detritus), that is 3·8-5·3% of dissolved quantity ($A \approx 4·5\%$). If the rate of protein hydrolysis on detrital surface is 33% per day and is a linear function of its total quantity, then under constant concentration of dissolved protein B, the flow is 1·5% of total dissolved protein. Thus the quantity of protein which may be transmitted to detritus and then to its consumers (C) is:

$$C = K(4·5\% - 1·5\%) = K\,3·0\%,$$

where K is a quotient reflecting the consumption of detritus by filter feeding animals. Not all dissolved macromolecular compounds in natural waters are proteins or polysaccharides. Most natural macromolecular surfactants are compounds of unknown chemical nature, but chemical analysis shows that these include both polysaccharides and proteins. The concentration of detritus is smaller too in natural conditions than in our experiments. Therefore our calculations suggest the magnitude of the maximum capacity of the chain. The real capacity will generally be smaller. However the total quantity of dissolved surfactants may be very great in some special periods and places—in phytoplankton blooms and in inshore waters. In such conditions this mechanism may be a significant way of inclusion of DOM into food chains through bacterial activity on the detritus surface and utilization of detritus by filter feeding animals.

In conclusion we shall emphasize that this external metabolic system is similar in many respects to the system described by Khailov and Gorbenko (1967). This last system included the association of periphytic micro-organisms (similar to that of detritus) and DOM of sea water.

REFERENCES

ANDREWS, P. 1965. The gel-filtration behaviour of proteins related to their molecular weights over a wide range. *Biochem. J.*, **96**, 595-606.

BARBER, R. T. 1966. Interaction of bubbles and bacteria in the formation of organic aggregates. *Nature, Lond.*, **211**, 257-58.

FLODIN, P. 1961. Methodological aspects of gel-filtration with special reference to desalting operations. *J. Chromatogr.*, **5**, 103-15.

JØRGENSEN, C. B. 1966. *Biology of suspension feeding*. New York, Pergamon Press. 357 pp.

KHAILOV, K. M. 1966. On the evolution of metabolic relations between marine organisms in societies. In *Physiology of marine organisms* (in Russian). Moscow, NAUKA.

KHAILOV, K. M., and BURLAKOVA, Z. P. 1965. On the use of gel-filtration in chemical oceangraphy (in Russian). *Okeanologia*, **5**, 739-48.

KHAILOV, K. M., and GORBENKO, Y. A. 1967. Regulation by external metabolites in the system—society of periphytic micro-organisms and dissolved organic matter of sea water (in Russian). *Dokl. Akad. Nauk SSSR*, **173**, 1434-37.

PORATH, J., and LINDNER, E. B. 1961. Separation methods based on molecular sieving and ion exclusion. *Nature, Lond.*, **191**, 69-70.

POSNER, A. M. 1963. Importance of electrolyte in the determination of molecular weights by "Sephadex" gel filtration with special references to humic acids. *Nature, Lond.*, **198**, 1161-63.

RILEY, G. 1963. Organic aggregates in sea water and the dynamic of their formation and utilization. *Limnol. Oceanogr.*, **8**, 372-81.

SKOPINTSEV, B. A. 1949. On the rate of decomposition of dead phytoplankton (in Russian). *Trudy vses. gidrobiol. Obshch.*, **1**, 34-43.

SUTCLIFFE, W. H., BAYLOR, E. R., and MENZEL, D. W. 1963. Sea surface chemistry and Langmure circulation. *Deep Sea Res.*, **10**, 233-43.

WALSH, G. E. 1965. Studies on dissolved carbohydrate in Cape Cod waters. I. General survey. *Limnol. Oceanogr.*, **10**, 570-76.

Energy flow
in a sand ecosystem

A. D. MCINTYRE, A. L. S. MUNRO and J. H. STEELE
Marine Laboratory
Aberdeen

ABSTRACT. Field studies on the interstitial ecosystem of a sandy beach provide the basis of a carbon budget, but some questions, such as how the requirement of the meiofauna is met, cannot be answered by direct study of the natural environment. Long-term laboratory experiments are described on columns of sand with natural unfiltered sea water flowing down them at a constant rate. The soluble organic carbon was removed in the columns but showed discrepancies when compared with the oxygen demand. Meiofauna populations, mainly nematodes, turbellarians and copepods, reproduced successfully in the columns, and the copepods were studied in detail. Effluent from one column was passed down through a second column treated in the same way. Input to this second column, having been filtered through the first, was largely of soluble material, suggesting as a main conclusion that the successful meiofauna populations depended on bacterial synthesis for the protein component of their food rather than directly on particulate matter.

INTRODUCTION

It has often been noted that "soluble" organic matter, usually defined as organic material passing a glass fibre filter, is about one order of magnitude greater than the particulate matter in sea water. However, the role of this reservoir of organic substrate in marine food chains is uncertain. Direct evidence (Degens *et al.*, 1964; Siegel and Degens, 1966; Chau and Riley, 1966) on the concentration of small molecules such as glucose, galactose and amino acids suggests that they contribute perhaps less than 10% to the total soluble concentration. Although direct observations are lacking on the rest of the "soluble" material this is likely to be composed of large molecules in solution (Khailov and Finenko, this volume) and small suspended particles.

These distinctions seem worth making when considering the possible means by which living organisms could utilize such materials. The soluble small

19

molecular weight fraction in sea water can be removed by the permease systems of bacteria (Cohen and Monod, 1957) which have been shown (Wright and Hobbie, 1966; Hamilton et al., 1967; Vaccaro and Jannasch, 1967) to work even at the low concentrations prevailing in sea water. Therefore, this fraction, where it exists, is likely to have a fairly fast turnover. The bacteria also posseses the capability of producing extracellular enzymes which break down large soluble molecules and particulate organic substrates, but when dealing with such heterogeneous material whose dimensions are smaller than a bacterium and which exist only at great dilution, the advantages of extracellular enzymes would seem diminished to a point where the return does not balance the outlay. It therefore seems likely that this fraction which may comprise the bulk of the "soluble" organic material would not be accessible to heterotrophic action, and this is borne out by the constant values for "soluble" carbon in deep water profiles (Duursma, 1961; Menzel, 1964). It has, however, been proposed that the "soluble" material may be taken up on existing particulate matter and also may be transformed into "flakes" when it then becomes available, perhaps through bacterial protoplasm, to higher levels in the food chain (Riley, 1963). Because this reservoir of "soluble" organic material is so large the possibility of its returning to the main food web is of considerable importance since this could form a significant addition to the energy available to higher levels of the food chain.

The main problem in investigating these factors in the open sea is the extreme dilution, not only of the substrates, but of the organisms, bacteria, phytoplankton, or animals, which would take part in these transformations. Benthic systems show much greater concentrations and so more easily measured changes. In particular, a sandy beach where there is continual exchange of the interstitial water can provide a suitable system to study heterotrophy both in the natural environment and experimentally in the laboratory.

THE BEACH ECOSYSTEM

Detailed studies have been made of the ecology of an exposed sandy beach in Loch Ewe on the west coast of Scotland (Steele and Baird, 1968; McIntyre and Eleftheriou, 1968). It has been found (Steele and Baird, 1968) that the organic material is predominantly attached to the sand grains and there is less than 5% detrital material in the interstitial water. On any tide the sand is mixed to a depth of between 2 and 10 cm depending on wave action, and there are also longer term movements of the sand level. As a consequence, the organic carbon per g of sand and the chorophyll per g of sand are uniform to a depth of about 20 cm. Estimates of the annual primary production are about 5 g of carbon per square metre. However, experiments on the rate of oxygen uptake of samples of this sand (Munro, unpublished) show that the equivalent carbon requirement is approximately 50 g of carbon per square metre. This oxygen requirement was 75-90% inhibited by azide and dinitrophenol indicating a biological oxygen demand. The oxygen requirement also showed a rapid increase on the addition of 5 and 30 mg/l of either glucose or acetate. Fluorochrome staining of the sand grains revealed approximately 10^8-10^9 bacteria-like objects per g of sand. Autoradiography using ^{3}H acetate or ^{3}H glucose consistently revealed label associated with the bacteria but never with the diatoms (Munro and Brock, 1968), indicating that heterotrophy is likely to be largely bacterial. This latter evidence and the primary production estimates based on ^{14}C data (less than 5 g carbon per square metre) indicate a low level of endogenous algal respiration (less than 0·1 mg C per kilo of sand per day). Observations on the other sand populations showed that protozoa were not abundant on this beach, and that the macrofauna, with a mean dry weight of 1·3 g/m², contributed only fractionally to the total oxygen demand. But the meiofauna—small metazoans passing a ½ mm sieve—were abundant, averaging over one million individuals per m² (McIntyre, unpublished). They were mostly true interstitial forms, largely copepods and nematodes, and although found to a depth of 20 cm, the main concentrations were in the surface layer. The mean dry weight was only 0·35 g/m² but their small size, and the presence of ovigerous females throughout most of the year suggested the possibility of a rapid turnover. The part they play in the sand ecosystem is uncertain.

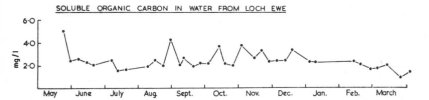

Fig. 1. Soluble organic carbon (mg/l) in water from Loch Ewe.

A consideration of the supply of nutrients for heterotrophy within the sand system led us to study the rate of flow of water into the sand and using techniques devised by Dr Graham Giese (University of Puerto Rico) it was shown that there was a net flow into the sand of approximately 5 cm per day occurring mainly around the time of low water. The concentration of particulate organic matter in the sea water of this area is approximately 200 µg/l, which would be insufficient to fuel the interstitial ecosystem. However, a contribution from the "soluble" organic carbon of 1 mg/l would be sufficient, a value which is much less than those observed in water from Loch Ewe (Fig. 1). As it was not possible to study the natural system in greater detail the work was transferred to the laboratory, where an artificial ecosystem in the form of sand columns was set up.

EXPERIMENTAL SAND COLUMNS

The construction of the experimental columns is shown in Fig. 2. The sand used was taken from low water mark on the beach. This sand is quartz material predominantly, and is well sorted with a median diameter of 250 µ. The sand is supported on a porous plastic sheet with average pore size of 50 µ.

Sea water from the bay was collected every two or three days and run through the columns at a rate between 6 and 7 litres per day, corresponding to an entry rate of 8 to 10 cm per day. This differs from the beach system since the rate of input was regular rather than pulsating as it appears to be on the beach.

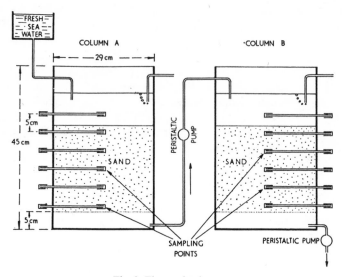

Fig. 2. The sand columns.

The effluent water from the first column (A) was passed through another column set up and sampled in exactly the same way as the first. The purpose of this second column (B) was to investigate the events, particularly the animal populations, that occur in sand where the entrant water had already been filtered through a sand column. The oxygen profiles within the sand columns were measured by sampling from the inlet points at the side, approximately once per week. Ammonia, nitrite and nitrate were sampled from the water entering and leaving both columns. Ammonia was measured using the method of Johnston (1966). The ammonia results averaged 1-2 $\mu M/l$ in the ingoing water with little difference in the water emerging from the base of column B. Similarly no difference was noted in the nitrite results and both nitrite and ammonia were discontinued. "Soluble" organic carbon was sampled from the top and bottom of each column using the method of Menzel and Vaccaro (1964). Cores of sand, approximately 1 cm^2 in area were taken vertically from the columns to provide estimates of the organic carbon and chlorophyll of the attached organic matter. Cores, 4 cm^2 in surface area, were taken for estimation of the animal populations in the columns. These animals were washed off the sand samples by elutriation and extracted after staining in rose bengal. The methods for estimating carbon and chlorophyll attached to the sand and in the interstitial water are those described by Steele and Baird (1968).

RESULTS

The oxygen data provide the most detailed picture of events within the sand column and Fig. 3 shows the decrease of saturation for both columns. After the first month column A showed significant differences between the decrease in the top 5 cm and that below. There appears to be a gradual increase of activity within this upper layer, whereas all the layers below it show a decrease and are relatively uniform after the first month. In column B there is some sign of an increase in the top 5 cm relative to the layers below but much less marked than in column A. This may be due to the fact that the water in the top of column B was aerated to ensure it was near saturation before entering the sand, thus possibly adding organic material or aggregating it into large pieces.

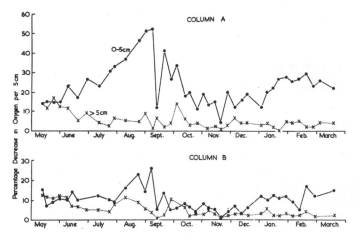

Fig. 3. Percentage decrease in oxygen per 5 cm in each column.

For comparison with the other parameters measured it has been necessary to consider the oxygen requirement of the column as a whole and in Fig. 4 this has been transformed into a carbon requirement using an RQ of 1, expressed as requirement per kilogram of sand. Fig. 4 also shows the data which were obtained in the uptake of "soluble" organic carbon (SOC) by the columns. It is apparent that this is significantly less than the estimated carbon requirements from the oxygen data. In column A there was a measurable decrease in SOC on nearly all occasions, but in B there is no significant reduction which, as previously pointed out, conflicts with the oxygen data. This point is discussed in detail later.

For column A there is a further source of organic material from the particulate matter in the water. The concentration of this is shown in Fig. 4, but it is apparently not enough to balance the difference between the oxygen and soluble estimates. Particulate carbon analyses from the water entering the top of B usually gave values between 10% and 25% of the inflow to A. It is not clear whether this is residual material from the inflow to A or material, such

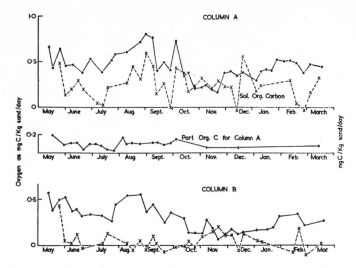

Fig. 4. Showing, for both columns, the carbon requirement calculated from oxygen data, the uptake of "soluble" organic carbon, and, for column A, the concentration of particulate organic carbon.

as bacteria, produced within the sand column. (It may also be partly produced by the aeration in the top of B.) As a loss from column A it is very small (0·01-0·02 mg C/kg sand, day) compared with the demand rate indicated by the oxygen data.

Finally it is necessary to consider whether the system can be fuelled by the breakdown of organic material present in the sand, and Fig. 5 shows the concentration of organic carbon from analysis of sand samples. The 95%

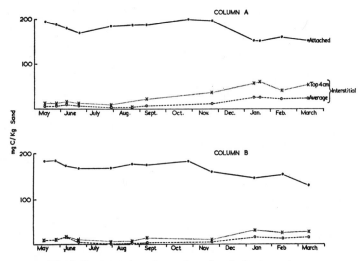

Fig. 5. Concentration of organic carbon in the two columns.

confidence limits for each value are ± 20 μg/g of sand. After the first three weeks the columns appear to settle down and there is no breakdown until about November. After November there is a significant decrease in attached carbon and a marked increase in the unattached interstitial material, particularly in the upper 4 cm of column A. Thus, until November, at least, there is no evidence of an overall decrease in organic matter within the column that could balance the energy requirements. After November, as will be described later, the carbon and fauna data suggest changes in the column system.

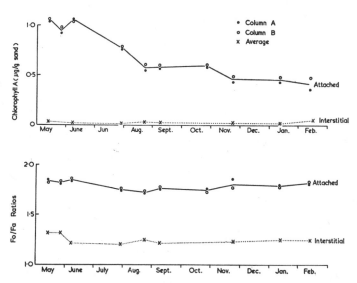

Fig. 6. Showing for attached and interstitial organic material the chlorophyll content (upper graphs) and the Fo/Fa ratios (lower graphs).

The chlorophyll *a* content of the attached organic matter shows a decline throughout the period (Fig. 6), but the pigment appears to remain undegraded as shown by Fo/Fa where a value of 1·7-1·8 probably indicates very low levels of phaeophytins in the sample (Yentsch, 1965). Microscopic examination shows that with time the chloroplasts appear to occupy an increasingly smaller proportion of the cells, suggesting that there is a decrease in chlorophyll per cell rather than a decrease in the number of cells. The interstitial pigments are at a very low level throughout the whole experiment and are considerably degraded, presumably representing old or digested material.

It has already been concluded that the attached diatoms are probably not heterotrophic, and so the main interest of the chlorophyll data is to show the survival of a diatom population in the dark for periods in excess of ten months. Their contribution to the oxygen demand cannot be assessed, but is assumed to be very low in view of their survival over such long periods. Their situation is analogous to that of algal cells found under ice (Rhode, 1955) or in the deep oceans (Wood, 1956) and suggests that a very low catabolic rate rather than heterotrophy is the determining factor.

B

Another indication of the transformations occurring is given by the production of nitrate on the passage of water through both columns (Fig. 7). The marked increase in nitrate without any change in ammonia or nitrite in the water shows that the microbial population of the column oxidizes organic nitrogen, the end result being nitrate. In column A this nitrogen is likely to be derived from entering particulate and soluble carbon but in column B it is likely to be largely derived from the soluble component despite the fact that the SOC method shows little or no change in the soluble carbon passing through the column.

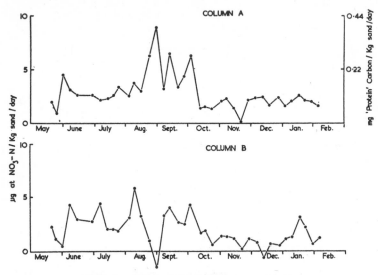

Fig. 7. Production of nitrate in the two columns.

When setting up the sand columns no precautions were taken to exclude larger animals, but there has been no indication that these were present, and since protozoa are not generally abundant on the beach, no attempt was made to estimate this group in the columns, although at least some ciliates were seen in the samples. The higher levels of the food chain were thus represented by meiofauna populations and for these, the columns were found to be qualitatively and quantitatively similar to the beach. Eight major taxonomic groups were represented—copepods, nematodes, turbellarians, gastrotrichs, tardigrades, polychaetes, archiannelids and coelenterates. The copepods were most numerous, followed by nematodes, and these two groups together usually made up more than 90% of the individuals. The sample cores were divided into 4 cm lengths to show vertical distribution, and this indicated that the animals were found throughout the whole length of the column, but with a tendency towards highest numbers in the top 4 cm.

Column A was studied by weekly samples for ten months. The total numbers of individuals per kilo of sand ranged from 1500 to 10 700, and dry weights from 0·6 to 2·4 mg (Fig. 8).

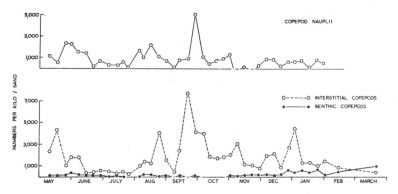

Fig. 8. Numbers of interstitial and benthic copepods, and of nauplii in column A.

A detailed examination was made of the dominant group, the copepods (Fig. 9), and the presence of nauplii throughout the whole period indicated successful breeding populations. For the first six months the copepods were mainly interstitial forms, harpacticoids of about 0·2 μg dry weight and up to 0·4 mm long, modified to move within the spaces between sand grains. Initially six species were present as breeding populations but from August onwards one, *Arenopontia subterranea* Kunz, became overwhelmingly dominant. These interstitial copepods are known to be browsers on sand diatoms and epigrowths (Remane, 1933). A benthic copepod, *Asellopsis intermedia* (T. Scott), which is normally found in intertidal sand, was present in the column during the first few months, but was later replaced by a number of typical subtidal species. These benthic animals, which are about five times the weight of the interstitial forms, were entirely confined to the sand surface. Although first detected in samples in July, they were not numerically significant until December, but thereafter the bulk of the nauplii belonged to these species. The increase in benthic copepods, which can utilize detritus on the bottom or in suspension (Gilat, 1967) may be associated with the breakdown of some of the attached organic material (Fig. 5) and the accumulation of organic debris. The column, after six months submersion, was thus becoming more subtidal in character.

The second major group, the nematodes, were studied in less detail, but at least 23 species were present and their average numbers showed a slow but steady decline from 3600 per kilo of sand in May to 600 in January, with a corresponding reduction in the number of species. Wieser (1960) has suggested that many nematodes previously thought to be predators are in fact not so, and the small or unarmed buccal cavities of most of the nematodes from the column place them in the category which Wieser assumed to be deposit feeders.

The remaining animal groups in column A were much less numerous (Fig. 9), but a small population of turbellarians persisted throughout the

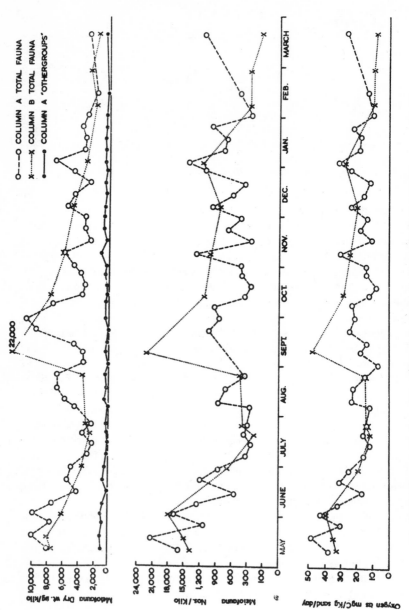

Fig. 9. Meiofauna—numbers, dry weights and carbon requirements calculated from oxygen data. Top figure shows also the numbers of animals other than copepods and nematodes.

whole period. Previous records (Hyman, 1951) agree that these are largely predators on the meiofauna, taking nauplii, copepods and small nematodes, and this is supported by an observation on one of the sand column turbellarians which contained copepod remains.

From respiration studies of the copepods by Dr Lasker, and from observations available in the literature, estimates have been made of the carbon requirements of the meiofauna in column A, ranging from 7 to 49 μg C/day/kilo of sand, with a mean of 21 μg C/day (Fig. 9).

Column B was studied in less detail, but the composition and fluctuations of the meiofauna (Fig. 9) were generally similar to those of column A, with a gradual decline in the nematodes and with the same species of harpacticoid copepod eventually dominating the population. The main difference was that although some benthic copepods appeared on the surface of column B, their numbers remained low, about one tenth of those in column A, a difference probably associated with the filtering effect of column A.

The metazoans of the columns thus consist of large numbers of deposit or epistrate feeders, together with a small population of predators which could account for the mortality.

DISCUSSION

While the division of the organic material in sea water into particulate and "soluble" fractions does define the large reservoir of non-living organic matter less than about 1 μ, it is rather artificial and does little to help understand the complex processes involved in the turnover of this material. The water in the bay, which was used in these experiments, although taken from close inshore, appears to be fairly typical of shelf conditions (Steele and Baird, 1968), and from this water the two sand columns taken together can extract between 1 and 4 mg of carbon per litre based on the oxygen results. There are some signs of a seasonal maximum in August and September, particularly in the breakdown of the nitrogenous component. Furthermore, this maximum is most pronounced in the top 5 cm of column A, suggesting that the organic material available at this time is easily assimilated by the bacteria in the same way as added glucose, which stimulates rapid oxygen uptake in the top few centimetres of a sand column. The deeper parts of column A and column B, especially during the winter, show a much less variable and much lower oxygen consumption, suggesting that there is a longer lasting or less easily removed component of the organic substrate which is broken down much more slowly by the bacteria, and it is this which provides the energy for the deeper parts of column A and for column B.

It is known that the sand column acts as a filter bed and it is probable that by non-specific adsorption it carries out a minor enrichment of large molecules and the small suspended particles on to the sand grain surfaces, thus allowing the extracellular enzymes of the bacteria to work to their own advantage. The bacteria in the surface layers of the sand column may remove the small soluble molecules and also the large particulate material filtered by the sand, thus explaining the large activity in the top 5 cm of A. The much lower activity of the deeper layers coupled with a slight decrease with depth in attached carbon suggests that any concentration of particulate carbon is being utilized forthwith. The fact that animal populations continue to survive and indeed multiply in these deeper layers also supports the belief that a viable bacterial population is maintaining itself on a substrate and against predation by these animals. The extent of nitrate production, especially the amounts produced in column B, supports the rate of oxygen consumption. However, it has not been possible to account quantitatively for the energy source of the oxygen demand in both columns. During the first six months in column A, 89% of the oxygen decrease can be accounted for by the particulate matter entering the column, the decrease of SOC within the column and the production of NO_3 from organic nitrogen. This is a reasonable agreement considering the accuracy of the methods. The organic matter attached to the sand grains does not show any change during this period in either column and so does not affect the calculations. In column B, assuming no entry of particulate to the column, the consumption of oxygen based on nitrate production and the oxidation of the SOC account for only 50% of the observed oxygen decreases. It is possible that some particulate matter may have been transferred from column A to B but, taking the columns together there would still be a deficit. Thus there is the possibility that the methods for particulate or soluble organic carbon are not measuring some component (Menzel and Ryther, 1968).

The carbon requirements of the meiofauna amount to about 5% of the total carbon utilization of the column. Since the animals are distributed throughout the length of the columns B as well as A, it would appear that they do not depend on the particulate carbon, which is largely filtered out in the upper part of A, and there is no indication of their feeding on the diatoms. The main source of energy for the interstitial fauna would seem to be the bacterial flora, which is in turn maintained by the "soluble" organic material.

It is apparent that the laboratory sand column is not an exact simulation of the beach environment. Some feature of the laboratory conditions resulted in the selection of one species of copepod which became dominant after three months. Furthermore, after seven months, the environment became "sublittoral" and conditions began to depart markedly from the initial setup. Thus, in the long term the column is not in a steady state but the changes, particularly in the first seven months, are small compared with the rates of breakdown of organic matter indicated by the oxygen data. On this basis, and with the assumption that effectively, all the organic matter passes through bacteria before being utilized by the meiofauna, the efficiency expressed as (meiofauna requirement)/(bacterial requirement) appears to be of the order of 5%.

Finally, it is suggested that the experimental sand ecosystem used is easier to study than the corresponding pelagic one, and provides insight, not only into events on the beach but also on the more general problems of transfer of organic matter up the food chain. It is concluded that the existing concentration and degree of dispersion of "soluble" organic matter in sea water is suboptimal for heterotrophic action, but that its concentration by sand allows its further utilization, first by heterotrophic bacteria, then by animal populations. In the sea larger organic or inorganic particles may perform the same role as the sand grains, although at a much greater dilution.

REFERENCES

CHAU, Y. K. and RILEY, J. P. 1966. The determination of amino-acids in sea water. *Deep Sea Res.*, **13**, 1115-24.
COHEN, G. N., and MONOD, J. 1957. Bacterial permeases. *Bact. Rev.*, **21**, 169-94.
DEGENS, E. T., REUTER, J. H., and SHAW, N. F. 1964. Biochemical compounds in offshore California sediments and seawaters. *Geochim. cosmochim. Acta*, **28**, 45.
DUURSMA, E. K. 1961. Dissolved organic carbon, nitrogen and phosphorus in the sea. *Neth. J. Sea Res.*, **1**, 1-141.
GILAT, G. 1967. On the feeding of a benthonic copepod, *Tigriopus brevicornis* O. F. Müller. *Bull. Sea Fish. Res. Stn Israel*, (45), 79-95.
HAMILTON, R. D., MORGAN, K. M., and STRICKLAND, J. D. H. 1966. The glucose uptake kinetics of some marine bacteria. *Can. J. Microbiol.*, **12**, 995-1003.
HYMAN, L. H. 1951. *The invertebrates. Platyhelminthes and Rhynchocoela, the acoelomate Bilateria.* **2.** New York, McGraw-Hill, 550 pp.
JOHNSTON, R. 1966. Determination of ammonia in seawater as subazoic acid. I.C.E.S., C.M. 1966, *Hydrogr. Comm.*, Pap. no. N:10.
KHAILOV, K. M., and FINENKO, Z. Z. 1970. Organic macromolecular compounds dissolved in sea water and their inclusion into food chains. This volume, pp. 6-18.
MCINTYRE, A. D., and ELEFTHERIOU, A. 1968. The bottom fauna of a flat fish nursery ground. *J. mar. biol. Ass. U.K.*, **48**, 113-42.
MENZEL, D. W. 1964. The distribution of dissolved organic carbon in the western Indian Ocean. *Deep Sea Res.*, **11**, 757-65.
MENZEL, D. W., and RYTHER, J. H. 1968. Organic carbon and the oxygen minimum in the South Atlantic Ocean. *Deep Sea Res.*, **15**, 327-37.
MENZEL, D. W., and VACCARO, R. F. 1964. The measurement of dissolved organic and particulate carbon in sea water. *Limnol. Oceanogr.*, **9**, 138-42.
MUNRO, A. L. S., and BROCK, T. D. 1968. Distinction between bacterial and algal utilization of soluble substances in the sea. *J. gen. Microbiol.*, **51**, 35-42.
REMANE, A. 1933. Verteilung und Organisation der benthonischen Mikrofauna der Kieler Bucht. *Wiss. Meeresunters., Abt. Kiel*, **21**, 161-221.
RHODE, W. 1955. Can plankton production proceed during winter darkness in subarctic lakes? *Verh. int. Verein. theor. angew. Limnol.*, **12**, 117.
RILEY, G. A. 1963. Organic aggregates in sea water and the dynamics of their formation utilization. *Limnol. Oceanogr.*, **8**, 372-81.
SIEGEL, A., and DEGENS, E. T. 1966. Concentrations of dissolved amino acids from saline waters by ligand-exchange chromatography. *Science, N.Y.*, **151**, 1098-1101.
STEELE, J. H., and BAIRD, I. E. 1968. Production ecology of a sandy beach. *Limnol. Oceanogr.*, **13**, 14-25.
VACCARO, R. F., and JANNASCH, M. W. 1967. Variations in uptake kinetics for glucose by natural populations in seawater. *Limnol. Oceanogr.*, **12**, 540-42.
WIESER, W. 1960. Benthic studies in Buzzards Bay. II. The meiofauna. *Limnol. Oceanogr.*, **5**, 121-37.
WOOD, E. J. F. 1956. Diatoms in the ocean deeps. *Pacif. Sci.*, **10**, 377-81.
WRIGHT, R. T., and HOBBIE, J. E. 1966. Use of glucose and acetate by bacteria and algae in aquatic ecosystems. *Ecology*, **47**, 447-64.
YENTSCH, C. S. 1965. Distribution of chlorophyll and phaeophytin in the open ocean. *Deep Sea Res.*, **12**, 653-66.

Particulate organic matter and its role in the productivity of the sea

Z. Z. FINENKO and V. E. ZAIKA
Institute of the Biology of the Southern Seas
Academy of Sciences of the Ukraine SSR
Sevastopol, USSR

ABSTRACT. The particulate organic matter content of different seas and oceans (in volume units) varied twenty–fortyfold in concentration but the amount under 1 m² of the surface in the euphotic zone varied by no more than fivefold.

Data from the Black Sea, the Sea of Azov, the Arabian Sea and the tropical part of the Atlantic Ocean show that the living part of the seston comprises 10-50% of the total and that this portion increases up to 100% in periods of algal bloom. The proportion of chlorophyll a in particulate organic matter is not constant. The relation of chlorophyll a to seston varied 1000 times and more.

By indirect methods of calculation it is demonstrated that the non-active part of the particulate organic matter comprises about 50% of the total quantity in the euphotic zone and the turnover rate of the biochemically active part is about once a month.

In the tropical part of the Atlantic Ocean and in the open Black Sea the food needs of the zooplankton is one and a half to four times greater than the primary production.

INTRODUCTION

The biological productivity of the seas and oceans is mainly a function of the small planktonic organisms which constitute the energetic basis for the existence of all other trophic levels, including fishes. A considerable amount of dead organic matter is accumulated in the ocean waters as a result of plankton mortality.

It is known that the suspended organic matter in sea water is composed of four main components: phytoplankton, zooplankton, bacteria and detritus. The detritus is a product of the destruction of the dead organisms of vegetable and animal origin. In recent years the importance of colloid aggregates arising in water from dissolved organic matter has also been emphasized.

It is interesting that living pelagic organisms form only a small fraction of the particulate organic material in sea water. The major proportion of particulate organic material is composed of dead organic matter which is an important source of energy for bacteria and for a significant number of animals. Thus an investigation of the living part of suspended organic matter is not enough if one is to understand the productivity of sea water. The methods for collecting particulate organic material now in use do not enable one to make a direct separation of the living and dead fractions or to evaluate the quantities of detritus in different stages of decomposition. The elaboration of such methods is essential for better work on the role of suspended organic matter in the cycles of transformation in the seas and oceans. At present one can only make mainly indirect quantitative estimations of the living and dead fractions of the particulate organic matter.

In the present report we give an account of the quantitative relationship between particulate organic matter, phyto- and zooplankton, pigments and primary production in the Black Sea, Azov Sea, Arabian Sea and the tropical part of the Atlantic Ocean. The data in the discussion were obtained in recent years by the research workers of the Institute of the Biology of the Southern Seas, Sevastopol.

MATERIAL AND METHODS

The method for the collection of organic particulate matter was as follows: a fine glass powder (3 parts by weight) was mixed with sea water (97 parts by weight, prefiltered through a membrane filter of $0\cdot2$-$0\cdot3$ μ pore size). This suspension (10 ml) was filtered through a membrane filter of 3-5 μ pore size. A thin sediment of glass powder remained on the filter. That filter was employed for the collection of particulate matter from samples of sea water. The filters were dried at 50-60°C and stored in a desiccator.

The glass powder (with seston) was easily separated from the membrane filter. For determination of the total particulate organic matter an oxidation with dichromate was employed as modified by Ostapenya (1965). The oxidation of the sample was carried out at 140°C for 15 minutes in a $0\cdot1$N solution of $K_2Cr_2O_7$ in concentrated H_2SO_4.

It should be noted that sea water chlorides react with dichromate and this can lead to errors. A correction was made from a blank using glass powder suspended in filtered sea water.

The results of the analyses were expressed in mgC/m³. It was assumed that 1 mg O_2 is the equivalent of $0\cdot67$ mg of organic matter, or $0\cdot33$ mg C. For the calculation of organic carbon from chlorophyll a data, it was assumed that in the Black Sea and Azov Sea 1 mg of chlorophyll a was equivalent to 35 mg C and in the Arabian Sea and the tropical part of the Atlantic Ocean 1 mg of chlorophyll a was equivalent to 75 mg C. These factors were assumed constant for all layers in the euphotic zone, irrespective of light adaptation of the algae. To calculate carbon units from wet weight data we assumed a ratio for zooplankton of $1:0\cdot065$ and for bacteria of $1:0\cdot09$.

RESULTS OF THE DETERMINATIONS OF THE TOTAL QUANTITY OF THE
PARTICULATE ORGANIC MATTER

The largest values for the concentration of particulate organic matter were found in the Sea of Azov, where the seston was between 0·7 and 1·4 mg C/l or 10 and 20 g C/m² in the summertime. It should be noted that variations in seston concentration were small during the course of the summer months although chlorophyll values varied by a factor of 10 times during this period. The same picture was noticed with the particulate organic matter concentrations in Sevastopol Bay during the course of the year, where the quantity of the particulate organic matter was within limits 6·2-9·8 g C/m² (mean 7·7 g C/m²) with the exception of short periods of spring and autumn phytoplankton blooming.

The very great quantity of particulate organic matter in the Sea of Azov is, of course, the result of the huge river discharge.

In the deeper parts of the Black Sea the mean concentration of particulate organic matter was 0·26 mg C/l (Finenko, 1965). This value changed no more than twice during the course of the year. In the coastal parts of the Black Sea the concentration of particulate matter increased to twice that of the open sea. In the Aegean Sea, the Adriatic Sea and the Ionic Sea quantities of particulate organic matter in the 0-50 m layer were 0·07-0·09 mg C/l (Sushchenya, 1961). In the Mediterranean Sea off Rhodes the quantity of particulate organic carbon (Sushchenya, 1961) was nearly the same as in waters off the African coast, i.e. 0·05-0·09 mg C/l (Ostapenya and Kowalewskaya, 1965).

Particulate organic matter has been investigated mainly in surface waters, which is why the vertical distribution of particulate matter and the total quantity under 1 m² of surface is not so well known. However, the data in hand shows that there may be no great difference in the quantity of the particulate organic matter in the different parts of the Mediterranean Sea. As a rough approximation it may be calculated that the quantity of the particulate organic carbon in the euphotic zone (0-150 m) is 7-15 g C/m².

In the tropical part of the Atlantic Ocean the concentration of the particulate organic carbon is considerably higher than in the Mediterranean Sea (Sushchenya and Finenko, 1966). In the Canary Current values of 0·53-1·30 mg C/l were found, values found only for waters of the highest productivity. Values of such magnitude were also reported in the Arabian Sea in an area near Karachi (Finenko and Zaika, 1968). In the equatorial waters of the Atlantic Ocean the quantity of particulate organic matter was somewhat lower (0·1-0·4 mg C/l). The lowest values in the tropical Atlantic water were found in the central part (0·1-0·25 mg C/l). The same values were found also in the central part of the Arabian Sea (Finenko and Zaika, 1968).

Although the data in Table 1 are approximate it is clear from these that the mean concentration of particulate organic matter in different seas varied 20-40 times, whereas quantities under 1 m² of the euphotic zone varied by no more than 5 times.

TABLE 1. Average quantities of particulate organic matter in the different seas

Sea	Thickness of the euphotic layer (m)	Particulate organic matter in euphotic zone in mg C/l	under 1 m^2 (gC)	Author
Azov Sea	10	0·7 -2·0	10-20	Finenko, 1965 a, b
Black Sea	50	0·2 -0·6	11-30	Finenko, 1965 a, b
Mediterranean Seas	150	0·05-0·10	7-15	Sushchenya, 1961; Ostapenya, Kowalewskaya, 1965
Atlantic Ocean	75	0·20-0·55	20-30	Riley, Wangersky, Hemert, 1964; Sushchenya, Finenko, 1966
Atlantic Ocean (oligotrophic waters)	100	0·10-0·25	7-18	Sushchenya, Finenko, 1966
North Sea	50	0·10-0·30	5-15	Krey, 1960; Steele, Baird, 1961; Ostapenya, Kowalewskaya, 1965
Pacific Ocean (oligotrophic waters)	100	0·10-0·15	10-15	McAllister, Parsons, Strickland, 1960
Arabian Sea	100	0·10-0·30	10-20	Finenko, Zaika, 1968

QUANTITATIVE RELATIONS BETWEEN THE PLANKTON AND PARTICULATE ORGANIC MATTER

Data concerning the relation between the quantities of the living and dead organic matter obtained in recent years have led to the rather unexpected results that the dead fraction comprises a great part of the total (Sushchenya, 1961; Jørgensen, 1962; Finenko, 1965). However, it must be noted that in these papers, the true role of the living organisms in seston is underestimated, as methods for plankton investigation now in use are not perfect. In fact, the role of the living fraction of the seston is greater than may be indicated from papers cited.

Data, obtained in the Black Sea, Azov Sea, the Arabian Sea, and in tropical

TABLE 2. Relation between the main components of particulate organic matter in some seas (average values for the euphotic zone)

Sea	Particulate organic matter mg C/m^3	Phyto-plankton	Zoo-plankton	Bacteria	Dead organic matter
		Percentage of total particulate organic matter			
Sea of Azov	750-1500	5-10	3-10	0·3-7	80-92
Arabian Sea	100-250	1-31	—	—	—
Black Sea	200-250	0·2-1	5-20	0·4	78-95
Tropical part of the Atlantic Ocean along 15th meridian	450-600	0·5-1·3	0·6	—	98-99
along 16th parallel	100-250	0·6-1·3	0·7	—	98-99

areas of the Atlantic Ocean, concerning the relation between the main components of the seston (Table 2), have no great precision or accuracy, but are sufficient to give the order of the values (Finenko, 1965a; Finenko and Zaika, 1968; Sushchenya and Finenko, 1966). The living part of the seston comprises 1-20% in the above areas. In coastal waters the living part of the seston comprises 10-50% and in the course of the spring bloom it may have a value of up to 100% (Finenko, 1965b).

The lack of data from the other seas makes it difficult to draw comparisons between the different areas. Nonetheless, we have attempted to find the relation between concentration of detritus and the total particulate organic matter. Detritus comprises about 90% of the seston, with the exception of the North Sea, where the living part is 30% or more of the seston (Fig. 1).

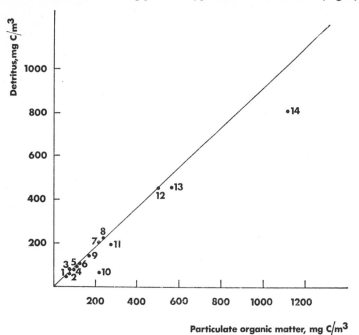

Fig. 1. Relation between the concentration of detritus and particulate organic matter (average values for the 1 m³ of the euphotic zone). 1—Aegean Sea, 2—the Mediterranean Sea, 3—the Ionic Sea, 4—the Adriatic Sea, 5—the Sargasso Sea, 6-8—different areas of the Atlantic Ocean, 9—the Pacific Ocean, 10—the North Sea, 11—the Black Sea, 12—the Riga bay of the Baltic Sea, 13—the Indian Ocean, 14—the Sea of Azov.

RELATIVE QUANTITY OF CHLOROPHYLL *a* IN PARTICULATE ORGANIC MATTER

In a number of investigations it has been shown that the quantity of chlorophyll *a* in total particulate organic matter may serve as one of the indices of the plankton abundance. In the open part of the Black Sea the ratio of chlorophyll *a* to the particulate organic matter is in the limits 1:5000-1:8000 in summer. At depths greater than 75 m this ratio sharply decreases (3-4 times; Finenko, 1964).

The same is characteristic of the equatorial parts of the Atlantic Ocean and Arabian Sea, where the ratio of chlorophyll *a* to the organic carbon of the seston in the 0-70 m layer is equal to 1:3000-1:10 000. This ratio is found mainly when chlorophyll *a* concentrations in the plankton have values below 0·1 mg/m³.

When the phytoplankton is abundant, as in the Azov Sea and in coastal waters of the Black Sea and Arabian Sea, 1 mg of the chlorophyll *a* corresponds to 35-500 mg C of the total particulate organic matter.

The following results were obtained from investigations in these different seas:

1. The ratio of chlorophyll *a* to the particulate organic matter is not constant and may vary a thousandfold or more.
2. The value of the ratio of chlorophyll *a* to seston is correlated with phytoplankton abundance and is greater the more phytoplankton.

When analysing this ratio, some authors assume that chlorophyll *a* and particulate organic matter are linearly related (Steele, 1964; Steele and Baird, 1961). This is indeed the case when phytoplankton concentrations are high. However, where the phytoplankton biomass comprises only a small part of

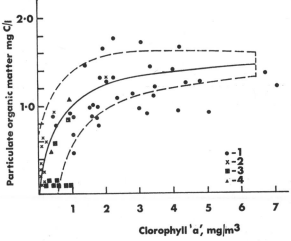

Fig. 2. Relation between particulate organic matter and chlorophyll "a" concentrations in plankton. 1—the Azov Sea, 2—the Black Sea, 3—north-west part of the Atlantic Ocean, 4—the Indian Ocean.

total seston, the relation is not linear. This is especially true in the Sea of Azov, where the chlorophyll concentrations in the different regions varied a hundredfold and more, but concentration of the particulate organic matter varied only threefold (Fig. 2).

From the data from the Azov Sea, we derive the following equation:

$$C = \frac{5x}{1+3\cdot5\ x\pm2}$$

where C is the concentration of the particulate organic matter in mgC/1 and x is the chlorophyll a concentrations in mg/m^3.

Fig. 2 also presents data from the Black Sea, the tropical and northwest parts of the Atlantic Ocean and the Indian Ocean (Finenko, 1965a; Ostapenya and Kowalewskaya, 1965; Sushchenya and Finenko, 1966; Kutjurin and Ulubekova, 1961). It may be seen that the data from different seas are in close correspondence with the theoretical curve obtained from data on the Azov Sea, although there are few points in the high chlorophyll region (more than 5 mg/m^3). It is seen that the ratio of chlorophyll a to the particulate organic matter is not constant and this ratio has not a straight-linear relationship with the chlorophyll concentration.

RELATION BETWEEN THE PRIMARY PRODUCTION AND PARTICULATE ORGANIC MATTER

A priori, the quantity of detritus must have a causal relation to the value of the primary production and it is of interest to compare primary production under 1 m^2 with the detritus quantity under 1 m^2. The relationship has no regularity. This is best explained by remembering that primary production is a function of the euphotic zone only but that the detritus thus generated is distributed over a thicker layer. There is some reason to suppose that detritus in deep water is not the same age as in a euphotic zone and has no direct causal relation with recent primary production.

TABLE 3. Average quantities of the particulate organic matter in different layers of the Arabian Sea

Station	Layer m	Particulate organic matter mg C/m^3	Percentage of the upper layer content
14°16′ N 51° 54′ E	0-200	70	100
	200-500	48	68
	500-1500	37	53
20° 26′ N 62° 21′ E	0-100	110	100
	100-500	64	58
	500-1500	51	46
	1500-3000	70	63
21° 23′ N 66° 30′ E	0-200	308	100
	200-500	181	42
	500-1500	236	54
16° 24′ N 70° 23′ E	0-100	370	100
	100-200	398	108
	200-500	217	58
	500-1000	220	59
	1000-3000	234	63

There is no direct evidence for the age of detritus nor method for deciding what fractions are biochemically reactive or nonreactive. We can only use an indirect method of calculation to obtain an order of magnitude for these values. It is known that the more reactive fraction of the organic matter is decomposed in the upper 200 m of the sea. Below 200 m the seston is mainly composed of more stable organic matter. Supposing that the concentration of this more stable organic matter is distributed uniformly in all depths, then the difference between the concentration of the particulate organic matter in the layer 0-200 m and in the layer below 200 m may give some idea about the quantity of the more reactive and younger organic matter. This calculation was made for the Arabian Sea where samples of the particulate organic matter have been collected in the entire water column (Table 3). Results show that the stable organic matter comprises an average of about 50% of the total particulate organic matter. This value is probably too large as we have not taken into account the role of the bathypelagic forms of the zoo- and ichthyo-plankton. Moreover, it is known that large dead plankton organisms reach the deep water relatively intact.

A knowledge of the decomposition rate of the more reactive organic matter permits us to calculate when some part of the detritus is oxidized (Skopintsev, 1949). After decomposition of the more reactive organic matter a stable part remains which is 6-21% of the initial quantity and which is then oxidized very slowly. If the rate-constant of decomposition is 0·05, as it is in the case of dead mixed plankton and some diatoms (Skopintsev, 1949), then 90% of the organic matter must be oxidized in 20 days at about 25°C. Now the concen-

TABLE 4. Quantities of active and stable particulate organic matter in the Arabian Sea (layer 0-200 m)

Station	Average daily value of the primary production g C/m^2 . day	Stable organic matter mg C/m^3	Active organic matter (mg C/m^3) values calculated from:	
			primary production data	ratio of organic matter in different layers
14° 16' N 51° 54' E	0·101	50	18	22
20° 26' N 62° 21' E	0·414	47	74	53
21° 23' N 66° 30' E	0·462	217	83	100
16° 24' N 70° 23' E	0·883	221	160	166

tration of organic matter in sea water stays relatively constant so that if all nonstable organic matter is oxidized in 20 days, this process must be equalled by the entry of material into the system. Hence, in 20 days there must be formed a quantity of organic matter equal to the quantity decomposing. The source of that matter is the primary production. That is, the quantity of re-active organic detritus formed is approximately equal to the primary pro-duction occurring over 20 days. Such calculations have been made for those stations where parallel determinations of the primary production and particulate organic matter have been made (Table 4). It may be seen that average quantities of the reactive organic matter calculated from primary production data are in rather good correspondence. Thus, approximate calculation leads to the conclusion that, in the Arabian Sea, biochemical cycles involved, about 50% of the particulate organic matter in the 0-200 m layer and about 4% in the whole water column.

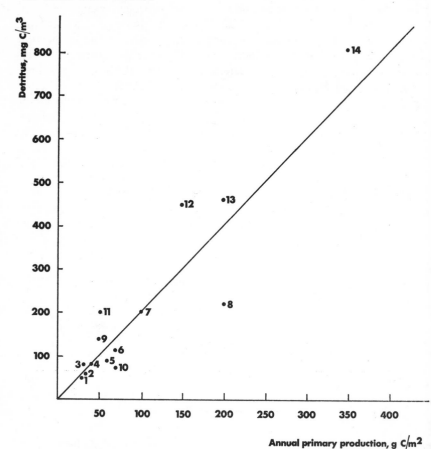

Fig. 3. Relation between the concentration of detritus in 1 m³ of the euphotic zone and annual primary production. Marks as in Fig. 1.

Proceeding from these data of the vertical distribution of the "young" and "old" seston, it may be supposed that there is a correlation between the average values of the annual primary production in the whole euphotic zone and the average detritus concentration in 1 m³ of this zone. The comparison of these values for the different seas shows straight-linear relation between two parameters (Fig. 3). It is seen from Fig. 3 that the average concentration of detritus in 1 m³ is more than twice the annual primary production under

1 m². In Fig. 4 is shown the relation between the annual primary production and the quantity of the detritus in the euphotic zone.

The linear relation of these parameters is seen only for the seas where the euphotic zone is thicker than 50 m. Apparently the main part of the unstable detritus is concentrated in this layer and we then compare the primary production and nearly the whole reactive detritus. In seas where the euphotic zone is small a great part of the detritus is outside the euphotic zone. In seas with deep euphotic zones, as is seen from Fig. 4, the detritus comprises about

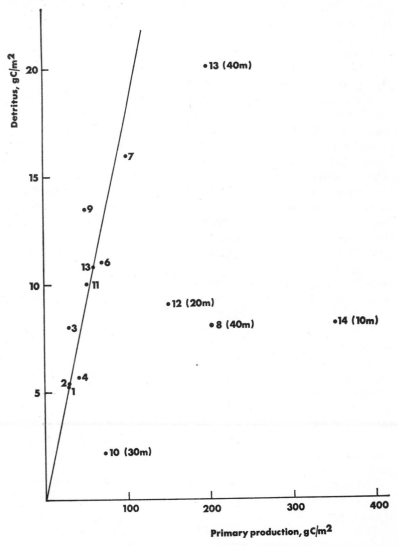

Fig. 4. Relation between the quantity of detritus in the euphotic zone and annual primary production. Marks as in Fig. 1. In brackets the thickness of the euphotic zone.

18% of the annual primary production. Hence, the detritus must turn over about 5 times in the year. If, as we suppose, in the upper layers of the seas about 50% of the detritus is unreactive, then 50% of the biochemically active detritus must turn over about 8-10 times in the course of the year. However, the detritus may also take part in biological cycles in the form of food for some detritophagous organisms and bacteria. This reasoning leads us to conclude that the reactive fraction of the detritus turns over about once a month or more.

The zooplankton is important in this process. From data on zooplankton respiration, we may calculate their food requirements. In oligotrophic waters of the tropical part of the Atlantic Ocean the biomass of the zooplankton in the 0-100 m layer is near to 3 g wet weight under 1 m^2 (Sushchenya and Finenko, 1966). The dry weight, taken as 15% of the wet weight, is about 450 mg/m^2. It is known that the average respiration of mixed marine zooplankton is about 0·125 mg C/mg of the dry weight a day (Menzel and Ryther, 1960). Hence, for the respiration of 450 mg of the zooplankton we need 56 mg C per day. The primary production has a value 250 mg C/m^2 (Sushchenya and Finenko, 1966), and about 22% of the total synthesized carbon in the euphotic zone is used for the respiration of the zooplankton. The synthesis of the organic carbon by the phytoplankton is the source of food for the zooplankton in a whole column of water. In the layer 0-500 m the zooplankton biomass is 6·5 g (wet weight) or about 975 mg (dry weight) under 1 m^2. Assuming the coefficient of utilization of assimilated food is 0·5 for these conditions, and the coefficient of the assimilation of the consumed food is 0·7, the food requirement of the zooplankton is about 390 mg C/m^2 per day. This calculation demonstrates that the food requirement of the zooplankton in the 0-500 m layer is one and a half times more than the primary production in this region. It is obvious that detritus and bacteria must have a great role in the feeding of the zooplankton.

In the Black Sea in the period June-September the food requirement of the zooplankton is 500-1000 mg C/m^2 per day and primary production has a value 120-250 mg C/m^2. It must be noted that in this case we have a potential value for the primary production in the absence of predation, as in bottle experiments on primary production determination there are very small quantities of phytoplankton predators. The net production of the phytoplankton in the sea is about 2-3 times smaller than the potential production (Kondratjeva, 1967). Hence, the same picture is seen in the Black Sea as is seen in the Atlantic Ocean, that is, detritus must play a great role in the feeding of zooplankton.

Such approximate calculations based on net zooplankton data clarify the role of detritus as the source of organic matter for heterotrophic micro-organisms.

To conclude we give the results of some observations on the micro-organisms living in the corpses of the dead copepod *Acartia clausi*. Crustacea were killed and placed in sea water from the Sevastopol Bay of the Black Sea, or with the "old detritus" from a flowing aquarium kept at about 20-25°C.

After one hour of exposure small flagellates were seen in the corpses of the dead crustacea. In this time the tissues of crustacea were partly disintegrated and micro-organisms had the opportunity to penetrate into the corpses. In the experiments with "old detritus", *Uronema*, small Euplotes and sometimes nematodes were seen in the animals after 3 hr. After penetrating the dead *Acartia*, small Flagellata and the Infusoria *Uronema* increase very rapidly in number. If the flagellates penetrated *Acartia* free from Infusoria, then about 300-500 flagellates were found in the corpses after 2 days. The flagellates then decreased in number and after 4 days 100-200 individuals only were present, many of them incysted. After 5 days of exposure there were no more than 40-50 flagellates.

When there is mixed population of Flagellata and *Uronema* in the corpses the former do not reach a high number. After 1 day exposure in such cases there are about 180-200 Infusoria and no more than 70-80 flagellates. In a separate experiment we determined that *Uronema acutum* and *U. marinum* divide every 2-3 hr. In the case of the mixed population of the flagellates and Infusoria the corpses of *Acartia* are free from tissue after 3 days exposure.

In these experiments there was no direct control of the bacterial number but we noted that when there were no Infusoria there were lots of bacteria. When Infusoria were present, the number of the bacteria was visibly lower. We think that *Uronema* are not histiophagous, but feed on the bacteria. Hence, in dead crustacea there may be a particular food chain: dead tissues-bacteria-Flagellata and Infusoria. Thus, in a summer period in coastal waters of the Black Sea one corpse of dead *Acartia* may give rise to a number of hundreds of small flagellates and Infusoria after 2 days. The bodies of *Acartia* are free of biochemically active organic matter after 3-4 days and then are a component of the more stable "old" detritus.

REFERENCES

FINENKO, Z. Z. 1964. Chlorophyll quantity in plankton of the Black Sea and Azov Sea. *Okeanologia*, 4 (3), 462-68. (In Russian).
FINENKO, Z. Z. 1965a. Quantity of the particulate organic matter in seston of the Black Sea and Azov Sea. In *Issledovanija planktona Chernogo i Azovskogo morey.* Ed. by W. A. Wodyanitzky. Kiev, Naukova dumka. (In Russian.)
FINENKO, Z. Z. 1965b. Relation between primary production, chlorophyll and organic carbon in the sea plankton. In *Vop. gidrobiol. Tes. dokl.* 1 *s'esda vses. gidrobiol. Obshch.* Moscow, Nauka. (In Russian.)
FINENKO, Z. Z., and ZAIKA, V. E. 1968. Particulate organic matter in the Arabian Sea waters. *Okeanologia* (in press). (In Russian).
JØRGENSEN, C. B. 1962. The food of filter feeding organisms. *Rapp. P.-v. Réun. Cons. perm. int. Explor. Mer,* 153, 99-107.
KONDRATJEVA, T. M. 1967. *Production and daily variations of the phytoplankton in the south seas.* Dissertation, Odessa University, pp. 36. (In Russian).
KREY, J. 1960. The role of detritus in the sea. *I.C.E.S., C.M. 1960, Plankton Comm.,* Pap. no. 47.
KUTJURIN, V. M., and ULUBEKOVA, M. V. 1961. Intensity of the photosynthesis of the algae and method determination of the photosynthesis from chlorophyll quantity. In *Pervichnaya produkciya morey i vnutrennich vod.* Ed. by G. G. Winberg, Minsk. Belorussk. Univ., 249-56. (In Russian.)

MCALLISTER, C. D., Parsons, T. R., and STRICKLAND, J. D. H. 1960. Primary productivity at station "P" in the north-east Pacific Ocean. *J. Cons. perm. int. Explor. Mer*, **25**, 240-59.

MENZEL, D. W., and Ryther, J. H. 1960. The annual cycle of primary production in the Sargasso Sea off Bermuda. *Deep Sea Res.*, **6**, 351-67.

OSTAPENYA, A. P. 1965. Oxidation completeness of the organic matter from water invertebrata when using the method of bichromate oxidation. *Dokl. Akad. Nauk belorussk. SSR*, **4** (4), 273-76. (In Russian.)

OSTAPENYA, A. P., and KOWALEWSKAYA, R. Z. 1965. Particulate organic matter in the upper layer of the sea. *Okeanologia*, **5** (4), 649-52. (In Russian.)

PARSONS, T. R., 1963. Suspended organic matter in sea water. In *Progress in oceanography*, 1. Ed. by M. Sears. London and New York, Pergamon Press, 205-39.

RILEY, G. A., WANGERSKY. P. J., and HEMERT, D. V. 1964. Organic aggregates in tropical and subtropical surface waters of the North Atlantic ocean. *Limnol. Oceanogr.*, **9** (4), 546-50.

SKOPINTSEV, B. A. 1949. Decomposition rate of the organic matter from plankton. *Trudy vses. gidrobiol. Obshch.*, **1**, 34-45. (In Russian.)

STEELE, J. H. 1964. A study of production in the Gulf of Mexico. *J. mar. Res.*, **22** (3), 211-22.

STEELE, J. H., and BAIRD, I. E. 1961. Relation between primary production, chlorophyll and particulate carbon. *Limnol. Oceanogr.*, **6**, 68-78.

SUSHCHENYA, L. M. 1961. Some data about organic seston in the Aegean Sea, the Ionic Sea, and the Adriatic Sea. *Okeanologia*, **1** (4), 664-70. (In Russian.)

SUSHCHENYA, L. M., and FINENKO, Z. Z. 1966. Suspended organic matter content of the tropical Atlantic waters and some quantitative ratios between its components. *Okeanologia*, **6** (5), 835-47. (In Russian.)

Some problems related to the food chain in a tropical estuary

S. Z. QASIM
Biological Oceanography Division
National Institute of Oceanography
Cochin—11, South India

ABSTRACT. An annual cycle of gross and net primary production was determined for the Cochin Backwater (a tropical estuary) along with phytoplankton and zooplankton crops. It then appeared that the bloom of phytoplankton and growth of zooplankton were somewhat out of phase. Throughout the year the phytoplankton production far exceeded the rate of consumption by the zooplankton herbivores.

If it is assumed that most of the organic material which is produced in an ecosystem is likely to be consumed by the animal communities, then there is every possibility that the production of large quantities of surplus food may lead to several "alternate pathways" in the food chain of the estuary. Some of these have been shown and discussed. The precise nutritive requirement, however, of any species or group is not yet known.

INTRODUCTION

The backwaters of Kerala (India) include a chain of shallow, brackish water lagoons and swamps of the tropical zone. These are hundreds of miles long and are located along lat. 9°58'N and long. 76°15'E. The northern part of the backwaters, adjacent to Cochin Harbour, is called the "Cochin Backwater". It has a permanent connection with the Arabian Sea which forms the main entrance to the Cochin Harbour. This region is subjected to regular tidal influence and has all the characteristics of a typical estuary.

In recent years the Cochin Backwater has been a centre of intensive studies on the hydrography and primary productivity. The following account of the food chain formed part of a more detailed investigation carried out recently on the organic production of this estuary (Qasim *et al.*, 1969).

METHODS

Fig. 1 shows the stations in the estuary from where sub-surface tows of

45

phytoplankton (with no. 25 net) and zooplankton (with no. 3 net) were made throughout the year. Measurements of primary production were carried out at station 2 by using oxygen and ^{14}C methods. All experiments dealing with the rate of photosynthesis were conducted *in situ* throughout the year, at different depths of the euphotic zone, which was predetermined from the compensation depth as 1 % of incident illumination.

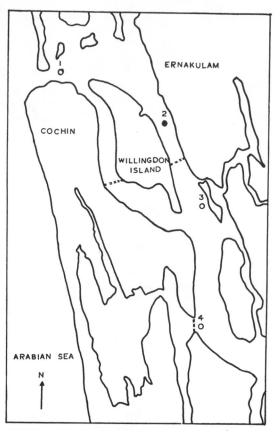

Fig. 1. Map showing a portion of the Cochin Backwater and the stations from where data on phytoplankton and zooplankton crops were obtained. Closed circle indicates station 2 where measurements on primary production were made.

The phytoplankton crop was estimated by taking settled volume and the zooplankton crop by taking displacement volume of the total organisms in relation to the volume of water filtered by the nets. Other parts of the same programme included studies on the microbial flora, fish and benthos.

Gross production was estimated by the oxygen method while the ^{14}C assimilation was taken as the net production. The average value of respiration from the oxygen decrease in the dark bottle during light-dark bottle experiments was taken as the respiratory loss which probably gave rise to the difference between gross and net production. This was done on the assumption that the rate of respiration occurring in the dark is the same as that during the corresponding periods of light in conjunction with photosynthesis. A further assumption that the rate of respiration remains constant throughout day and night made it possible to estimate the 24 hr (daily) net-production, after subtracting the respiratory loss occurring during the night from the daytime production. This 24 hr net production was taken to be the potential source of organic matter which is transferred to the next trophic level and has been mentioned here as the basic food.

RESULTS AND DISCUSSION

Seasonal changes in the column production (gross and net) have been shown in Fig. 2, together with the average respiration for the euphotic zone. As can

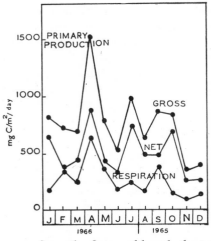

Fig. 2. Annual cycle of gross and net primary production in the Cochin Backwater together with the seasonal changes in community respiration. The values of production and respiration refer to euphotic zone.

be seen from the figure, although the annual cycle of primary production forms three small peaks, the fluctuations hardly exceed twofold. It can thus be concluded that primary production in the backwater goes on at a uniform level, with little seasonal increase, throughout the year.

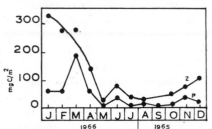

Fig. 3. Relation between phytoplankton (P) and zooplankton (Z) taken from parallel hauls on the same day.

Using the conversion factor, 96 mg C for each ml of zooplankton by displacement (Cushing *et al.*, 1958) and 1·4 mg C for each ml of settled phytoplankton (Sverdrup *et al.*, 1942) the zooplankton and phytoplankton crops have been shown in Fig. 3 in terms of their carbon equivalents. It is clear from the figure that maximum zooplankton in the estuary occurred during Jan.-April (pre-monsoon months) when the salinity in the backwater is high. During this period the estuary becomes virtually an arm of the adjoining sea. From May, with the onset of monsoon rains, the salinity in the estuary begins to fall and during the following months (June-Sept.), typically brackish water occupies the entire euphotic zone. These changes in the salinity are accompanied by a marked increase in the turbidity of water throughout the monsoon months. The zooplankton crop during the monsoon months remains at a low level and does not show any appreciable change until December.

The ratio of gross photosynthesis to the carbon content of phytoplankton has been shown in Fig. 4 as production coefficient (Riley *et al.*, 1949), from which it can be seen that the range in the ratio is very large (3·7-119). This indicates that the phytoplankton crop sampled by the plankton-net included only a small fraction of the total carbon assimilation (1-28%), as measured by the rate of photosynthesis. During the period when the ratio was large (monsoon and post-monsoon months), the crop was mainly dominated by such "small forms" (nanoplankton) which were not retained by the net; and similarly the low ratio obtained from November to March signifies that the net sampled nearly one-quarter of the standing crop consisting mainly of larger marine diatoms and dinoflagellates. Fig. 4 also includes the ratios

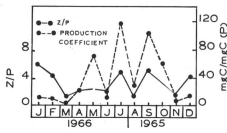

Fig. 4. Relation between zooplankton: phytoplankton (Z/P) and production coefficient (photosynthesis:carbon content of phytoplankton).

of zooplankton to phytoplankton (Z/P). It can be seen from the figure that the seasonal fluctuations in Z/P fall in accordance with a similar increase and decrease in the production coefficient. This indicates that the type of phytoplankton bloom in the estuary is so adjusted that, with the seasonal changes in the zooplankton composition, it provides the specific food requirement of the herbivores.

Taking the daily metabolic requirement of tropical zooplankton in terms of carbon as 12% of its dry weight (Menzel and Ryther, 1961), the seasona, changes in the rate of consumption of the basic food have been given in Fig. 5, along with gross and net primary production. The rate of consumption works out as 10% of the production by plants during the monsoon months, 20% during the post-monsoon months and 40% during the pre-monsoon months.

The estimated gross production in the estuary ranges from 270-295 g C/m²/year (average = 280 g C/m²) while the net production, for days only, is 180-200 g C/m²/year (average = 195 g C/m²). The average daily net production (24 hr) is approximately 124 g C/m²/year. The estimated annual con-

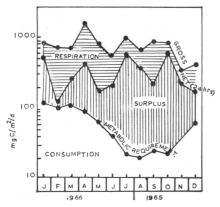

Fig. 5. Relation between metabolic requirement of zooplankton and gross and net primary production. The lower portion shows the approximate consumption of primary production by zooplankton herbivores; the middle portion shows the approximate surplus of basic food and the upper portion indicates the average respiration (24 hrs.).

sumption by the zooplankton herbivores is only about 30 g C/m². This indicates that there is a large surplus of basic food in the estuary. A general lack of zooplankton herbivores seems to be the main factor responsible for such an inefficient utilization of the primary production, which in turn may be due to changing environmental features of the estuary. From the general composition of the zooplankton crop, however, there seems no evidence to suggest that the herbivore population is kept at a minimum because of the presence of a large number of carnivorous forms.

The "apparent wastefulness" of the primary production, however, may not remain so well marked, if we consider that the unconsumed material may perhaps be used more efficiently by the other members of the estuarine ecosystem. This raises the possibility of several "alternate pathways" in the food chain. One such pathway may be directly linked with the herbivorous fishes (mullets) which are always present in the estuary in appreciable numbers. Another direct link from the basic food may be with the omnivorous prawn and shrimp populations through detritus, as the backwater is well known for its rich crustacean fishery (see Fig. 6). Finally, a substantial portion of the basic food which keeps sinking to the bottom may be utilized directly by the sedentary animal communities. The small depth of the euphotic zone greatly increases the rate of loss from this zone and thus prevents an excessive outburst of the phytoplankton bloom at any one time of the year. The unconsumed food which, on reaching the bottom becomes a part of the sediment, may be chemically and biologically altered, but remains in the system to take part in the cycle of events.

In a typically marine environment it has been suggested that nearly 80% of the primary production is consumed by the zooplankton hervibores, while the remaining 20% goes to the demersal animal communities (Steele, 1965). In the oceanic environment of the Sargasso Sea off Bermuda, the utilization of the basic food by the zooplankton herbivores is even more efficient and works out nearly 100% (Menzel and Ryther, 1961). These figures seem certainly true of such environments where the primary production is spread over a large column of water and has to travel a long distance, even below the

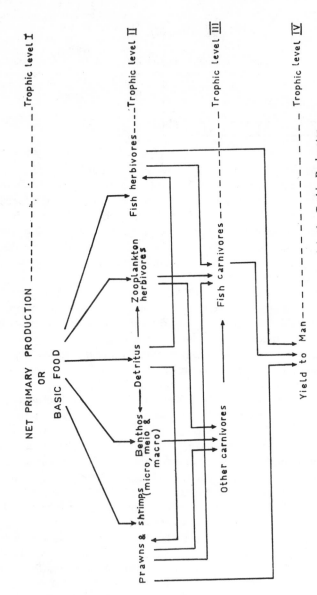

Fig. 6. Generalized representation of a food web in the Cochin Backwater.

euphotic zone, before it decays and gets chemically altered. Thus its chances of being grazed are invariably high. In a shallow estuary an increase in the benthic herbivorous forms thus compensates for the lack of zooplankton, and probably makes the system just as efficient. It is therefore less surprising to find the primary production in the estuary far exceeding the rate of consumption by the zooplankton herbivores.

Based on the above discussion, a diagrammatic and rather simplified representation of the food chain in the estuary is given in Fig. 6. The web has been constructed according to the generalized scheme enunciated by Steele (1965). It is not intended to give any final answer to the problems of food chains in the estuary, but to give an overall picture of the transmission of food to different trophic levels, I-IV. The representation, though to a large extent arbitrary, indicates some possible links within each trophic level and from one to the next higher trophic level. The several links within the same trophic level, as shown in Fig. 6, recall Steele's (1965) hypothesis of a greater "ecological cannibalism" which leads to competition and a decrease in the amount of food transmitted to the next higher trophic level. Such a situation is to be expected in a tropical estuary where, to start with, a fairly uniform rate of production is maintained throughout the year, which probably induces a large assemblage of herbivorous feeders.

ACKNOWLEDGEMENTS

I wish to thank Dr A. D. Ansell and Dr Ann Trevallion for their very helpful discussions.

REFERENCES

CUSHING, D. H. (and others). 1958. Report on the committee on terms and equivalents. *Rapp. P.-v. Réun. Cons. perm. int. Explor. Mer*, **144**, 15-16.
MENZEL, D. W., and RYTHER, J. H. 1961. Zooplankton in the Sargasso Sea off Bermuda and its relation to organic production. *J. Cons. perm. int. Explor. Mer*, **26**, 250-58.
QASIM, S. Z. (and others). 1969. Organic production in a tropical estuary. *Proc. Indian Acad. Sci.*, **59**, 51-94.
RILEY, G. A., STOMMEL, H., and BUMPUS, D. F. 1949. Quantitative ecology of the plankton of the western North Atlantic. *Bull. Bingham oceanogr. Coll.*, **12**, 1-169.
STEELE, J. H. 1965. Some problems in the study of marine resources. *Spec. Publs int. Commn N.W. Atlant. Fish.*, **6**, 463-76.
SVERDRUP, H. U., THOMSON, M. W., and FLEMING, R. H. 1942. *The Oceans: their physics, chemistry and general biology.* N. Y., Prentice-Hall, Inc. 1087 pp.

Food transfer through the lower trophic levels of the benthic environment

NELSON MARSHALL
Professor of Oceanography
Graduate School of Oceanography
University of Rhode Island
Kingston, R.I. 02881, U.S.A.

ABSTRACT. At representative shoal benthic environments in southern New England the measured organic input from primary production sources approximates 300 g C/m²/yr and the total may be greater. This organic matter is utilized rapidly in the upper layers of the sediments. At maximum rates, respiration by the interstitial microbiota and meiofauna would account for all of this depletion; however the demand of the interstitial community is not obligatory. Also many of the larger consumers compete directly for basic organic food available from the sediment environment and make little use of the trophic pathway from microbiota to meiofauna.

It is suggested that these shoal areas exhibit an intricate dynamic balance. If a portion of the organic input is carried away by currents or is utilized more directly by the macrofauna, utilization by the interstitial community may be curtailed. If organic matter tends to accumulate, interstitial demand may increase and offset this. Such a system, in balance, is not optimal for harvests beneficial to man. Aquaculture efforts should be directed to well-conceived manipulative measures, both biological and non-biological, which would disrupt the balance.

The role of various larger consumers is discussed. Particular attention is directed to the functional adaptations of benthic filter feeders, some of which are edible forms and are effective in competing directly for the organic input. More than is commonly realized, these filter feeders utilize benthic as well as planktonic food sources and enrich their own environment with their deposition.

INTRODUCTION

The harvests in shellfish, fin fish, etc., surveys of the biota, and some initial studies of rates of production have directed attention to the production potential of the shoal, benthic, sediment environments of coastal and estuarine areas. My interest was focused on such environments on noting harvests

52

of bay scallops, *Aequipecten irradians*, as high as 20 g C/m²/yr from a shoals area of about 2 km² in the Niantic River estuary in southern New England.

For an analysis of trophic relationships in such habitats I will consider sites within a 60 kilometre stretch along the southern shore of New England, U.S.A., Fig. 1. The data presented apply to shoals behind inlets connecting

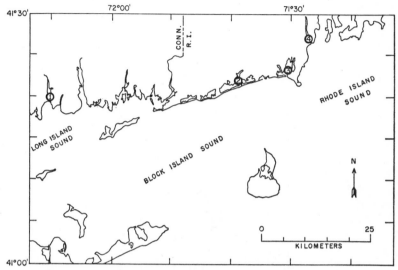

Fig. 1. The location of the principal shoals environments referred to: 1. Niantic River, 2. Charlestown Pond, 3. Point Judith Pond, 4. Pettaquamscutt River. Three of these are estuarine environments; Charlestown Pond is a lagoon area. All shoals considered are close to inlets.

with the ocean or larger sounds, Fig. 2. Sediments are primarily of siliceous sand, with a small silt-clay component and not infrequently with miscellaneous larger components including shell fragments and gravel. Depths over the shoals are generally less than a metre at mean low water. Tidal currents range up to one-half knot, and the tidal range is about one metre. These are lagoon sites or are at the mouths of estuaries of little tributary flow, and salinities generally range between 20 and 30‰. Vegetation cover varies from dense stands of *Zostera marina* to the relatively open areas which usually develop mixed algal stands in mid-summer. These shoals habitats are not to be confused with intertidal flats, which are also characteristic of this region of coastal submergence.

Such shoals habitats have been studied rather intensively by myself, my colleagues, and students at the University of Rhode Island. Thus some of the data extend considerably beyond the literature, being available through the courtesies of my associates.

In seeking general concepts, data obtained from different locales are applied broadly to others, assuming a uniformity which, critically considered, does not exist. An effort is made to qualify interpretations and to offer only such generalizations as seem warranted in view of this liberal application of the data.

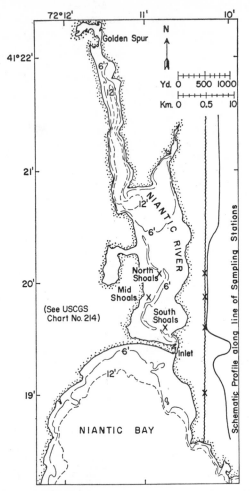

Fig. 2. The Niantic River estuary, a representative site. The schematic profile portrays the gross physiographic form of the shoals.

SOURCES OF ORGANIC MATTER

The first step in considering the trophic relationships is to estimate the input of organic matter to the benthic or sediment enviroment of the shoals. Sources include:

1. Decay of *Zostera*, its aufwuchs community, and of mixed stands of algae; also deposition by grazers on these populations.
2. Photo- and chemo-synthesis of the microflora of the sediments.

3. Settling of phytoplankton by sinking or via the faeces and pseudofaeces of suspension feeders.
4. Settling out of allochthonous matter draining from tidal marshes, from the tributaries, or from offshore and other source areas.
5. Processes whereby additional organic matter present in a dissolved state may be transferred in an available state to the sediments.

Data on the input of the benthic macroflora are available for Charlestown Pond from J. T. Conover's studies (1964), and from personal communication with Conover. The maximum standing stock of this vegetation was 87 g C/m² (average of three stations in August). Most of the algal component disintegrates into the sediments each year and Conover estimates that approximately two-thirds to three-quarters of the *Zostera*, the blades of which tend to break off in the growing season, also decays into the sediments annually. The addition from the substantial aufwuchs of the *Zostera* is estimated at 20 g C/m²/yr, based on Conover's data on primary production of the benthic plant community, an assumption that approximately half of this is *Zostera*, and Conover's interpretation that aufwuchs production is about one-sixth that of the host plants (slightly higher than indicated by Brown, 1962). Finally, in considering the input derived from the macroflora, one must include the feeding and subsequent deposition by herbivores throughout the growing season. Thus I suggest (Table 1) 125 g C/m² as an estimate of the total yearly addition to the sediments from the macroflora community.

TABLE 1. Estimates of organic carbon contributed to estuarine shoals in southern New England

	gm C/m²/yr
Macroflora, including *Zostera* and its aufwuchs, also macroscopic algae	125
Benthic microflora	90
Deposition of phytoplankton	50
Allochthonous matter	0-10
Conversion of dissolved organic matter (aggregates)?	No estimate
	265-275 plus unknowns

Data on the rate of production of the benthic microflora, and thus on the organic input from this source, are available from my observations in which small, *intact* and undisturbed samples of the sediment were exposed to ^{14}C and photosynthesis was calculated from scintillation counts indicating the isotope uptake. The results of the latest observations (Table 2), which also includes some specifics on procedures, indicate a net production of 30 mg C/m²/hr at two representative stations. Preliminary supplementary observations suggest that rates might be less in the shade of a dense vegetation cover. An earlier series of ten sets of observations made at the Pettaquamscutt estuarine site, for all seasons over a two-year period starting November 1964, averaged 39 mg C/m²/hr. These earlier sets involved only three light and three dark flasks and the counting was for beta radiation. The earlier work also indicated an average of 8 mg C/m²/hr for seven sets of observations representing all seasons at the north shoals of the Niantic estuary and 9 mg C/m²/hr for six sets of observations over the year at the south shoals station there. I have reservations as to whether the slow rates for the Niantic habitat are characteristic. They may reflect inadequate sampling and difficulties in the early techniques.

TABLE 2. Rate of production of the benthic microflora of the sediments of estuarine shoals habitats, as ascertained using intact core samples 0·9 cm in diameter and 0·5 cm thick. Replicates were placed in six light and four dark flasks containing filtered water from the shoals with ¹⁴C added, and held at the sediment surface for 3½ hours during mid-day. Organic carbon was removed from samples by wet oxidation, trapped in NaOH and counted by liquid scintillation.

		mg C/m²/hr		
		East station in Pettaquamscutt River		Station in Point Judith Pond*
February	1967	32		
March	,,	12		
April	,,	27		
June	,,	75	{ 34 / 158 / 32	
July	,,	55	{ 38 / 72	42
August	,,	74	{ 43 / 106	34
September	,,	47		
October	,,			39
November	,,	13		19
December	,,	1		28
January	1968	27		
February	,,			11
March	,,	3		19
April	,,	15		
Average		32		27
Average of two stations				30

* This station is about 0·4 km north of the inlet (see Table 3 for the Pettaquamscutt River station).

Taking nine hours as an estimate of the productive period of an average day and a rate a little less than 30 mg C/m²/hr as representative of microflora production in these habitats, 90 g C/m²/yr is suggested as the entry for Table 1. This is lower than Grøntved's (1960) suggestion of 116 g C/m²/yr for the microflora in sediments of shoals in fjords in Denmark that, from the author's descriptions and my own observations (Marshall and Wheeler, 1965), seem very similar to the environment in southern New England. Since Grøntved worked with sediment samples dispersed and separated into fractions, he had to estimate corrections to allow for important natural environmental limitations not operative in his procedure. He did attempt a correction for the limited light available to microflora in the sediments but did not correct for other space limitations that might be operative in natural compact sediment conditions.

There are three sources of data on the net rate of phytoplankton production in the locales being discussed, all based on oxygen changes in light and dark bottles. Rates determined by R. L. Davis (reported in Marshall, 1967) on nine observation dates (all seasons) on samples from over the Niantic shoals average 41 g $C/m^2/yr$. Rates by Conover (1964) for Charlestown Pond average 69 g $C/m^2/yr$ for gross production, which suggests a net rate similar to that for the Niantic. Smayda (1962) reported a net rate of phytoplankton production of 51 g $C/m^2/yr$ for Charlestown Pond waters.

For use in this discussion it is necessary to estimate how·much such net production adds to the shoals. Since Marshall and Wheeler (1965) could not discern changes in phytoplankton counts in upstream-downstream samples taken along the course of the tidal flow over the shoals of the Niantic estuary, it is suggested that the losses from the phytoplankton population equal the growth of the population. These losses involve the settling of this organic input onto the sediments through death, sinking and decay, and through the biodeposition (after respiratory utilization) by zooplankton and benthic filter feeders. As a generalization for the phytoplankton addition by such means, 50 g $C/m^2/yr$ is entered in Table 1.

There are no data on allochthonous organic matter carried to the shoals. Probable sources include transport from tributary drainage basins, from bordering marshes, and excesses from biological production taking place inshore and offshore from the shoals. Like the phytoplanton, allochthonous material must reach the sediments either by sinking or biodepositon. Ash free dry weight determinations were made on the suspended matter concentrated on filters from water taken over the shoals in the Niantic estuary (Marshall and Wheeler, 1965). As mentioned for the phytoplankton, an upstream-downstream station pattern was used, sampling as the tidal currents crossed the area. The suspended components included, of course, both the non-living seston and the plankton, but, as just noted, the latter did not change; futhermore the total weight did not change sufficiently to show differences under the technique used. While such negative observations do not preclude a net contribution of allochthonous matter, they do suggest that additions from such sources may be small; so a value of 0-10 g $C/m^2/yr$ is entered in Table 1.

No determinations were made on dissolved organic matter in the waters over the shoals. My colleague, John McN. Sieburth, is currently conducting experiments which indicate (personal communication) that representative littorine phaeophytes may exude over 2000 g C/yr for every m^2 of dense beds. He further suggests that these dissolved organic compounds may condense into compounds of higher molecular weights and precipitate as organic aggregates. Possibly there is a substantial input through such routes since dissolved organic matter may be derived from populations outside the area as well as from macroflora over the shoals. (A dissolved organic and aggregate contribution from the microflora of the sediment would not be in addition to production estimates presented for that community.)

DISPOSITION OF ORGANIC MATTER

The disposition of the organic matter contributed to the shoals may be as follows:

1. Some may be oxidized in the metabolism of the interstitial community of bacteria, microalgae, protozoans, meiofauna, etc., within and on the surface of the sediments.
2. Some may be taken up by larger consumers, infauna and epifauna, and, after respiratory loss, may be returned via the excretion, biodeposition and the ultimate death and decay of these consumers (thereby re-entering these first steps).
3. Some that is consumed may, after respiratory loss, be channelled into trophic pathways leading to the export of organic matter from the area by natural routes, such as migration, or via the fisheries.
4. Some may be carried away from the area by currents.
5. Some may accumulate in the sediments.

Data showing low levels of organic matter in the sediments at the shoals sites, Table 3 (from Tietjen, 1966), Table 4 from recent studies on Point Judith Pond, and Fig. 3 from Conover (1964) indicate that very little accumulates. The living interstitial community comprises part of this benthic organic content and the balance must include highly refractory compounds. Some organic matter buried in the sediments may, through the action of the infauna (part of step 2 above) or by storm effects, be returned to the sediment surface and used later if it is not highly refractory.

TABLE 3. Organic carbon (by dichromate oxidation) and percent silt-clay at estuarine shoals habitats as reported by Tietjen. Mean values for the top 5 cm from a monthly year around sampling programme (1964-5)

	Stations in the Niantic River		Stations in the Pettaquamscutt River	
	*North	South	†East	West
Organic carbon,‡ % by wt	1·3	0·8	1·2	1·0
Silt-clay content, % by wt	15·5	4·1	12·1	1·3

* North station is about 1·3 km north of inlet; South station is about 0·4 mm from the inlet.
† East and West stations are on opposite sides of estuary about 2 km from the inlet.
‡ For use in this table, Tietjen's entries, reported as organic matter, are divided by two as an estimate of organic carbon.

The obvious effects of tidal currents, wind-driven currents and waves suggest that water transport is paramount in preventing accumulation of organic matter, generally sufficiently buoyant to be carried off by moderate currents. The contrast (Table 3) between the low silt-clay content at the south shoals station as compared to the high value at north shoals in the Niantic estuary where the maximum currents are 47 and 18 cm/sec, respectively, illustrates the work of the currents. The usual, though moderate built-up of organic matter in deeper areas adjacent to shoals (as shown in Fig. 3), is the typical end result. This should not be interpreted as just scour and fill, however, for with reduced circulation, a tendency toward stagnation, and salinity inter-

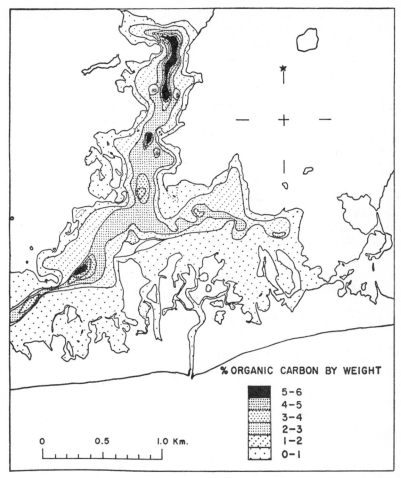

Fig. 3. Central area of Charlestown Pond showing per cent of organic carbon in the surface layers of the sediments. Depth and percent of organic matter are closely correlated: the area with less than 1% C is almost entirely less that 0·5 m deep at mean low water; the area with 5-6% C is almost entirely over 2 m deep. (From Conover, 1964.)

actions between the deeper and the surface waters, there are many accumulation forces affecting the depressions and basins.

As to the conditions on the shoals, it is important not to overlook evidence that the biological demands of such environments may equal or nearly equal the organic matter input. This, also, is demonstrated in Charlestown Pond for there are extensive protected shoal areas with persistent low organic levels where the fetch is too limited for vigorous wave action and the maximum tidal currents are less than 10 cm/sec (Conover, 1964).

Table 5 lists the organic matter demand calculated from the decrease in dissolved oxygen in dark B.O.D. bottles containing small, intact sediment samples (see the table for specifics on procedure). Since these results exceed the estimates of the organic matter input given in Table 1, questions arise as to whether the reported input suggestions are too low or whether unknown organic sources, such as from dissolved organic matter, are substantial. The results of the demand observations may be too high, due perhaps to an accelerated oxidation of reduced compounds in the sediment even though care was taken to keep the samples intact, or due to distortion from the bacteria wall effect on the flasks or from other unknown technique complications.

At this symposium McIntyre, Munro and Steele, reporting laboratory work with columns simulating a sandy beach ecosystem, also mention oxygen demands in excess of the organic input they can acount for. On the other hand Wieser and Kanwisher (1961) reported lesser oxygen requirements in the surface muds low in the intertidal zone at a nearby marsh site on Cape Cod, Massachusetts. Their data, which indicate a demand of 19 and 14 mg C/m²/hr for an observation in the fall and one in the spring respectively, are about half the average for fall and spring dates in Table 5.

Though our observations on the oxygen demand of the interstitial community may involve errors of consequence, it is probably meaningful that the highest demands reported were for the north shoals station of the Niantic

TABLE 4. Organic carbon (by combustion in a CHN analyser), per cent silt-clay, and E_h from two adjacent sampling sites on the shoals of the Point Judith Pond estuary. Number of samples indicated in brackets.

	No Zostera Cm beneath sediment surface				
	Top	2½	5½	8½	12½
Org C, % by wt					
Mid-July	0·9 (9*)	0·9 (10)	—	—	0·9 (9)
Mid-August	0·9 (5)	0·8 (5)	1·9 (1)	1·4 (1)	0·6 (5)
% silt-clay					
Mid-July	4 (5)	—	—	—	—
Mid-August	5 (8*)	—	—	—	—
E_h (mv)					
Mid-July	− 58 (10)	− 124 (9)	− 152 (9)	− 153 (9)	− 122 (9)
Mid-August	− 22 (5)	− 79 (5)	− 163 (5)	− 144 (5)	− 100 (5)
	Dense Zostera cover Cm beneath sediment surface				
	Top	2½	5½	8½	12½
Org C, % by wt					
Mid-July	1·6 (10)	1·3 (10)	—	—	0·7 (7)
Mid-August	1·2 (5)	1·2 (5)	1·6 (1)	1·4 (1)	0·7 (5)
% silt-clay					
Mid-July	13 (9)	—	—	—	—
Mid-August	14 (5)	—	—	—	—
E_h (mv)					
Mid-July	− 166 (10)	− 225 (10)	− 216 (10)	− 196 (10)	− 164 (10)
Mid-August	− 89 (5)	− 145 (5)	− 186 (5)	− 182 (5)	− 174 (5)

* One extreme outlier value not averaged in.

estuary where the organic content is relatively high and the quantity of silt-clay is considerably greater than at other stations. Also, if the respiratory rates of the interstitial community are high as the foregoing comments suggest, one would expect the organic carbon content to be lowered immediately at and just beneath the surface layers of the sediment as is demonstrated by the data of Table 4 and Fig. 3. Such a high respiratory demand at the surface zone of the sediments is also suggested by Tietjen (1968). The sharp drop in E_h in the top cm of the sediments, illustrated in Table 4, is undoubtedly associated with anaerobic respiration. Anaerobic conditions may suggest an accumulation of organic matter; however, Tietjen (1966) reported only minor differences in organic levels in the upper and lower portions of the 5-cm cores he studied from the shoal habitats. Apparently, in these environments the sharp drop to anaerobic conditions reflects the very high organic demand of the interstitial bacteria operative below levels where oxygen is available from photosynthesis or from exchange with the overlying waters.

TABLE 5. Utilization of organic carbon by the interstitial community (microbiota and meiofauna) of estuarine shoals habitats as determined from the oxygen demand of sediment samples 0·9 cm in diameter and 0·5 cm thick kept intact and held in dark B.O.D. containers for 24 hr. Except for losses noted, entries before November 1964 are averages of three replicate observations; entries thereafter are averages of four observations

| | | Stations in the Niantic River | | East station in the Pettaquamscutt River |
		North mg C/m²/hr	South mg C/m²/hr	mg C/m²/hr
July	1964	52		
August	,,			16
September	,,	85		57*
October	,,	15		
November	,,	69		38
December	,,		29	
January	1965	43		1
February	,,		13	
March	,,	53		42*
April	,,		9	18
May	,,	75*		
June	,,		60	48
July	,,	119		
August	,,		40	58
September	,,	98		
Average		596 g C/m²/yr	263 g C/m²/yr	305 g C/m²/yr
Average for three stations: 425 g C/m²/yr				

* One sample discarded because of a procedure difficulty.

From the foregoing, and irrespective of any criticism that I may be extending limited data beyond reason since generalizations for several areas are based on observations from relatively few sites, one point seems persistent and consistant, namely the end result, the low organic content of these sediments. A simple interpretation ties together both the foregoing observations, which suggest that the low organic content is the result of a high interstitial demand and the well-established fact that many of these shoals are highly productive of shellfish and are important nursery grounds for young fish which later migrate to deeper waters. While the interstitial community is potentially very active and must be important in releasing nutrients which sustain productivity, its requirements are not obligatory and the larger consumers can, at least in part, circumvent this community in their feeding. If some organic matter is carried off with the currents or is utilized directly by macrofauna, as will be discussed below, interstitial respiration may be curtailed. On the other hand if organic matter should tend to accumulate, interstitial utilization, particularly that of the fast-growing bacteria populations, might increase rapidly and counteract this (a plausible explanation for high demands observed in the B.O.D. flasks). A system in dynamic balance is portrayed here, with a variable organic input, some export of this organic supply by currents, some utilization by macrofauna, and a high loss through the respiration of the interstitial biota. It should be noted that, as a system in balance with high demands by the interstitial community operative as soon as the organic matter becomes available, the amount left for larger consumers is not optimal.

PRACTICAL IMPLICATIONS

There are several practical points to be derived from such an interpretation:
1. Aquaculture research should focus on manipulative measures, biological and non-biological, designed both to augment rates of production and to disrupt the balance in the shoals environment in order to counteract the tendency for available organic matter to be lost immediately via the high respiratory demand of the interstitial community. Such an emphasis would parallel the fundamental technique of agriculture wherein relatively closed, balanced, often climax, systems are disrupted to obtain a yield.
2. Trawling, the digging and raking common on shoals, and other fisheries practices which disturb the sediments may not only free the environment of some of the demands of the balanced state but may augment production. When, for example, we have dispersed sediment samples and measured production by the microflora, the rates are much higher than in the intact and naturally compacted sediments.
3. If such shoals were not balanced systems and were not fished, organic matter might accumulate in excess of utilization and lead to an adverse eutrophication of the environment. However, it might be suggested, in view of low organic carbon levels commonly encountered in marine environments (for example, the levels are not very high even in the deposition areas shown in Fig. 3), that high utilization by the interstitial community and an approach to balance is widespread in the marine environment.
4. While the idea of deliberately disrupting the environmental system is most readily considered for shallows where aquaculture seems feasible, the basic premise may have application in the future management of larger bodies of water. For example, pursuing Odum's suggestions (also presented in this symposium) that, through aquaculture, we might take advantage of the fact mullet feed directly on benthic organic matter,

would be to apply this thinking to fin-fish, to other locales, another climate and perhaps to very extensive bodies of water.

MEIOFAUNA CONSUMERS

Further data are available on the meiofauna fraction of the total interstitial community. Tietjen (1966) reported on a year around, detailed monthly sampling of these populations at four shoals stations. He estimated the organic matter demands of the meiofauna by applying respiration rate data obtained by other investigators, most notably the rates indicated by Wieser and Kanwisher (1961) for nematodes. For the north and south stations on the Niantic shoals and the east and west stations of the Pettaquamscutt, he calculated that meiofauna respiration amounts, respectively, to 52, 58, 50 and 74 gm $C/m^2/yr$. This is between one-fifth and one-tenth of the total utilization of the interstitial community, about as expected for a group which, according to the review of their habits by Tietjen (1966), is in part competing with the interstitial microbiota for the utilization of sediment organic materials and is in part predatory upon those smaller forms.

The suggestion is often made that the meiofauna, utilizing both organic sources and the microbiota in the sediments, being larger in size than the microbiota, and being very abundant, must constitute a highly significant food source for the bottom macrofauna. If this is truly the major route to higher trophic levels, without significant alternatives, the suggestion above that interstitial demands may be curtailed by competing macrofauna utilizing sediment organic sources more directly, is not valid.

Tietjen (1966), noting the paucity of meiofauna, particularly of the nematodes, in stomach analyses which have included bottom feeding fishes and other representative benthic consumers, thinks the emphasis on meiofauna as food is exaggerated. In discussing trophic relationships on the Fladen ground of the North Sea, McIntyre (1961) noted that non-selective deposit feeders of the macrofauna population must ingest meiofauna present but the meio- and macrofauna also compete for the limited supply of organic matter, thus he states "a lower meiobenthos population might lead to a more efficient production of fish food on these grounds". Without very complete feeding habit studies over a broad representation of browsing groups, the importance of meiofauna to higher trophic levels will not be fully clarified. If the meiofauna are important, the chain seems to require intermediate consumers, an annelid worm as a link between the meiofauna level and the browsing fish for example. As interpreted in this paper and in keeping with McIntyre's suggestion, there are more efficient pathways.

COMPETING MACROFAUNA CONSUMERS

Thus it is appropriate to direct attention next to the larger consumers which in part feed on the interstitial community but also compete with this community for the basic organic matter available in these environments. As will be noted, filter feeders as well as grazers and deposit feeders utilize the sediment resources. Utilization rates are not available for the macrofauna populations as a whole but the pace may be suggested by the observations of Gordon (1966), indicating that the polychaete, *Pectinaria gouldii*, may assimilate all nutritionally available organic matter in its immediate surroundings (a column $3 \cdot 5$ cm $\times 6$ cm) in less than a month. These worms occur where studied in Barnstable Harbor, Cape Cod, in number ranging from 10 to 40 per m^2. Such populations are but a fraction of the total macrofauna, with numbers which are almost always far in excess of 1000 per m^2 as the smaller forms are counted (see, for example, Phelps (1964) for details on Charlestown Pond populations and for comparative data). Thus it seems highly probable that observations on the total assimilation would show populations taking in organic carbon at rates many times the annual input rate, while in an overall situation, total respiration per unit of time obviously cannot exceed the rate of input of organic matter, the excess assimilation is possible through rapid recycling or turnover.

I will discuss at some length unique features of the trophic role of filter feeders, at least as interpreted through our studies of representative bivalve molluscs. Contrary to a common assumption, the suspension feeding bivalves of the benthic environment may feed largely on benthic materials. In analysing the gut contents of bay scallops, *Aequipecten irradians*, Davis and Marshall (1963) found both benthic and planktonic diatoms, with the former types outnumbering the latter nine to one. Tenore (1967), using detritus tagged with radioactive zinc and phosphorus, observed that the marsh-clam, *Rangia*, gets a major share of its organic foods from the sediments. During this symposium G. Thorsson pointed out that *Tellina* and *Macoma* feed on material in and on the sediments. Scallops, in an aquarium with diatomaceous earth spread over the bottom, will suck up this pinkish film very rapidly. Scallops will also draw in diatomaceous earth sprinkled on the upper shell, which in nature serves as an aufwuchs garden.

Recognizing that an average-sized scallop lies with its intake approximately 1 cm off the bottom, one must assume that many other filter feeders, with intake apertures that are not well clear of the sediment surface, probably also draw in benthic food materials. The turbulent boundary layer of reduced current velocities, forming a somewhat distinct epibenthic environment (Marshall, 1967) into which light sediment materials may be sucked up and held in suspension, should enhance such bottom feeding.

This concept, that filter feeders of the benthos may depend largely on food in the sediment environment, can, of course, be over-stated. If some of the benthic microflora is being ingested, defaecated and reingested, gut counts will exceed utilization. The fact that the production rates of the benthic microflora and of the plankton over the shoals are about equal, yet the scallop gut contains nine times as many benthic diatoms as it does plankton forms may reflect such recycling of the benthic populations. Furthermore, some of the bottom organisms seem primarily adapted to feeding on plankton and seston, for example, oysters, which prosper on their own reefs, and mussel populations, living on rocks in the swift tidal currents or in wave zones. It has proved feasible to raise many shellfish on plankton and positive correlations between phytoplankton concentrations and the growth have been shown in nearby areas for quahogs, *Mercenaria mercenaria* (Pratt and Campbell, 1956) and for soft-shell clams, *Mya arenaria* (Matthiessen, 1960). Finally, to overlook the role of filterers in concentrating plankton and suspended matter would obscure their significance in the deposition of seston taken from the large volumes of water that cross the shoals.

Several important trophic functions of benthic filter feeders merit comment. First the varied and abundant bivalves, annelids and other forms which filter organic matter from the sediment environment serve as a direct and thus as an efficient trophic link from these organic sources to predator populations. In the trophic scheme these are competing with the meiofauna, as discussed above. The rate of growth and ultimate size of these organisms must be augmented appreciably by their capability of drawing on both the planktonic and benthic food resources. Several of the suspension feeders, most notably the edible shellfish, constitute efficient pathways from basic organic matter sources to harvests utilized by man. These are species to consider for aquaculture efforts designed to draw heavily on the organic supply to the sediments in a manipulated environment, as suggested above.

On analysing the rate of waste output, particularly the pseudofaeces and faeces of the marsh mussel, *Modiolus demissus*, Kuenzler (1961) concluded that they "seem to be more important as biogeochemical agents than as energy consumers in the salt marsh ecosystem". A similar role has been demonstrated for the common oyster, *Crassostrea virginica*, and two other bivalves, barnacles and a tunicate by Haven and Morales-Alamo (1966). Data from Moore (1966) on bay scallops from the shoals we are studying in southern New England indicate that, in areas of moderate abundance ($20/m^2$), scallops incorporate 123 g $C/m^2/yr$ in biodeposits.

Much of the deposition by bottom forms involves returning and thus recycling organic matter already on the bottom, an action which may not alter the sources appreciably, either as a food supply or as a reserve from which nutrients are released. On the other hand the quantities of phytoplankton and the non-living seston incorporated in the biodeposition of suspension feeders may, in this habitat of high oxidation by the microbiota, be subject to accelerated mineralization. For example Moore (1966) reports that the biodeposition of scallops solubilizes in fifteen days, and Seki's (1968) recent data on slightly deeper waters suggest that mineralization in the muds may exceed that of the overlying waters by two orders of magnitude (a point in harmony, incidentally, with foregoing comments that organic matter may be quickly used up in the sediments). It follows that deposition functions, thereby, as a food substrate for the interstitial biota, as nourishment for the macrofauna, and as a reservoir of nutrients (see Tietjen (1967) for a note on biodeposition serving as food for nematodes, Jeffries (1962) for comments on benthic nutrient release, and Moore (1966) for initial work, which is being pursued further by R. L. Davis, on regeneration rates from bivalve biodeposits). Thus the benthic filter feeders, many of which are highly adapted to benefit from the shoal habitats and the water transport across the shoals, effectively nourish their own environments or, to quote Moore (1966), "fertilize their own pastures".

REFERENCES

BROWN, CHARLES LOGAN, Jr. 1962. *On the ecology of aufwuchs of* Zostera marina *in Charlestown Pond, Rhode Island.* Ph.D. thesis, Univ. of Rhode Island.

CONOVER, J. T. 1964. Environmental relationship of benthos in salt ponds. *Tech. Rep. publ. Health Serv., Grant No. WP-0023,* (3) processed.

DAVIS, R. L., and MARSHALL, N. 1963. The feeding of the bay scallop, *Aequipecten irradians. Proc. natn. Shellfish Ass.,* 1961, **52,** 25-29.

GORDON, D. C., Jr. 1966. The effects of the deposit feeding polychaete, *Pectinaria gouldii,* on the intertidal sediments of Barnstable Harbour. *Limnol. Oceanogr.,* **11,** 327-32.

GRØNTVED, JUL. 1960. On the productivity of microbenthos and phytoplankton in some Danish fjords. *Meddr Danm. Fisk.-og Havunders.,* N.S. 3, 55-92.

HAVEN, D. S., and MORALES-ALAMO, R. 1966. Aspects of biodeposition by oysters and other invertebrate filter feeders. *Limnol. Oceanogr.,* **11,** 487-98.

JEFFRIES, HARRY P. 1962. Environmental characteristics of Raritan Bay, a polluted estuary, *Limnol. Oceanogr.,* **7,** 2-31.

KUENZLER, EDWARD J. 1961. Phosphorus budget of a mussel population. *Limnol. Oceanogr.,* **6,** 440-25.

MARSHALL, N. 1967. Some characteristics of the epibenthic environment of tidal shoals. *Chesapeake Sci.,* **8,** 155-69.

MARSHALL, N., and WHEELER, B. M. 1965. Role of coastal and upper estuarine waters contributing phytoplankton to the shoals of the Niantic estuary. *Ecology,* **46,** 665-73.

MATTHIESSEN, GEORGE C. 1960. Observations on the ecology of the soft clam, *Mya arenaria,* in a salt pond. *Limnol. Oceanogr.,* **5,** 291-300.

MCINTYRE, A. D. 1961. Meiobenthos of sub-littoral muds. *J. mar. biol. Ass., U.K.,* **44,** 665-74.

MOORE, J. K. 1966. *Biodeposition by the bay scallop,* Aequipecten irradians. Ph.D. thesis, Univ. of Rhode Island.

PHELPS, D. K. 1964. *Functional relationships of benthos in a coastal lagoon.* Ph.D. thesis, Univ. of Rhode Island.

PRATT. DAVID M., and CAMPBELL, DONALD A. 1956. Environmental factors affecting growth in *Venus mercenaria. Limnol. Oceanogr.,* **1,** 2-17.

SEKI, H. 1968. Relation between production and mineralization of organic matter in Aburatsubo Inlet. *J. Fish. Res. Bd Can.,* **25,** 625-37.

SMAYDA, THEODORE J. 1962. Some quantitative aspects of primary production in a Rhode Island coastal salt pond. *Proc. 1st natn. coastal and shallow Water Res. Conf., 1961.*

TENORE, KENNETH, 1967. *Some effects of the bottom substrate on the ecology of* Rangia cuneata *in the Pamlico River estuary.* M. S. thesis, N.C. State University.

TIETJEN, J. H. 1966. *The ecology of estuarine meiofauna with particular reference to the class nematode.* Ph.D. thesis, Univ. of Rhode Island.

TIETJEN, J. H. 1967. Observations on the ecology of the marine nematode, *Monhystera filicaudata* Allgen, 1929. *Trans. Am. microsc. Soc.,* **86,** 304-6.

TIETJEN, J. H. 1968. Chlorophyll and pheo-pigments in estuarine sediments. *Limnol. Oceanogr.,* **13,** 189-92.

WIESER, WOLFGANG, and KANWISHER, JOHN. 1961. Ecological and physiological studies on marine nematodes from a small salt marsh near Woods Hole, Massachusetts. *Limnol. Oceanogr.,* **6,** 262-70.

Part Two

PELAGIC FOOD CHAINS

Introduction

Any supermarket shows the complexity of the simplest link in a food chain. Despite the textbook simplification of algae/*Calanus*/herring, Hardy's (1924) original study revealed an array of food items for *Calanus* and a second array, including *Calanus*, for herring. In this symposium, Petipa has simplified such networks by grouping animals, but she has revealed that some animals proceed from level to level during their life histories and that there is consequently a number of interlocking trophic levels. As a study becomes more penetrating, it necessarily becomes more complex. But before the methods of complex study can be exploited, their bases must be established. For example, as Petipa has shown, the trophic levels may not always be completely identified. So, although plants and animals can be caught quantitatively, the quantities may be wrongly allocated. Furthermore, it is of little value to sample the quantities of plants and animals in the sea, if their growth rates and death rates are not sampled at the same time.

The algae can be counted and measured, estimated as chlorophyll, carbon or as particles and rather roughly the methods correspond. There is an enormous size range of algae and many are colonial. They provide food for a broad size range of animals, from nauplii and protozoa to copepods, euphausiids and appendicularians. It is likely that small animals feed on small algae and big ones on large cells (Mullin, 1963). So estimates of carbon or chlorophyll are perhaps less useful than size distributions of particles. However, because algae divide once or twice a day, information on their division rates is always needed. Ideally, they should be given by species in terms of environmental conditions, but in general such information is lacking. The radiocarbon method estimates increments to a population and can be used with methods of determining the quantity of carbon in the population. The Coulter counter can be used to estimate division rates by size groups. At the present time analysis of a food chain is limited at the primary level by our ability to measure division rates in the field.

The herbivores can be sampled with water bottles (for small nauplii and protozoa), fine nets (for smaller copepodites), coarse nets (for larger copepods) and with high-speed nets (for euphausids and fish larvae). Lohmann (1908) and Hentschel (1937) made good estimates of the smallest animals. The problem of escape from the mouth of the net can be solved by comparing catches of different sized nets (McGowan and Fraundorf, 1966). The animals are distributed in a patchy way. Because the variances are no greater than in other environments, the patchiness is probably of biological origin, due to variance in space and time of rates and quantities. In larval plaice studies carried out at Lowestoft, differences due to patchiness are very much less than differences due to the loss rate in time (Harding, in press).

If the quantities of herbivorous animals in the sea can be determined, their growth rates and death rates are not well known. The work of Mullin in this symposium is a valuable pointer to the type of information needed. There are two ways in which production of an animal can be estimated. First, if the duration of stages with temperature is known, the quantity at a given time divided by the duration gives the rate of production, which can be integrated over the life of that stage, the weight of which is known. Although laborious and needing specialized information, the method is complete and estimates production in weight throughout the life history, that at each stage being a unit. Growth rates and mortality rates can be estimated from the material. A second method is to multiply average weights by average numbers over a time period, which is sometimes not specified in terms of the life history of the animal. The first method is really preferable because the times are dictated by how long the stages are present in the water.

Networks proliferate in the secondary and tertiary levels. Indeed it is difficult to classify some animals as herbivores or carnivores unless their gut contents are continuously examined throughout the period of study. But many carnivores are not identified and indeed some may remain beyond the reach of planktonic capture; midwater trawls and acoustic methods might become necessary. Although the quantities of some carnivores are known, those of other carnivores remain unknown. Their growth rates and death rates are also unknown. So the work of Reeve in this symposium on the growth of *Sagitta hispida* is a considerable step forward.

For a fisheries biologist the purpose of food chain studies is to estimate ecological efficiency (Slobodkin, 1959). Primary production and secondary production can be estimated, and if the ecological efficiency is known, so can the fish production in unexploited areas of the world. Slobodkin's estimate of about 10% for *Daphnia* populations was based on a number of generations, and in the sea the coefficient should be based on at least one generation. Petipa's high coefficient for her bathyplankton community is based on a 20-day series of observations; perhaps a rapidly growing stock of young grazers was destroying an accumulated stock of algae.

Steele (1965) suggested that ecological efficiencies might be considerably greater than 10%. In temperate waters, during the spring, populations increase by many times from generation to generation. So we are concerned less with growth than with survival. When food is abundant, survival is high and the population increases; when food is scarce, survival is low and the population decreases. Under a regime of increasing population, the yield is greater and vice versa. But is the yield as ratio to the food intake of the population constant through these changes? Steele suggested that if herbivores could increase their populations in rich food supply and be harvested themselves at the end of their growth phase, ecological efficiencies of 25% might be obtained.

It is possible to examine the question in a study of upwelling areas. An upwelling area can be defined physically by the distribution of surface isotherms by months (U.S. Hydrogr. Office, Publ. 225, 1944). Because of divergences beyond about 100 km from shore, biological activity extends about 250 km offshore, from zooplankton and phosphorus observations; because of the positive correlation between phosphorus and zooplankton, it is likely that most of the phosphorus is regenerated. So the area and season of upwelling can be defined.

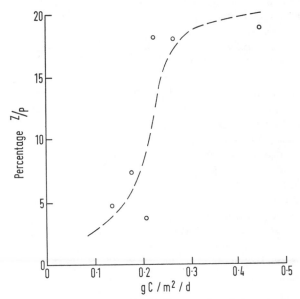

Fig. 1. Ratio of production of zooplankton to that of phytoplankton as a function of phytoplankton production.

The primary production is given by g $C/m^2/d$ × area (m^2) × season (in days). Koblentz-Mishke (1965) has summarized all observations in the Pacific and has expressed surface observations and *in situ* observations in the same way. Observations are available from Chile, Peru, the Costa Rica dome, the eastern tropical divergences, California, the divergences off New Guinea and north of the Marquesas.

Zooplankton observations as ml/1000 m³ from 100-300 m depth in the Pacific are given in Reid (1962); they can be converted to g C/m². There are some data on the duration of copepodite stages with temperature (Marshall and Orr, 1955; McLaren 1965) and it is possible to construct a rough relation between generation time and temperature. Then the production can be estimated from the observations of stock and the number of generations during the upwelling season.

Because most of the phytoplankton and most of the zooplankton are eaten, the ratio of production in each trophic level is very nearly an ecological coefficient. The ratio of zooplankton to phytoplankton (Z/P) can be plotted on the average productivity in g C/m²/d, as in Fig. 1. The three high figures are from Chile, Peru and California; the three low ones are from the zones of divergence off New Guinea, north of the Marquesas and in the eastern tropical Pacific. The data are taken from many workers and are averaged over areas and seasons. There is a suggestion that the Z/P ratio, as an estimate of ecological efficiency, increases with food concentration, thus supporting Steele's contention.

Like temperate production, an upwelling system is really a cycle of production. As algal numbers increase, herbivore survival improves, increasing the herbivore population by many times. Predation on the herbivores is low because the predators are few. Consequently the food is taken at maximum efficiency by the young stages, the effect of which is manifest in increased survival or more herbivores. The yield of herbivores is taken later when fish aggregate. Ecological efficiency here is the ratio of food uptake at two separate times and at different parts of the life history. Where food is less abundant, the cycle is less pronounced and survival is less. There is then less increase of population between generations and so food may remain unexploited. So ecological food conversion is less efficient.

There is little known about ecological efficiency in pelagic food chains. However, because of conveniences in sampling and because much of the network has been worked out, the pelagic food chain remains an attractive one to work on. But we need to solve the remaining sampling problems, particularly the problem of escape from the mouth of a net, and to estimate the duration of stages with temperature. The survival of young stages can be estimated under different conditions. The problem studied is then how a population regulates itself naturally, which is becoming the central problem of fisheries biology itself.

REFERENCES

HARDING, D. W. (in press) *The biology of larval fish*. Volume presented to Michael Graham. Logos Press.

HARDY, A. C. 1924. The herring in relation to its animate environment I. The food and feeding habits of the herring with special reference to the east coast of England. *Fishery Invest.; Lond.*, Ser. 2, 7 (3), 53 pp.

HENTSCHEL, E. 1937. Zur quantitativen Plankton Methodik. *Kieler Meeresforsch.*, 1, 322-6.

KOBLENTZ-MISHKE, O. I. 1965. Primary production in the Pacific. *Oceanology*, 5, 104-16.

LOHMANN, H. 1908. Untersuchungen zur Feststellung des vollständigen Gehaltes des Meeres an Plankton. *Wiss. Meeresunters., Abt. Kiel*, 10, 129.

MCGOWAN, J. A., and FRAUNDORF, V. J. 1966. The relationship between size of net used and estimates of zooplankton diversity. *Limnol. Oceanogr.*, 11, 456-69.

MCLAREN, I. A. 1965. Some relationships between temperature and egg size, body size, development rate and fecundity of the copepod *Pseudocalanus*. *Limnol. Oceanogr.*, 10, 528-38.

MARSHALL, S. M., and ORR, A. P. 1955. *Biology of a marine copepod*. Edinburgh, Oliver and Boyd. 188pp.

MULLIN, M. M. 1963. Some factors affecting the feeding of marine copepods of the genus *Calanus*. *Limnol. Oceanogr.*, 8, 239-50.

REID, J. L., Jr. 1962. On circulation, phosphate phosphorus content and zooplankton volmes in the upper part of the Pacific Ocean. *Limnol. Oceanogr.*, 7, 287-306.

SLOBODKIN, L. B. 1959. Energetics in *Daphnia pulex*. *Ecology*, 40, 232-43.

STEELE, J. H. 1965. Some problems in the study of marine resources. *Spec. Publs int. Commn N.W. Atlant. Fish.*, (6), 463-76.

U.S. Hydrographic Office 1948. World atlas of sea surface temperature. 2nd ed. *Publ. Hydrogr. Off.*, no. 225.

Growth and metabolism of two planktonic, marine copepods as influenced by temperature and type of food *

MICHAEL M. MULLIN and ELAINE R. BROOKS
Institute of Marine Resources
University of California
San Diego, P.O. Box 109
La Jolla, California

ABSTRACT. The planktonic copepods, *Rhincalanus nasutus* and *Calanus helgolandicus*, were raised from eggs to adults at 10° and 15°C on the diatoms, *Ditylum brightwellii* and *Thalassiosira fluviatilis*, as food. Rates of growth and ingestion were measured in terms of carbon per unit time, and the gross efficiency of growth was computed from these measurements. The laboratory-raised copepods were compared with wild copepods in terms of their sizes and respiratory rates.

INTRODUCTION

The particle-grazing, calanoid copepods, *Calanus helgolandicus* (*pacificus*) and *Rhincalanus nasutus*, occur commonly in the California Current (Fleminger, 1964). *Calanus* is the more abundant of the two in local waters off La Jolla, California. Both may be raised from egg to adult under controlled conditions in the laboratory (Mullin and Brooks, 1967) and are therefore suitable species with which to measure directly the gross efficiency of growth as affected by temperature and by the type of food which is available to the animals. The ratio of growth (net production) to ingestion will be termed the gross efficiency of growth, while net efficiency of growth refers to the ratio of growth to that fraction of ingestion which is actually assimilated. Gross and net efficiencies are often referred to in the Russian literature as the first and second order coefficients of utilization (K_1 and K_2) respectively. Since the ultimate goal of this research is to apply measurements made in the laboratory to the natural populations, comparisons between "wild" and "laboratory-raised" animals have also been made. Also of interest are metabolic differences which may be related to the distinctness of the ecological niches of the two species, since they are potential competitors.

* Supported by U.S. Atomic Energy Commission Contract No. AT(11-1) GEN 10, P.A. 20.

METHODS

Unialgal cultures of the planktonic diatoms, *Ditylum brightwellii* and *Thalassiosira fluviatilis*, were grown at 20°C in semi-continuous culture under constant light for use as food in the experiments. The culture vessels were 1-l gas washing bottles fitted to permit sterile withdrawal of cells and periodic introduction of Guillard's f-1 medium (Guillard and Ryther, 1962) from a reservoir. We felt that such cultures would give cells of more constant size and chemical composition than would a series of batch cultures (*cf.* Myers, 1962). Both diatoms grew as single cells in our cultures, rather than chains, and were thus easy to count electronically. *Ditylum* is the larger of the two, containing an average of 1700 picograms of carbon per cell (range 800-2900 pg C/cell in our experiments), while *Thalassiosira* contains 270 pg C/cell (range 130-380 pg C/cell). The ratio of cellular carbon to nitrogen is about 5:1 in both species.

Seawater from the Scripps seawater system was filtered through rinsed, 0·8-μ porosity membrane filters and stored in 20-l glass carboys in a cold room at the experimental temperature. About 100 l were filtered at a time and the supply was replenished approximately every two weeks.

Twenty-four hours before cells were to be used, they were suspended at the desired concentration in 8 l of filtered seawater plus an antibiotic (alternating 50 mg/l penicillin G or streptomycin sulphate) and kept in the dark at the experimental temperature. A fresh suspension was made up every 2 days. The pre-conditioning was an attempt to minimize changes in cellular carbon and in numbers of cells during the 2-day intervals of the experiments, since the cells could complete their last division before use.

The amount of carbon per cell changed considerably during the course of a complete experiment, and also decreased significantly within any 2-day interval in one of the two experiments tested. The concentration of cells also changed significantly within the 2-day intervals of sampling in 3 of the 4 experiments tested. These changes in cellular carbon and concentration were not large, however, and it was assumed that for any 2-day interval of an experiment the mean of initial and final control values gave a satisfactory estimate of the average conditions during the interval.

Gravid female *Calanus helgolandicus* and *Rhincalanus nasutus* were collected 4 km offshore from Scripps Institution of Oceanography, placed in filtered sea water at 12°C in small containers, and fed mixed phytoplankton and *Artemia* nauplii. Female *Calanus* were always available, and when female *Rhincalanus* were rare in the plankton, juveniles were grown to adults in 20-l carboys and allowed to breed, and the females produced were placed in smaller containers (Mullin and Brooks, 1967). All containers were checked daily and any eggs present were removed for hatching. The resulting nauplii were used to start the experiments.

A water bath containing a rack to hold 1-l beakers and enclosed by a dark cover was placed in a walk-in cold room. Both the bath and the cold room were held at the experimental temperature so that all manipulations could be carried out without subjecting the animals to temperature shock. The experimental containers were 1-l beakers fitted with plexiglass covers and small plexiglass plungers driven by a 3 r.p.m. motor, following the principle of the plunger jar (Browne, 1898). The plungers provided gentle agitation which kept the food in suspension and facilitated moulting of the animals.

An experiment was started by distributing several hundred stage I and II nauplii between 3 or 4 experimental beakers containing 1 l of the suspension of pre-conditioned cells. Every 2 days the water was changed. The animals were pipetted from the beakers and counted and identified under a dissecting microscope in the cold room, and dead animals were discarded. The beakers were rinsed with distilled water and filtered sea water and filled with a new suspension of pre-conditioned cells, and the animals were returned to them. The whole process of changing required less than 20 minutes per beaker. Two control beakers containing only the suspension of cells were treated in the same way.

Each time the water was changed, two initial samples of 450 ml each from the new cell suspension for the subsequent 2 days, and two similar samples from each of the control beakers for the previous 2 days, were filtered onto heat-cleaned (450°C for 4 hours) GF/C glass fibre filters, washed with about 15 ml of Na_2SO_4 (60 mg/l), and stored over silica gel in an evacuated desiccator for subsequent analysis of organic carbon. About 100 ml from each control and experimental beaker and from the new suspension were reserved for counts of cells. Counting was done with a Model A Coulter Counter (Hastings, Sweeney, and Mullin, 1962), with at least six counts made on each sample. The coefficient of variation on counts was 2% for *Thalassiosira* and 4% for *Ditylum*.

Organic carbon on the filters was measured as CO_2 on a Beckman Model 15A Infrared analyser after the method of Menzel and Vaccaro (1964) adapted for particulate samples. The amount of organic carbon was determined by comparison against dextrose standards. The coefficient of variation for measurements of carbon was 10%.

The amount of carbon ingested per copepod was calculated for each 2-day interval by the following expression:

$$\frac{(N_0 - N_2) \cdot C_0/N_0}{A} \qquad [1]$$

where N_0 is the mean of initial and final control cells/l; N_2 is the final cells/l in the experimental beaker containing copepods; C_0 is the mean of initial and final control μg C/l; and A is the number of animals alive in the experimental vessel. Ingestion for each 2-day interval was summed to give total ingestion over longer periods. The mean number of cells in an experimental beaker during the 2-day interval was calculated as:

$$\frac{N_0(1-e^{-G})}{G} \qquad [2]$$

where G is ln (N_0/N_2). Use of this expression assumes that the copepods feed at a constant rate during the 2 days. The mean amount of cellular carbon present in an experimental beaker was calculated by multiplying expression [2] by the amount of carbon per cell, C_0/N_0.

Dead animals were treated in the calculation as if they had not fed during the preceding 2-day interval. This is one of several possible assumptions concerning the feeding of such animals. The transfer of animals to fresh suspensions of cells every 2 days seemed the most likely time for fatal injuries, and we have assumed that all dead animals were too seriously injured or sick to feed during the 2-day interval in which they died.

The juvenile life of an individual copepod was divided into three developmental periods of approximately equal duration. The entire naupliar life up to the first copepodite stage, the first three copepodite stages, and copepodite stages IV and V. The first of these divisions is separated from the others by a major morphological change, but the second division is arbitrary. Copepodite stages I and IV were moved to separate beakers when they appeared, and when animals reached the adult stage they were removed from the experiment. Some copepodite stages were removed for measurement of respiratory rate, dry weight, and bodily carbon and nitrogen (see below). By removing or redistributing animals (in addition to unplanned mortality) we attempted to keep the total requirement for food by all the animals in a beaker constant through the course of an experiment. If the number of animals in a beaker had been kept constant as the animals grew, the older stages would have depleted a supply of food which had been quite adequate for the same number of nauplii.

Respiratory rates were measured by a modification of the method used by Conover (1960). Laboratory-raised copepodite stages IV and V and females were taken directly from the beakers in which they had been reared since acclimated respiration was measured at the same temperature as that of rearing. "Wild" animals captured with a plankton net offshore from La Jolla were kept overnight at the temperature to be used during the measurement of respiration to prevent respiratory overshoot but still give non-acclimated respiratory rates (cf. Halcrow, 1963). Three to 15 animals were enclosed in 130-ml glass-stoppered bottles which were completely filled from a large reservoir of filtered sea water of the correct temperature to which 50 mg of streptomycin had been added per litre, and bottles without animals were used as controls. The experiments were run in the dark at 10°C or 15°C (± 0.5°C). The bottles were attached to a slowly rotating wheel to provide a constant, although unknown, amount of mechanical stimulation to the animals. The respiratory rates thus are probably intermediate between resting or basal rates and normal, active rates. Oxygen content of aliquots from control and experimental bottles was determined by the Winkler titration after 22-29 hours. The precision of the method, including variability in taking of aliquots as well as in titration, was about 0·6% (mean coefficient of variation for 4 replicate controls, based on 6 trials). Animals were rinsed very briefly in distilled water, dried at 60°C, and weighed. The carbon and nitrogen content of the animals were measured with an F and M Carbon-Hydrogen-Analyzer, Model 185. The coefficient of variation of this measurement was

TABLE 1. Conditions during the six experiments

Inclusive dates	Copepod	Temperature	Food	Cells/ml during exp. Mean	Cells/ml during exp. Range	Cellular µg C/l during exp. Mean	Cellular µg C/l during exp. Range	Total unplanned mortality
21 XII 66 – 26 I 67	*Rhincalanus*	15° C	*Ditylum*	82	43-161	148	76-288	23%, mostly as nauplii
11 II – 21 III 67	*Rhincalanus*	15° C	*Thalassiosira*	830	360-1760	196	99-335	56%, mostly as nauplii
14 II – 6 III 67	*Calanus*	15° C	*Thalassiosira*	910	510-1720	177	103-275	47%, mostly as C I and C II
5 IV – 25 V 67	*Calanus*	10° C	*Thalassiosira*	760	400-1610	226	101-399	75%, mostly as C I and C II
7 V – 6 VII 67	*Rhincalanus*	10° C	*Thalassiosira*	878	470-1760	352	143-614	90%, mostly as nauplii
8 VII – 29 VIII 67	*Rhincalanus*	10° C	*Ditylum*	134	48-312	200	98-327	60%, mostly as nauplii

3·4%, including variability in weighing. Respiratory rate was converted to loss of carbon in μg by multiplying the μl of oxygen used per animal per day by 0·43. This assumes a respiratory quotient of 0·8.

Table 1 shows the experiments which were used to test the effects of temperature and type of food on growth and metabolism of the two species of copepods. We attempted to keep the cellular carbon (calculated from expression [2] and the mean carbon per cell) at a constant level during all experiments, but were not entirely successful. The level of food available was higher than the animals normally encounter in nature, but not higher than reported for phytoplankton blooms. We were unable to raise *Calanus* on *Ditylum* at either temperature because the nauplii did not feed, supporting the finding of Marshall and Orr (1956) that these nauplii do not ingest large phytoplankton cells. Mullin and Brooks (1967) found that *Rhincalanus* nauplii fed quite well on *Ditylum*.

RESULTS

Laboratory-raised animals

Fig. 1 indicates that the rate of growth decreased with increasing age except in the experiments in which *Rhincalanus* was fed *Thalassiosira* at 15°C. This is also shown by the coefficients of exponential growth (k in the expression $W_t = W_0 e^{kt}$, where W is the animal's weight in μg of carbon and t is time in days) given in Table 2 (*cf*. Mullin and Brooks, 1967). *Thalassiosira* is apparently a rather unsatisfactory food for *Rhincalanus* nauplii, since their rates of growth were retarded relative to those of the copepodites on the same food. The rates of growth of the copepodites at each temperature were similar on both types of food, but the nauplii grew less rapidly on *Thalassiosira* than on *Ditylum*, and ingested *Thalassiosira* at lower rates. Mortality was also considerably higher among animals raised on *Thalassiosira* (Table 1).

Both species grew more slowly at 10° than at 15°C, and mortality was also higher at 10°C (Table 1), perhaps only because of the longer time required for development. In the experiments where *Rhincalanus* was fed *Thalassiosira*, a poor food for the nauplii, the effects of temperature on the rate of growth were much less marked than where *Rhincalanus* was fed *Ditylum*, a more satisfactory food (values for Q_{10} given in Fig. 1). The effects of temperature on growth were pronounced when *Calanus* was fed *Thalassiosira*.

A classical generalization, based largely on field studies (e.g. Deevey, 1960), is that copepods growing at low temperatures will be larger in size at any developmental stage than animals of the same species growing at high temperatures. The amount of carbon in the copepod was used as a measure of size. A combined rank sum test on the bodily carbon of females and copepodite stages I,I V, and V (see Fig. 1 and Table 3) showed a difference of doubtful significance between the *Calanus* raised at 15°C and those raised at 10°C, the animals raised at 10°C containing more carbon ($p \sim 0·05$ for a 1-tailed test). There was no significant difference between the *Rhincalanus* raised at the two temperatures on *Thalassiosira*, although the animals raised

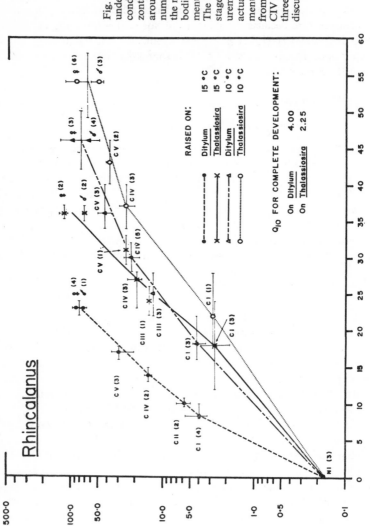

Fig. 1. Growth of copepods under different experimental conditions. Vertical and horizontal lines indicate the ranges around each point and the number in parentheses indicates the number of measurements of bodily carbon, each measurement involving 3 to 15 animals. The bodily carbon of naupliar stage I is from previous measurements rather than animals actually used in these experiments. Lines have been drawn from NI to CI, CI to CIV, and CIV to adult representing the three developmental periods discussed in the text.

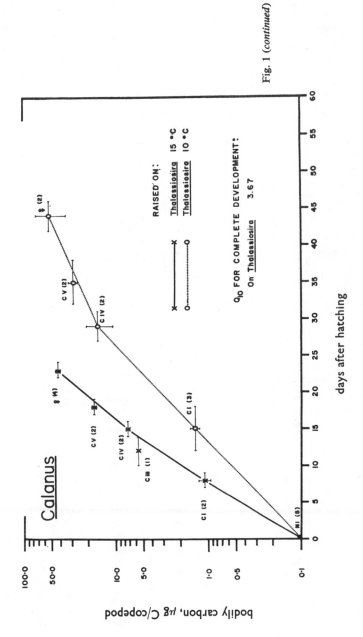

Fig. 1 (*continued*)

TABLE 2. Ingestion and growth of laboratory-raised copepods. See Fig. 1 for the durations of the developmental periods. In each row, $\bar{x}$ is the mean, w is the range, and n is the number of replicates

Period	N I to C I	C I to C IV	C IV to Adult	Total, N I to Adult
Rhincalanus feeding on *Ditylum* at 15° C				
Total µg C ingested	$\bar{x}$ = 9·6 w = 8·1-10·8 n = 4	$\bar{x}$ = 32·4 w = 23·4-43·7 n = 5	$\bar{x}$ = 136·2 w = 122·2-159·9 n = 5	178·2
Total µg C as growth	$\bar{x}$ = 3·7 w = 3·0-5·4 n = 4	$\bar{x}$ = 10·5 w = 9·6-11·3 n = 2	$\bar{x}$ = 65·5 w = 56·4-69·2 n = 5	79·7
Gross growth efficiency	39%	32%	48%	45%
Coefficient of exponential growth	0·64 per day	0·22 per day	0·17 per day	
Rhincalanus feeding on *Thalassiosira* at 15° C				
Total µg C ingested	$\bar{x}$ = 11·7 w = 10·2-12·9 n = 3	$\bar{x}$ = 51·6 w = 47·9-53·7 n = 3	$\bar{x}$ = 195·3 w = 158·1-255·0 n = 4	258·6
Total µg C as growth	$\bar{x}$ = 2·5 w = 1·6-3·5 n = 3	$\bar{x}$ = 16·4 w = 13·9-20·9 n = 3	$\bar{x}$ = 77·7 w = 71·5-83·9 n = 4	96·6
Gross growth efficiency	21%	32%	40%	37%
Coefficient of exponential growth	0·15 per day	0·22 per day	0·18 per day	
Calanus feeding on *Thalassiosira* at 15° C				
Total µg C ingested	$\bar{x}$ = 5·7 w = 5·4-6·1 n = 4	$\bar{x}$ = 24·6 w = 22·4-27·4 n = 4	$\bar{x}$ = 96·8 w = 93·7-102·7 n = 3	127·1
Total µg C as growth	$\bar{x}$ = 1·0 w = 0·8-1·2 n = 2	$\bar{x}$ = 6·3 w = 6·0-6·6 n = 2	$\bar{x}$ = 35·8 w = 33·4-38·0 n = 4	43·1

				Total
Total µg C ingested	$w = 8.2\text{-}12.2$ $n = 4$	$w = 41.2\text{-}77.1$ $n = 6$	$w = 155.6\text{-}180.7$ $n = 3$	
Total µg C as growth	$\bar{x} = 3.9$ $w = 3.1\text{-}4.4$ $n = 3$	$\bar{x} = 18.0$ $w = 13.3\text{-}24.3$ $n = 5$	$\bar{x} = 56.3$ $w = 38.1\text{-}81.9$ $n = 7$	78.2
Gross growth efficiency	39%	31%	34%	34%
Coefficient of exponential growth	0·18 per day	0·14 per day	0·08 per day	

Rhincalanus feeding on Thalassiosira at 10° C

				Total
Total µg C ingested	$\bar{x} = 12.0$ $w = 10.1\text{-}13.2$ $n = 3$	$\bar{x} = 40.9$ $w = 28.9\text{-}52.3$ $n = 5$	$\bar{x} = 177.7$ $w = 115.2\text{-}238.6$ $n = 4$	230.6
Total µg C as growth	$\bar{x} = 2.6$ $w = 2.6$ $n = 1$	$\bar{x} = 22.6$ $w = 174\text{-}26.1$ $n = 3$	$\bar{x} = 44.1$ $w = 30.4\text{-}64.1$ $n = 9$	69.3
Gross growth efficiency	22%	55%	25%	30%
Coefficient of exponential growth	0·13 per day	0·15 per day	0·06 per day	

Calanus feeding on Thalassiosira at 10° C

				Total
Total µg C ingested	$\bar{x} = 6.1$ $w = 5.7\text{-}6.5$ $n = 4$	$\bar{x} = 19.6$ $w = 17.0\text{-}23.9$ $n = 5$	$\bar{x} = 127.8$ $w = 100.4\text{-}155.2$ $n = 2$	153.5
Total µg C as growth	$\bar{x} = 1.3$ $w = 1.1\text{-}1.4$ $n = 3$	$\bar{x} = 14.1$ $w = 9.3\text{-}18.9$ $n = 2$	$\bar{x} = 38.0$ $w = 18.8\text{-}57.2$ $n = 2$	53.4
Gross growth efficiency	21%	72%	30%	35%
Coefficient of exponential growth	0·18 per day	0·17 per day	0·08 per day	

on *Ditylum* at 10°C contained more carbon ($p \sim 0.04$ for a 1-tailed test) than those raised at 15°C on this food. Thus our data provide no clear support for the generalization.

The type of food was not shown to affect size, since *Rhincalanus* raised on *Ditylum* were not significantly different in amount of bodily carbon from those raised on *Thalassiosira* at the same temperature. Neither the type of food nor temperature had any demonstrable effect on the ratio of bodily carbon to nitrogen in either species, although the species themselves differed significantly (Table 3). Most zooplankters contain only a small amount of carbohydrate (Raymont and Conover, 1961), so the C/N ratio in these species probably reflects the ratio of lipid to protein.

Rates of feeding increased as the animals grew, and Fig. 2 shows the rate of ingestion plotted logarithmically as a linear function of time. The equations given are computed linear regressions. However only the data for *Rhincalanus* feeding on *Thalassiosira* satisfy the assumption of linearity, so the other equations have no more validity than equations for straight lines fitted by eye. Inspection of Figs 1 and 2 suggests that when the animals metamorphosed from nauplii to copepodites a distinct increase in rate of feeding occurred, and also that the rate of feeding of any one developmental stage was nearly constant.

The rate of ingestion increased less rapidly with time in all experiments than did the weight of carbon in the individual animals, so that the rate of ingestion of carbon per unit bodily carbon decreased with time. For example, 5-day-old *Rhincalanus* feeding on *Ditylum* at 15°C would ingest 1·2 μg C per μg bodily C per day, while the same animals at 10 days would ingest 0·45 μg C per day and at 20 days would ingest 0·30 μg C per μg C per day (see also Table 2 of Mullin and Brooks, 1967). These data support the classical generalization for many metabolic processes that smaller animals metabolize faster per unit weight than larger animals under identical conditions.

The effects of temperature on the rates of feeding of herbivorous copepods are shown most clearly by Anraku (1964). In the present study, late copepodite stages of *Rhincalanus* ingested *Ditylum* and *Thalassiosira* at approximately the same rates, but these rates were lower at 10° than at 15°C (Fig. 2). The rate of ingestion of *Thalassiosira* by *Calanus* was also lower at 10° than at 15°C.

Combined rank sum tests were used to determine from the data summarized in Table 2 whether the copepods ingested different amounts of carbon under different experimental conditions during any given developmental period, without regard to the duration of the period. *Rhincalanus* ingested significantly less carbon during any developmental period when feeding on *Ditylum* than when feeding on *Thalassiosira* at 15°C, but there was no

---

Fig. 2. Rate of ingestion of carbon (left ordinate) as a function of age in the six experiments. In each graph, the circles and upper line represent data for 15° C while the crosses and lower line represent data for 10° C. In the equations, $\overline{Y}_x = \log_{10}$ (μg C ingested/copepod-day) and X = age in days. The right ordinate shows the equivalent filtering rate at a concentration of food equal to 150 μg C/l.

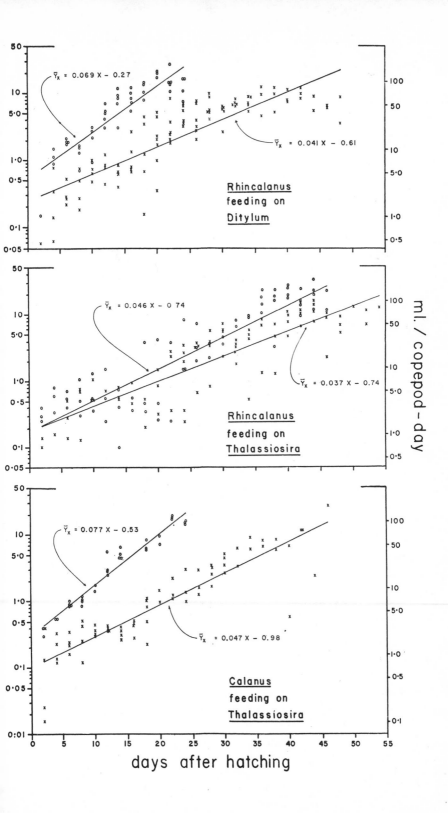

$\bar{Y}_x = 0.069 X - 0.27$

$\bar{Y}_x = 0.041 X - 0.61$

Rhincalanus
feeding on
Ditylum

$\bar{Y}_x = 0.046 X - 0.74$

$\bar{Y}_x = 0.037 X - 0.74$

Rhincalanus
feeding on
Thalassiosira

$\bar{Y}_x = 0.077 X - 0.53$

$\bar{Y}_x = 0.047 X - 0.98$

Calanus
feeding on
Thalassiosira

ml. / copepod-day

days after hatching

TABLE 3. Bodily carbon and the ratio of bodily carbon to bodily nitrogen of wild and laboratory-raised copepods. In each row $\bar{x}$ is the mean, w is the range, and n is the number of measurement, each measurement involving 3 to 15 animals

Stage	15° C, Ditylum	10° C, Ditylum	15° C, Thalassiosira	10° C, Thalassiosira	Wild animals
		Bodily carbon, μg C/copepod, *Rhincalanus*			
C IV	$\bar{x} = 14\cdot4$ $w = 13\cdot5\text{-}15\cdot2$ $n = 2$	$\bar{x} = 22\cdot1$ $w = 17\cdot4\text{-}25\cdot4$ $n = 5$	$\bar{x} = 18\cdot9$ $w = 16\cdot4\text{-}23\cdot4$ $n = 3$	$\bar{x} = 25\cdot4$ $w = 20\cdot2\text{-}28\cdot9$ $n = 3$	$\bar{x} = 25\cdot2$ $w = 16\cdot1\text{-}35\cdot1$ $n = 11$
C V	$\bar{x} = 29\cdot6$ $w = 19\cdot9\text{-}35\cdot2$ $n = 3$	$\bar{x} = 42\cdot2$ $w = 35\cdot7\text{-}50\cdot7$ $n = 3$	$\bar{x} = 25\cdot2$ $w = 25\cdot2$ $n = 1$	$\bar{x} = 39\cdot0$ $w = 38\cdot6\text{-}39\cdot5$ $n = 2$	$\bar{x} = 56\cdot0$ $w = 33\cdot2\text{-}77\cdot0$ $n = 12$
♀	$\bar{x} = 87\cdot4$ $w = 69\cdot2\text{-}94\cdot6$ $n = 4$	$\bar{x} = 95\cdot6$ $w = 72\cdot7\text{-}130$ $n = 3$	$\bar{x} = 122\cdot0$ $w = 112\text{-}133$ $n = 2$	$\bar{x} = 88\cdot7$ $w = 67\cdot5\text{-}120$ $n = 6$	$\bar{x} = 109\cdot2$ $w = 90\cdot6\text{-}138$ $n = 19$
		Bodily carbon, μg C/copepod, *Calanus*			
C IV			$\bar{x} = 7\cdot4$ $w = 7\cdot1\text{-}7\cdot7$ $n = 2$	$\bar{x} = 15\cdot5$ $w = 10\cdot7\text{-}20\cdot3$ $n = 2$	$\bar{x} = 13\cdot6$ $w = 10\cdot1\text{-}21\cdot6$ $n = 6$
C V			$\bar{x} = 17\cdot3$ $w = 164\text{-}18\cdot2$ $n = 2$	$\bar{x} = 29\cdot2$ $w = 26\cdot5\text{-}31\cdot8$ $n = 2$	$\bar{x} = 31\cdot6$ $w = 18\cdot7\text{-}44\cdot0$ $n = 12$
♀			$\bar{x} = 43\cdot1$	$\bar{x} = 53\cdot5$	$\bar{x} = 67\cdot8$

$w = 4\cdot4\text{-}5\cdot4$ $n = 2$	$w = 5\cdot1\text{-}5\cdot9$ $n = 5$	$w = 4\cdot6\text{-}5\cdot0$ $n = 3$	$w = 3\cdot9\text{-}6\cdot3$ $n = 3$	$w = 3\cdot3\text{-}6\cdot4$ $n = 11$
C V $\bar{x} = 5\cdot8$ $w = 4\cdot9\text{-}7\cdot4$ $n = 3$	$\bar{x} = 6\cdot3$ $w = 6\cdot0\text{-}6\cdot8$ $n = 3$	$\bar{x} = 5\cdot0$ $w = 5\cdot0$ $n = 1$	$\bar{x} = 6\cdot0$ $w = 5\cdot8\text{-}6\cdot2$ $n = 2$	$\bar{x} = 6\cdot9$ $w = 5\cdot9\text{-}8\cdot2$ $n = 12$
♀ $\bar{x} = 6\cdot7$ $w = 6\cdot5\text{-}7\cdot0$ $n = 4$	$\bar{x} = 6\cdot7$ $w = 6\cdot4\text{-}7\cdot4$ $n = 3$	$\bar{x} = 7\cdot0$ $w = 6\cdot6\text{-}7\cdot4$ $n = 2$	$\bar{x} = 6\cdot5$ $w = 4\cdot8\text{-}7\cdot9$ $n = 5$	$\bar{x} = 7\cdot1$ $w = 6\cdot3\text{-}7\cdot7$ $n = 18$

Bodily μg C/μg N, *Calanus*

C IV		$\bar{x} = 3\cdot6$ $w = 3\cdot6$ $n = 1$	$\bar{x} = 4\cdot4$ $w = 4\cdot0\text{-}4\cdot8$ $n = 2$	$\bar{x} = 3\cdot6$ $w = 3\cdot3\text{-}3\cdot8$ $n = 6$
C V		$\bar{x} = 3\cdot7$ $w = 3\cdot7\text{-}3\cdot7$ $n = 2$	$\bar{x} = 3\cdot8$ $w = 3\cdot0\text{-}4\cdot5$ $n = 2$	$\bar{x} = 4\cdot1$ $w = 3\cdot4\text{-}6\cdot1$ $n = 12$
♀		$\bar{x} = 3\cdot8$ $w = 3\cdot7\text{-}3\cdot9$ $n = 4$	$\bar{x} = 3\cdot4$ $w = 2\cdot8\text{-}4\cdot1$ $n = 2$	$\bar{x} = 3\cdot5$ $w = 3\cdot4\text{-}3\cdot5$ $n = 14$

difference in ingestion of the two foods at 10°C. *Rhincalanus*, when feeding on *Ditylum*, ingested significantly more carbon at 10°C than at 15°C, but there was no significant difference in ingestion of *Thalassiosira* at the two temperatures by either *Rhincalanus* or *Calanus*. Thus, for either species of copepod fed on *Thalassiosira*, the same amount of carbon was ingested to attain a particular developmental stage at both temperatures, in spite of the effect of temperature on rates of growth and ingestion.

Table 2 gives the gross efficiencies (K_1) during the various periods of growth and over the whole of juvenile life. Exuviae (cast exoskeletons) were treated as detritus rather than growth. Gross efficiency of growth from nauplius I to adult was 30-45%, agreeing with previous measurements or estimates for *Calanus* (Conover, 1964; Corner, Cowey and Marshall, 1965, 1967; Petipa, 1966). The range of efficiencies for the various developmental periods was larger, but there was no regular decrease in efficiency with increasing age as might have been expected by analogy to vertebrates. There was no significant difference (by signed rank test) in gross efficiency between *Calanus* and *Rhincalanus* feeding on *Thalassiosira*; at 15°C, animals of both species were most efficient during the last period of growth, while at 10°C the animals were most efficient between copepodite stages I and IV. Although *Rhincalanus* appeared to be more efficient when feeding on *Ditylum* than when feeding on *Thalassiosira*, the difference was not statistically significant by signed rank test. Nor could it be shown that gross efficiency was significantly affected by temperature. Reeve (1963) found that gross efficiency of *Artemia* feeding on *Phaeodactylum* varied directly with temperature.

Our finding that growth efficiency was unaffected by temperature suggests that under the conditions of our experiments the effects of temperature on ingestion and respiration were equal, so that the same fraction of ingested carbon appeared as growth. This is the simplest explanation in the absence of evidence (Conover, 1966) that the efficiency of assimilation of carbon also changes with temperature. The respiration of fully acclimated *Rhincalanus* was more affected by temperature (mean $Q_{10} = 8\cdot3$) than was that of *Calanus* (mean $Q_{10} = 3\cdot1$) in the range of temperature tested. Type of food had no demonstrable effect on rate of respiration. The values for Q_{10} suggest that *Rhincalanus* is a very poor regulator of metabolic rate at different temperatures, since *Calanus* and other copepods from the colder North Atlantic generally have Q_{10}'s of less than 3·0, even when unacclimated (Anraku, 1964; Clarke and Bonnet, 1939; Conover, 1956, 1962; Halcrow, 1963; Haq, 1967, Marshall *et al.*, 1935).

In summary, the effects of temperature were most marked on the rates of growth, ingestion of a suitable food, and respiration, and were less apparent on the size of the copepod and the total ingestion of food during development to any stage. Rates of growth and ingestion were also greatly affected by the nature of the available food, illustrated by the effects of *Thalassiosira* on *Rhincalanus* during the naupliar stages. The gross efficiency of growth and the ratio of bodily carbon to nitrogen were not shown to be affected by either of the two variables tested.

One other observation was of interest. A total of 32 adult females and no males was produced in the cultures of *Calanus*. If the eggs used to start the cultures had equal probabilities of becoming females or males, resulting in a normal sex ratio of 1:1, then the production of 32 females by chance alone is extremely improbable ($p = 2^{-32}$). This suggests that the conditions of culture were either highly selective against potential males or else that sex determination is environmentally influenced (*cf.* Conover, 1965) and the conditions of culture favoured "femaleness". Forty-five females and 41 males were produced in all the cultures of *Rhincalanus*. Neither the individual experiments nor the pooled result were significantly different from a sex ratio of 1:1 by χ^2 test ($p > 0.1$ in all cases).

Comparisons with wild animals

It is necessary to know how similar the metabolism of laboratory-raised copepods is to that of the wild animals before one can with any confidence use the laboratory-raised animals as "models" of the wild populations. We used size (measured as bodily carbon), ratio of bodily carbon to nitrogen, and respiratory rate as measures of metabolic similarity.

Calanus raised at 10°C were not significantly different in size or in carbon/nitrogen ratio from the wild copepods (combined rank sum tests on data for females and copepodite stages IV and V summarized in Table 3). Those raised at 15°C were significantly smaller and had a higher C/N ratio than the wild copepods ($p < 0.05$ in both cases). Of the laboratory-raised *Rhincalanus*, only those raised at 15°C on *Thalassiosira* did not differ significantly from wild animals in the two measurements. The copepods raised at 15°C on *Ditylum* were significantly smaller than the wild animals and had a lower C/N ratio. The same trends were found in the animals raised at 10°C on *Ditylum*, but the differences were of questionable significance ($p \simeq 0.05$ in both cases). The copepods raised at 10°C on *Thalassiosira* were significantly smaller than the wild animals but did not differ from them in the ratio of bodily carbon to nitrogen.

Differences in respiratory rates per animal were also tested by combined rank sum tests comparing acclimated, laboratory-raised copepods with unacclimated, wild animals at the same temperature (data summarized in Table 4). There was no difference ($p > 0.20$) between laboratory-raised and wild *Calanus* at either temperature. *Rhincalanus* raised on either food at 15°C had significantly lower respiratory rates than wild copepods at the same temperature ($p \leqslant 0.05$ in both cases), but there were no significant differences at 10°C ($p > 0.50$).

Respiratory rates of individual animals could of course differ either because of fundamental differences in rates per unit size (bodily carbon) or simply because the animals being compared were of different sizes even though being of the same developmental stage. Hence, respiratory rates per unit bodily carbon were compared. These rates decreased as bodily size increased, as shown by the following regression equations for the wild copepods: *Rhincalanus* at 15°C, $y = -0.26\,x - 0.86$; *Rhincalanus* at 10°C, $y = -0.17\,x - 1.50$;

TABLE 4. Respiratory rates of wild and laboratory-raised copepods. In each row, $\bar{x}$ is the mean, w is the range, and n is the number of measurements, each measurement involving 3 to 15 animals

Stage	$\mu l\ O_2$/copepod-day			$\mu g\ C$/bodily $\mu g\ C$-day		
	Wild animals	Ditylum-raised	Thalassiosira-raised	Wild animals	Ditylum-raised	Thalassiosira-raised
Rhincalanus at 15° C						
C IV	$\bar{x} = 3\cdot7$ $w = 1\cdot8\text{-}4\cdot5$ $n = 5$	$\bar{x} = 2\cdot1$ $w = 2\cdot1,\ 2\cdot1$ $n = 2$	$\bar{x} = 4\cdot2$ $w = 2\cdot6\text{-}5\cdot7$ $n = 5$	$\bar{x} = 0\cdot053$ $w = 0\cdot045\text{-}0\cdot060$ $n = 5$	$\bar{x} = 0\cdot063$ $w = 0\cdot059\text{-}0\cdot067$ $n = 2$	$\bar{x} = 0\cdot086$ $w = 0\cdot067\text{-}0\cdot105$ $n = 2$
C V	$\bar{x} = 6\cdot8$ $w = 5\cdot4\text{-}9\cdot2$ $n = 5$	$\bar{x} = 4\cdot5$ $w = 4\cdot5,\ 4\cdot5$ $n = 2$	$\bar{x} = 5\cdot1$ $w = 5\cdot1$ $n = 1$	$\bar{x} = 0\cdot048$ $w = 0\cdot037\text{-}0\cdot058$ $n = 5$	$\bar{x} = 0\cdot064$ $w = 0\cdot064,\ 0\cdot064$ $n = 2$	$\bar{x} = 0\cdot087$ $w = 0\cdot087$ $n = 1$
♀	$\bar{x} = 11\cdot4$ $w = 8\cdot6\text{-}14\cdot9$ $n = 9$	$\bar{x} = 8\cdot4$ $w = 5\cdot1\text{-}12\cdot0$ $n = 4$	$\bar{x} = 5\cdot9$ $w = 5\cdot6\text{-}6\cdot2$ $n = 2$	$\bar{x} = 0\cdot045$ $w = 0\cdot031\text{-}0\cdot066$ $n = 8$	$\bar{x} = 0\cdot043$ $w = 0\cdot023\text{-}0\cdot058$ $n = 4$	$\bar{x} = 0\cdot021$ $w = 0\cdot018\text{-}0\cdot024$ $n = 2$
Calanus at 15° C						
C IV	$\bar{x} = 3\cdot5$ $w = 3\cdot1\text{-}4\cdot3$ $n = 3$		$\bar{x} = 1\cdot8$ $w = 1\cdot5\text{-}2\cdot1$ $n = 2$	$\bar{x} = 0\cdot121$ $w = 0\cdot110\text{-}0\cdot132$ $n = 2$		$\bar{x} = 0\cdot109$ $w = 0\cdot089\text{-}0\cdot130$ $n = 2$
C V	$\bar{x} = 5\cdot0$ $w = 3\cdot2\text{-}6\cdot5$ $n = 7$		$\bar{x} = 4\cdot2$ $w = 4\cdot1\text{-}4\cdot2$ $n = 2$	$\bar{x} = 0\cdot88$ $w = 0\cdot062\text{-}0\cdot111$ $n = 7$		$\bar{x} = 0\cdot105$ $w = 0\cdot101\text{-}0\cdot109$ $n = 2$

$w = 0.7\text{-}1.9$ $n = 4$	$w = 0.9$ $n = 1$	$w = 1.4\text{-}1.6$ $n = 3$	$w = 0.015\text{-}0.035$ $n = 4$	$w = 0.018$ $n = 1$	$w = 0.025\text{-}0.030$ $n = 3$
C V $\bar{x} = 2.1$ $w = 1.2\text{-}3.4$ $n = 7$	$\bar{x} = 2.6$ $w = 2.2\text{-}3.1$ $n = 3$	$\bar{x} = 2.3$ $w = 2.0\text{-}2.5$ $n = 2$	$\bar{x} = 0.017$ $w = 0.009\text{-}0.026$ $n = 7$	$\bar{x} = 0.026$ $w = 0.025\text{-}0.027$ $n = 3$	$\bar{x} = 0.025$ $w = 0.022\text{-}0.028$ $n = 2$
♀ $\bar{x} = 4.7$ $w = 1.0\text{-}18.3$ $n = 13$	$\bar{x} = 3.5$ $w = 2.8\text{-}4.4$ $n = 3$	$\bar{x} = 3.4$ $w = 0.7\text{-}5.2$ $n = 4$	$\bar{x} = 0.019$ $w = 0.003\text{-}0.057$ $n = 11$	$\bar{x} = 0.016$ $w = 0.014\text{-}0.020$ $n = 3$	$\bar{x} = 0.017$ $w = 0.004\text{-}0.028$ $n = 4$

Calanus at 10° C

C IV $\bar{x} = 2.6$ $w = 2.3\text{-}3.0$ $n = 4$	$\bar{x} = 3.3$ $w = 2.6\text{-}4.0$ $n = 2$	$\bar{x} = 0.078$ $w = 0.061\text{-}0.092$ $n = 4$	$\bar{x} = 0.095$ $w = 0.085\text{-}0.106$ $n = 2$
C V $\bar{x} = 4.6$ $w = 3.8\text{-}5.4$ $n = 5$	$\bar{x} = 4.0$ $w = 3.2\text{-}4.7$ $n = 2$	$\bar{x} = 0.048$ $w = 0.043\text{-}0.054$ $n = 5$	$\bar{x} = 0.060$ $w = 0.044\text{-}0.076$ $n = 2$
♂ $\bar{x} = 9.0$ $w = 8.3\text{-}9.7$ $n = 6$	$\bar{x} = 4.8$ $w = 2.6\text{-}6.9$ $n = 2$	$\bar{x} = 0.051$ $w = 0.047\text{-}0.060$ $n = 6$	$\bar{x} = 0.037$ $w = 0.032\text{-}0.041$ $n = 2$

Calanus at 15°C, $y = -0.31 x - 0.62$; *Calanus* at 10°C, $y = -0.26 x - 0.83$; where in each equation $y = \log_{10}$ (μg C respired per μg bodily C per day) and $x = \log_{10}$ (μg bodily C). Analyses of covariance were used to compare the respiration of laboratory-raised copepods to the appropriate equation for wild animals; in no case was there a significant difference ($p > 0.25$ in all cases). This means that the respiration per unit bodily carbon of wild and laboratory-raised copepods of the same size was the same (as far as our data can tell) under similar conditions of measurement.

These results indicate that the conditions of rearing did not have a marked effect on respiratory rate, although bodily carbon and chemical composition were somewhat altered. In the balance, the copepods raised at 10°C seem to be more like the wild animals in metabolism than those raised at 15°C.

Differences between the species

Several metabolic differences between *Calanus* and *Rhincalanus* have been noted above, but those which may affect the ability of the two species to coexist are worth emphasizing since so little is known about the separation of ecological niches in the planktonic environment. Adults and late copepodite stages of both species feed preferentially on large particles (compare Mullin, 1963, with Mullin and Brooks, 1967), but the naupliar stages appear to differ considerably in selective feeding. Naupliar *Calanus* are not able to feed on particles as large as *Ditylum* (Marshall and Orr, 1956, and the present study), while naupliar *Rhincalanus* feed and grow more rapidly on *Ditylum* than on smaller cells such as *Thalassiosira* (Mullin and Brooks, 1967, and the present study) which are good food for *Calanus*. Hence, separation of niches with regard to food is probably most marked for the naupliar stages. The two species were not shown to differ in efficiency of growth on the same food.

The bodily C/N ratio of *Rhincalanus* was significantly higher (by rank sum test) than that of *Calanus*, even when both were raised on the same food (data in Table 3). This probably means that *Rhincalanus* has greater reserves of lipid than *Calanus*. Extraction of lipid of copepodite Stage V and females with acetone in a Soxhlet apparatus supported this.

Conover (1960) plotted *Rhincalanus* and *Calanus finmarchicus* on a common regression line for log (respiratory rate at 5-10°C per unit dry weight) against log (dry weight). However, analyses of covariance on similarly log-transformed data for wild copepods (regression equations in preceding section) showed that at 15°C and at 10°C *Rhincalanus* had lower respiratory rates per unit bodily carbon than *Calanus* of the same bodily carbon content.

DISCUSSION

There were several potential sources of bias which might have affected the results presented above. It was pointed out that the amount of carbon available as food was not the same in the different experiments (Table 1), but we have assumed that this did not cause the observed differences between experimental results. Different broods of nauplii were used to start different experiments, so that some of the effects which have been attributed to temperature or nature of food could conceivably be due to inherent differences in the nauplii. Also, the animals removed from the experiments for measurement of respiratory rate and content of carbon were not selected in a truly random manner, although removal was not systematic. Finally, the wild animals were collected at different times over 9 months, but since both size (bodily carbon) and respiratory rate per unit size are known to vary seasonally (Marshall and Orr, 1958) the animals do not necessarily represent the "average" state of the wild population.

The comparisons between wild and laboratory-raised animals were chosen more for convenience than for desirability These comparisons indicated that the animals raised at 10°C were similar to those taken from the wild population, but this obviously does not mean that the metabolic measurements made on the laboratory-raised animals are directly applicable to the wild animals. The most obvious problem, recently stressed by Petipa (1966), is that the wild animals almost certainly spend more energy in searching for food and in diurnal vertical migration than do the animals in the laboratory. Conover (1966) showed that the efficiency of assimilation is not significantly affected by the concentration of food. Hence, even if the wild animals ingest as much carbon per day as did the laboratory-raised animals, the gross efficiency of growth in the field is probably lower since the respiratory costs are higher.

SUMMARY

1. *Rhincalanus* grew less rapidly on *Thalassiosira* than on *Ditylum*, apparently because the nauplii did not feed readily on the former. Naupliar *Calanus* were unable to grow on *Ditylum*.

2. Both species ate and grew more slowly at 10° than at 15°C, but did not necessarily grow to larger size.

3. Rates of ingestion per unit bodily carbon decreased as bodily carbon increased.

4. When either species was raised on *Thalassiosira*, the amount of carbon ingested during the attainment of a particular developmental stage was not affected by temperature.

5. The gross efficiency of growth over the whole juvenile life was 30-45%. This efficiency did not decrease with increasing age, was not affected by temperature, and did not differ between the two copepods when they were raised on the same food.

6. Respiration per unit bodily carbon decreased as bodily carbon increased. *Rhincalanus* respired at a lower rate than *Calanus* of the same size, and its respiration was more affected by temperature.

7. The laboratory conditions markedly altered the sex ratio of *Calanus* but not of *Rhincalanus*.

8. The respiratory rates of laboratory-raised copepods were similar to those of wild animals when corrected for differences in size of the copepods, but both size and bodily C/N ratio of some of the laboratory-raised copepods differed from the wild animals.

ACKNOWLEDGEMENTS

Dr P. M. Williams of the Institute of Marine Resources, U.C.S.D., provided the infrared CO_2 analyser, and Drs R. Lasker and P. Smith of the Bureau of Commercial Fisheries, La Jolla, provided the F and M model 185 analyser. Dr E. W. Fager of Scripps Institution of Oceanography gave us space in a temperature-controlled room.

Mrs S. Davidor assisted in several phases of the experiments, and suggestions on the manuscript were made by Drs R. Lasker, R. W. Eppley, S. Richman, and J. D. H. Strickland. The research was supported by U.S. Atomic Energy Commission Contract No. AT(11-1)GEN 10, P.A. 20.

REFERENCES

ANRAKU, M. 1964. Influence of the Cape Cod Canal on the hydrography and on the copepods in Buzzards Bay and Cape Cod Bay, Massachusetts. II. Respiration and feeding. *Limnol. Oceanogr.*, 9, 195-206.

BROWNE, E. T. 1898. On keeping medusae alive in an aquarium. *J. mar. biol. Ass. U.K.*, 5, 176-80.

CLARKE, G. L., and BONNET, D. D. 1939. The influence of temperature on the survival, growth, and respiration of *Calanus finmarchicus*. *Biol. Bull. mar. biol. Lab. Woods Hole*, 76, 371-83.

CONOVER, R. J. 1956. Biology of *Acartia clausi* and *A. tonsa*. *Bull. Bingham oceanogr. Coll.*, 15, 156-233.

CONOVER, R. J. 1960. The feeding behaviour and respiration of some marine planktonic Crustacea. *Biol. Bull. mar. biol. Lab. Woods Hole*, 119, 399-415.

CONOVER, R. J. 1962. Metabolism and growth in *Calanus hyperboreus* in relation to its life cycle. *Rapp. P.-v. Réun. Cons. perm. int. Explor. Mer*, 153, 190-7.

CONOVER, 1964. Food relations and nutrition of zooplankton. *Occ. Publ. Proc. Symp. exp. mar. Ecol.*, 2, 81-91.

CONOVER, R. J. 1965. Notes on the moulting cycle, development of sexual characters and sex ratio in *Calanus hyperboreus*. *Crustaceana*, 8, 308-20.

CONOVER, R. J. 1966. Factors affecting the assimilation of organic matter by zooplankton and the question of superfluous feeding. *Limnol. Oceanogr.*, 11, 339-45.

CORNER, E. D. S., COWEY, C. B., and MARSHALL, S. M. 1965. On the nutrition and metabolism of zooplankton. III. Nitrogen excretion by *Calanus J. mar. biol. Ass. U.K.*, 45, 429-42.

CORNER, E. D. S., COWEY, C. B., and MARSHALL, S. M. 1967. On the nutrition and Metabolism of zooplankton. V. Feeding efficiency of *Calanus finmarchicus*. *J. mar. Ass. U.K.*, 47, 259-70.

DEEVEY, G. B. 1960. Relative effects of temperature and food on seasonal variations in length of marine copepods in some eastern American and western European waters. *Bull. Bingham oceanogr. Coll.*, 17, 54-86.

FLEMINGER, A. 1964. Distributional atlas of calanoid copepods in the California Current region, Part I. *Calif. coop. ocean. Fish. Invest. Atlas*, (2), 313 pp.

GUILLARD, R. R. L., and RYTHER, J. H. 1962. Studies of marine planktonic diatoms. I. *Cyclotella nana* Hustedt, and *Detonula Confervacea* (Cleve) Gran. *Can. J. Microbiol.*, 8, 229-39.

HALCROW, K. 1963. Acclimation to temperature in the marine copepod, *Calanus finmarchicus* (Gunner). *Limnol. Oceanogr.*, 8, 1-8.

HAQ, S. M. 1967. Nutritional physiology of *Metridia luceus* and *M. longa* from the Gulf of Maine. *Limnol. Oceanogr.*, 12, 40-51.

HASTINGS, J. W., SWEENEY, B. M., and MULLIN, M. M. 1962. Counting and sizing of unicellular marine organisms. *Ann. N.Y. Acad. Sci.*, 99, 280-9.

MARSHALL, S. M., and ORR, A. P. 1956. On the biology of *Calanus finmarchicus*. IX. Feeding and digestion in the young stages. *J. mar. biol. Ass. U.K.*, 35, 587-603.

MARSHALL, S. M., and ORR, A. P. 1958. On the biology of *Calanus finmarchicus*. X. Seasonal changes in oxygen consumption. *J. mar. biol. Ass. U.K.*, 37, 459-72.

MARSHALL, S. M., NICOLLS, A. G., and ORR, A. P. 1935. On the biology of *Calanus finmarchicus*. VI. Oxygen consumption in relation to environmental conditions. *J. mar. biol. Ass. U.K.*, **20**, 1-28.

MENZEL, D. W., and VACCARO, R. F. 1964. The measurement of dissolved organic and particulate carbon in seawater. *Limnol. Oceanogr.*, **9**, 138-42.

MULLIN, M. M. 1963. Some factors affecting the feeding of marine copepods of the genus *Calanus*. *Limnol. Oceanogr.*, **8**, 239-50.

MULLIN, M. M., and BROOKS, E. R. 1967. Laboratory culture, growth rate, and feeding behavior of a planktonic marine copepod. *Limnol. Oceanogr.*, **12**, 657-66.

MYERS, J. 1962. Laboratory cultures. In *Physiology and biochemistry of algae*, edited by R. A. Lewin. New York and London, Acad. Press, Ch. 39, 603-15.

PETIPA, T. S. 1966. On the energy balance of *Calanus helgolandicus* (Claus) in the Black Sea. In *Physiology of marine animals*, Moscow, Oceanogr. Commn scient. publ. Ho., 60-81. (Eng. trans.)

RAYMONT, J. E. G., and CONOVER, R. J. 1961. Further investigations on the carbohydrate content of marine zooplankton. *Limnol. Oceanogr.*, **6**, 154-64.

REEVE, M. R. 1963. Growth efficiency in *Artemia* under laboratory conditions. *Biol. Bull mar. biol. Lab. Woods Hole*, **125**, 133-45.

The influence of temperature on the transformation of matter in marine invertebrates

I. V. IVLEVA

Institute of the Biology of the Southern Seas
Academy of Sciences of the Ukraine SSR
Sevastopol, USSR

INTRODUCTION

The influence of temperature on the main elements of energy budgets in marine animals is of importance for evaluating biological productivity in aquatic systems. The transformations of energy and matter in a single link of food chains, are established ordinarily in the narrow temperature range, 15-20°C, which reflect the features normally present in one season of the year for a single geographic region. Obviously, the general processes in aquatic systems at natural temperature may be obtained only if the temperature dependence of parameters is known.

Therefore, the influence of temperature on parameters of the balance equation, $A = P + T$, was studied. This equation represents the relation between the assimilated part of food (A), the expenditure on energy metabolism (T) and plastic metabolism (P) in marine invertebrates.

The animals investigated were: *Nereis diversicolor* O. F. Muller and *Leander adspersus* (Rathke), species which are common in the Black Sea and play a significant part in the production in neritic waters.

METHODS

The animals were sampled near Sevastopol and acclimated to temperatures from 3 to 30°C in chambers of a polythermostat (*N. diversicolor*) and in flowing aquaria with regulated temperature (*L. adspersus*) (Semionov and Ivleva, 1968). Young animals of approximately equal size (live weight of 200 to 250 mg) were used in experiments on *L. adspersus*. Four age groups of *N. diversicolor* having different rates of growth were investigated at each temperature. Live weight of the individuals in the first group varied from 10 to 50 mg, in the second group—from 50 to 100 mg, in the third—from

96

100 to 200 mg, and in the fourth—from 200 to 350 mg. The animals were cultivated for up to 60 days, the accuracy of temperature control was ±1·0°C. In accordance with temperature tolerance, the temperature range in the experiments on *L. adspersus* was from 7 to 25°C and in experiments on *N. diversicolor*—from 3 to 25°C.

The measurements of nutrition, respiration and growth rate were carried out at each temperature as the animals were growing. All the animals were fed regularly once or twice a day. *N. diversicolor* were fed with live *Enchytraeus albidus* Henle, and *L. adspersus* were fed with gonads of *Mytilus galloprovincialis*. In both experiments the proportion of unconsumed food was evaluated two hours after feeding. The proportion of consumed food was measured in a special series of experiments where animals were kept in small crystallizers in cells of the polythermostat. In each experiment faeces of 5-10 animals were gathered for several days. The quantities of consumed food and faeces were accurately measured. Faeces were dried to constant weight at 60°C. In *N. diversicolor* the assimilated part of diet was determined from the total matter consumed by growth and metabolism.

The intensity of plastic metabolism was evaluated from the growth rate. The initial and the final weight of animals were measured by means of an analytical balance after a short period of drying between two sheets of filter paper. Dry weight was determined after drying in a desiccator at 60°C to constant weight. Dry samples of animals, their food and excreta were used for determination of calorific value. This was carried out by wet oxidation using chromic acid. Dichromate remaining after oxidation of a sample was titrated by ferrous ammonium sulphate in the presence of phenylauthranilic acid (Ostapenia, 1965).

Expenditure on metabolism was calculated from the rate of consumption of oxygen by animals, placed in closed respirometers. The initial and the final concentration of oxygen in the water in respirometers was determined by Winkler's method. Metabolism was measured in animals of various size after acclimation to the experimental temperatures had been completed (Ivleva and Popenkina, 1968 a, b). In each experiment the equation for metabolism as a function of weight was obtained. The coefficients in this equation, calculated by the method of least squares were then used to determine the amount of oxygen consumed by animals that took part in nutrition and growth experiments. The results obtained on *L. adspersus* allowed the calculations to be carried out not only in weight, but also in calorific units.

RESULTS

Temperance tolerance

Despite the fact that *N. diversicolor* and *L. adspersus* are eurythermal species, they have rather different temperature tolerances. *N. diversicolor* are more tolerant of low temperatures and at 5°C they have normal activity, growth, feed, and in adults reproductive products ripen under such conditions. *N. diversicolor* can also easily tolerate lower temperatures. At 0°C they show good survival and can even tolerate negative temperature. The upper temperature limit is 28-30°C in this species. At 30° they can survive only for a limited period of time: after 15-20 days the mortality reaches 80%. Temperature range in *L. adspersus* was found to be more restricted. *L. adspersus* easily tolerate temperatures of 27-28° and at 30° can survive for 10-15 days. 5°C should be assumed as the lower temperature limit of this species. At this temperature they can survive for about 20 days in laboratory; however, they completely lose activity, often remaining lying on their sides, their appendages hardly moving, and scarcely responding to food. At 7°C they feed, but still show reduced activity. In the natural environment, long before the temperature drops to this level, they migrate to the depths where water temperature is never lower than 10°C. Reproduction in *L. adspersus* occurs at temperatures above 12°C. The survival of *L. adspersus* during cultivation in the laboratory is illustrated in Fig. 1.

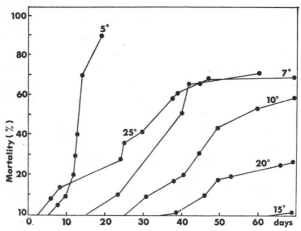

Fig. 1. The mortality in *L. adspersus* at various temperatures.

The relation between dry and wet weight

This was determined from a large number of samples obtained in investigations of natural populations of *N. diversicolor* and *L. adspersus*, consisting of various sized individuals. Live weight of *N. diversicolor* varied between 0·004 and 1·850 mg and that of *L. adspersus*—from 0·080 to 2·800 mg. As the content of dry matter in the body of each species remained constant irrespective of temperature, the data obtained in the laboratory under different temperature conditions were put into the general results. Within the size range investigated, the relationship between wet (W_{wet}) and dry (W_{dry}) matter is described by the regressions:

$$W_{dry} = -0·002 + 0·143\ W_{wet} \qquad \text{for } N.\ diversicolor$$
$$W_{dry} = -0·011 + 0·255\ W_{wet} \qquad \text{for } L.\ adspersus$$

Correlation (r) between dry and wet weight was found to be very close: in *N. diversicolor* $r = 0·895 \pm 0·039$ and in *L. adspersus* $r = 0·001 \pm 0·017$.

Food-consumption rate

This was determined for groups and individuals. The duration of experiments was as follows: 14 days for *N. diversicolor* and 57-60 days for *L. adspersus*. In each case specific consumption rate ($\bar{C}_R$) was determined, which is expressed in per cent and is equal to the ratio between the diet (R) for a definite period of time (t_2-t_1) and the average weight of an animal ($W_1 + W_2$)/2, which feeds during this period:

$$\bar{C}_R = \frac{2R}{(W_2 + W_1)(t_2 - t_1)} \cdot 100.$$

The value of daily diet in *N. diversicolor* in four age groups as a function of temperature is illustrated in Fig. 2. The consumption rate increased with temperature and reached a maximum value at the upper temperature limit. In *L. adspersus* the dependence was similar (Fig. 3).

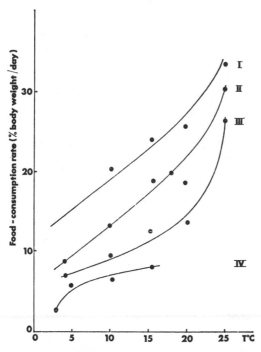

Fig. 2. The nutrition rate in four (I-IV) size groups of *N. diversicolor* as a function of temperature.

Excretion rate

In *L. adspersus* the excretion rate was determined at temperatures of 7°, 10°, 13°, and 20° from the quantity of faeces released by a single animal daily. The quantity of faeces was related to the average body weight of the animals that took part in these experiments. The results obtained are shown in Fig. 4.

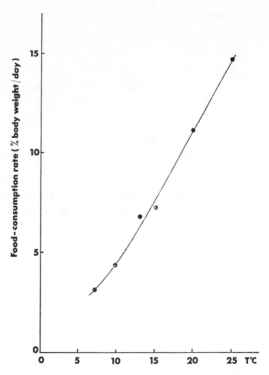

Fig. 3 The nutrition rate in *L. adspersus* as a function of temperature.

Under the influence of increasing temperature the excretion rate increased from 0·2 to 1·6%.

Assimilation

Excretion accounted for 2 to 8% of the total weight of food consumed. This proportion kept stable within the limits mentioned above, irrespective of conditions under which the animals were cultivated. This allowed the use of an average assimilation coefficient of 94% in dry matter for the full scale of temperatures investigated. This value was somewhat larger if calculated in calories. The calorific value of faeces showed a slight, but quite regular, decline in assimilation with increasing temperature (Table 1). However, these differences affected only slightly the assimilation coefficients calculated from calories. The calorific value of gonads of *Mytilus galloprovincialis* which served as food for *L. adspersus*, was determined to allow corresponding calculations. By analysis of a great number of samples 1 mg of dry matter was found to contain 4·05±0·36 cal. The values of assimilation obtained in this case are shown in Table 2. Considerable spread of points shows that there is no correlation with temperature. Assimilation coefficients ranged from 96 to 99% and on the average was 97·7%. In *N. diversicolor* the assimilated part of food was calculated from expenditure on growth and metabolism. The results

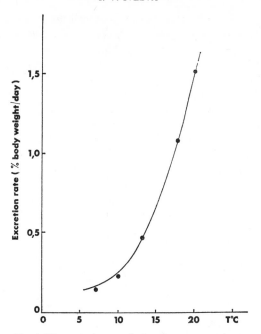

Fig. 4. The excretion rate in *L. adspersus* at various temperatures.

TABLE 1. The calorific values of excreta of *L. adspersus*, cultivated at various temperatures (cal/mg)

$7°$ C $n = 10$	$10°$ C $n = 15$	$13°$ C $n = 14$	$20°$ C $n = 17$
0.72 ± 0.26	1.24 ± 0.13	1.74 ± 0.26	1.87 ± 0.32

TABLE 2. The values of assimilation of the food in *L. adspersus* at various temperatures

$T°$ C n	7 3	10 3	13 6	20 5
Assimilation %	99.4 ± 0.2	98.9 ± 0.2	96.8 ± 0.3	98.1 ± 0.9

TABLE 3. The values of assimilation of food $(P+T)$ in four size groups in *N. diversicolor* at various temperatures

$T°$ C		4	10	15	20	25
size-groups	I	32.0	38.1	53.6	45.3	40.6
	II	38.1	51.1	47.8	60.9	37.7
	III	35.1	66.3	58.7	75.1	41.4
	IV	38.1	43.6	40.9	—	—

are illustrated in Table 3. The data obtained indicate that assimilation varied not only with temperature, but also depended on the size of the experimental animals. In the smallest *N. diversicolor* assimilation was lower than in middle-sized ones and never exceeded 55%. Assimilation coefficient increased with increase in temperature, but only to a certain limit: in the first age group the maximum occurred at 15°C. In the two other age groups assimilation coefficient reached a maximum at 20°C. At 25°C the assimilation coefficient dropped in all animals. Large nereids showed maximal assimilation at low temperatures.

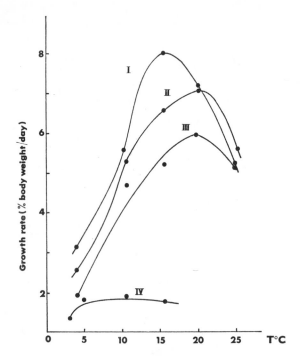

Fig. 5. The growth rate in four (I-IV) size groups of *N. diversicolor* as a function of temperature.

Growth rate

Growth rate which reflects the rate of plastic metabolism, was determined in both species within a short period of time. The experiments on *N. diversicolor* lasted for 15-20 days. In *L. adspersus* growth rate was very low, in consequence of which the observations were prolonged up to 60 days. Inasmuch as the growth in shrimps proceeds by moulting, it was of importance to evaluate matter and energy expenditure in moulting. At each temperature the interval between moults and the weight of the moulted skin relative to the body weight were measured.

In each case the growth rate was investigated, calculated with the following formula (Winberg, 1968):

$$\bar{C}_w = \frac{2(w_2 - w_1)}{(w_2 + w_1)\,(t_2 - t_1)}$$

where w_1 and w_2 are the initial and the final weight within the time $t_2 - t_1$.

Fig. 5 illustrates the growth rate in the four age groups of *N. diversicolor* at various temperatures. The average daily growth in small and middle-sized animals was considerable at all temperatures. In the smallest animals it ranged between 3·5 and 8% of body weight. In large individuals it remained invariable within temperature range 5-15°C and was about 2% of body weight. It is significant that in the smallest individuals maximum growth rate occurred at lower temperatures than in older age groups. This temperature was 15°C, while in the second and third groups the fastest growth occurred at 20°C. Nereids' response to the temperature of 25°C should also be noted: in the first three groups growth rate at this temperature fell to the same level as at 10°C; the animals lost weight and soon died, in spite of intensive feeding.

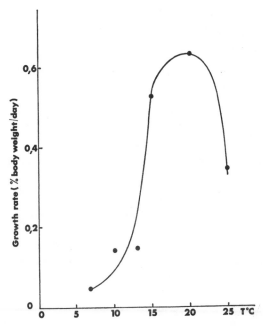

Fig. 6. The growth rate in *L. adspersus* as a function of temperature.

Fig. 6 illustrates the dependence between growth rate and temperature in *L. adspersus*. The diagram clearly shows that there also exists a range of temperature within which the most intensive growth occurs. Even under optimum conditions, however, shrimps showed low daily growth, which never exceeded 0·64% of body weight.

Growth rate proved to be higher if expenditure on moulting was taken into account. The weight of moulted skins accounted for 11 to 18% of body weight in each moulting. This proportion remained within given limits at all temperatures and averaged 15·4%. Temperature, however, affected the frequency of moulting. As illustrated in Fig. 7 the moulting process slowed down considerably as temperature decreased. The duration of the period between moults at 7°C was not defined exactly. It was assumed to be of about 50 days. Extended keeping at 25°C caused irregularities in periodicity of moults in shrimps. Expenditure of matter on moulting increased from 0·3% of body weight at 7°C to 1·1% at 25°.

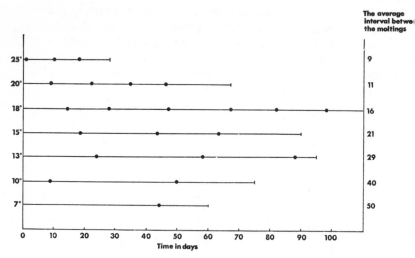

Fig. 7. The intermoulting period in *L. adspersus* at various temperatures.

The expenditure on plastic metabolism (growth + moulting) was highest at 20°C and reached 2% of body weight.

It is known that exoskeletons of Crustacea contain a considerable proportion of mineral matter and the calorific value may be very low. Our analyses found that the calorific value of moulted skins does not change with temperature and averages 1·14±0·37 cal/mg (Table 4). Losses on moulting grew with the rise in temperature from 0·205 to 1·129 cal/day (Table 5).

TABLE 4. The calorific values of moulted skin of *L. adspersus*, cultivated at various temperatures (cal/mg)

7° C n = 3	10° C n = 3	13° C n = 3	18° C n = 7	20° C n = 10
1·15 ± 0·47	1·17 ± 0·02	1·08 ± 0·37	1·16 ± 0·02	1·33 ± 0·25

TABLE 5. The losses on moulting in *L. adspersus* at various temperatures (% body weight/day)

7° C	10° C	13° C	15° C	20° C	25° C
0·30	0·39	0·51	0·58	0·76	1·13

The energy metabolism rate

Preliminary investigations were carried out to define the effect of temperature on oxygen consumption rate in *N. diversicolor* and *L. adspersus* of a wide range of weight (Ivleva and Popenkina, 1968 a, b). At each temperature the dependence of metabolism (Q) on weight (W) was expressed as:

$$Q = AW^k.$$

The temperature dependences of parameters A and k are given in Table 6 These parameters were used for calculations of average metabolic rate in the nutrition and growth experiments, with the exception of data on metabolism in *N. diversicolor* at 20°C, obtained by direct measurement of oxygen consumption rate.

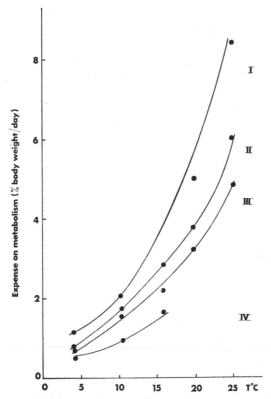

Fig. 8. The expenditure on metabolism in four (I-IV) size groups of *N. diversicolor* as a function of temperature.

All the values are expressed in weight and energy units. It was assumed that 1 ml of oxygen consumed is expended in oxidation of 1 mg of dry matter and the oxycalorific coefficient is equal to 4·86 cal/ml O_2 (Winberg, 1968).

The comparison of the parameter k in *N. diversicolor* shows a more significant reaction in large-sized animals to temperature changes. At the same

TABLE 6. Temperature dependence of *A* and *k* in *N. diversicolor* and *L. adspersus*

$t°$ C		·5	10	15	20	25	28
N. diversicolor	*A*	0·80	0·85	0·80	—	1·356	—
	k	0·69	0·79	0·73	0·73	0·73	—
L. adspersus	*A*	0·50	1·62	2·38	3·35	4·11	5·19
	k	0·91	0·78	0·81	0·81	0·81	0·80

time, in all age groups investigated the expenditure on metabolism increased with the rise in temperature and became maximal at 25°C (Fig. 8). The expenditure on respiration was 0·6-8·4% of body weight. In *L. adspersus* a rise in temperature from 10°C up to 28°C did not change the character of dependence of metabolism rate on body weight. Within this interval the exponent *k* remained practically constant (Table 6), and increased noticeably only near the lower and the upper temperature limits. Expenditure on metabolism in the groups of shrimps investigated increased with the rise in temperature from 1·1% at 7°C up to 4·2% at 25°C (Fig. 9).

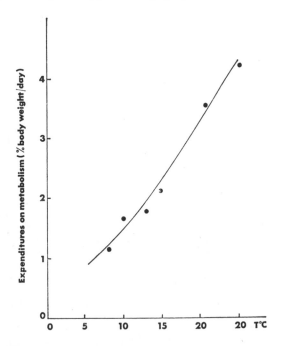

Fig. 9. The expenditure on metabolism in *L. adspersus* as a function of temperature.

The general energy balance

As a result of our investigations, the general energy balance was calculated for *L. adspersus* at various temperatures. The data on chemical composition of the body of *N. diversicolor*, obtained by I. A. Stepanuk (1967), and that on the composition of the body of *E. albidus* by E. M. Malikova (Ivleva, 1961) were used to evaluate the energy expenditure on growth in *N. diversicolor*. In the calculations the calorific equivalents were assumed to be 5·65 kcal/g for proteins, 9·45 kcal/g for fats and 4·10 kcal/g for carbohydrates (Ostapenia, 1968). The values obtained are given in Table 7.

TABLE 7. The chemical composition (in per cent) and the calorific value (in cal/mg) of the body of *Nereis diversicolor* and *E. albidus*

	Water	Dry matter	Protein	Fat	Carbo-hydrates	Ash	Caloricity
N. diversicolor	84·3	15·7	40·3	14·8	33·2	11·7	5·05
E. albidus	82·7	17·3	64·3	19·8	9·3	6·6	5·88

The energy balances of the four age groups of *N. diversicolor* and *L. adspersus* are listed in Tables 8 and 9. In the two bottom lines the values of the first and the second order growth quotients are given. K_1—gross growth efficiency—represents growth as a percentage of all food consumed, and K_2—net growth efficiency—represents growth as a percentage of the assimilated part of the food (Ivlev and Ivleva, 1948).

Temperature differently affected the processes of growth and metabolism, in consequence of which the relation between expenditures on energy metabolism and on plastic metabolism did not remain constant. Also, the two species investigated differed in the character of the changes occurring.

It is seen from the tables that in both *N. diversicolor* and *L. adspersus* the energy consumed as food does not fit the energy expended in plastic and energy metabolism. In each case the quantity of consumed food was about twice as much as required for growth and respiration. The chief cause of errors in calculations of this kind might be the lack of data on expenditure of energy on moving. It has been found, for instance, that in *L. adspersus* at 20°C the expenditure on active metabolism accounted for 120% of standard metabolism (Ivlev, 1963). On the other hand, it is known that any intensification of functional activity, mainly muscular activity, is associated with a considerable increase in metabolism. It has been proved that in *Calanus helgolandicus* the expenditure of energy in the course of daily migrations may increase by several times (Petipa, 1964). There is no evidence, however, of changes in expenditure as a result of movement under the influence of temperature changes. The activity of animals increases with increasing temperature and this is reflected, to some extent, by standard metabolism. Only at low temperatures, where activity is reduced, the measured metabolism may be assumed to be near to its true value but, in this case, the excess may also be considererable in the period of acclimation. Since the calculation of exact values of general metabolism (total expenditure on basal and active metabolism, Ivlev, 1963) is well known to be difficult, we calculated the energy growth coefficient (K_2) using the ratio between daily growth and the part of food consumed by growth and basal metabolism instead of the ratio of growth to assimilated food (A).

As illustrated in Tables 8 and 9, the relationship between energy expenditure on plastic and energy metabolism did not remain invariable within the temperature range of 4°-25°C. The energy growth coefficient in all age groups of nereids was very high. With the rise in temperature it fell regularly from 70-80% at 4°C to 60-70% at 15°-20°C and 40-50% at 25°C. In *L. adspersus* K_2 was much lower. It amounted to 7-22%, if the total expenditure on the process of growth (growth + moulting) was taken into account. K_2 was maximal in the range of temperature within which maximal growth occurred.

TABLE 8. The total energy balance in four size groups of *N. diversicolor* (cal/day)

$T°$ C	Size group	Consumption food	Expense on metabolism	Expense on growth	K_1	K_2
4·1	I	4·43	0·29	0·92	20·5	76·0
	II	7·67	0·55	1·74	22·3	75·6
	III	8·72	0·69	1·95	22·4	73·8
	IV	17·86	1·43	4·44	24·9	75·6
10·3	I	7·95	0·71	1·91	24·0	72·9
	II	12·20	1·08	4·31	35·3	80·0
	III	14·45	2·01	6·25	43·3	75·6
	IV	14·45	2·11	3·30	22·8	61·0
15·7	I	11·23	1·09	3·27	29·1	63·2
	II	18·43	2·22	5·38	29·2	70·8
	III	23·19	3·52	8·20	35·3	70·0
	VI	20·52	3·51	3·65	17·8	50·0
19·6	I	13·77	2·22	3·13	22·7	58·5
	II	20·95	3·71	7·27	34·7	66·2
	III	27·81	5·98	11·97	43·1	66·7
25·0	I	11·62	2·45	1·66	14·2	40·4
	II	23·57	3·64	3·77	16·0	50·9
	III	27·02	4·94	4·61	16·9	48·2

TABLE 9. The total energy balance in *L. adspersus* at various temperatures (cal/day)

$T°$ C	7	10	13	15	20	25
Consumption of food	7·48	9·61	16·11	19·30	29·50	34·51
Losses as faeces	0·13	0·16	0·27	0·39	0·50	0·59
Assimilation of food	7·36	9·45	15·84	18·93	29·00	33·92
Expense on metabolism	3·26	4·48	4·96	6·92	11·46	10·93
Expense on growth	0·05	0·30	0·32	1·37	1·62	0·72
Expense on moulting	0·21	0·24	0·33	0·57	1·13	0·68
K_1	3·4	5·6	4·0	10·0	0·9	3·4
K_2	7·4	10·8	11·6	21·9	19·4	7·4

DISCUSSION

In spite of availability of a considerable number of works devoted to the influence of temperature on the rate of physiological reaction, there are no data enabling simultaneous comparison of the character of temperature influence on various processes occurring in the bodies of invertebrates. In this respect it was of particular interest to obtain the temperature dependence of distinct parameters of energy balance and to evaluate the relation between them quantitatively. The rates of respiration, growth and nutrition in *L. adspersus* and in the two younger groups of *N. diversicolor* are given in Figs 10 and 11 as a function of temperature. Analysis of temperature dependence of re-

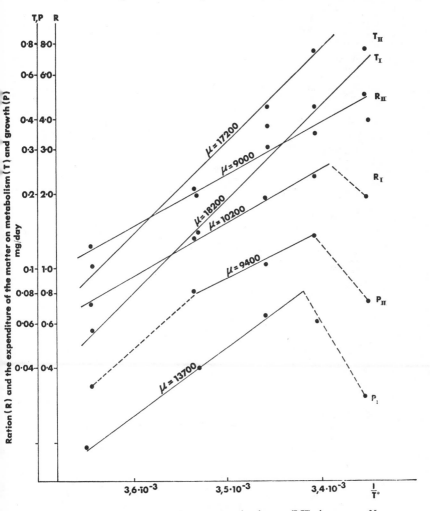

Fig. 10. The parameters of the balance equation in two (I-II) size groups *N. diversicolor* as a function of temperature.

action rates was performed with parameters of the Arrhenius equation. If the ordinate is the logarithm of rate and the abscissa is the inverse of absolute temperature, then the values obtained fall on a straight line, with the exception of the data on growth rate which are represented in the diagram by two sections of a straight line. The break occurs as temperature passes the optimal values. There is also a noticeable departure in values of respiration and growth obtained at 25°C which is the temperature that affects animals adversely. In each case the effect of temperature on the rate of processes was evaluated by parameter μ known as "temperature characteristic". It was calculated with the following formula:

$$\mu = R\frac{(\log V_2 - \log V_1)}{1/T_1 - 1/T_2}$$

where $V_2 - V_1$ is the change in process rate as temperature changes from T_1 to T_2, and R is the gas constant, equal to 1·98.

The values obtained are shown in Figs 10 and 11. These data give evidence of quite different response to temperature variations in the two species. The rates of respiration, nutrition and growth in *L. adspersus* rose with tempera-

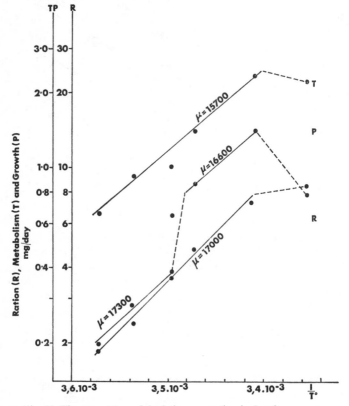

Fig. 11. The parameters of the balance equation in *L. adspersus* as a function of temperature.

ture to approximately the same degree. For all these processes the value of μ was almost the same, equal to 16 000-17 000 cal/mol. In *N. diversicolor* the acceleration of the processes did not correlate so well. The increase in metabolic rate was almost the same in this species as in *L. adspersus* (in the first age group $\mu = 17\,000$ cal/mol and in the second $\mu = 18\,000$ cal/mol), while the acceleration of nutrition and growth occurred more slowly. For nutrition rate $\mu = 9000$-$10\,000$ cal/mol, for growth rate $\mu = 9000$-$13\,000$ cal/mol. Additional explorations seem to be indicated to clarify the cause of this difference. It is significant, however, that the variations in processes investigated had quite similar characters in the two age groups. At the same time, the two groups showed considerable deviations in the character of growth acceleration. The acceleration of growth was extremely irregular within the temperature range investigated. In small nereids regular acceleration of growth occurred within the limits of 4-15°C. The temperature increase up to 20°C slightly affected the growth rate, and prolonged exposure at 25°C resulted in its decline. In the second age group of *N. diversicolor* the temperature range of 10-20°C should be noted within which the values obtained fall on a straight line. As temperature drops to 4°C or rises up to 25°C growth rate considerably decreases. The parameter μ which represents the acceleration of growth in both groups of nereids within the mentioned interval of temperatures, ranged from 13 500 to 9000 cal/mol. Thus, the increase in temperature did not always cause equal acceleration of processes. Most deviations occurred in growth, the rate of which at different temperatures depended not only on the age of animals, but also on availability of optimal temperature for growth and differences in temperature tolerance between species.

Variations in parameter μ and energy growth coefficient K_2 reflect particularities in distribution of energy consumed with food at each temperature. Comparison of these values shows that in *N. diversicolor* which are able to grow intensively within a wide range of temperatures, the degree of food consumption for growth was very high at low temperatures. The increase in temperature affected the oxidative processes to a greater extent than the synthetic ones, in consequence of which the ratio between expenditures on plastic and energy metabolism ($P:T$) dropped regularly from 2·5-3·8 at 4-10°C to 0·6-1·0 at 25°C. In *L. adspersus* the oxidative processes exceeded the synthetic ones at all temperature ranges and the effect of temperature changes on this ratio was much lower. In this case $P:T$ ratios increased from 0·3 at 7°C and 10°C to 0·6 at 15°C and 20°C. Prolonged exposure at 25°C caused breakdown in oxidative processes and in protein metabolism, more pronounced in older size groups of *N. diversicolor*.

The results obtained reflect specific features of temperature changes of plastic and energy metabolism in animals fed with a particular kind of food. Low growth rate in *L. adspersus* points to the fact that food such as gonads of *Mytilus galloprovincialis* was of low value for this species. This, undoubtedly, reflected the values of energy growth coefficient, which, so far as is known, is strongly dependent on the kind of food. It should be noted, however, that low growth rate also occurred if *L. adspersus* were fed with larvae of Chironomidae (Karpevitch and Bogorad, 1940).

REFERENCES

IVLEV, V. S. 1962. The active energy exchange in young salmon (*Salmo salar*). *Vop. Ikhtiol.*, **2** (I) (22), 158-68.

IVLEV, V. S. 1963. The energy expended by moving shrimps. *Zool. Zh.*, **42** (10), 1465-71.

IVLEV, V. S., and IVLEVA, I. V. 1948. Energy transformations in the growth of birds. *Bull. Mosc. Soc. Nat. biol. Div.*, **53** (4), 23-37.

IVLEVA, I. V., 1961. Feeding of *Enchytraeus albidus* during its cultivation. *Byull. vses. Inst. rỹb. Khoz. Okeanogr.*, **51**, 96-117.

IVLEVA, I. V., and POPENKINA, M. I. 1970. On temperature dependence of metabolism in poikilothermic animals. In *The physiological principles of aquatic animals ecology*. Kiev "Naukova Dumka".

IVLEVA, I. V., and POPENKINA, M. I. 1968. Temperature influence on respirative intensity of shrimps *Leander adspersus* Rathke. In *Problems of energetic metabolism of aquatic animals*. Moscow, "Nauka" Publ. House.

KARPEVITCH, A., and BOGORAD, G. 1940. The consumption of food by the shrimp *Leander adspersus*. *Zool. Zh.*, **19**, (1), 134-8.

OSTAPENIA, A. 1968. Caloricity of the body substance in aquatic animals and methods for its determination. In *The methods for the estimation of production of aquatic animals*, Part I, ch. 2. Minsk, "Higher School" Publ. House.

PETIPA, T. S. 1964. Diurnal rhythm of the consumption and accumulation of fat in *Calanus helgolandicus* (Claus) in the Black Sea. *Dokl. Akad. Nauk SSSR*, **156**, (6), 1440-3.

SEMENOF, J. V., and Ivleva, I. V. 1970. The automatic thermoregulation arrangement for flowing aquaria. In *Problems of energetic metabolism of aquatic animals*. Moscow, "Nauka" Publ. House.

STEPANUK, I. A. 1967. *The biochemical composition of some bottom invertebrates of the north-western Black Sea*. Master's thesis. Odessa University Edit., 1-22.

WINBERG, G. G. 1968. Methods for the estimation of production in aquatic animals. In *The methods for the estimation of production of aquatic animals*. Part I, ch. I. Minsk, "Higher School" Publ. House.

Variations of length, weight, respiratory rate, and chemical composition of Calanus cristatus in relation to its food and feeding

MAKOTO OMORI
Ocean Research Institute
University of Tokyo
Nakano, Tokyo, Japan

ABSTRACT. Influence of the quantity of food available in the sea for *Calanus cristatus* was studied by determining the variations of its body length, weight, respiratory rate, and the hydrogen, carbon, nitrogen and ash contents. Because of the change of its feeding habit, the effect was most significant in copepodite stage V. Positive correlations were observed between body length and the quantity of food available during its growth. The average wet weight of animals decreased to approximately ⅖, and that of dry weight to ⅓ of maximum when food became scarce. Respiratory rate was reduced to less than half due to adaptation to the unpleasant condition. Carbon contents ranged from 58·9% to 50·3% of the dry weight. Nitrogen showed negative correlation with carbon content and varied from 8·3% to 10·6%. The chemical composition of the copepod varied with its sex and developmental stage. Generally, the carbon content of the dry weight was largest in the adult female and smallest in the adult male, while the situation was the reverse with the nitrogen content. It was observed that the scarcity of food influences not only the occurrence of adults but also their sex ratio.

Conditions of the species preserved with different methods were compared. Usually, loss of the dry weight and the hydrogen, carbon, and nitrogen contents was greatest in formalin-preserved materials. The carbon: nitrogen ratio was also changed considerably.

INTRODUCTION

Zooplankton is carried from one place to another by movement of water-masses. Animals are always influenced by changes of the environment and as they attempt to adapt to their new circumstances, their condition varies from time to time by being affected by the environment. Such influence caused by the environment has been described as the seasonal or geographical variations of relative abundance, species composition, or chemical composition of zooplankton. However, most studies discuss nothing but the relation between

113

the environment and the quantitative or qualitative variation of the total zooplankton or some large taxonomic groups. There are only a few data from studies showing how the fluctuation of the environment influences a given species. Recently, several workers have attempted to study the dynamics of plankton production through the measurement of chemical composition of plankton organisms. But these studies are again limited to the analysis of the net plankton or major groups in the zooplankton. A knowledge of the influence of the environment and the process of such variations as body length, weight, and chemical composition found in a selected species must also be important to the understanding of the production in the sea.

Among various environmental factors, the quantity of food available in the sea must be one of the most effective components. It has been known, for instance, that there is good agreement between reproductive rates of some copepods and rotifers and the small phytoplankton increases (Marshall and Orr, 1955, 1964; Edmondson, 1962).

Calanus cristatus, a typical subarctic copepod, is abundant and one of the most important members of the total zooplankton biomass in the Bering Sea and in the northern North Pacific. In the Pacific water of Japan, it is most abundant in the upper 100 m of the Oyashio area where it reproduces. In this area, the species feeds on diatoms densely distributed in shallow layers and stores great reserves of fat in copepodite stage V, the most prolonged of all the stages. The population in the waters south of 40°N in general is transported down from the north with the submerged Oyashio water. The species sinks deeper and decreases in concentration with increasing distance from the Oyashio area. In the southern waters, as the shallow layer is covered with the unfavourably warm Kuroshio water, distribution of the species is restricted to below 400 m. Since diatoms, their most important food, become very scarce there, the species cannot store nutrition in their bodies during the copepodite V, but have to use it for survival (Omori, 1967; Omori and Tanaka, 1967).

In the present study, body length, wet and dry weights, respiratory rate, and total carbon, nitrogen, hydrogen and ash contents of *Calanus cristatus* distributed in the various depths and localities with different nutrient conditions have been measured. Through the relation between these variations and the quantity of food available in the sea, it was proposed to study how the condition of zooplankton is influenced by fluctuations of the environment.

Variations of the weight and chemical composition caused by the different methods of handling and preservation were also determined.

MATERIALS AND METHODS

Except for the respiration experiments, the species was collected at 9 stations located in the Pacific water off eastern and central parts of Honshu (Stas. 83 and 191-9) (Fig. 1). Sampling was made by oblique or horizontal tow with ORI-200 net (Omori, 1965) on three cruises of RV *Tansei Maru*. At Station 83 the sampling was made on 23 October 1965, but at others it was carried out during the period from 27 September to 17 October 1966. Samples were all preserved in 5% formalin-seawater solution neutralized by hexamine, and brought back to the laboratory. In the laboratory, *C. cristatus*, consisting of adult and copepodite stage V, was sorted completely from the original samples.

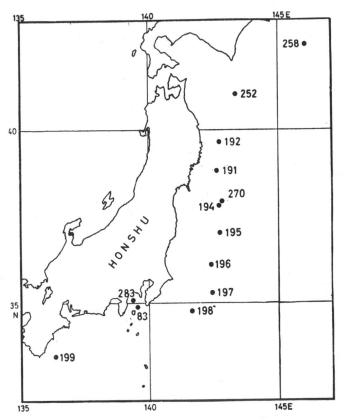

Fig. 1. Sampling stations of *Calanus cristatus*.

Determinations of body length, weight, and chemical composition were carried out about a month after the sampling. Generally, unless the number exceeded 100 individuals, all individuals found at each station were used for analyses. The body length was measured from dorsal view along midsagittal plane: from base of crest to end of right fucal ramus. For the wet weight measurement, samples were rinsed with distilled water to remove adhering interstitial formalin-seawater, and then rolled on filter paper. Following this, each group was placed on a glass ring with a silk netting XX13 base, and dried in an oven to a constant weight at 60°C. The dry material was then stored in a desiccator.

Total carbon, nitrogen, and hydrogen as well as ash content were determined for the specimens obtained at Stations 192, 195, 197, 83, and 199 using a Yanagimoto CHN Corder, Type MT-1. In one analysis, the samples of approximately 1·80-2·95 mg in dry weight were heated at 500°C and com busted at 800°C.

Respiratory measurement was made using the method given by Anraku (1964). *C. cristatus*, copepodite V, was collected with an ORI-100 net at stations (Stas. 252, 270, and 283) between 30 July and 8 August 1967. The animals were transferred to 2-l glass containers with filtered seawater as soon after capture as possible. These were immersed in a bath with a constant temperature of 8°C, the temperature which was closest to that of the seawater at the point of capture, for at least 24 hours to permit acclimatization. Three series of 100-ml glass bottles were filled with 10 ml of seawater filtered by Millipore type HA filter. Then, 3 individuals were inserted with a pipette. Filtered seawater, with 1 mg/litre of Kanamycin sulphate, was next flushed twice through the bottles by a siphon arrangement. Controls without animals were prepared in exactly the same way. All bottles were wrapped with aluminum foil and immersed in the bath at 8°C. Experiments were run for 12 hours. After the run, the water in the bottles was transferred for titration to 50-ml glass bottles through another siphon system. Titrations were by the Winkler method with 0·005 N thiosulphate solution. The animals were checked for their activity. They were then rinsed with distilled water and dried at 60°C for the determination of dry weight and chemical composition. All these procedures were made on the ship.

In addition to the work on the ship, variation of the respiratory rate of the species under starvation conditions was observed in the laboratory. *C. cristatus*, copepodite V, was obtained by the ORI-100 net surface tow at 43°52′N, 149°58′E on 7 December 1967. Thirty animals were placed in 2-l glass containers with one individual in each container, and kept in the filtered seawater (Millipore type HA filter) at a temperature of 8°C. Every 3 days, the animals were changed to fresh filtered seawater and clean receptacles. The respiration measurement was made once every 5 days until all individuals had died.

As already mentioned, except for the specimens used for the respiration measurement, the chemical analysis was done for those preserved in formalin-seawater. Thus, the present values obtained might be somewhat different from that of fresh specimens. To clarify this question, three series of samples preserved by different methods were analysed and compared. The female, male, and copepodite V of *C. cristatus* were collected at Station 258 on 30 July 1967. Immediately after sampling, 150 individuals of each group were sorted, and divided into three series. The specimens of the first series were then preserved in exactly 5% formalin-seawater neutralized by hexamine for 1 month. After that, they were rinsed with distilled water and dried at 60°C in the manner already described. All specimens of the second and third series were rinsed briefly with distilled water and placed on the glass ring with silk netting base. Subsequently, the second ones were frozen at −20°C, while the third ones were dried at 60°C at first and then stored in the freezing box (−20°C) for a month. The animals of the second series were melted at 5°C, and then dried like others in the laboratory. All were then

lesiccated for 2 days and analysed. In the present paper, these three preserving methods are named tentatively as formalin-preserving (Ser. 1),
reezing (Ser. 2), and drying (Ser. 3).

RESULTS

The regional variations of the body length distribution of the species is shown
in Fig. 2. The difference was not very significant in adult female and male,
but was remarkable in copepodite V. Proceeding south-westward, the
average body length of copepodite V was quite definitely observed to be reduced from 8·6 mm to 7·8 mm (Table 1).

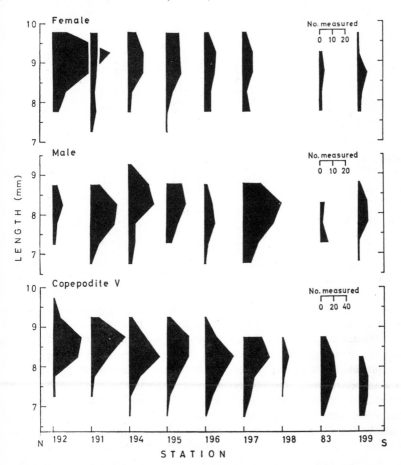

Fig. 2. Histograms showing the body length distributions of *Calanus cristatus*.

Both the wet and dry weight of the species were larger in the northern
waters than in the southern waters. The maximum decrease of the wet weight
was about 36% in copepodite V. The difference was more distinct in the dry
weight. The decrease ranged from 72% in copepodite V to 39% in the male.

TABLE 1. Regional variations in body lengths, wet weights, and dry weights of *Calanus cristatus*

Station	Number of individuals measured	Mean body length (mm)	Wet weight per 1 copepod (mg)	Dry weight per 1 copepod (mg)	Dry weight % on wet weight
Female					
192	50	9·0	24·0	4·46	18·6
191	22	8·8	21·9	3·50	16·0
194	35	9·0	20·9	3·47	16·7
195	40	9·2	23·3	4·00	17·2
196	30	8·8	21·1	3·20	15·2
197	27	·8·8	22·4	3·82	17·0
198	—	—	—	—	—
83	10	8·5	17·8	2·24	12·6
199	14	8·6	15·4	2·15	14·0
Male					
192	15	8·2	17·1	2·77	16·2
191	55	8·1	16·6	2·77	16·7
194	50	8·3	16·6	3·15	19·0
195	40	8·2	17·0	2·79	16·4
196	20	8·0	16·2	2·17	13·4
197	85	8·0	16·0	2·55	15·9
198	—	—	—	—	—
83	11	7·7	14·5	1·69	11·7
199	19	7·9	14·8	1·89	12·7
Copepodite V					
192	100	8·6	19·6	2·48	12·3
191	100	8·5	19·5	2·11	10·8
194	100	8·3	15·8	1·25	7·9
195	100	8·4	15·4	1·28	8·3
196	100	8·1	14·4	1·23	8·6
197	100	8·2	14·5	0·92	6·3
198	15	8·2	11·6	0·70	6·0
83	70	7·9	12·1	0·73	6·0
199	38	7·8	12·5	0·90	7·2

The dry to wet weight ratio also decreased with increasing distance from the Oyashio area. The decrease was most significant in copepodite V.

Variation of the total carbon, nitrogen, and ash contents of the dry weight are shown in Table 2. In the female, the regional variation as seen in the weight determination was not found in both the carbon and nitrogen contents. The carbon:nitrogen ratio (C/N) was usually greater than 7·3. Ash content was almost constant ranging between 2·3% and 2·6% of the dry weight. In the male, variation of the carbon content was not closely related to the locality, but the nitrogen content increased towards the south-west. Consequently, the ratio C/N decreased gradually from 5·3 to 4·6. A marked change of the ratio was seen in copepodite V. As the nitrogen content showed a negative correlation with the carbon content, the ratio decreased from 7·1 to 4·8.

Generally, at the same location, the carbon content of the dry weight was largest in the adult female and lowest in the adult male, whereas the situation was the contrary with nitrogen. The values of copepodite V were found be-

ween these two. The ratio C/N was usually largest in the female and smallest
n the male.

The respiratory rate again changed with locality (Table 3). The difference
n the oxygen consumption per animal between the Oyashio area (Sta. 252)
and Kuroshio area (Sta. 283) was nearly 3:1. The oxygen consumption per
ng dry weight was also slightly lower in the southern waters. The variation

TABLE 2. Regional variations of total carbon, nitrogen and ash contents
(% of dry weight) and the carbon:nitrogen ratio of *Calanus cristatus*

| Station | Total dry weight analysed (mg) | % of dry weight | | | C:N ratio |
		C	N	Ash	
Female					
192	37·522	60·9	7·5	2·4	8·1
195	52·665	60·0	8·2	2·5	7·3
197	23·799	61·8	6·8	2·3	9·1
83	14·461	62·6	7·4	2·6	8·5
199	12·755	62·7	8·6	2·5	7·3
Male					
192	15·863	55·9	10·5	2·2	5·3
195	17·500	56·0	10·8	2·7	5·2
197	15·863	56·1	11·2	2·4	5·0
83	12·645	52·4	11·5	2·7	4·6
199	10·130	54·1	11·9	2·6	4·6
Copepodite V					
192	12·254	58·9	8·3	2·2	7·1
195	10·359	58·3	9·5	2·8	6·1
197	7·389	56·8	10·3	2·9	5·5
83	12·645	53·9	10·7	2·9	5·0
199	10·130	50·3	10·6	2·9	4·8

TABLE 3. Regional variations of respiratory rate, total hydrogen, carbon,
nitrogen, and ash contents (% of dry weight) and the carbon:nitrogen ratio
of *Calanus cristatus*, copepodite V

| Station | Dry weight per 1 copepod (mg) | Average respiratory rate | | % of dry weight | | | | C:N ratio |
		μl/animal ×day	μl/mg dry weight ×day	H	C	N	Ash	
252	4·44	33·93	7·64	9·1	55·9	6·2	2·0	9·0
270	2·47	18·57	7·52	8·7	54·0	9·0	4·3	6·0
283	2·40	12·22	5·09	8·2	48·6	9·3	4·4	5·2

TABLE 4. Variation of respiratory rate of *Calanus cristatus*, copepodite V,
under starving condition. (Species was obtained on 7 December 1967)

Date of experiment	Average respiratory rate (μl/animal ×day)
8 Dec., 1967	29·18
13 Dec., 1967	16·30
19 Dec., 1967	14·28
24 Dec., 1967	13·89

of chemical composition including hydrogen showed a similar tendency to that found in the previous determination of copepodite V, although the present C/N ratio was slightly higher. Hydrogen decreased gradually from north to south.

The decrease of the respiratory rate of animals under starvation condition was ascertained in the laboratory (Table 4). The oxygen consumption per animal per day dropped suddenly in the first 5 days after capture. Thereafter the rate of the decrease was slow.

Table 5 shows the summary of the comparison in the wet and dry weight and the chemical composition of *C. cristatus* preserved by three different methods. Except for wet weight, there was considerable variation among the three methods. The dry weight of the species was usually largest after freezing followed by drying. The great decrease of the dry weight occurred after formalin-preserving. Generally, the hydrogen and carbon contents of the dry weight were largest in the frozen materials and smallest in the formalin-preserved materials. For the female, the values ranged from 10·9% to 8·3% in hydrogen and from 61·7% to 54·8% in carbon. On the contrary, the nitrogen varied from a maximum 7·4% of the dry weight in formalin-preserving to a minimum 6·6% in freezing. It can be seen that the C/N ratio definitely changed. The variation was largest for the female, ranging from 9·4 (freezing) to 7·4 (formalin-preserving). Ash content of the dry weight was generally increased after drying.

With a decrease in dry weight, quantities of the total carbon, nitrogen, and hydrogen in the specimens were decreased. The differences among the three preserving methods became clearer when the quantity of each element was calculated. The values obtained after freezing have been rated as 100 for convenience. The loss from initial value was usually greatest in the formalin-preserved material; it was up to approximately 65% in hydrogen, 59% in carbon, and 48% in nitrogen. The difference between the freezing and formalin-preserving was most significant in the female.

DISCUSSION

Under normal conditions, the Oyashio area extends as far south as Station 192. The mixing of the Oyashio and Kuroshio waters is found in the area between Stations 191 and 196. Station 197 and its southern stations are located in the Kuroshio area. Generally, with the submergence of the Oyashio water *C. cristatus* starts to sink at around Station 192. The temperature of the water where the species is most numerous, is between 3 and 8°C. The centre of distribution of copepodite V is found at depths greater than 500 m at Station 195, 600 m at Station 197, 700 m at Station 83, and 1000 m at Station 199. The adult is usually found below the 400 m level.

With the deepening of the layers containing *C. cristatus*, the quantity of food available for the species in the sea decreases. Since diatoms,* their most

* Three species of large diatoms, *Coscinodiscus asterompharus, C. lineatus,* and *Denticula seminae,* constituted the major portion of the recognizable stomach contents in summer. Some tintinnids were also present.

TABLE 5. Variations of wet and dry weight, total hydrogen, carbon, nitrogen, and ash contents (% of dry weight) and the carbon: nitrogen ratio in *Calanus cristatus* by three different methods of preservation. Numbers in parentheses in the chemical composition mean the ratio of the quantity among three methods of the preservation. The values obtained after freezing have been rated as 100

Method of preservation	Wet weight per 1 copepod (mg)	Dry weight per 1 copepod (mg)	Total dry weight analysed (mg)	% of dry weight				C:N ratio
				H	C	N	Ash	
Female								
Freezing	25·5 (100)	8·94 (100)	44-252	10·9 (100)	61·7 (100)	6·6 (100)	2·3 (100)	9·4
Drying	—	8·70 (97)	41-536	10·0 (89)	59·5 (94)	7·2 (106)	3·0 (127)	8·3
Formalin-preserving	26·3 (102)	4·13 (46)	27-401	8·3 (35)	54·8 (41)	7·4 (52)	2·6 (53)	7·4
Male								
Freezing	17·9 (100)	5·96 (100)	27-947	10·0 (100)	57·3 (100)	7·2 (100)	2·5 (100)	8·0
Drying	—	5·46 (92)	27-111	9·1 (83)	55·9 (89)	7·8 (99)	3·3 (121)	7·2
Formalin-preserving	18·1 (101)	3·51 (59)	19-649	8·4 (49)	53·0 (54)	8·4 (69)	2·8 (66)	6·3
Copepodite V								
Freezing	20·8 (100)	3·47 (100)	20-385	9·8 (100)	60·9 (100)	6·3 (100)	2·1 (100)	9·7
Drying	—	3·09 (89)	21-488	9·0 (82)	56·4 (83)	6·3 (89)	3·5 (148)	9·0
Formalin-preserving	21·6 (103)	1·85 (53)	17-557	10·8 (59)	58·8 (52)	6·7 (57)	2·5 (63)	8·8

important food, are numerous in the shallow layers in the northern waters, the animal must find its nutrient source in something else in the southern deep water. The influence of the scarcity of food appeared clearly on the condition of the animal. In fact, both wet and dry weights of the species decreased south-westward. The variation of the dry to wet weight ratio indicates that the organic substance of the species was reduced and its water content was increased gradually in the unfavourable condition. The situation can be observed more distinctly from the variations of chemical composition and respiratory rate of copepodite V. With increasing distance from the Oyashio area, the total carbon content of the dry weight was decreased, while the total nitrogen content was increased. This means that in the southern waters, due to hunger, the species loses its nutrient balance, and the consumption of fat and carbohydrate increases rapidly. Conover (1964) found in *Calanus hyperboreus* that up to 92% of the total loss of the weight by starvation was due to a loss of fat. To avoid this dangerous situation, the species tends to reduce its metabolic rate. For *C. hyperboreus*, the respiratory rate was lowered seasonally by two-thirds when food became scarce (Conover, 1962). In the present experiment, the reduction of the respiratory rate of *C. cristatus* due to adaptation to a scarcity of food exceeded half the initial rate. Probably, the species occasionally eats other food in the southern waters. In fact, the stomachs of the species collected south of Station 195 were entirely empty.

As seen in Fig. 2, the body length of copepodite V decreased with de-creasing quantity of food available. As the size of copepod is fixed by moult-ing, the present smallness of the body means insufficient increase of the length at the time of several moults during the growth of the species. There is no doubt that such variation of the length was caused by the influence of the environment in which the species had spent its early developmental stages. The species found in the southern stations must have lived in the water where the food was very scarce from an earlier period of its younger stages than that obtained in the northern stations. Such relations between the body length of the oceanic copepod and the quantity of food available during its growth has been observed in the Sargasso Sea (Deevey, 1964). According to her, there were significant positive correlations between the mean lengths of *Pleuro-mamma abdominalis* and *P. piseki* and the quantity of phytoplankton during the previous month.

The reason why the regional variation of the body length was not very significant in the adult might be explained partly by the characteristic feeding habits of the species, i.e. the adult does not eat. As shown in Fig. 3, when the species becomes an adult, the cutting edge of its mandible is reduced greatly, and it is probable that the species almost stops feeding. Even so, however, if a majority of the copepods could reach adulthood in each station, the length distribution of adults would show patterns more similar to copepodite V. The present result seems to indicate that in such an unfavourable environment, the number of copepodite V which can moult into adults is quite limited. Actually, the occurrence of adults, especially the female, in the southern waters was usually scarce. This circumstance is defined from the difference of extreme

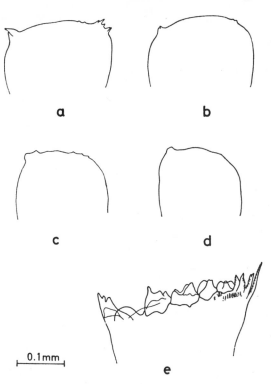

Fig. 3. Variation of cutting edge of mandible in *Calanus cristatus*.
a and b, female; c and d, male; e, copepodite V.

values of the ratio C/N for each group of the species (Fig. 4). At Stations 192 and 195, the extreme values among individuals of copepodite V were found in the range of these measured in the female and the male. However, the values were then lowered and separated gradually from those of the female towards the south-west. Since the recovery of the C/N ratio is impossible under starvation conditions, the animal may not be able to become female. Probably almost all of the females were transported from the favourable northern waters after they matured there. The similarity of the range of adult male to copepodite V may indicate some possibility of the growth from copepodite V to adult male in the southern waters. In fact, comparing the trends in regional variation of all items in the female and the male, that of the male rather resembled that of copepodite V. The sex ratio was biased toward males in the southern waters. From these findings, it can be said that the scantiness of food influences not only the occurrence of adults, but also their sex ratio. Conover (1965) pointed out that the adult males of *Calanus hyperboreus* appeared in winter when their food, phytoplankton, was scarce. It was also shown that the sex ratio of the same species established from laboratory moults indicated an appreciably higher percentage of males than was observed in nature (Conover, 1965). These facts seem to support the present conclusion,

124   PELAGIC FOOD CHAINS

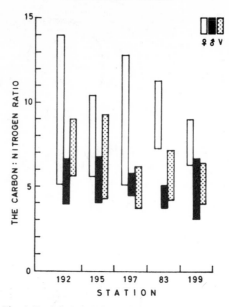

Fig. 4. Regional variations of extreme values of the
carbon:nitrogen ratio in *Calanus cristatus*.

and suggest that the sex ratio of copepods is influenced by the quality of
food available in the sea. Regional variation of the sex ratio has been found
in several species of copepod. According to Mednikov (1961), the occurrence
of the males of *Pseudocalanus elongatus* was larger in the North Sea than in
the White Sea. This might be the effect of the availability of food as suggested
by him.

The chemical composition as well as the dry weight of the species varied
considerably with the method of preservation. Even in the same species,
since the chemical composition is different in the sex and developmental stage,
the rate of loss in each element caused by formalin-preserving was not the
same. At present, zooplankton biomass is generally reported as wet weight or
displacement volume. Since the difference in the wet weight between the
frozen materials and the short-term formalin-preserved materials was not
serious, the latter is useful for the study of species identification or the wet
weight determination. However, when estimation of zooplankton production
is attempted by the dry weight or contents of some chemical elements, certainly we have to pay more attention to the effect of the fixing and preserving
methods of the specimens.

There was some reduction of dry weight and chemical contents in the
drying method, too. However, the fluctuation of the values among individuals of the same species was smallest in this method, while it was largest after
the freezing. It is frequently experienced that the condition of frozen zooplankton varies greatly with freezing temperature and with handling at the
time of freezing and melting. The preservation of samples by freezing will be

frequently employed in the future study of zooplankton production in the sea. Thus, it is necessary to establish a standard method for plankton freezing.

We now know, as a result of the present study of *C. cristatus*, that the total carbon, nitrogen, and hydrogen contents of zooplankton can vary considerably with the quantity of food available in the sea. The difference of the nutrient condition becomes one of the main factors in causing variations of the chemical composition within a species. For instance, nitrogen levels reported for *Calanus finmarchicus* vary markedly from 10·21% (Vinogradov, 1933, cited in Vinogradov, 1953) to 4·7-5·9% (Curl, 1962) for dry weight. In the present study, the hydrogen, carbon, and nitrogen contents of the dry weight measured in *C. cristatus* which seemed to be distributed in the most favourable environmental conditions were, 10·9%: 61·7%: 6·6% in the female, 10·0%: 57·3%: 7·2% in the male. In the Sargasso Sea, the total carbon content of copepods varied from a maximum 47·6% of the dry weight to a minimum 35·2%, while the nitrogen content ranged from 11·17% to 8·16% (Beers, 1966). The above values of *C. cristatus* are apparently higher in carbon and lower in nitrogen than the range of the Sargasso Sea copepods. The copepods in the northern North Pacific must have more fat than those in the Sargasso Sea. The nitrogen content of the dry weight of *C. cristatus*, copepodite V, was previously determined using the formalin-preserved material (Nakai, 1955). His value, 8·54%, is close to the present one found at Station 192.

ACKNOWLEDGEMENTS

The author would like to express his sincere thanks to Dr John D. H. Strickland for his helpful criticism and kind reading of the manuscript. Thanks are also due to staff members of the Plankton Laboratory of the Ocean Research Institute who assisted the author either in the field or in the laboratory.

REFERENCES

ANRAKU, M. 1964. Influence of the Cape Cod Canal on the hydrography and on the copepods in Buzzards Bay and Cape Cod Bay, Massachusetts. II. Respiration and feeding. *Limnol. Oceanogr.*, 9, 195-206.

BEERS, J. R. 1966. Studies on the chemical composition of the major zooplankton groups in the Sargasso Sea off Bermuda. *Limnol. Oceanogr.*, 11, 520-8.

CONOVER, R. J. 1962. Metabolism and growth in *Calanus hyperboreus* in relation to its life cycle. *Rapp. P.-v. Réun. Cons. perm. Int. Explor. Mer*, 153, 190-7.

CONOVER, R. J. 1964. Food relations and nutrition of zooplankton. *Occ. Publ., Proc. Sym. exp. mar. Ecol.*, 2, 81-91.

CONOVER, R. J. 1965. Notes on the molting cycle, development of sexual characters and sex ration in *Calanus hyperboreus. Crustaceana*, 8, 308-20.

CURL, H., Jr. 1962. Standing crops of carbon, nitrogen, and phosphorus and transfer between trophic levels in continental shelf waters south of New York. *Rapp. P.-v. Réun. Cons. perm. int. Explor. Mer*, 153, 183-9.

DEEVEY, G. B. 1964. Annual variations in lengths of copepods in the Sargasso Sea off Bermuda. *J. mar. biol. Ass. U.K.*, 44, 581-600.

EDMONDSON, W. T. 1962. Food supply and reproduction of zooplankton in relation to phytoplankton population. *Rapp. P.-v. Réun. Cons. perm. int. Explor. Mer*, **153,** 137-41.

MARSHALL, S. M., and ORR, A. P. 1955. *The biology of a marine copepod*, Calanus finmarchicus *Gunnerus*. Edinburgh, Oliver and Boyd.

MARSHALL, S. M., and ORR, A. P. 1964. Grazing by copepods in the sea. in *Grazing in the terrestrial and marine environments*. Oxford, Blackwells Sci. Publs., 227-38.

MEDNIKOV, B. M. 1961. On sex ratio in deep water Calanoida. *Crustaceana*, 3, 105-9.

NAKAI, Z. 1955. The chemical composition, volume, weight, and size of the important marine plankton. *Spec. Publ., Tokai reg. Fish. Res. Lab.* 5, 12-24.

OMORI, M. 1965. A 160-cm opening-closing plankton net. I. Description of the gear. *J. oceanogr. Soc. Japan*, 21, 212-20.

OMORI, M. 1967. *Calanus cristatus* and submergence of the Oyashio water. *Deep Sea Res.,* 14, 525-32.

OMORI, M., and TANAKA, O. 1967. Distribution of some cold-water species of copepods in the Pacific water off east-central Honshu, Japan. *J. oceanogr. Soc. Japan*, 23, 63-73.

VINOGRADOV, A. P. 1953. The elementary chemical composition of marine organisms *Mem. Sears Fdn mar. Res.*, (2), 647 pp.

Food rations, metabolism and growth of crustaceans

L. M. SUSHCHENYA
Institute of the Biology of the Southern Seas
Academy of Sciences of the Ukraine SSR
Sevastopol, USSR

ABSTRACT. Some results of food, growth and metabolism experiments are discussed in relation to the studies of energy transformation by planktonic and bottom Crustacea. General equations for food intake of these animals are obtained, based on relations between body weight, food concentration and ration value. Some quantitative relations between food consumption and food assimilation were also found. The available empirical data show that the absolute quantity of food assimilated, like the ration consumed, increases with increasing food concentration and tends to some maximum point which may be called the maximum value of assimilation of a given type of food by these animals. The metabolism of Crustacea is expressed through the rate of oxygen consumption and in the equivalent energy units. The energy expenditure for the growth of animals was estimated from the experimental data on weight increase converted to units of energy. Based on these data, the growth coefficient of the first and of the second order was calculated. To study quantitative relationships between food intake, metabolism and growth of Crustacea the equations of Winberg, Paloheimo and Dickie were used.

INTRODUCTION

Among aquatic animals the Crustacea play an important part in determining the productivity of water basins. The efficiency of matter and energy transformation between the autotrophic and first heterotrophic level in aquatic ecosystems depends largely on the productivity of these animals. In this connection the estimation of quantitative relations between some ecological-physiological parameters on the one hand and growth of Crustacea on the other hand is most important.

In energy units, the quantitative relations between food intake, total metabolism and growth must balance as in equation [1]:

$$R = G + T + N \qquad [1]$$

127

where R is the energy in the food intake, G the energy of growth, T the metabolic expenditure (or the energy of respiration) and N the energy of non-assimilated part of food, whence

$$G = R-(T+N) \text{ or } \frac{dW}{dt} = AR-T \qquad [2]$$

where dW is the increment of growth in energy during any short period of time, dt, and A is the coefficient of assimilation. Now we shall consider each of these parameters separately.

FOOD CONSUMPTION

Relation between rations value and food concentration

The literature on the feeding of Crustacea shows that the quantity of food consumed by animals per unit of time or ration (R), depends first of all upon two factors—food concentration (K) and weight of the animal (W). In considering this problem for fishes, Ivlev (1955) found that as food concentration increased, the increased ration was less than expected, suggesting a progressive drop in grazing efficiency. A number of experimental data (Lucas, 1936; Richman, 1958; Sushchenya, 1958, 1962, 1966; Petipa, 1959; Monakov, Sorokin, 1960) allowed us to extend this conclusion to Crustacea.

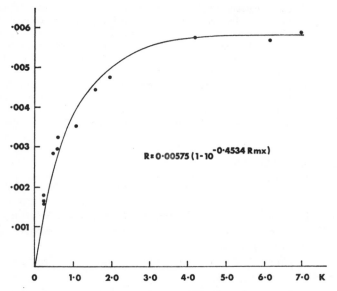

Fig. 1. Relation between rations value (R) and food concentration (K) in the *Diaphanosoma brachyurum*. R in mg of food per animal per day; K is the biomass of *Chlorella* in mg wet weight per litre (see Table 1 for definition of terms).

TABLE 1. Value of maximum ration (R_{mx}) and of the coefficient p in equation [4] for different species of Crustacea

Species	Food		R_{mx} per animal per day	p	Weight of animal: w—wet d—dry	$t°C$	Author
	Type	Concentration					
		mln. cells per ml	mln. cells				
Artemia salina juv.	Yeast	0·1 -1·6	10·08	0·680	0·175d	20	Sushchenya, 1962
Artemia salina ad.	Yeast	0·1 -1·6	33·60	0·550	0·590d	20	Sushchenya, 1962
Bosmina longirostris	Chlorella	0·03-0·90	1·05	1·600	0·024w	20	Sushchenya, unpubl.
Simocephalus vetulus	Chlorella	0·02-1·20	1·80	2·300	0·900w	20	Sushchenya, unpubl.
Eurytemora hirundoides	Nitzschia	0·25-2·00	0·55	0·125	?	summer	Lucas, 1936
Neomysis vulgaris	Nitzschia	0·01-2·00	57·60	0·180	?	summer	Lucas, 1936
		biomass, mg wet weight per litre	mg of wet weight				
Artemia salina	Dunaliella	20-1000	2·15	0·0022	4·200w	25	Khmeleva, unpubl.
Daphnia longispina	Various algae	1-12	0·0325	0·1986	0·027w	20	Monakov et al., 1960
Daphnia pulex	Chlamydomonas	0·5 -11	0·0525	0·1345	0·050d	20	Richman, 1958
Acartia clausi	Various algae	0·05-0·5	0·0011	0·5163	0·040w	20	Petipa, 1959

From Fig. 1 it appears that there is an asymptotic relation between ration and food concentration as ration tends to its maximum, R_{mx}. Therefore at sufficiently high food concentration $R = R_{mx}$.

Then, $$\frac{dR}{dK} = \xi(R_{mx} - R)$$

whence $$R = R_{mx}(1 - e^{-\xi K}) \qquad [3]$$

where ξ is a coefficient of proportionality.

In common logarithms,

$$R = R_{mx}(1 - 10^{-pK}) \qquad [4]$$

where p is the coefficient of proportionality.

So $$\xi = p\, 2{\cdot}3026$$

In equation [4] p is estimated by least squares:

$$p = \frac{1}{n} \sum_{i=1}^{n} \frac{\log R_{mx} - \log(R_{mx} - R_i)}{K_i}.$$

The results of calculations of R_{mx} and p for some species of Crustacea are shown in Table 1. It is evident that numerical values of these differ markedly by species of animals and type of food. The relationship between body weight and the maximum ration R_{mx} has not been determined. The effect of temperature and the effect of food searching capacity has not been analysed.

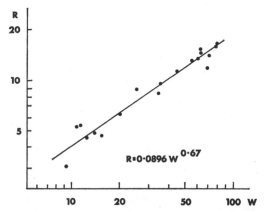

Fig. 2. Relation between rations value (R) and body weight (W) in the amphipoda, *Orchestia bottae*. R in mg of food per animal per day; W the wet weight of animal in mg. Correspondings value in the equation are expressed in g of wet weight.

Relation between ration and body weight of animals

A review of literature and of our own experimental data shows that ration

(R), like metabolism, is proportional to body weight (W) and can be described by a relation of the form

$$R = \alpha W^{\beta}$$

whence $$\log R = \log \alpha + \beta \log W \qquad [5]$$

where W is the weight of the animal, α is a coefficient defining the level of food consumption per unit time when $W = 1$, and β defines the rate of change of food consumption with body weight, i.e. the slope of the log line (Fig. 2). This relation is true for Crustacea of various types of feeding: filterers (Ryther, 1954; Richman, 1958; Monakov and Sorokin, 1960); "scavengers" (Sushchenya, 1966; and unpublished data on feeding of *Orchestia bottae*); and predators (Karpevich and Bogorad, 1940; Kurenkov, 1958). Table 2 compares numerical values of coefficients α and β from the equation [5] calculated by the method of least squares for different species.

TABLE 2. Numerical values of coefficients α and β from equation [5] relating the daily ration and body weight of some Crustacea. R and W in g

Species	Type	Food Concentration mg/l	α	β	Author
Daphnia pulex	*Chlamydomonas*	0·62	0·032	0·83	Richman, 1958
Daphnia pulex	*Chlamydomonas*	1·24	0·064	0·83	Richman, 1958
Daphnia pulex	*Chlamydomonas*	1·86	0·080	0·82	Richman, 1958
Daphnia pulex	*Chlamydomonas*	2·48	0·092	0·81	Richman, 1958
Daphnia pulex	Natural phyto-plankton	2·00	0·095	0·88	Monakov and Sorokin, 1960
Daphnia magna	*Chlorella*	4·00	0·085	0·88	Ryther, 1954
Orchestia bottae	Decomposed *Cystozeira*	—	0·090	0·67	Sushchenya, non-published data
Leander adspersus	*Chironomus tummi*	—	0·046	0·57	Karpevich and Bogorad, 1940
Leander modestus	*Heterocypris* and *Daphnia*	—	0·037	0·47	Kurenkov, 1958
Palaemonetes sinensis	*Heterocypris* and *Daphnia*	—	0·044	0·60	Kurenkov, 1958
Menippe mercenaria	Fish *Opistonema oglinum*	—	0·084	0·66	Sushchenya, 1966

There is no doubt that the available data are insufficient for a general conclusion. However, values of the coefficient β of different species of Crustacea are very similar or even identical with the coefficient b in the equation [8] relating rate of respiration and body weight of these animals. In addition to this, Richman's data clearly indicate that the coefficient α is a constant only at a particular concentration of food and increases with the increase of food concentration. We have found that this relation can be expressed by the equation

$$\alpha = \alpha_1 K^q \qquad [6]$$

where α_1 is a constant defining the level of food consumption per unit time when $K = 1$, and q defines the rate of change of coefficient α with food concentration.

In a numerical expression the equation [6] according to Richman's data for *Daphnia pulex* which consumed *Chlamydomonas* may be written in this form

$$\alpha = 1 \cdot 37 \times 10^{-5} K^{0 \cdot 77} \qquad [6a]$$

where K is quantity of algae per ml. It is obvious that an equation like this can be calculated when K is expressed in weight units of biomass, B:

$$\alpha = 9 \cdot 82 \ B^{0 \cdot 77} \qquad [6b]$$

where B is biomass of *Chlamydomonas* in mg of dry weight per ml.

In predatory Crustacea the same relation may be observed. But, in scavengers in excess food conditions the coefficient α is a constant. In all cases the value of coefficient α is affected by temperature, but there are no data to evaluate this point.

ASSIMILATION OF FOOD

For a determination of energy expenditure on respiration and growth the estimation of the assimilated or physiologically useful part of a ration is most important. The experimental data on this question are not numerous. Table 3 gives certain values of food assimilation by some species of Crustacea.

TABLE 3. Percentage of assimilation of various food organisms by some species of Crustacea, A in % of R

Species	Food	A	Author
Artemia salina	'Yeast	33-78	Sushchenya, 1962
Artemia salina	*Dunaliella*	66	Khmeleva, unpubl.
Artemia salina	*Ankistrodesmus*	57-70	Khmeleva, unpubl.
Daphnia pulex	*Chlamydomonas*	7-24	Richman, 1958
Calanus helgolandicus	*Seston*	76-93	Corner, 1961
C. hyperboreus	*Exuviella* sp.	55-85	Conover, 1965
C. hyperboreus	Various algae	18-92	Conover, 1965
Asellus aquaticus	Leaf of trees	68	Levanidov, 1949
Euphausia pacifica	*Dunaliella*	81-97	Lasker, 1960
Pontogammarus maeoticus	*Enteromorpha*	82-88	Karpevich, 1946
Orchestia bottae	*Cystozeira*	30-58	Sushchenya, unpubl.
Menippe mercenaria	Fish *Opistonema*	97-99	Sushchenya, 1966

These data show that the percentage of assimilation of food varies from 7-24% in *Daphnia pulex* to 97-99% in *Menippe mercenaria*. In most cases a high value of assimilation (from 60 to 80-90%) was obtained. Apparently in filter-feeding animals at the highest food levels the percentage of assimilation may be lower.

The data from our laboratory allowed us to establish some quantitative relation between food consumption and food assimilation. Fig. 3 shows that

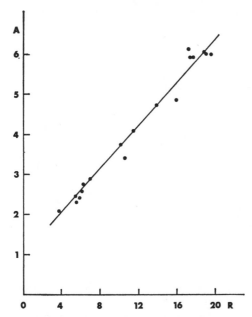

Fig. 3. Relation between food assimilation (*A*) and food consumed (*R*) by *Orchestia bottae* ranging in weight from 9 to 80 mg. The change of *R* depends upon different weight of animals. *A* and *R* are expressed in calories per animal per day.

there is a straight-line relation between ration (*R*) and quantity of food assimilation (*A*) by *Orchestia bottae* ranging in weight from 9 to 80 mg. The relation is:

$$A = 0.906 + 0.274\ R. \qquad [7]$$

From Fig. 3 the percentage of assimilation decreases regularly from 50% in young animals (with small rations) to 32% in older groups (with large rations).

Data of great interest were taken in our laboratory by N. Khmeleva on feeding of adult *Artemia salina*. The experiment was carried out using *Dunaliella salina* labelled with ^{14}C as food. The concentration of algae ranged from 0.005 to 0.3 million cells per ml (biomass correspondingly from 0.02 to 1.0 mg per ml) and was equivalent to 0.1-2.0 mg per animal. The percentage assimilation was practically constant despite the significant change of food concentration. Between *A* and *R* there was a straight-line relation described by the simple regression equation (Fig. 4):

$$A = 0.66\ R \qquad [7a]$$

whence percentage of assimilation is equal to 66% of ration value. The

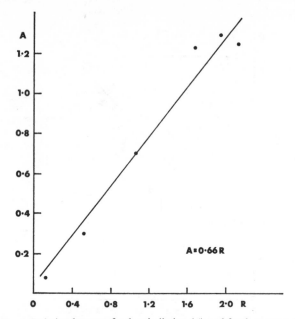

Fig. 4. Relation between food assimilation (A) and food consumed (R) by *Artemia salina* of definitive weight (wet W = 4·2 mg). The change of R defines by change of food concentration. A and R are expressed in mg of wet weight of Dunaliella per animal per day (by N. Khmeleva, nonpublished data).

change of absolute amount of food assimilation [1] in relation to food concentration (K) can be expressed in this case by equation [4]:

$$A = 1·3 \, (1 - 10^{-0·00294K}).$$

Suitable empirical data plotted in Fig. 5 show that the absolute quantity of food assimilated, like the ration consumed, increases with increasing food concentration and tends to a maximum, the maximum value of assimilation [A_{mx}] for this animal with a given type of food.

METABOLISM AND GROWTH

The metabolism of Crustacea as reflected in the rate of oxygen consumption has been extensively studied by many investigators. It was found to be well described by a relation of the form

$$Q = a \, W^b$$

whence $\qquad\qquad\qquad \log Q = \log a + b \log W \qquad\qquad$ [8]

where Q is the rate of oxygen consumption per animal per unit time (ml/hour), W is the body weight (g), a is a constant defining the level of metabolic

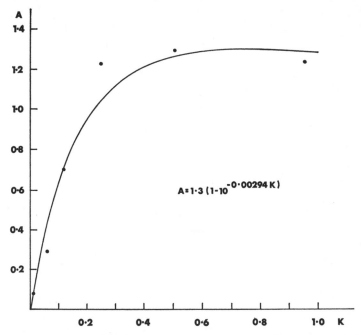

Fig. 5. Relation between amount of food assimilated (*A*) and food concentration (*K*) in the adult *Artemia salina*: *A* in mg of food per animal per day, *K* if g wet weight of *Dunaliella* per litre (by N. Khmeleva, nonpublished data).

expenditure per unit time, and *b* defines the rate of change of metabolism with body weight.

The numerical values of *a* and *b* in equation [8] have been calculated using many experimental data for Crustacea by Weymouth *et al.* (1944) and by Winberg (1950). According to studies by these authors the coefficient *a* is between 0·165 and 0·170, the coefficient *b* between 0·826 and 0·810 at 20°C. Preliminary re-examination of all data published up to the present time corroborates these values very well. From our calculations for a number of species the relation between rate of respiration and body weight of Crustacea is adequately described by the equation

$$Q = 0.174 \ W^{0.75}. \tag{8a}$$

Equation [8a] and the calorific equivalent of oxygen (5 cal per ml O_2), permits us to express the expenditure for respiration in energy units. After multiplying coefficient *a* in equation [8a] by 5 we find that

$$Q = 0.87 \ W^{0.75} \ \text{cal (animal} \times \text{hour)}^{-1} \tag{8b}$$

or

$$Q = 20.9 \ W^{0.75} \ \text{cal (animal} \times \text{day)}^{-1}. \tag{8c}$$

It is necessary to note that interactions between the rate of respiration and temperature may be expressed by temperature coefficients of the Krogh correction curve with the necessary degree of precision (Winberg, 1956).

The energy expenditure for growth of animals can be estimated by empirical data of weight increase and caloricity per unit body weight. This quantity may be calculated from equation [2] if we know the value of the remaining parameters (R, T and N). Then it is possible to calculate the growth coefficients of the first and of the second order (Ivlev, 1938):

$$K_1 = \frac{G}{R}, \; K_2 = \frac{G}{T+G} \qquad [9, 10]$$

where K_1 has been called by Ivlev the energy coefficient of growth of the first order (the gross efficiency of growth), K_2 the energy coefficient of growth of the second order (the net efficiency of growth). The efficiency, measured as K_1 and K_2 depends on whether total R, or assimilated rations, AR, are used.

In the very interesting work by Winberg (1966) it has been shown that there are some quantitative relations between metabolism and growth of animals. On the basis of Winberg's equations we can calculate the growth curve of a given animal under given temperature conditions using the rate of respiration and some other indices (maximum value of K_2, initial and maximum weight of animal and its caloricity).

TABLE 4. Values of energy coefficients K_1 and K_2, and percentage of energy expenditure for respiration (T), in % of AR

Species	$t°$ C	K_1	K_2	T	Author
Artemia salina	25	18·5	23·6	76·4	Sushchenya, 1962
Artemia salina	25	13·0	26·5	73·5	Sushchenya, 1962
Artemia salina	25	9·0	27·4	72·6	Sushchenya, 1962
Daphnia pulex	20	13·2	55·4	44·6	Richman, 1958
Daphnia pulex	20	9·1	57·6	42·4	Richman, 1958
Daphnia pulex	20	4·8	56·9	43·1	Richman, 1958
Daphnia pulex	20	3·9	58·7	41·3	Richman, 1958
Calanus helgolandicus	10	48·9	52·5	47·5	Corner, 1961
Calanus helgolandicus	10	37·2	47·0	53·0	Corner, 1961
Calanus helgolandicus	10	41·9	46·2	53·8	Corner, 1961
Asellus aquaticus	14-19	28·3	40·4	59·6	Levanidov, 1949
Asellus aquaticus	14-19	23·3	33·3	63·7	Levanidov, 1949
Asellus aquaticus	14-19	20·6	30·0	70·0	Levanidov, 1949
Asellus aquaticus	14-19	18·7	26·7	73·3	Levanidov, 1949
Asellus aquaticus	14-19	16·3	23·3	76·7	Levanidov, 1949
Euphausia pacifica	10	7·1	7·4	92·6	Lasker, 1960
Euphausia pacifica	10	31·8	39·2	60·8	Lasker, 1960
Euphausia pacifica	10	14·2	14·9	85·1	Lasker, 1960
Euphausia pacifica	10	28·0	29·0	70·0	Lasker, 1960
Leander adspersus	16	3·35	—	—	Karpevich and Bogorad, 1940
Orchestia bottae	20	8·0-31·0	26·5-64·0	36·0-73·5	Sushchenya, unpublished
Menippe mercenaria	25	44·0-78·0	45·0-80·0	20·0-55·0	Sushchenya, 1966

Table 4 gives a comparison of experimental data and values of K_1, K_2 and energy expenditure for the respiration of different species of Crustacea.

According to Winberg, if during the growth process, animals of any species reach a maximum size, their absolute growth w to a given moment of time t may be described by the following equation:

$$w_t = W^{1-b} - (W^{1-b} - w_0^{1-b}) \, e^{(1-b)kt} \, 1/(1-b) \qquad [11]$$

where w_t is the absolute increase of body weight, t is time of growth in days, W the maximum weight of the animal, w_0 the initial weight of the animal, b is the same as in equation [8] and k is another constant. The value of k depends on relations between metabolism, weight and maximum size of K_2 for the species studied, and can be expressed by the equation:

$$k = \frac{T_1}{W^{1-b} - w_0^{1-b}} \cdot \frac{K_2}{1 - K_2} \qquad [12]$$

where T_1 shows the level of metabolic expenditure per unit time when body weight of animal is equal to 1.

Relation between ration, metabolism and growth of Crustacea

From the foregoing we have a number of expressions for the calculation of any parameter in the original energy equation [1] but none of them describes the simultaneous functional relationship between all the above-mentioned quantities. In this connection the papers by Paloheimo and Dickie (1965a, 1965b, 1966) on growth and feeding of fishes are of great interest. These authors show that when fish are fed on one type of food, the logarithm of the gross growth efficiency (log K_1) decreases with increase in ration. For a number of species this relation is adequately described by the linear equation

$$\log K_1 = \log (dW/Rdt) = -1 - b'R \qquad [13]$$

whence
$$dW/dt = R \, e^{-a' - b'R} \qquad [14]$$

where dW is the growth, R the ration intake during the period time dt, and a' and b' are the parameters fitted to the linear form of the equation.

Equation [14] implies that $e^{-a'}$ is equal to a maximum value of K_1, $(K_1)_{mx}$, i.e. from the maximum value at low ration, $[(K_1)_{mx} = e^{-a'}]$, the growth efficiency decreases by constant fraction, $e^{-b'}$, for each unit increase in amount consumed per unit time.

From the original energy equation [2] we write

$$T = AR - dW/dt.$$

Substituting the value of dW/dt according to equation [14] we have

$$T = AR - R \, e^{-a' - b'R}$$

whence
$$T = R \, (A - e^{-a' - b'R}). \qquad [15]$$

We have thus, on the basis of equations [14] and [15], a relationship between ration, metabolic expenditure and growth of animals. This system of equations may be applied to investigations of quantitative relationship between food ration, metabolism and growth of Crustacea. Thus, for example, all the terms of equations [14] and [15] have been calculated by Richman's data on the feeding of *Daphnia pulex* at a concentration of *Chlamydomonas* of 10^5 cells per ml. For computation of energy expenditure for respiration (T) the coefficient of food assimilation adduced by Richman for pre-adult

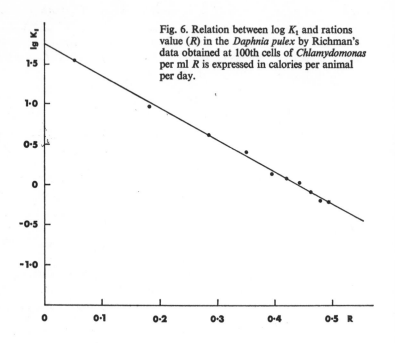

Fig. 6. Relation between log K_1 and rations value (R) in the *Daphnia pulex* by Richman's data obtained at 100th cells of *Chlamydomonas* per ml R is expressed in calories per animal per day.

animals at given concentration of food (6·6% or 0·066) was used. The computed value of $(K_1)_{mx}$ is equal to 0·60 or 60%. The relation between log K_1 and ration consumed by *Daphnia* of various weight (K-line) is shown in Fig. 6. Coefficient b' of the slope of regression line may be determined by the equation,

$$K_1 = (K_1)_{mx} \, e^{-b'R}$$

whence
$$\log_e K_1 = \log_e (K_1)_{mx} - b'R.$$

Paloheimo and Dickie (1965a) give the following equation:

$$a'W^\gamma = (1 - e^{-a'-1})/b'$$

whence
$$b' = \frac{(1 - e^{-a'-1})}{a'W^\gamma}$$

where $e^{-a'}$ is the maximum K_1, $a'W^\gamma$ the respiration of animals at the maximum rate of growth.

The following numerical expressions have been obtained for *Daphnia pulex*:

$$\frac{dW}{dt} = R\,e^{-0\cdot58-8\cdot91\;R} \qquad [14a$$

$$T = R\,(0\cdot066-e^{-0\cdot58-8\cdot91\;R}) \qquad [15a]$$

where R, T and dW/dt are expressed in calories per animal per day.

The relationship between daily ration in calories and dry body weight of *Daphnia* was found to be well described by a relation of the form

$$R = 3\cdot74\,W^{0\cdot88}. \qquad [16]$$

The growth curve calculated from equation [14a] is in good agreement with the experimental growth curve by Richman (Fig. 7). Respiration values

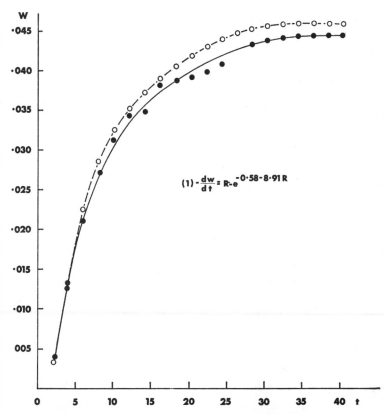

Fig. 7. Comparison of growth curves of *Daphnia pulex* derived from experimental data by Richman (2) and calculated by equation according to Paloheimo and Dickie (1): W is the dry weight of animal in mg, t the time of growth in days. All experimental data obtained by Richman at 100th cells of *Chlamydomonas* was used.

computed according to equation [15a] was also fitted by the experimental data:

by Richman $\qquad -T = 0{\cdot}525 \ W^{0{\cdot}88 *}$

by equation [15a] $\qquad -T = 0{\cdot}489 \ W^{0{\cdot}89}$

where T is the daily expenditure for respiration in calories per animal, W the dry weight of animal in mg.

It must be noted that our equation [15] differs from that of Paloheimo and Dickie (1965a, equation [6]). These authors erroneously took the value of ration (R) without a coefficient of assimilation and assumed that

$$T = R(1 - e^{-a' - b'R}).$$

CONCLUSION

The experimental data available demonstrate that some success was reached in the investigation of quantitative relationships between such important parameters as body weight and ration of different food, metabolism and growth of Crustacea. But these are only the first steps in the examination of the problem. At the moment there is a need for extensive accumulation of data on the rations of the most abundant species of Crustacea as well as on relation between the latter and food concentration, temperature and body weight of the animals. These investigations must be conducted simultaneously with the study of metabolism and growth of Crustacea. The results of such studies may increase our understanding of the role of Crustacea in natural production processes.

REFERENCES

CONOVER, R. J. 1966. Assimilation of organic matter by zooplankton. *Limnol. Oceanogr.*, **11**, 338-45.

CORNER, E. D. S. 1961. On the nutrition and metabolism of zooplankton. I. Preliminary observations on the feeding of the marine copepod, *Calanus helgolandicus* (Claus). *J. mar. biol. Ass. U.K.*, **41**, 5-16.

IVLEV, V. S. 1938. Sur la transformation de l'énergie pendant la croissance des invertébrés. *Byull. mosk. Obshch. Ispÿt. Prir., Otd. Biol.*, **47**, 267-78.

IVLEV, V. S. 1955. *Experimental ecology of nutrition of fishes.* Moscow, Pishchepromizdat. Transl. D. Scott, New Haven, Yale University Press. 1961.

KARPEVICH, A. F., and BOGORAD, G. 1940. Feeding of the crayfish, *Leander adspersus. Zool. Zh.*, **19**, 134-8. (In Russian).

KARPEVICH, A. F. 1946. Feeding of *Pontogammarus maeoticus* in the Caspian Sea. *Zool. Zh.*, **25**, 517-22. (In Russian).

KURENKOV, I. I. 1958. Some data to ecology of fresh-water shrimps of the Far East in connection with the possibility of their acclimatization. *Trudÿ mosk. tekhnol. Inst. rÿb. Prom. Khoz.*, **9**, 80-101. (In Russian).

LASKER, R. 1960. Utilization of carbon by a marine crustacean. Analysis with Carbon-14. *Science*, **131**, 1098-1100.

* Richman's equation showing the relation between respiration and body weight of *Daphnia pulex* contains the following error: value of coefficient a should be $4{\cdot}5$ instead of $0{\cdot}0014$ as he points out.

LEVANIDOV, V. I. 1949. The significance of allochtonous material as food resource in a water-basin and its consumption by *Asellus aquaticus*. *Trudy vses. gidrobiol. Obshch.*, **1**. (In Russian).

LUCAS, C. E. 1936. On certain inter-relations between phytoplankton and zooplankton under experimental conditions. *J. Cons. perm. int. Explor. Mer*, **11**, 343-62.

MONAKOV, A. V., and SOROKIN, Yu. I. 1960. Experimental study of feeding of *Daphnia* by means of the method C-14. *Dokl. Akad. Nauk SSSR*, **135**, 1516-18. (In Russian). English trans. in *Dokl. (Proc.) Acad. Sci. USSR*, **135**, 925-6, 1961.

PALOHEIMO, J. E., and DICKIE, L. M. 1965a. Food and growth of fishes. I. A growth curve derived from experimental data. *J. Fish. Res. Bd Can.*, **22**, 521-42.

PALOHEIMO, J. E., and DICKIE, L. M. 1965b. Food and growth of fishes. II. Effects of food and temperature on the relation between metabolism and body weight. *J. Fish. Res. Bd Can.*, **23**, 869-908.

PALOHEIMO, J. E., and DICKIE, L. M. 1966. Food and growth of fishes. III. Relations among food, body size and growth efficiency. *J. Fish. Res. Bd Can.*, **23**, 1209-48.

PETIPA, T. S. 1959. Feeding of the copepod, *Acartia clausi*. *Trudy sevastopol' biol. Sta.*, **11**, 72-100. (In Russian).

RICHMAN, S. 1958. The transformation of energy by *Daphnia pulex*. *Ecol. Monogr.*, **28**, 273-91.

RYTHER, J. H. 1954. Inhibitory effects of phytoplankton upon the feeding of *Daphnia magna* with reference to growth, reproduction and survival. *Ecology*, **35**, 522-633.

SUSHCHENYA, L. M. 1958. Quantitative data on the filtration feeding of plankton Crustacea. *Nauch. Dokl. vyssh. Shk. biol. Nauk*, **1**, 16-20. (In Russian).

SUSHCHENYA, L. M. 1962. Quantitative data on the feeding and energy balance of *Artemia salina* (L.). *Dokl. Akad. Nauk SSSR*, **143**, 1205-7. (In Russian). English trans. in *Dokl. (Proc.) Acad. Sci. USSR*, **143**, 329-30, 1962.

SUSHCHENYA, L. M. 1966. Quantitative regularities of feeding and its connection with energy balance of the edible crab, *Menippe mercenaria* (Say). *Trudy Sov.-Kub. morsk. Eksped.*, **1**. (In Russian).

WEYMOUTH, F. W., and others. 1944. Total and tissue respiration in relation to body weight *Physiol. Zool.*, **17**, 50-71.

WINBERG, G. G. 1950. The intensity of metabolism and total size of Crustacea. *Zh. obshch. Biol.*, **11**, 367-80. (In Russian).

WINBERG, G. G. 1956. Rate of metabolism and food requirements of fishes. *Nauch. Trudy Beloruss. gos. Univ. Lenin*, 253 pp. (*Fish. Res. Bd Can. Transl. Ser.*, no. 194).

WINBERG, G. G. 1966. Rate of growth and intensity of metabolism in animals. *Usp. sovrem Biol.*, **61**, 274-93. (In Russian).

The food web structure, utilization and transport of energy by trophic levels in the planktonic communities

T. S. PETIPA, E. V. PAVLOVA and G. N. MIRONOV
Institute of the Biology of the Southern Seas
Academy of Sciences of the Ukraine SSR
Sevastopol, USSR

ABSTRACT. This paper presents a qualitative and quantitative description of the food web system within the planktonic communities of the Black Sea. The utilization of energy by the association of the organisms in the different ecological groups composing one or another trophic level is estimated.

The analysis of the transformation of the energy flow through the planktonic communities of different ecological types is given.

The results are discussed in terms of further studies of the ecological role of the heterotrophic levels in the general productivity of the ocean.

To investigate the matter and energy flow through the trophic levels of any community, one must first study the structure of food chains or food webs. We now wish to discuss the schemes of food interrelations in the epiplankton and bathyplankton communities of the Black Sea. The epiplankton community of the Black Sea usually occupies the surface strata above the thermocline while the bathyplankton community inhabits the water space under the thermocline. We investigated the specific rates of uptake, accumulation and expenditure of matter and energy by the main ecological groups of each trophic level of the two communities. With these data we determined the flow of matter and energy in each ecosystem, and the efficiency of transfer of matter and energy from one trophic level to another.

Our experiments were carried out in the most stable area of the western Black Sea *galistaza* during the early summer period, when almost all species of plankton-eating fishes feed and spawn in coastal waters. Fish larvae were not present at this early season and the pelagic communities consisted of phytoplankton and zooplankton only.

In the observed area phytoplankton consisted largely of Dinoflagellata (including Noctiluca) and Diatomea, while zooplankton included eight

species of Copepoda, two species of Chaetognatha, one species of Cteno-
phora, one species of Appendicularia and various species of Lamellibranchiata, Gastropoda and Polychaeta larvae. These plankton are abundant in
the Black Sea during the longest season of the cold spell (the upper temperature
limit of the surface water is 17-18°C). During the warm seasons the smallest
phytoplankton forms as well as bacteria and infusorians play a great part.
In summer, unlike winter, thin filter-feeders (Cladocera mainly) play a
significant role in zooplankton.

An account is given here of preliminary results of the first analysis of the
materials obtained at anchor stations lasting many days. The results of the
investigation of different processes observed in the Black Sea ecosystems, represent average data from 10-14 field observations or laboratory experiments.

MATERIALS AND METHODS

Together with catching the organisms on four 24-hour stations in the observed area at different depths, we obtained measurements on such environmental factors as light intensity, temperature, salinity, currents, oxygen,
phosphate, nitrate and detritus.

Phytoplankton was caught by a Nansen 1-litre water bottle and zooplankton with nets of 36 and 80 cm diameter in the mouth opening; the diameter
of the mesh size is 130μ.

The reproductive rate of algae (unpublished data of L. A. Lanskaya) and
the animals' respiration (Pavlova, 1967, 1968, unpublished data; Petipa, 1966,
1966a) were determined experimentally with bottles at different depths in the
sea. The weight increment of the animals was computed from the duration
of stages as a function of temperature (Sadgina, 1960, 1961) and the difference
in weight from stage to stage (Petipa, 1966 a, b). The fecundity of the animals
was determined experimentally (Sadgina, personal communication). The
values for diurnal rations were determined primarily by gut contents and rate
of digestion, secondly from losses in growth and energy metabolism, taking
faeces into account. The amount of faeces was determined from the difference between the ration obtained by gut contents and that from the sum of the
weight increment and the energy metabolism, expressed in the form of food.
Losses by liquid excretion were not taken into account.

The total diurnal increment, respiratory losses and rations of all the
organisms of any trophic level were computed by multiplying the data for one
organism by its average number during 24 hours. The diurnal rate of production of matter in the trophic levels of each community was determined as
a sum of organisms eaten from a given level, of dead organisms and the
difference between standing stock at the beginning and at the end of the 24-
hour period and organisms transferring to the next trophic level. The diurnal
rate of production of organic matter was determined as a sum of the diurnal
increase in standing stock of both living and dead organisms in all the trophic
levels. The energy metabolism of the total community was determined by
summing the energy dissipated by respiration which was determined for each
trophic level by the method mentioned above.

In the area we investigated the distribution of large detritus from the death of organisms, taking into account moulting and defaecation in particular. The amount of detritus contained in a cubic metre of water was determined in the following way. The daily mortality equals the number of dead and unbroken forms found in the plankton samples divided by the number of days during which dead organisms were present unbroken. At the present time the rate of destruction of dead organisms as well as of moults and faeces is not known. However, some experiments showed that faeces and algae are present more or less unbroken for a day, dead animals for 2-4 days, and the chitinous exoskeletons of planktonic crustaceans for not less than 4-5 days.

The rate of moult of crustaceans by duration of stage development was determined experimentally together with the rate of defaecation of some typical groups. Knowing the amount of detritus formed daily for all the groups and the duration of their presence unbroken in the sea, we can compute for each group the instantaneous amount of detritus contained, approximately, in the layer of water. For this purpose the daily amount of detritus formed for a definite group must be multiplied by the number of days, during which the detritus group is present unbroken in the sea.

RESULTS

All the organisms composing the investigated planktonic communities are divided into the following six trophic levels:
1. primary producers and saprophagous organisms (phytoplankton),
2. herbivorous organisms,
3. mixed-food (plant and animal material) consumers,
4. primary carnivores,
5. secondary carnivores,
6. tertiary carnivores.

Primary producers and saprophagous organisms

The phytoplankton saprophagous organisms which eat non-living particulate and dissolved organic matter, are included in the first trophic level, because they are sometimes identical with the producers. Both the producers and saprophagous organisms feed upon the dead elements of the environment. The particulate organic matter as well as dissolved organic matter are considered as transition products of dead organisms decaying, while biogenous salts, that producers consume, are taken as the final products of decaying process. However, up to now for many algae, nobody can separate completely autotrophic nourishment from heterotrophic. It is known that most of the Dinoflagellata can be both autotrophic and heterotrophic.

The first trophic level was divided into three groups:
1. small-sized phytoplankton from 7 μ to 25 μ;
2. medium-sized phytoplankton from 30 μ to 50 μ;
3. large-sized phytoplankton of more than 50 μ.
The ratio between these groups was in general, the same in both epiplank

on and bathyplankton communities and the first phytoplankton group, the smallest algae, predominated in numbers in both communities. In the epiplankton community the first phytoplankton group amounted to 86%; the second, 12%; and the third, 42% of total phytoplankton. In the bathyplankton community at the depth of 12-100 m the number of the first, second and third groups amounted respectively to 92·6, 3·9, and 3·5%. However, the proportion of larger algae increases with depth; from 12-50 m, the first group comprised 90-96% of the total phytoplankton in number and the third group, 2-3%; from 50-75 m, the first group comprised 57 per cent in number, while the third group comprised 30%. Margalef (1967) observed a similar picture of changes in phytoplankton number in the Mediterranean Sea.

The ratio of standing stock between the three groups in both communities is directly opposite to their ratio in numbers: the first two groups constitute 1·2-1·7% and the third 97-99% of the total phytoplankton standing stock in each layer. On the whole, the greater part of the organisms is concentrated in the epiplankton ecosystem and in the upper layer of the bathyplankton ecosystem.

The species composition of each phytoplankton group in both communities varies slightly with depth. In the epiplankton at temperatures of 16-18°C and in the upper layer of the bathyplankton at temperatures of 12°C, small and medium Dinoflagellata (*Exuviaella cordata, E. compressa, Peridinium steinii, Prorocentrum micans*) and the chains of small diatoms (*Cyclotella caspia, Chaetoceros insignis, Coscinosira oestrupii*) are abundant. Their number is 0·5-4·5 million cells/m³ and the standing stock 1-4 mg/m³. Besides these dominant species, a considerable proportion of large Dinoflagellata, such as *Ceratium furca* (0·13 million cells/m³; 4·6 mg/m³), *Noctiluca miliaris* (0·019-0·043 million cells per m³; 2390-5103 mg/m³) and large diatoms, *Rhizosolenia calcaravis* (0·006-0·014 million cells/m³; 2-5 mg/m³), are present. All these algae in the upper layers, where light intensity is usually more than 50 luxes, possess high division rates of one or two times per 24 hours.

In the deeper layers of the bathyplankton (up to 75 m), some common representatives of the upper layers are present. Among them the most conspicuous are: small *Chaetoceros insignis, Coscinosira oestrupii*, some *Peridinium* sp., medium Dinoflagellata—*Glenodinium* sp. and *Dinophysis acuta* etc.; and large Dinoflagellata—*Noctiluca miliaris*. Some algae are typical of rather deep waters and are not found in the surface layers. For example, in the layer, 25-100 m, the most typical are small diatoms, shade-loving *Nitzschia seriata* and small Silicoflagellata—*Distephanus speculum*. Also, large Dinoflagellata —*Ceratium tripes, C. fusus*—and large diatoms—*Coscinodiscus yanischii, Rhizosolenia alata, Cerataulina bergonii* are present.

In the layers from 25 to 100 m the temperature is constant at 7-8°C, but the light intensity varies from a few units of lux to a few tens of luxes. So the division rate does not usually exceed 0·4-0·5/24 hours. *Nitzschia seriata* is the only algal species which divides three times a day at any depth.

The most interesting algal species from the third group of the first trophic level is *Noctiluca miliaris*, which may be considered mainly as a saprophagous organism. Detritus of various sizes comprises 70-90% of its ration. The daily ration is rather low, 3-5% of the wet body weight. *Noctiluca miliaris* reaches 43-260/m³ and 5-100 mg/m³ in the bathyplankton community, and in the epiplankton community its number is 19 000/m³ and the standing stock is 2390 mg/m³ wet weight.

ZOOPLANKTON

In each community of the Black Sea zooplankton, we may distinguish five trophic levels (see above). Six ecological groups were distinguished amongst the herbivores in both epiplankton and bathyplankton communities. One ecological group in each level was distinguished in mixed-food consumers, primary, secondary and tertiary carnivores in both epiplankton and bathyplankton communities. All the groups are dominant in number and standing stock and represent the main mass of zooplankton. In general, the epiplankton community comprises non-migratory and slightly-migratory organisms and the bathyplankton community comprises migratory organisms. The values for number, standing stock, growth increment, energy metabolism and daily rations of the ecological groups in both communities are given in Tables 1 and 2.

Herbivores. There are the following ecological groups of herbivores in the epiplankton community:

1. Stages III-VI nauplii of copepods, *Oithona minuta*, *Paracalanus parvus*, *Acartia clausi*, Pontellidae and nauplii of the bathyplanktonic species. Nauplii consume, mainly, small and medium phytoplankton of 15 to 50 μ in diameter. The daily ration of nauplii in the epiplankton is rather high, 140% body weight.

2. Stages I-III copepodites of copepods. Younger copepodites consume small, medium and, partly, large algae from 55 to 60 μ in diameter. Their daily ration is also high, 100-125% of the body weight.

3. Stages IV-V copepodites, males and females of *Paracalanus parvus*. They consume small algae from 20 to 30 μ. Their diurnal ration is rather low, 45% of the body weight.

4. *Oikopleura dioica* may be segregated into an independent group. I consumes only small algae and small detritus. The daily ration is 60-70% of the body weight.

5. Larvae of molluscs and polychaetes during the cold seasons are no abundunt, but they are always present. This ecological group may be divided into feeding and non-feeding forms. The feeding forms consume small, medium, large algae and detritus. The daily ration ranges from 2 to 30% of the body weight.

6. Stages IV-VI copepodites of *Pseudocalanus* and *Calanus*. Their number and standing stock in the epiplankton community are very low. They are not typical representatives of the surface waters and can get there only by chance migration, passing through the thermocline. Nevertheless, if they enter the surface layer their ration doesn't vary.

TABLE 1. Consumption and expenditure of matter by ecological groupings of the epiplankton community (16-17°, on field and experimental data)

Ecological groupings	Number no/m²	Standing stock mg/m²	P Increment in body weight and reproduction	T Energy of respiration	R Daily ration	R (from food eaten) in % of wet body weight
			% of wet body weight			
Phytoplankton—producers and saprophages						
1. Small-sized forms	113 868 000	190·8	115·3	20·3*		
2. Medium-sized forms	16 212 000	243·6	60·9	10·7*		
3. Large-sized forms	2 592 000	28 815·0	74·1	7·0*		
including Noctiluca	260 500	28 658·0	74·1	6·9	93·0	5·0
Herbivores						
1. Nauplii	61 908	33·0†	9·4	36·8	56·9	129·3
2. I-III copepodites	83 904	184·7	14·4	57·1	82·4	194·2
3. IV-VI Paracalanus	28 296	320·7	8·0	30·0	45·0	35·1
4. Oikopleura	35 940	215·6	16·0	30·0	60·0	16·2
5. Larvae of molluscs and polychaetes	372	2·4	3·0	9·0	15·0	10·0
6. IV-VI Pseudocalanus, Calanus‡	396	35·5	0·4	4·0	4·8	9·4
Mixed-food consumers IV-V Acartia, Oithona; males, females Acartia	33 252	193·5	7·0	68·5	82·8	88·2
Primary carnivores Males, females Oithona minuta, Oithona similis	37 932	175·2	4·8	150·0	164·7	169·0
Secondary carnivores Sagitta setosa	3 984	297·3	18·0	48·2	85·2	37·5
Tertiary Carnivores Pleurobrachia, Medusae	1 836	263·2	4·0	140·0	332·5	20·0

* Algal respiration is taken 15% of the value of the photosynthesis.
† Feeding and non-feeding nauplii.
‡ Data are given on half an hour nutrition while present in the epiplankton system.

TABLE 2. Consumption and expenditure of matter by ecological groupings of the bathyplankton community (7-12°, on field and experimental data)

Ecological groupings	Number no/m²	Standing stock mg/m²	P Increment in body weight and reproduction	T Energy of respiration	R Daily ration	R (from food eaten) in % of wet body weight
			% of wet body weight			
Phytoplankton—producers and saprophages						
1. Small-sized forms	202 669 000	218.5	84.7	14.9*	—	
2. Medium-sized forms	8 474 000	212.7	11.7	2.1*	—	
3. Large-sized forms	7 664 000	83 080.2	65.3	2.8*	—	
including Noctiluca	739 800	81 379.2	66.2	2.8	86.4	4.6
Herbivores						
1. Nauplii	249 109	167.8†	6.3	12.5	21.3	58.0
2. I-II copepodites	76 726	385.0	18.8	33.5	66.4	90.3
3. IV-VI Paracalanus	13 578	136.7	4.0	15.0	22.0	16.0
4. Oikopleura	52 196	313.3	4.0	20.0	30.0	8.0
5. Larvae of molluscs and polychaetes	1 767	34.5	1.5	4.0	7.0	8.5
6. IV-VI Pseudocalanus, Calanus	35 878	2765.0	13.6	99.2	121.9	198.7
Mixed-food Consumers						
IV-V Acartia, Oithona; males, females Acartia	18 533	122.0	3.0	34.7	48.1	69.0
Primary Carnivores						
Males, females Oithona minuta, Oithona similis	33 768	187.8	8.5	70.0	87.0	95.0
Secondary Carnivores						
Sagitta setosa	402	837.3	9.0	12.3	65.4	20.0
Tertiary Carnivores						
Pleurobrachia, Medusae	1 122	18 775.0	2.0	24.3	80.2	3.5

* Algal respiration is taken 15% of the value of the photosynthesis.
† Feeding and non-feeding nauplii.

Of the six ecological groups of herbivores in the bathyplankton, three groups are the same as in the epiplankton community. They are: *Oikopleura dioica*, stages of VI-YI of *Paracalanus parvus* and the larvae of molluscs and polychaetes. The maximum number and the standing stock of these groups is 0·67 that of the epiplankton community. Their number decreases with depth, and their daily rations are 0·5-0·33 that in the epiplankton community. Low temperature accounts for this.

The other ecological groups of herbivores in bathyplankton are rather peculiar. Stages III-VI of nauplii and stages I-III of copepodites in the bathyplankton differ in their species composition from those epiplankton. Besides epiplankton species, young *Oithona similis, Pseudocalanus elongatus* and *Calanus helgolandicus* are present. The number of the organisms in these groups is, in general, lower than in the epiplankton but standing stock is, approximately, the same. By depth their number and standing stock decrease. The daily ration of nauplii and copepodites in the bathyplankton is much lower (18-20% and 60-70% of the body weight correspondingly), than in the epiplankton. Organisms of both groups consume various algae up to 60-70 μ in diameter.

Copepodite stages IV-VI of *Pseudocalanus* and *Calanus* are characterized by low numbers, the highest standing stock, high daily rations (100-140% of the body weight), intensive diurnal vertical migrations and high energy metabolism. The adult stages of *Pseudocalanus* and *Calanus* eat all sizes of phytoplankton. Their feeding is characterized by a distinct diurnal rhythm. On the whole, in the period of development from nauplii to adult, one can notice a constant increase in the daily ration of the migratory bathyplankton species and quite the opposite for non-migratory organisms of the epiplankton species (Petipa, 1966 a, b).

Mixed food consumers. The third trophic level includes consumers of plant and animal material and consists of only one ecological group. It includes stages IV-VI copepodites of *Acartia clausi, Centropages ponticus*, Pontellidae, *Oithona minuta, Oithona similis*, male and female *Acartia* and *Centropages*. All these organisms consume small and medium algae, nauplii of copepods and small copepodites. In the epiplankton, *Oithona similis* is almost absent, whereas it is the main representative of the bathyplankton community. *Acartia clausi* is abundant, but all other epiplankton species decrease in number and standing stock by a factor of 10-15. The daily rations of mixed-food consumers in the epiplankton and bathyplankton communities amount respectively to 40-144% and 20-70% of the body weight.

Primary carnivores. Only one ecological group was distinguished in the trophic level of primary carnivores, the predatory Copepoda. In the epiplankton this group consists of adult *Oithona minuta* and Pontellidae. All the representatives are active carnivores, feeding on small Copepoda and Cladocera up to 1·5 mm in size. The daily ration of Pontellidae is 40-80μ%, the daily ration of *Oithona minuta*, 158% of the body weight. In the bathyplankton the predatory group of Copepoda consists of *Oithona similis*. At a low temperature the daily ration decreases to 79% of body weight.

Secondary carnivores. The trophic level of secondary carnivores consists o one ecological group in each community. In the epiplankton this grou includes the small *Sagitta setosa*. Its daily ration is about 85% of the bod weight. *Sagitta setosa* consumes adult copepods (0·8-1·5 mm), *Oikopleur* and its own juveniles. In the bathyplankton this group of secondary carni vores consists, mainly, of large *Sagitta* sp. Its daily ration amounts to 65% of the body weight. Of all the organisms, the Copepoda is of the most im portance in the ration of *Sagitta*.

Tertiary carnivores. The trophic level of tertiary carnivores consisted mainly, of *Pleurobrachia pileus*. Medusae of all species were found ver rarely. In the epiplankton only small *Pleurobrachia pileus* were met. Thei daily ration is 330% of body weight. *Pleurobrachia* consumes animals o various size, largely Copepoda and *Sagitta*. In the bathyplankton communit large individuals of *Pleurobrachia pileus* become of great importance. Thei daily ration decreases to 80% of the wet body weight. Food compositio does not vary very significantly.

That was the structure of the trophic levels in the epiplankton and bathy plankton communities of the Black Sea.

Detritus

Detritus, i.e. particles of dead organic matter in suspension, is of great im portance for food interrelations in the sea. On the basis of the above con siderations, organic detritus is not included in any trophic level of livin organisms. Side by side with such elements of the environment, as nutrien salts and dissolved organic matter, detritus is considered as an independen group. The data on the distribution and rate of detritus formation by th epiplankton and bathyplankton ecosystems are given below.

Plant detritus. The distribution of dead phytoplankton, or plant detritus is the converse of the distribution of the living phytoplankton. In the epi plankton and in the upper layer of the bathyplankton down to 25 m, dea algae, except *Noctiluca miliaris*, are almost absent; dead cells usually consti tute 1-2%, and sometimes 8% of the living cells. In the deeper layers of th bathyplankton the number of dead, medium and large diatoms and Dino flagellata gradually increases from 8-20% of the number of the living organ isms, to 80%. The number of dead *Noctiluca* is distributed differently. In th epiplankton the number of dead *Noctiluca* is 0·42 that of the living organisms In the bathyplankton the number of dead cells decreases with depth and th percentage of dead cells to living ones increases from 12% in the layer 12-25 m to 60% in the layer 75-100 m. Dead algae are present in the sea unbroken fo 24 hours, so the daily mortality of cells must be equal to the number of dea organisms. The death rate of phytoplankton in ecological groups is given i Tables 3 and 4. The highest mortality rate is found for the large phytoplank ton, *Noctiluca miliaris* in particular. Total absolute number of dead phyto plankton cells during the period of observation decreased with depth.

Animal detritus. The distribution and rate of formation of animal detritu are characterized by the following peculiarities. As for algae, the absolute mas

TABLE 3. Rate of detritus formation by the epiplankton community (mg/m² . 24 hours)

Life forms	Standing stock of the living organisms B	Standing stock of the dead organisms B'	Death Rate	Defaecation	Moulting	Total increase of detritus D	$\frac{D}{B}$ %
Phytoplankton—producers and saprophages	29 249·6	12 593·8	12 593·8	183·4		12 777·2	4·4
1. Small-sized forms	190·8	4·1	4·1			4·1	2·1
2. Medium-sized forms	243·6	5·2	5·2			5·2	2·1
3. Large-sized forms	28 815·2	12 584·5	12 584·5			12 584·5	43·7
including Noctiluca	28 658·0		12 581·0	183·4		183·4	0·6
Herbivores	792·0	11·6	3·9	80·2	8·2	92·3	11·7
1. Nauplii	33·1	0·48	0·2	6·5	1·0	7·7	23·3
2. I-III copepodites	184·7	1·66	0·5	47·5	3·1	51·1	27·7
3. IV-VI Paracalanus	320·7	3·85	1·3	17·6	4·0	22·9	7·1
4. Oikopleura	215·6	5·61	1·9	8·2	—	10·1	4·7
5. IV-VI Pseudocalanus, Calanus	35·5	—	—	0·3	0·1	0·4	1·1
6. Larvae of molluscs and polychaetes	2·4	—	—	0·05	—	0·05	2·1
Mixed-food consumers	193·5	3·48	1·9	15·1	2·4	19·4	10·0
Primary carnivores	175·2	3·85	1·3	17·9	—	19·2	10·9
Secondary carnivores	297·3	8·32	2·8	24·9	—	27·7	9·3
Tertiary carnivores	263·2	—	0	29·7	—	29·7	11·3
Sum on life forms	30 970·8	12 621·05	12 603·0	351·2	10·6	12 965·5	41·9

* The weight of moulted exoskeletons of crustaceans constitutes 5% of the weight of moulting organisms (data of Khmeleva N. N. for Artemia).

TABLE 4. Rate of detritus formation by the bathyplankton community (mg/m² . 24 hours)

Life forms	Standing stock of the living organisms B	Standing stock of the dead organisms B'	Death Rate	Defaecation	Moulting	Total increase of detritus D	$\frac{D}{B}$ %
Phytoplankton—producers and saprophages	83 511·0	12 821·0	12 821·0	756·8		13 577·8	16·3
1. Small-sized forms	218·5	38·0	38·0			38·0	17·4
2. Medium-sized forms	212·7	32·0	32·0			32·0	15·0
3. Large-sized forms	83 079·7	12 751·0	12 751·0			12 751·0	15·3
including Noctiluca	81 379·0		12 451·0	756·8		756·8	0·9
Herbivores	3 802·3	199·17	66·2	502·1	25·1	593·4	15·6
1. Nauplii	167·8	0·67	0·2	9·8	8·3	18·3	10·9
2. I-III copepodites	385·0	7·32	2·4	73·8	6·7	82·9	21·5
3. IV-VI Paracalanus	136·7	1·78	0·6	2·9	1·7	5·2	3·8
4. Oikopleura	313·3	12·20	4·0	5·0	—	9·0	28·7
5. Pseudocalanus, Calanus	2 765·0	177·20	59·0	410·0	8·4	477·4	17·3
6. Larvae of molluscs and polychaetes	34·5	—	—	0·6	—	0·6	1·7
Mixed-Food consumers	122·0	4·76	1·6	18·2	1·5	21·3	17·5
Primary carnivores	187·8	7·51	2·5	17·4	—	19·9	1·7
Secondary carnivores	837·3	242·82	81·0	112·9	—	193·9	23·2
Tertiary carnivores	18 775·0	142·30	47·4	441·2	—	488·6	2·6
Sum of life forms	107 235·4	13 417·56	13 019·7	1848·6	26·6	14 894·9	13·9

of dead zooplankton decreases with depth, but as a percentage of living organisms it increases. In the epiplankton the number of dead animals increases from 2·5% in the layer 12-25 m to 10% in the layer 100-150 m. The total number of dead animals in the bathyplankton constitutes 2·5% of the number of the living organisms. The maximum number of dead organisms is characteristic of secondary carnivores (Chaetognatha), and of tertiary carnivores (*Pleurobrachia*). The daily animal mortality is one third of the standing stock of dead organisms (Tables 3 and 4).

Faeces of heterotrophs and moulted exoskeletons of Crustacea are distributed in the water column, much in the same way as the living organisms. In the epiplankton the total weight of faeces constitutes 1·2%, and the weight of exoskeletons 0·03% of the standing stock of all living organisms. In the bathyplankton faeces and exoskeletons constitute respectively 1·7 and 0·02% of the standing stock of the living organisms. In the epiplankton the daily production of faeces and moulted exoskeletons is 351·2 and 10·6 mg/m² respectively. In the bathyplankton there is a daily production of 1848·6 mg/m² of faeces and 26·6 mg/m² of exoskeletons (Tables 3 and 4).

On the whole, the daily intensity of detritus formation in weight is three times greater in the epiplankton (42%) than in the bathyplankton (14%). The main mass of detritus in both systems is formed at the expense of dead *Noctiluca* and faeces (Tables 3 and 4).

The transfer of matter and energy

We shall now consider the ways in which matter and energy are transferred and changed in the planktonic communities of the Black Sea. To determine the direction of matter and energy flow in the observed ecosystems, data on food composition in the daily rations of animals from various ecological groups were used. The daily rations were estimated from gut contents. Knowing the quantity of consumed material, the food chains, or flows of matter and energy transfer, for the Black Sea epiplankton and bathyplankton communities were worked out. The main direction of matter and energy flow in the epiplankton is mediated through the small and medium-sized forms of all the trophic levels (Fig. 1). The standing stock of the algae up to 50 μ in diameter and of the animals up to 1·5 mm in length is very small, but these algae and animals are very active relatively in producing matter and in metabolizing energy. In the epiplankton these organisms comprise most of the plankton and are used as food. In particular, the main groups, which transfer matter and energy, are the small and medium Peridinians and diatoms, nauplii and young copepodite stages of small copepods, the thin filter feeder *Paracalanus*, the mixed-food consumers (IV-V copepodite stages of *Acartia clausi* and *Oithona minuta*) and small primary, secondary, tertiary carnivores (males and females of *Oithona minuta*, *Sagitta setosa*, *Pleurobrachia pileus*). The large migratory organisms appearing here for a very short period, are of small significance in matter and energy transfer.

Epiplanktonic forms, especially filter-feeders such as *Oikopleura* and *Paracalanus*, consume a lot of detritus (Fig. 1). The daily rate of detritus

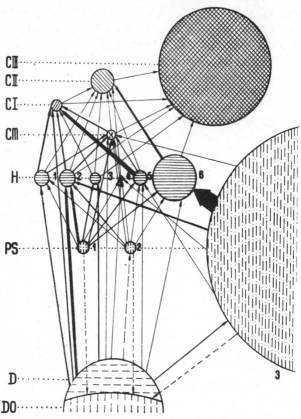

Fig. 1. Food webs in the epiplankton community. C III—tertiary carnivores;
C II—secondary carnivores; C I—primary carnivores; CM—mixed-food (plant
and animal) consumers; H—herbivorous organisms; PS—producers and
saprophagous organisms (phytoplankton); D—detritus; DO—dissolved organic
matter; 1, 2, 3 etc., written near the circles, are the ecological groups of the
tropic level. The level of producers: 1—small-sized forms; 2—medium-sized
forms; 3—large-sized forms. The level of herbivores: 1—*Oikopleura*; 2—II-III
Copepodites; 3—IV-VI copepodites of *Paracalanus*; 4—larvae of molluscs and
polychaetes; 5—naupliuses; 6—IV-VI copepodites of *Pseudocalanus* and
Calanus. Areas of circles are proportional to the average daily standing stock
of the ecological groups. Thickness of solid pointers is proportional to daily
specific rate of this or that food consuming. Dotted pointers mean the processes
of dissolved organic matter excretion and the processes of detritus consumption,
where no quantitative data were obtained.

consumption in the epiplankton is 20% of standing stock of the living
organisms.

The main direction of matter and energy flow in the bathyplankton, in contrast to the epiplankton, occurs through the larger animals of the trophic

levels. Of all the producers and saprophagous organisms, large diatoms, *Noctiluca*, and small Peridinians and diatoms, play an important role in matter and energy transfer (Fig. 2).

At the herbivorous level the most powerful flow of matter and energy is through the migratory copepodites and adult large-sized copepods (*Pseudocalanus* and *Calanus*). At the level of primary, secondary and tertiary carnivores, the flow is through the males and females of *Oithona similis* and large individuals of *Sagitta* and *Pleurobrachia* (Fig. 2). The bathyplanktonic

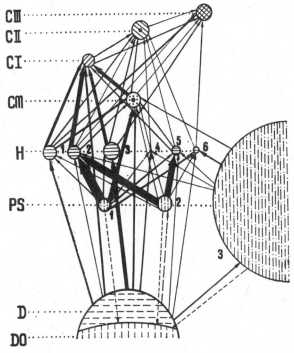

Fig. 2. Food webs in the bathyplankton (signatures are the same as on Fig. 1).

organisms use detritus more actively than the epiplanktonic ones. The daily rate of detritus consumption in the bathyplankton is 45% of standing stock of the living organisms.

Before giving a quantitive estimate of energy flow in both communities, let us consider the character of changes in matter and energy, consumed by each trophic level.

From experimental work the data on food consumption, increment in body weight, rate of reproduction and respiration for all the main ecological groups of the trophic levels were obtained. These materials are given in Tables 1 and 2. Summing the quantitative indices for each process on all the ecological groups composing the trophic level, we obtain values for each level. So, food consumption, energy accumulation (increment in body weight and eggs production expressed in calories), energy metabolism (respiratory losses)

TABLE 5. Rate of consumption, increment in body weight and respiratory losses (cal/m²) by the trophic levels of the epiplankton community at 16-17°

Trophic levels	B Average standing stock	P Diurnal increment in body weight, re-production	T Diurnal energy of respi-ration	R Diurnal ration	$\frac{P}{B}$ %	$\frac{T}{B}$ %	$\frac{R}{B}$ %
Producers and saprophages (phytoplankton)	2709	2072	165	—	76	6	—
Herbivores	871	62	260	448	7	30	51
Mixed-food consumers	213	15	146	176	7	68	83
Primary carnivores	192	9	288	317	5	150	165
Secondary-tertiary carnivores*	113	18	66	127	16	58	112

* Levels are added, as all the animals of both levels, practically, consumed one and the same food in the investigated period.

TABLE 6. Rate of consumption, increment in body weight and respiratory losses (cal/m²) by the trophic levels of the bathyplankton community at 7-12°

Trophic levels	B Average standing stock	P Diurnal increment in body weight, re-production	T Diurnal energy of respi-ration	R Diurnal ration	$\frac{P}{B}$ %	$\frac{T}{B}$ %	$\frac{R}{B}$ %
Producers and saprophages (phytoplankton)	7944	4984	241	—	63	3	—
Herbivores	4182	514	3228	4096	12	77	98
Mixed-food consumers	134	4	46	64	3	34	48
Primary carnivores	206	18	145	181	9	70	88
Secondary-tertiary carnivores*	1183	43	253	907	4	21	77

* Levels are added, as all the animals of both levels, practically, consumed one and the same food in the investigated period.

TABLE 7. Dry matter and caloricity of the main representatives or groupings of the plankton

Plankton groupings	Dry matter and caloricity	Dry matter %	Caloricity of 1 mg dry mat-ter, g-cal
Peridinea		14·5	3·5
Diatomea		12·2	2·5
Noctiluca		2·0	4·4
Copepoda		20·0	5·5
Oikopleura		10·0	4·0
Sagitta		8·6	3·9
Pleurobrachia		2·4	2·0
Detritus		40·0	4·2

by total quantity of animals in each of the trophic levels, as well as specific rates of all these processes are given in Tables 5 and 6 and Figs 3 and 4. The illustrations show that in the epiplankton there is a tendency for the average value of daily energy metabolism and daily food consumption to increase from the first trophic level of the community to the levels of primary, secondary and tertiary carnivores. In the bathyplankton, this tendency appears less as a consequence of the high specific rates amongst the herbivores. The daily specific rate of body weight increment and reproduction, on the contrary, declines in both communities from the first trophic level to the higher levels of the carnivores.

The comparison of the specific rates of the processes calculated on wet weight, is very difficult, as the organisms of various trophic levels differ in their water contents. Similarly the caloricity of dry matter differs (Table 7) in the organisms of the various trophic levels.

A great range in data on the specific rates of all three processes in caloricity and wet weight is characteristic of the secondary and tertiary carnivores—*Sagitta* and *Pleurobrachia*. These animals contain much water and so have a high weight and low caloricity. However, they consume small quantities of highly calorific food (mainly crustaceans) enough to cover their requirements (Tables 1 and 2). The tertiary carnivores (*Pleurobrachia* in particular) are characterized by a decline in the rates of respiratory loss. It is not surprising as these predators not only include much matter of little importance in their metabolism but also they are less mobile than other plankton organisms

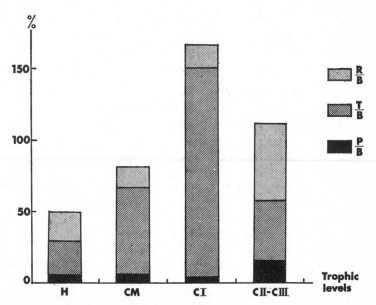

Fig. 3. Specific rates of uptake, accumulation and expenditure of energy by the trophic levels of the epiplankton.

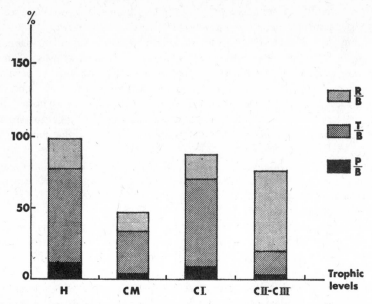

Fig. 4. Specific rates of uptake, accumulation and expenditure of energy by the trophic levels of the bathyplankton.

(Tables 5 and 6, Figs 3 and 4). The lowest rate of respiratory loss is found in the secondary and tertiary carnivores of the bathyplankton (Fig. 2).

The increase in the rate of matter and energy uptake and expenditure from the lowest heterotrophic level to the highest is an effect of the increasing role of predation as an active search for food. A very high rate of food consumption and energy metabolism is typical of the smallest and most active primary carnivores in both communities—males and females of *Oithona minuta* and *Oithona similis*.

The comparison of two processes, the accumulation of matter and energy (as increment of weight and gonad) and the metabolism of energy showed that for all heterotrophic levels, the energy dissipated by respiration is 4-10 times as much as the accumulation of matter and energy in the animal's body.

As the increments of matter and energy were estimated experimentally in all the trophic levels from the growth and reproduction of individuals, the production for all the trophic levels was found, estimated from the organisms taken from each level. The production of a trophic level is the increase of organic matter in that level per unit of time.

It was noted in the section on method that the production of a given trophic level was estimated as the sum of organisms eaten from the level, dead organisms, organisms transferring to the upper level and the difference in standing stock at the beginning and at the end of the day. Or, formally, the production of a given level (P') will be:

$$P' = G + M + L + (B_1 - B_0),$$

TABLE 8. Production of the trophic levels of the Black Sea epiplankton community (mg/m² . 24 hours)

Trophic levels	G Food amount, eaten from a given level	M Death rate	$B_1 - B_0$ Difference between final and initial standing stock	L Transport to the following level	P' Production
Producers, saprophages	478·5	12 593·8	5397·2	—	18 469·5
Herbivores	361·2	4·0	−6·3	156·0	515·0
Mixed-food consumers	67·0	1·2	20·0	82·2	170·4
Primary carnivores	53·5	1·3	27·4	—	82·2
Secondary-tertiary carnivores	33·0	2·8	−1·1	—	34·7

TABLE 9. Production of the trophic levels of the Black Sea epiplankton community (cal/m² . 24 hours)

Trophic level	G Food amount, eaten from a given level	M Death rate	$B_1 - B_0$ Difference between final and initial standing stock	L Transport to the following level	P' Production
Producers, saprophages	213·5	1 112·4	474·9	—	1800·8
Herbivores	316·6	3·0	33·8	171·6	525·0
Mixed-food consumers	73·7	1·3	22·0	90·4	187·4
Primary carnivores	58·9	1·4	30·1	—	90·4
Secondary-tertiary carnivores	11·1	9·4	12·9	—	33·4

TABLE 10. Production of the trophic levels of the Black Sea bathyplankton community (mg/m² . 24 hours)

Trophic level	G Food amount, eaten from a given level	M Death rate	$B_1 - B_0$ Difference between final and initial standing stock	L Transport to the following level	P' Production
Producers, saprophages	37 415·6	12 821·0	22 273·0	—	72 509·6
Herbivores	967·4	66·2	−59·7	90·0	1063·9
Mixed-food consumers	21·3	1·6	11·4	60·0	94·3
Primary carnivores	47·2	2·5	10·4	—	60·1
Secondary-tertiary carnivores	0	128·4	1289·1	—	1417·5

TABLE 11. Production of the trophic levels of the Black Sea bathyplankton community (cal/m² . 24 hours)

Trophic level	G Food amount, eaten from a given level	M Death rate	$B_1 - B_0$ Difference between final and initial standing stock	L Transport to the following level	P' Production
Producers, saprophages	3581·6	1198·5	1960·0	—	6740·1
Herbivores	1033·0	70·0	−63·6	99·7	1139·1
Mixed-food consumers	23·4	1·8	12·5	66·1	103·8
Primary carnivores	51·9	2·8	11·4	—	66·1
Secondary-tertiary carnivores	0	29·5	46·3	—	75·8

where G = the number or weight of the organisms eaten from that level
       per day;

   M = the number or weight of the organisms dying per day;

   L = the number or weight of the organisms, transferring to the next
       level per day;

$(B_1 - B_0)$ = difference in standing stock at the beginning and at the end of
       the day.

The availability of data on nutrition, mortality and standing stock in the sea at different times of day for all the main phytoplankton and zooplankton groups made it possible to obtain values of each parameter of the equation and to calculate the production of each level (Tables 8-11).

For the following three trophic levels—producers, primary carnivores, secondary and tertiary carnivores, $L = 0$. It means that no organisms leave for the higher trophic level.

For the herbivores and mixed food consumers a different result is observed The level of mixed-food consumers consists of copepodite states IV-VI of

TABLE 12. Specific rate of production and accumulation of matter and energy by the trophic levels of the epiplankton community

Trophic levels		$\frac{P'}{B}$ %	$\frac{P' - G - L}{B}$ %	$\frac{dB}{B}$ %
Producers and saprophages	on wet weight	63·0	61·5	20·3
	on caloricity	66·0	58·7	17·5
Herbivores	on wet weight	64·0	−0·3	−0·8
	on caloricity	60·0	4·2	3·9
Mixed-food consumers	on wet weight	88·0	10·9	10·3
	on caloricity	88·0	10·9	10·3
Primary carnivores	on wet weight	46·9	16·4	15·6
	on caloricity	46·9	16·4	15·6
Secondary-tertiary carnivores	on wet weight	6·2	0·3	−0·2
	on caloricity	29·6	19·7	11·4
Total community	on wet weight	—	58·2	17·6
	on caloricity	—	41·5	13·9

TABLE 13. Specific rate of production and accumulation of matter and energy by the trophic levels of the bathyplankton community

Trophic levels		$\frac{P'}{B}$ %	$\frac{P' - G - L}{B}$ %	$\frac{dB}{B}$ %
Producers and saprophages	on wet weight	87·0	42·0	26·7
	on caloricity	84·8	39·8	24·7
Herbivores	on wet weight	28·0	0·2	−1·6
	on caloricity	27·0	1·8	−1·5
Mixed-food consumers	on wet weight	77·5	10·7	9·3
	on caloricity	77·5	10·7	9·3
Primary carnivores	on wet weight	32·0	6·9	5·5
	on caloricity	32·0	6·9	5·5
Secondary-tertiary carnivores	on wet weight	7·2	7·2	0·7
	on caloricity	6·4	6·4	3·9
Total community	on wet weight	—	34·1	21·9
	on caloricity	—	23·9	14·4

Arcartia and copepodite stages IV-VI of *Oithona*. Their daily growth increment in weight is not very high. The production of herbivores includes not only the increase in standing stock, as well as organisms eaten from that level and dead organisms, but also organisms transferring to the level of mixed-food consumers. In its turn the production of the level of mixed-food consumers consists of their increase in standing stock, animals eaten from the level and dead animals, and animals transferring to the level of primary carnivores. Copepodites, stage V of *Oithona*, becoming adult, transfer to the level of primary carnivores. Although males and females of *Oithona*, comprising the level of primary carnivores, produce eggs, they do not grow at all. So the level of primary carnivores increases solely at the expense of the preceding level of mixed-food consumers.

Eggs of organisms of all trophic levels, together with increment in body weight are the main sources of increase of herbivores.

A comparison of production of all the trophic levels showed that it decreases by 2-10 times from the level of producers to the tertiary levels of carnivores (Tables 8-11). A decrease in the same direction is noted for the daily specific rate of production—P'/B (Tables 12 and 13) (except for the mixed-food consumers, for which P'/B is the highest value).

Now let us give quantitative characteristics to the energy flow in both communities. The flow of energy passing through any trophic level is the amount of energy received by a level per unit of time, expressed in terms of calories, utilized for the increase of organic matter of that level and dissipated by respiration. If there were no transfer of organisms from a lower level to a higher one, all the energy fixed by that level would be nothing more than assimilated from the consumed food.

To estimate the value of energy flow through a trophic level one must have data on total energy uptake by each level, in terms of the assimilated part of the food.

TABLE 14. Value and specific rate of energy flow through the trophic levels of the communities

Trophic levels	Epiplankton community $P'+T = U$ $(cal/m^2 . 24\ hours)$	$\dfrac{U}{B}\%$	Bathyplankton community $P'+T = U$ $(cal/m^2 . 24\ hours)$	$\dfrac{U}{B}\%$
Producers and saprophages (phytoplankton)	1966	73	6981	88
Herbivores	785	90	4367	104
Mixed-food consumers	333	156	150	110
Primary carnivores	378	197	211	102
Secondary-tertiary carnivores	99	87	329	28

The values of energy flow for each level can be obtained from Tables 9 and 11 by summing the daily production of that level and the energy dissipated by respiration. Table 14 shows that in both communities the absolute value for energy flow decreases about 20 times while passing from the first trophic level of the producers to the higher levels of secondary and tertiary carnivores.

The average intensity of energy flow, or daily specific rate, changes in a somewhat different way. In the epiplankton, at the level of primary carnivores, it is thrice as much as in the level of the producers and twice as much as in the herbivores. A sharp decline in the rate of energy flow through the higher levels of secondary and tertiary carnivores is observed. A rather low rate of growth and energy metabolism may account for this. In the bathyplankton the specific rate of energy flow from the first trophic level to the higher one averages 86%.

TABLE 15. Transfer efficiency of matter and energy by the heterotrophic level of the epiplankton community

Trophic level		G Org. eaten from the level	R Consumption by the level	$\frac{G}{R}$ %	$G+L$ Extraction from the level	$R+L_0$ Receipt to the level	$\frac{G+L}{R+L_0}$ %
Herbivores	mg/m²	361·2	544·0	66	517	569	91
	cal/m²	316·6	448·0	71	488	465	105
Mixed-food consumers	mg/m²	67·0	171·0	39	149	327	45
	cal/m²	73·7	176·0	42	164	348	47
Primary carnivores	mg/m²	53·5	296·0	18	53·5	296	18
	cal/m²	58·8	317·0	19	58·8	317	19
Secondary-tertiary carnivores	mg/m²	33·0	164·1	20	33	164	20
	cal/m²	11·1	155·0	7	11	155	7

TABLE 16. Transfer efficiency of matter and energy by the heterotrophic level of the bathyplankton community

Trophic level		G Org. eaten from the level	R Consumption by the level	$\frac{G}{R}$ %	$G+L$ Extraction from the level	$R+L_0$ Receipt to the level	$\frac{G+L}{R+L_0}$ %
Herbivores	mg/m²	967·0	5972·0	16	1057	6085	17
	cal/m²	1033·0	4096·0	25	1133	4220	27
Mixed-food consumers	mg/m²	21·3	84·0	25	81	174	46
	cal/m²	23·4	64·0	36	89	164	54
Primary carnivores	mg/m²	47·2	179·7	26	47·2	179·7	26
	cal/m²	51·9	181·0	29	51·9	181·0	29
Secondary-tertiary carnivores	mg/m²	0	824·6	0	0	824·6	0
	cal/m²	0	907·0	0	0	907·0	0

In order to judge the efficiency of matter and energy transfer from one trophic level to another, the coefficient of Slobodkin (1962) was used. It was called the ecological efficiency of the trophic level. This coefficient is the ratio of the amount of matter and energy (in terms of calories) extracted from a lower level by the upper levels in the form of food to the amount of food consumed by the lower level. Calculations showed that the ecological efficiency (G/R) in the epiplankton gradually decreases from the first trophic level to the highest one (Tables 15 and 16). The ecological efficiency of the producers is 71%, but that of primary carnivores and of secondary and tertiary carnivores constitute 19 and 7% respectively. In the bathyplankton this coefficient for all the trophic levels, with the exception of the highest averages 30%. From the present data the organisms of the highest level were not consumed, so the ecological efficiency of that level equals zero.

If it is necessary to know the total transfer of matter and energy from one trophic level to another, one must also take into account the data on the direct transfer of organisms from the lower level to the next one. Then the efficiency of matter and energy transfer by the trophic levels will be: $(G+L)/(R+L_0)$ (Tables 15 and 16). $(G+L)$ represents the amount of matter or energy, extracted from a lower level to the higher one; $(R+L_0)$ represents the amount of matter or energy, which was received by the lower level. For the herbivores we have eggs added in the denominator, produced by several trophic levels of the observed communities, as the larvae hatched out of these eggs are herbivorous.

The calculations showed (Tables 15 and 16) that the trend of changing efficiency for total matter and energy transfer from one level to another in both communities is the same as it was for matter and energy transfer at the expense of food only.

A summary of the accumulation of matter and energy in epi- and bathy-planktonic communities of the Black Sea is given in Tables 12 and 13. It has been already noted in the section on methods that the daily production of matter and energy (in terms of calories) by the total community was defined as a sum of standing stock increases in both living and dead organisms of all the trophic levels per day. The standing stock increase of the living organisms of all the levels is nothing more than the accumulation of the living matter in the community, while the standing stock increase of dead organisms is the accumulation of dead matter.

Calculations showed that the average daily specific rate of accumulation of total dead and living matter in the epiplanktonic community (41·5%) was twice as high as that in the bathyplanktonic community (23·9%). However, the average daily specific rate of accumulation of the living matter only, was, on the contrary, a little bit higher in the bathyplankton (14·5%) than in the epiplankton (13·9%).

A comparison of respiratory losses in total energy in both communities showed, that the specific rate of energy metabolism per day in the epiplankton (23%) is 6% lower than in the bathyplankton (29%).

DISCUSSION

An attempt is made to show matter and energy transfer through the trophic levels of the Black Sea planktonic communities as well as to compare two different methods in defining the production of each level.

While analysing the results, some questions arise and must be discussed. It is known that the notion of "trophic level" means that organisms with identical manner of feeding (autotrophy, herbivoriness, predation) are grouped independently of their systematic position. All the trophic levels are considered as the links of one and the same food chain of the community, where each link feeds upon the preceding link and the lower has a greater stock of energy than the higher. But any higher level may feed upon several lower levels.

Following this principle the main trophic levels in the Black Sea planktonic communities were segregated. On account of the definition of the trophic level, the same population of any species may be a component of several trophic levels, if its stages in age are characterized by different feeding habits and different food composition. Thus, the level of the mixed-food consumers included only adult copepods, consuming both plant and animal material. Their juveniles are included in the level of the herbivorous organisms (Tables 1 and 2).

The use of the ecological groups in the trophic levels made it possible to reduce the number of experimental and field observations to estimate uptake, accumulation and expenditure of matter and energy by all organisms of various species composing a level.

The composition of the trophic level is closely connected with the method of the calculation of the production of the level. The daily production of each level was determined by two methods: first by multiplying the average daily growth increment in body weight (including reproduction) of any organism by their average number. And secondly by summing all the individuals extracted from the level in the form of food, with losses from mortality and transport to the next level and with the difference in standing stock at the beginning and at the end of the day.

In both cases the calculated production is more or less close to the actual production (Petipa, 1967a), which is why both values for each level can be compared. Such a comparison, however, must take into account the structure of the trophic levels and the means of their supply. In other words, it is necessary to take into account which part of the production is formed from growth and which from the transfer of organisms from the lower level.

Comparison of values of production obtained by both methods showed that for most levels the production determined by the method of estimating the extraction of food from a given level proves to be very high. The greatest difference between parallel estimates for production is found for the herbivores in the epiplankton (Tables 5 and 9). The main reasons are as follows:

1. Escape of small copepods and their larvae through the meshes of the nets. According to the observations of V. E. Zaika the plankton nets used in the investigation catch only $\frac{1}{10}$ of the copepod nauplii and half of all the copepodites.

2. In laboratory experiments the values for the growth and for the fecundity of copepods are probably lower than that in the natural environment as a consequence of an insufficient quantity of suitable food, restricted living space, and of the unfavourable influence of excretory products.

3. Lack of estimation of the rhythms in the processes of nutrition, mortality and reproduction.

The results of both methods of estimating production in each trophic level were corrected for escape and for the low fecundity observed in experimental data.

Moreover, the definition of production by the method of estimating the extraction of food from the trophic level and by changes in the number of animals in the sea for the period of investigation must lead to more reliable values. The reason for this is that the rhythms of food consumption, morality, reproduction and the transfer of animals to the higher level are all taken into account. The first method of defining the production (increment of weight multiplied by numbers) does not take into account these factors. The point is well illustated by comparing values for phytoplankton production obtained by both methods (Tables 5, 6, 9 and 11).

As noticed earlier (Petipa, 1967a), the rhythms of consumption of the phytoplankton by the bathyplankton animals promotes a more intense production of organic matter by the phytoplankton in the bathyplankton, regardless of the fact that the potentional capacities of phytoplankton production in this system is lower than that in the epiplankton. Hence the second method of estimating the production at a trophic level is the better method.

CONCLUSIONS

A preliminary consideration of the main results on the investigation of an ecosystem allows us to make the following generalizations.

1. The specific environmental conditions and behaviour of the main ecological groups of each community define the means of transfer and the character of the production of matter and energy by the trophic levels, as well as the character of their utilization.

2. The main flow of matter and energy in the epiplankton goes through the small forms of all the trophic levels. Juveniles of copepods and thin filter-feeders play a significant role here. In the bathyplankton community the main matter and energy flow goes through the large forms of all the trophic levels.

3. The method of defining the production of matter for all the trophic levels by summing the increase of standing stock for the investigated period of time and the number of organisms eaten from the level gives us the production which is the most close to actual production of matter in the sea, because this method takes into account indirectly the influence of the rhythms of extraction and other factors.

4. As a result of differing rhythms of grazing on phytoplankton by each community, the daily specific rate of matter produced by the phytoplankton proved to be higher in the bathyplankton community (85%), than in the epiplankton (66%) as defined by the second method.

5. The predominance of a passive state in migratory animals in deeper layers during the day time and their capacity for forming an energy store in the form of fat every day gives the bathyplanktonic animals an economic use of energy. On the whole, the energy dissipated by respiration for all the bathyplanktonic community is only 6% higher than that for the total epiplankton, in spite of the active migration of the bathyplanktonic animals.

6. The daily specific rate increase of the standing stock of living organisms

in the bathyplankton is somewhat higher (14·5%) than for the epiplankton (13·9%). The specific rate of production of living and dead organic matter by the bathyplankton is half that by the epiplankton.

7. The production and use of detritus differs in the two communities. The epiplankton produces detritus three times as quickly as the bathyplankton, while the rate of detritus utilization in the epiplankton is half that in the bathyplankton.

8. The absolute value for energy flow in the first trophic level of both communities is 20 times as much as that in the highest trophic level. The daily specific rate of energy flow through the trophic levels of the epiplankton decreases from the lowest levels to the highest levels, whereas in the bathyplankton it is more or less of constant value.

9. The ecological efficiency can be considered as the efficiency of transfer of matter and energy from one trophic level to another at the expense of food. In the epiplankton this coefficient gradually decreases from the lowest heterotrophic level (i.e. the herbivores) (71%) to the highest (7%). In the bathyplankton the efficiency of matter and energy transfer is of rather constant value (30%). Similar changes are found for the efficiency of total matter and energy transfer.

A high and changeable ecological efficiency indicates a great use of matter in proportion to its consumption and, therefore, strained interrelations between trophic levels. A relatively constant low level of ecological efficiency indicates stable food interrelations and is possibly an effect of constant environmental conditions as well as constant physiological characteristics.

10. For the successful study of the structure, dynamics and production of planktonic communities one must take into account:

(a) the capture of small (<7-100 μ) phytoplankton, zooplankton and detritus;

(b) that fecundity, respiration and growth increment in body weight must be defined under conditions as close to natural conditions as possible;

(c) the influence of rhythm on production processes.

REFERENCES

MARGALEF, D. K. 1967. The food web in the pelagic environment. *Helgoländer wiss. Meeresunters.*, 15, 548-59.
MIRONOV, G. N. 1954. Nutrition of planktonic carnivores. I. Nutrition of *Noctiluca. Trudȳ sevastopol. biol. Sta.*, 8, 320-40.
PAVLOVA, E. V. 1967. Food consumption and energy metabolism of Cladocera populations in the Black Sea. In *Structura i dinamika vodnykh soobshchestv i populatsii*, Ser. "Biol. Morya". Kiev, Akad. Nauk, UkSSR, Nauk. dumka, 66-85.
PAVLOVA, E. V. 1968. Respiration of the planktonic animals of the Black Sea. In *Issledovanie Chernogo Morya i ego promyslovykh resursov.* Moscow, Nauka Press.
PETIPA, T. S. 1966. Oxygen consumption and food requirements in copepods *Acartia clausi* Giesbr. and *A. latisetosa* Kritcz. *Zool. Zh.*, 45 (3), 363-70.
PETIPA, T. S. 1966a. Energy balance of *Calanus helgolandicus* (Claus) in the Black Sea. In *Fiziologiya Morskykh Zhivotnykh.* Moscow, Nauka Press.

PETIPA, T. S. 1966b. The relation between growth increment, energy metabolism and the ration in *Acartia clausi* Giesbr. In *Fiziologiya Morskikh Zhivotnykh*. Moscow, Nauka Press.

PETIPA, T. S. 1967. On the life forms of pelagic copepods and structure of the trophic levels. In *Structura i dinamika vodnykh soobshchestv i populatsii*. Ser. "Biol. Morya". Kiev, Akad. Nauk, UkSSR, Nauk. dumka, 108-19.

PETIPA, T. S. 1967a. On the efficiency of energy utilization in the pelagic ecosystems of the Black Sea. In *Structura i dinamika vodnykh soobshchestv i populatsii*, Ser. "Biol. Morya". Kiev, Akad. Nauk, UkSSR, Nauk. dumka, 44-65.

SADGINA, L. I. 1960. Development of the Black Sea Copepoda. I. Nauplial stages of *Acartia clausi* G., *Centropages kröyeri*, Oithona minuta *Kritsz. Trudÿ sevastopol. biol. Sta.*, **13**, 49-67.

SADGINA, L. I. 1961. Development of the Black Sea Copepoda. II. Nauplial forms of *Calanus helgolandicus* (Claus). *Trudÿ sevastopol. biol. Sta.*, **14**, 103-9.

SLOBODKIN, L. V. 1962. *Growth and regulation of animal populations.* New York, Holt, Rinehart and Winston. 184 pp.

The biology of Chaetognatha
I. Quantitative aspects of growth
and egg production in Sagitta hispida

Michael R. Reeve
Institute of Marine Sciences
University of Miami

ABSTRACT. The growth of *Sagitta hispida* from immature juveniles to mature egg producing adults was studied in terms of the relative rates of maturation of the gonads and seminal vesicles. Large eggs appear in the ovary at the same time that seminal vesicles become full of spermatophores, and as the animal reaches its maximum size. The species has a wide temperature tolerance (15°C) and over the whole of its temperature range has a growth rate directly proportional to temperature. Growth and egg production are dependent on feeding level over a wide range, and starved animals can survive for short periods by using reserves; this causes shrinking. The changes in gonads and other parameters over the period of maturity were studied and it was shown that eggs could be laid over several days during which time the ovary length and maturity stage fluctuated with egg-laying. The maximum number of eggs laid by an individual was 420, and towards the termination of egg-laying death was heralded by a decrease in total and ovary length, and a decline in egg production. In mature animals, survival was inversely related to temperature, and egg production increased except at the highest temperature. Short term gross growth efficiency experiments were performed at three temperatures in terms of both dry weight and total nitrogen. Populations which had not reached maturity had efficiencies ranging between 19-50% with a mean of 34·5%. In mature egg-producing populations values in the region of zero were recorded on a dry weight basis but in terms of nitrogen were in the region of 41%. Egg production was taken into account in both cases. The discrepancy was considered to be due to utilization of stored carbon reserves at this age for energy metabolism while nitrogen was being incorporated into the eggs. Populations always lost weight over this period. In terms of nitrogen efficiency over the whole period studied values were not related to age and were at a mean level of 36·0%. This data was compared with the few other values available for marine zooplankton and was shown to be very closely comparable to calanoid copepods and euphausids. There was no suggestion that the carnivorous chaetognaths operated at higher efficiency levels than the largely herbivorous copepods. All this information was related to the marine environment by illustrating from the literature that chaetognaths appeared to be second in order of importance only to the copepods in the marine macroplankton community, possibly exceeding 20-30% of the latter in biomass units. This suggests that a high percentage of copepods must be converted into chaetognath tissue before being utilized by higher elements of the food chain.

INTRODUCTION

During 1962-63, as a visiting investigator at the Institute of Marine Sciences of the University of Miami, my attention was drawn to the fact that there occurred in adjacent waters a species of planktonic chaetognath which did not die within 24 hours of being taken from the sea. This species is almost unique within the phylum in this respect, as indicated by the virtual lack of physiological information to be found in such reviews as that of Hyman (1959), Alvariño (1965), and Nicol (1967). I was able to obtain some preliminary data, largely concerned with feeding in *Sagitta hispida*, which was reported in two papers (Reeve 1964a, 1966). Having subsequently returned to the Institute I have been able to begin a more detailed programme of research, in which I hope to cover many aspects of the biology and physiology of this and other chaetognath species. Some of the first results of this study are presented below.

Murakami (1959, 1966) has also reported success in keeping the planktonic *Sagitta crassa* alive, in the case of the adults for up to three months and up to three weeks in the case of the larvae. He was, however, largely concerned with the effects of various environmental parameters on the morphology of structures used in taxonomic identification of different races.

Observations on the nature of the food of chaetognaths based on gut contents of preserved or freshly-caught material indicated that they were voracious carnivores (e.g. Lebour, 1922; Furnestin, 1953). They have most frequently been observed to take copepods, as well as euphausids, tunicates, siphonophores, amphipods, medusae, fish larvae and other prey. Less well established is their ability to feed on phytoplankton, reported by Burfield (1927) and Wimpenny (1936). Aspects of digestion have been studied by Parry (1944) and David (1955), and Beers (1964) did some experiments on ammonia and phosphorus excretion. Horridge (1966), and Horridge and Boulton (1967), investigated the functions and structure of the sensory cilia of the benthonic *Spadella*. Hyman (1958) reported that their teeth and hooks were of chitin, and Curl (1962) determined the carbon and phosphorus content of the body. The vast body of chaetognath literature (not covered here) concerns observations drawn from analyses of preserved plankton samples and, occasionally, histological sections.

Sagitta hispida Conant is confined in its distribution to the epiplankton of tropical-equatorial Atlantic waters (Alvariño, 1965). There appears to be considerable doubt as to its occurrence, at least in any numbers, in mid-Atlantic waters, and Tokioka (1955) considered it to be confined to coastal waters, although Alvariño mapped a trans-oceanic distribution for the species.

It is certainly far more common in the coastal waters around Florida than it is in the oceanic Gulf Stream water of the Florida Straits (Owre, 1960) and in Biscayne Bay, adjacent to the Institute, is often the only chaetognath species present in the plankton.

The experiments to be described below concern studies on growth rate, survival, maturation and egg production, together with short-term studies on gross growth efficiency in *Sagitta hispida*.

METHODS AND RESULTS

Animals were collected by making net tows using 12 and 20 inch diameter nylon nets with a mesh-opening size of 300 μ. Making use of the tidal flow beneath the laboratory dock, which extended into Bear Cut (an entrance to Biscayne Bay) nets were hauled every five minutes until sufficient animals had been obtained. On being taken up to the laboratory the mixed plankton sample was selectively diluted by use of a suitable filter-cloth over a siphon. In this way the vast majority of small copepods and assorted larvae were removed with little loss of the larger chaetognaths. Even some of the smaller *Sagitta* (capable of passing through the filter) would be retained by virtue of their avoidance reaction to currents of the sort occurring at the filter surface. Chaetognaths so obtained would be held in plastic 5-gallon containers of bay sea water under gentle aeration until required, usually within three days.

In all the experiments, only water collected from the Florida Current was used. This water is typically oceanic in nature varying only a few tenths of a part per thousand in salinity at the surface seasonally with a mean of 36·0%. This water also shows relatively little seasonal variation and is very low in the amounts of dissolved nitrate and phosphate, phytoplankton, and zooplankton it contains, when compared to the neritic inshore and bay water (Bsharah, 1957). Bay water, for instance, can vary more than 10% seasonally, and fluctuate in zooplankton numbers by a factor of at least 20. Water from the Florida Current, therefore, may be expected to approach far more nearly to a constant medium in terms of physical and biochemical characteristics and composition. This water was collected from a point well within the edge of the Current at a depth of about 5 metres at a station approximately five miles east of the coast-line. The pump-head, transporting containers, and 50-gallon storage laboratory tank were all non-metallic, and collections were timed so that water was at least two weeks old before experimental use. Food for experimental animals was supplied in the form of newly-hatched nauplii of *Artemia* (the brine shrimp) of California origin.

Growth

The studies reported below are concerned with animals taken from the environment of a total length no less than 3-4 mm, after which point they begin a gradual process of attainment of sexual maturity. The maturation process in this hermaphrodite phylum could be followed in three ways, which are detailed in Table 1. Numbers were assigned to various arbitrarily defined

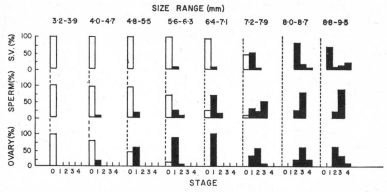

Fig. 1. The percentages of animals with ovary, sperm and seminal vesicles of a
given stage for 8 population size ranges. The unblocked portion of the histograms
represents zero development. The dashed vertical lines indicate the origin of
each histogram.

tages of development of the testes, seminal vesicles, and ovaries. In some
ases, these stages correspond to those designated by other workers (see
)iscussion).

In most of the experiments animals were inspected microscopically to
letermine total length (anterior tip of the head to posterior tip of the tail
xcluding the tail fin), ovary length (where present), and the three maturity
tage parameters. Some of these data were abstracted by setting up 8 total-
ength ranges of 0·8 mm interval and tabulating observations under these until
ach contained data for 30 animals. The information was translated into
ercentages and plotted in Fig. 1.

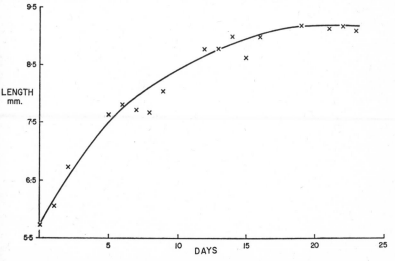

Fig. 2. Growth rate of *Sagitta hispida* in terms of length increase with time
at 21° C.

This figure contains the histograms describing, for each size range, the per
centage spread of maturity stages for ovary, sperm and seminal vesicles
The unblocked portion of the histograms indicate the original stage, in each
case, where there was no development (zero stage). The vertical dashed line
provide a reference to indicate the origin of each group of histograms. The
histograms indicate that the ovaries appear first, followed by signs of sperm
production in the tail, and lastly the seminal vesicles develop. The mean
length at which 50% of the samples first show these signs of approaching
maturity are 5·2, 6·8, and 7·6 mm respectively. The ovaries, although in
evidence earliest, do not show indications of small eggs (2) until a mean
length of 7·6 mm is reached, by which time over 50% of the animals have their
tail coelom full of testicular products and show thickenings indicative of the
emergent seminal vesicles. By the time a mean length of 8·4 mm is reached
animals are not again found without ovaries containing eggs (at least in this
sample-size of 30), with empty tail coeloms, or without some sign of seminal
vesicle production.

Fig. 1 offers no indication of the time scale involved in the attainment of
maturity. This is provided by Fig. 2 which plots the mean change in length of
three individual animals on 17 out of the 23 days of the experiment, at a
temperature of 21°C. Unlike the most abundant representatives of the net
plankton (Copepoda), growth in *Sagitta* does not proceed through a serie
of moults, but is a relatively continuous process. Another point to be noted
from Figs 1 and 2 is that by the time animals reach a size at which an appreci
able number of large eggs are visible in the ovaries, their growth rate is
beginning to fall off and rapidly reaches a plateau thereafter between 8·5 and
9·5 mm.

The effect of a wide range of temperatures (12-33°C) on the growth and
survival of *Sagitta hispida* was studied. Fig. 3 contains average values from

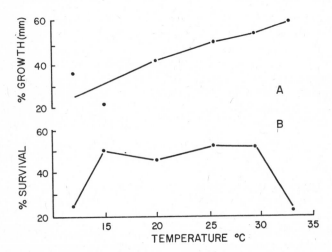

Fig. 3. The percentage increase in length (A) and survival (B) over the
experimental period as a function of temperature.

four series of experiments conducted between September and November when the ambient temperature of the natural environment ranged between 28-22°C. The basic experimental unit for each temperature in a series was a polystyrene box divided into 24 compartments into each of which was placed a single animal in 100 ml of sea water. Animals were pre-adapted to experimental temperatures for at least 24 hours. The experiments ran for 12 days, and animals were transferred into clean boxes with fresh medium and food three times per week. They were measured at the beginning and end of the experiment, and their size range initially was chosen so that their ovaries did not contain large eggs by the end of the period (i.e. they would not have reached the point in Fig. 2 when their growth rate was slowing down appreciably). Facilities were available for only four different temperatures simultaneously, and all the points in Fig. 3 do not represent the mean value for an equal number of initial animals. The lowest number was 24 at 12°C, followed by 48 at 29 and 33°C, the others being 72 and 96.

Curve A in Fig. 3 plots mean length increase over the experiment expressed as a percentage of the initial length. There appears to be an overall trend in which increasing temperature and growth rate are positively related over the entire temperature range. In curve B, plotting percentage survival, the highest and lowest temperatures (12 and 33°C) are clearly not within the preferred range of this species. On the other hand, there is a wide range of temperature (15-29°C) over which survival is independent of temperature.

The effect of food concentration upon growth, survival and maturity was examined by setting up a series of five compartmented boxes containing animals at 21°C. The experiment continued over 15 days and *Artemia* nauplii were added to the compartments of four of the boxes at 10, 20, 50 and 100 nauplii/compartment respectively. The animals in the fifth box were not fed. Animals were fed and/or changed daily except at the weekend when double the amount of food was added. *Artemia* were dispensed by estimating numbers per unit volume in the stock suspension by microscopically counting a sub-sample, and diluting accordingly to obtain the different concentrations required in the four boxes.

The results of this experiment are shown in Fig. 4. The percentage change in length over 15 days (curve C) bore a linear relationship with food concentration from 0 to 20 nauplii/day, animals with no food having shrunk in length by about 10% over the period. From 20 to 100/day there was little further growth rate increase. Curve B, the percentage of animals at each food level with some large eggs in their ovaries at the end of the experiment, did not follow this pattern. This figure rose from 10% to 76% between feeding levels of 20 and 100. Survival (curve A) rate was 32% over 15 days even for the starved animals. It appeared to have increased to 70% at 50 and decreased again to 53% at 100 nauplii/day.

Egg production

Some statistics of egg production were obtained by selecting animals for daily observation which contained some large eggs in their ovary and placing them

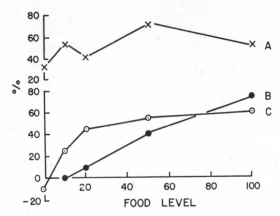

Fig. 4. The percentage survival (A), number of animals with mature eggs (B) and length increase over the experimental period (C) as a function of feeding level.

individually in dishes containing 50 ml of medium. Each day, animals were changed into fresh medium with new food, and at the same time their total length and ovary length were measured, the stages of the three maturation parameters (Table 1) were noted, and any eggs produced were counted. A series of 35 animals was thus observed daily until they died, although no more than 12 individuals were involved at any one time. The temperature was 21°C. A total of 14, 16, 3, and 2 animals lived from 2-5, 6-10 11-15, and 16-19 days respectively, producing from 6 to 420 eggs. Any animals which did not live to produce any eggs are not included in the original 35. In Fig. 5 all of the observations made for three of these animals have been recorded. They were chosen on the basis of being the three best egg producers (420, 186, 186), though not necessarily the longest lived. Curves A and B are the ovary and total length. Curves C, D, and E represent ovary, sperm and seminal vesicle stages respectively. A point has sometimes been plotted intermediate be-

TABLE 1. Descriptions of maturity stages for ovary, testis and seminal vesicles

	Ovary	Testis	Seminal vesicles
0.	Juvenile. Ovaries absent.	Complete absence of any developing sperm in tail coelom.	Not developed.
1.	Stage I. Ovaries very short, narrow, no eggs.	A few sperm balls free in the tail coelom.	Thickenings on tail body-wall.
2.	Stage II. Ovaries short, wider, with small eggs.	Tail coelom at least 50% occupied by testicular products.	Small, but distinctly hollow; empty.
3.	Stage III. Ovaries longer, with eggs of various diameters.	Tail coelom packed full with testicular products.	Large, with some spermatophores.
4.	Stage IV. Ovaries elongate, all eggs large.		Large, entirely filled.

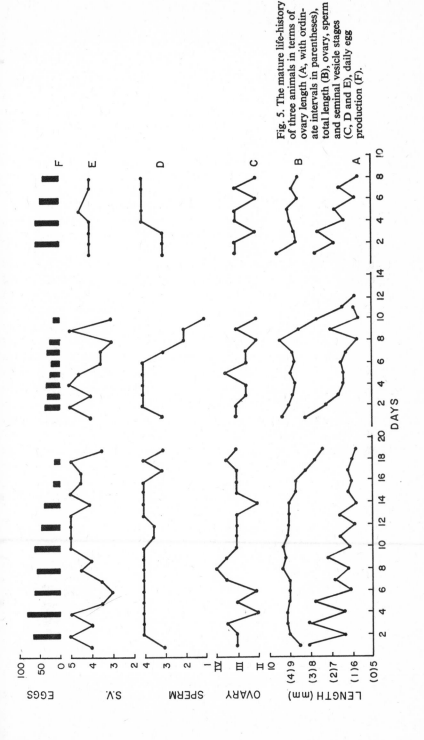

Fig. 5. The mature life-history of three animals in terms of ovary length (A, with ordinate intervals in parentheses), total length (B), ovary, sperm and seminal vesicle stages (C, D and E), daily egg production (F).

tween two stages because it was originally recorded in that manner. Histogram F refers to daily egg production.

Animal number one, which was the longest lived, produced 9 separate batches of eggs which were laid every second day. This pattern was shared by no. 3, while no. 2 produced smaller batches every day for 7 days. In all three, the number of eggs produced in each batch gradually diminished. This decline is mirrored in the overall decrease in ovary length (curve A). Superimposed on this is a day-to-day fluctuation in ovary length in animals no. 1 and 3 which is exactly related to egg batch production. On days when eggs were found in the dish, the ovaries had decreased in length from their previous value, the greatest reduction being almost $\frac{2}{3}$ on the basis of length on the day before and the day after. This fluctuation in ovary length is somewhat less clearly followed in curve C (ovary stage). On some days when the ovaries had contracted and eggs had been laid, stage II (all small eggs) was recorded. On one occasion only, however, was stage IV (all large eggs) seen. On most occasions, some condition between these extremes occurred when small and large eggs were present in variable proportions. It must be noted that in these three animals and in all the others, ovaries having once contained eggs never became completely empty. The same could be said of the tail coelom and its spermatogenic products (curve D) which rarely fell below stage 3 (moderately full) except on certain occasions such as exampled by animal no. 2. Death in this animal was preceded by progressive atrophy of the tail, cessation of spermatogenesis, and consequent rapid decline in total body length. This tail damage (or perhaps degeneration or disease) could be seen in animals of all sizes though it was certainly not an invariable precursor to death. When manifested, however, animals rarely lived for more than three days.

Seminal vesicles usually remained moderate to very full (curve E) and although they sometimes became empty and consequently smaller in size, they were never re-absorbed back into the tail body-wall. Although fluctuating in capacity more than the tail coelom, the data for seminal vesicles do not permit any clear correlation with egg-laying. Only two of the 35 animals increased in length from the first to the last day before they died, and most shrank somewhat. This is illustrated for the three animals of Fig. 5 by curve B.

The relationship between egg production and temperature was investigated by performing experiments exactly similar to those just described at 16, 26, and 31°C, with the exception that the only parameter measured was egg production.

The data were analysed, together with those already obtained at 21°C and are represented in Fig. 6. In most cases, successive batches of eggs were separated by at least one day on which no egg-laying occurred, as in the cases of animals nos. 1 and 3 in Fig. 5. The maximum separation was 9 days. On rare occasions, eggs were laid on more than two consecutive days, as in animal no. 2 of Fig. 5. On a few more occasions, eggs were laid on two consecutive days followed by a day or more of no egg laying. There was often considerable disparity between numbers laid on the two days (e.g. 33, 16; 7,

51). It seemed that counts may have been made on the first day while the animal was in process of laying eggs.

Mean values for 2 parameters are plotted in Fig. 6 for four temperatures. These are (A) total eggs produced per animal and (B) survival in days. Total number of eggs produced per animal (A) rose to a maximum at 26°C, falling at the highest temperature. Survival (B) was inversely related to temperature over the whole range.

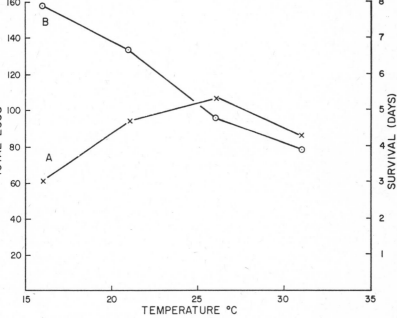

Fig. 6. The effect of temperature on total egg production per animal (A, left-hand scale), and survival (B, right-hand scale).

Efficiency

Gross growth efficiency based on dry weight and total nitrogen was measured in animals at various ages and at three temperatures (16, 21, and 26°C). Dry weighing was done with a Cahn Gram Electrobalance. Organic nitrogen was first converted to ammonia by digesting with 1 ml of concentrated sulphuric acid, with selenium metal powder added as a catalyst. The material to be digested, usually of the order of 1 mg, was flushed into the 30 ml Kjeldahl flask using de-ionized water which had been passed through an exchange column with at least 20 inches of unexhausted granules. After all the water had been driven off, the acid was refluxed for 2 hours. The resulting ammonia was determined according to the method of Newell (1967), modified to handle solutions originally more acid than the desired reaction pH. Spectrophotometric determinations were made at 650 mμ, with a light path of 19 mm or less, depending on the optical density of the solution.

Eight experiments were performed each lasting for 5 days on consecutive weeks. Plankton tows were made and the size range of animals available noted. Animals were isolated in small dishes, and 40-45 of these were selected on the basis of all being within 1·0-1·5 mm of each other in total length. Each animal was re-measured and 20 were randomly selected for the purpose of making initial dry weight and nitrogen determinations. These animals were quickly rinsed in de-ionized water, the excess being removed on absorbent tissue. They were then distributed with forceps onto pre-weighed aluminum foil pans. Up to five pans were used, depending on the size of the animals. They were oven-dried at 70°C for 48 hours and weighed.

The remaining animals (20-25) were placed in individual dishes with lids containing 50 ml of sea water and fed. Food, in the form of *Artemia* nauplii, were hatched out every day so that they were always used within 12-24 hours of hatching. Food was counted out by spotting a plastic tray with drops of medium containing the nauplii which were then counted microscopically. It was arranged that no more than about five nauplii were swimming in each spot of liquid, since above this number counting became unreliable. Nauplii so counted were washed into the dishes containing *Sagitta* and the exact number recorded. Numbers added on the first day were on the basis of information from Reeve (1964b). Thereafter nauplii were added on the basis that they were in excess of double the amount consumed on the previous day. This was averaged out so that approximately equal numbers were added to each dish. On the second day, animals were changed into dishes containing fresh medium and fed again. The old dishes and medium were warmed in an oven to 50°C. The nauplii remaining in these dishes were thus killed and sank to the bottom, which facilitated counting. Any eggs produced the day before were also counted. This daily procedure was repeated up to the fifth day, when the remaining animals were prepared for dry weight and nitrogen determinations.

At a later date dry weight and nitrogen determinations were made on samples of *Artemia* nauplii and *Sagitta* eggs. Both of these were collected on Millipore Corp. "Mitex" 13 mm diam. filters which had been pre-weighed together with control filters. "Mitex" filters do not change weight significantly at 70°C and have a non-stick surface. About 150 nauplii were counted in liquid spots and collected on each filter. *Sagitta* eggs were treated slightly differently because they are not laid singly but in clumps bound together by a mucous secretion. The eggs in a clump were counted and the clump picked up at the tip of a very fine insect pin and scraped directly onto the filter. About 300 eggs were placed on each filter. In the case of both nauplii and eggs, de-ionized water was drawn through the filters to remove salt before placing in the oven.

The results of this series of experiments are tabulated in Tables 2 and 3 and Fig. 7. The experiments are arranged in order of ascending temperature with 2 at 16°C and 3 each at 21 and 26°C. It was intended to utilize a temperature of 31°C also, but survival was so poor at this temperature that it was abandoned. At each of the temperatures, the animals in one experiment were mature and producing eggs. These were experiment nos. 2, 5, and 8. The data

1 Experiment no.	2 Temp. C	3 Total length mm Start	4 Total length mm Finish	5 No. Artemia consumed/animal Total (mean)	6 Per day (mean)	7 Range/day	8 No. eggs produced/animal Total (mean)	9 Per day (mean)	10 Range/day	11 Overall survival %	12 Overall wt. increase %
1	16	6·2	7·2	39·9	8·0	0·22	—	—	—	56	48·4
2	16	7·8	8·7	68·1	13·6	0·33	63·3	12·7	0·100	68	0
3	21	5·3	6·8	62·6	12·5	2·36	—	—	—	65	84·9
4	21	6·6	8·2	85·6	17·1	0·51	—	—	—	65	96·2
5	21	9·2	9·5	63·1	12·6	3·34	51·5	10·3	0·75	60	0
6	26	6·9	7·6	40·2	8·0	0·31	—	—	—	52	1·1
7	26	5·5	7·6	114·6	22·9	0·73	—	—	—	52	80·3
8	26	8·3	8·7	122·7	24·5	0·56	88·0	17·6	0·103	60	0

TABLE 3. Data and steps in the calculation of efficiencies in terms of dry weight and nitrogen with and without egg production taken into account

1 Experiment no.	2 Temp. C	3 Mean Dry wt./animal	4 Mean Nitrogen/animal	5 Change in wt. wt.	6 Change in wt. or N, N	7 Eggs produced wt.	8 Eggs produced N	9 Total food consumed/animal wt.	10 Total food consumed/animal N	11 Efficiency without eggs wt.	12 Efficiency without eggs N	13 Efficiency with eggs wt.	14 Efficiency with eggs N
1A	16	67·8	5·50	32·8	3·70	—	—	70·6	7·70	46·5	48·1	—	—
1B		100·6	9·20	-37·3	1·20	47·35	5·87	120·5	13·15	-31·0	9·1	8·3	53·8
2A	16	227·8	17·15	24·1	3·35	—	—	110·8	12·09	21·8	27·7	—	—
2B		190·5	18·35	75·1	5·75	—	—	151·5	16·53	49·6	34·8	—	—
3A	21	28·4	5·15	-54·0	-0·22	38·52	4·78	111·7	12·19	-48·3	-1·8	-13·9	37·4
3B		52·5	8·50	1·20	-1·65	—	—	71·2	7·77	1·7	-21·2	—	—
4A	21	78·1	8·30	59·0	4·12	—	—	202·8	22·13	29·1	18·6	—	—
4B		153·2	14·05	-57·6	-0·67	65·82	8·16	217·2	23·70	-26·5	-2·8	3·8	31·6
5A	21	302·0	35·78										
5B		248·0	35·56										
6A	26	112·7	13·85										
6B		113·9	12·20										
7A	26	73·5	5·73										
7B		132·5	9·85										
8A	26	252·2	25·60										
8B		194·6	24·93										

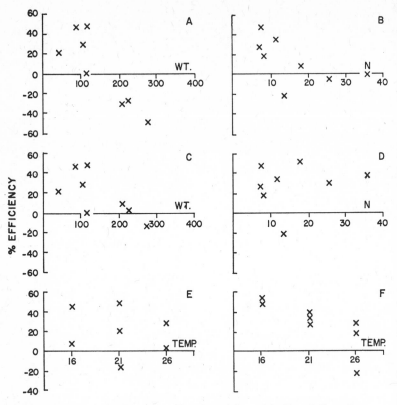

Fig. 7. Gross growth efficiency as a function of dry weight (A and C) total nitrogen (B and D) and temperature (E and F). See text for detailed description.

are separated into two tables largely from space considerations, although that contained in Table 3 is specifically confined to the derivation of gross growth efficiency. Table 2 contains such information as the total number of nauplii consumed over 5 days per animal and the total number of eggs produced (if any), the mean per day and the variation per day. It also lists the change in length survival, and weight increase (if any) over the 5 days. Table 3 contains the mean data for dry weight and nitrogen determinations and the changes in these between the start and finish of each experiment (columns 3-6). Minus signs indicate a decrease over the period of the experiment. Columns 7-10 contain the mean dry weight and nitrogen of any eggs produced and the nauplii consumed per animal during the experiment. These were calculated from mean values for 5 determinations each of dry weight and percentage total nitrogen of dry weight for nauplii and eggs. These figures were 1·77 μg and 10·91 % of nauplii and 0·748 μg and 12·40 % for eggs. In columns 11-14 gross growth efficiency is tabulated on the basis of weight and nitrogen changes of the animal only (11, 12) and on the basis of changes of the animal to which are added any eggs produced (13, 14). Negative numbers in these columns indicate that efficiency is zero and that there was an overall weight or nitrogen loss. In these cases the loss of material by the animal was expressed as a percentage of food consumed as in the case of material gained, but is preceded by a minus sign. In this way the extent of the loss is more clearly indicated.

The data of the last 4 columns of Table 3 are plotted in Fig. 7. In each case, gross growth efficiency constitutes the ordinate axes of the graphs, and includes the hypothetical negative values which were explained in the preceding paragraph. Graphs A and B plot efficiency in terms of weight and nitrogen respectively excluding any eggs produced (Table 3, columns 11 and 12) as a function of mean weight or nitrogen content for the experiment. Graphs C and D contain the same information except that egg production has been included in the three experiments concerned (Table 3, columns 13 and 14). Graphs E and F plot the same data as C and D although as a function of temperature rather than animal size. In graphs A to D, the three points on each, nearest to the right-hand side of the graphs are for experiment nos 2, 5, and 8, which were those in which there were some eggs produced, and which contained, not unnaturally, the largest animals.

Before commenting on these graphs, I wish to except from my remarks experiment no. 6. From reference to Tables 2 and 3 it can be seen that this population, although small in size, was not rapidly growing. There was almost no change in weight or nitrogen over the 5 days. Compared with experiment no. 7, which grew over the same range at the same temperature, it consumed less than half the amount of food of the latter. Though not being dismissed as due to experimental error, its behaviour was not normal when compared to other populations of a similar size range in other experiments.

The relationship between efficiency and animal size is affected very much on the basis of whether or not egg production is taken into account. When not taken into account, efficiency decreases, passing down very far into negative values on a dry weight basis (graph A), and marginally so on a nitrogen basis (graph B). In animals which were not mature, there are no obvious differences between efficiency on a weight and nitrogen basis, the figures ranging between 22-50% and 19-48% respectively excluding experiment no. 6. In egg-producing experiments, the lower end of the ranges were −48 and −21%. Taking into account egg production, efficiency at the higher end of the scale is raised so that on a dry weight basis (graph C) the points fall about zero, but on a nitrogen basis are as high as in younger animals, suggesting that there is no change in gross growth efficiency over the whole range of sizes under consideration. The efficiency ranges observed now become −48 to 50% and 19 to 54% on a dry weight and nitrogen basis respectively.

The effect of temperature on these efficiency values appears to be of little significance on a dry weight basis (graph E). Its effect on a nitrogen basis is clearer, especially if experiment no. 6, the only one to give a negative value, is excluded. Efficiency drops from a mean of 51% at 16°C to 25% at 26°C.

Inspection of Table 2 indicates that feeding rate is also closely correlated with temperature even without taking into account differences in size of populations. Again excluding experiment no. 6, mean daily feeding rate rises from 11 to 24 nauplii between 16 and 26°C.

DISCUSSION

The Chaetognatha are a phylum entirely confined to the marine environment, and, excepting the genus *Spadella*, are considered to be wholly planktonic in habit. Their importance within the plankton is implied by the very large literature available (e.g. Alvariño, 1965) resulting from studies of animals taken in expeditions. In a search of 11 such studies the following information was derived. In every case, copepods were recorded as the most abundant macro-plankton animal group, and in 5 out of 12 cases chaetognaths were next in order of importance. If one excluded the seasonal swarms of great numbers of temporary plankton typical of coastal regions, the figure would rise to 8. When quantity of chaetognaths was expressed as a percentage of the quantity of copepods, values varying very widely between 2-62 occurred when comparing total numbers. Three observations based on volume, all in the North Atlantic were much closer to each other, with a mean of 22%. In the work of Grice and Hart (1962), percentages on the basis of both numbers and volume are available which indicate a very serious underestimation of relative importance using numbers rather than some estimation of biomass. The chaetognaths were 25% by volume rather than 1·6% by number of the copepods. Bsharah (1957), who compared both numbers and dry weight, showed that the percentage was increased in the latter case by a factor of 3. From the data on numbers in the reports cited above the mean percentage of chaetognaths was 16% of that of copepods. On any other more equitable basis the percentage would presumably have to be at the very least doubled.

Many authors have assigned numbers to arbitrarily defined stages of maturity of the gonads and their products in this hermaphrodite phylum. With regard to the ovaries, which develop as paired tubes from the posterior end of the trunk expanding forward, the stages used here correspond essentially to those of Thompson (1947) and Ghirardelli (1961). Since the seminal vesicles also develop as part of the maturation sequence, these have also qualitatively been separated into five stages. In Fig. 1, the initial (zero) stage in all these parameters (unshaded on the histogram) is the juvenile condition where no development is visible using a stereoscopic microscope at a magnification of ×25.

Both Hyman (1959) and Alvariño (1965), in reviewing the literature, considered that chaetognaths are protandric, in that although ovary and testis may begin development at the same time, the tail first becomes full of sperm, and the eggs mature subsequently. This pattern would appear to hold for *Sagitta hispida* although the time difference is very slight. At a mean length of 7·4 mm, 50% of the animals have tail coeloms full of sperm, whereas less than 10% of the ovaries have any large eggs. In the next size group 55% of the ovaries contain some large eggs. Bearing in mind that less than 5% of the seminal vesicles, which are the last organs to develop, have reached a stage where they contain any spermatophores, the male products cannot be considered functionally mature at an earlier stage than the female. This will be referred to again below in reference to fertilization.

Although growth rate was related to temperature, day to day measurements of growing animals were made only at 21°C. In Fig. 2, the length variations are recorded for three animals over a period of 23 days. It is clear that this species has a maximum size of about 9·5 mm under these experimental conditions, and not until this size is approached does egg-laying commence.

The initial size of animals in this figure is about 5·5 mm, which obviously does not permit any definite observations to be made of the time required for one complete generation from hatching to egg production. Nevertheless, the generation time is to be measured in weeks rather than months. It will not be possible to be more precise on this point until the young have been reared from the egg. On the basis of inspection of size frequency histograms derived from seasonal samples from the natural environment, I concluded that the species Reeve, 1964b, 1966) required 3-12 months to complete its life-cycle in the natural environment. It is not clear at this point whether this interpretation was due to inadequacies in the data, or whether for reasons such as lack of food, growth rate was retarded in the natural habitat. If a short generation time did occur in nature, it would fit in with the concept of a "gradient of breeding habit" (Dunbar, 1941) in which the numbers of generations produced appear to increase from one per year near the poles to possibly five in the English Channel (Russell, 1932), to an indeterminate number hroughout the year at low latitudes. This latter situation was suggested by Owre (1960) for several tropical oceanic species off Miami. I found (Reeve, 1964b) less than three months of the year when animals from the bay did not show any ovary development.

From Fig. 3, which indicates that growth rate is directly proportional to temperature, one may postulate the shortest generation times in the summer. This growth rate pattern agrees with the relationship between feeding rate and temperature (Reeve, 1966), as does the fact that 33°C is close to the lethal temperature. Since the temperature in Biscayne Bay regularly approaches and in some shallow areas exceeds this in the height of the summer, it is not surprising that the *Sagitta hispida* population is very severely reduced over this period (Reeve, 1964b). A similarly reduced natural population was also found during the coldest month and may be related to survival at the lowest experimental temperature in Fig. 3, and with the fact that the northern extreme of distribution along the United States Atlantic coast is determined by the influence of the Gulf Stream in coastal waters as far north as Cape Hatteras (Pierce, 1953). Apart from the extreme ranges, however, this species appears to enjoy a very wide degree of temperature independence with regard to its survival (15-29°C).

The feeding rate experiment, the results of which are depicted in Fig. 4, was continued for a period of time such that the animals at the higher feeding levels developed large eggs in their ovaries. It has already been shown above that animals would have reached almost their maximum length by this time. This probably explains why curve C suggests that feeding levels above 20 nauplii per day produced no further growth, whereas curve B shows that the percentage of animals with mature eggs was low at 20 and rose progressively in feeding levels up to 100 nauplii per day. Growth was being manifested largely in terms of egg production. Reeve (1964a) showed that for a given size of *Sagitta* there was a preferred maximum ingestion rate, which for the largest animals was in the region of 100 nauplii per day. In the present experiment this was the maximum number offered, which may have slightly depressed egg development, since animals would have been unlikely to have been able to capture every nauplius. From curve A of Fig. 4 it can be seen that over 30% of the animals were able to survive for two weeks without any food. They shrank in length and showed no tendency to advance in maturity. Although this survival in the absence of food is not lengthy, it may be significant if measured in terms of their total life-span.

The three animals whose reproductive history is detailed in Fig. 5 were those, it must be emphasized, which produced the most eggs. I assume that they are more representative of the natural environment than the animals which produced the average amount and were shorter-lived, because the daily handling in the laboratory must have tended to increase mortality and depress egg production.

Kuhl (1938) assumed that chaetognaths died after egg-laying, although several workers before and since (e.g. Owre, 1960) have seen small animals of a particular species with mature ovaries and larger specimens in an immature stage. Owre suggested that several of the species which she studied in the Florida Current may have produced more than one brood. Alvariño urged caution in respect to these observations which she warned could be the result of confusing two or more populations originating from different geographical areas which matured at different sizes. It is clear that *Sagitta hispida* does lay eggs on more than one occasion in its lifetime. It may be a matter of definition however, as to whether this animal has more than one brood. Most animals laid a succession of "batches" separated by at least one day on which no eggs were laid, though a few deposited some eggs on several consecutive days. The eggs were not all matured and deposited at the same time, but the process was staggered. The period of egg-laying to death could be considered one extended brood. None of the animals lived long enough to indicate that they might pass through a non-egg-laying period followed by another egg-laying period. Since most animals did not produce eggs every day, their ovaries contained differing percentages of small and large eggs on successive days, which had a direct effect on ovary size, and maturity stage. It would, therefore, be possible for animals which showed any egg development in the ovary not to have any relationship between size and egg development, especially since total length itself fluctuated and tended to decrease before death (Fig. 5, curve B). Ovary maturity stages II-IV, therefore, do not provide any real information concerning the degree of maturity of this species unless it were known that the animal had not previously passed through any of these stages. This could only be ascertained by rearing the individual from stage I. This stage, in which the ovaries are tiny and show no sign of egg development, was never reverted to in animals which had previously attained maturity.

Owre (1960), in attempting to relate ovary maturity stages and total length, found that the size ranges of animals with stages II, III, and IV ovaries were very similar. She also showed that the numbers of animals with "fully mature" (stage IV) ovaries were very low. These observations correspond to the experimental results for *Sagitta hispida*, although it is considered here that "fully mature" ovaries can range between II and IV. Only once was a stage IV recorded in Fig. 5, but eggs were being laid regularly despite this.

During the egg-laying period, seminal vesicles rarely became empty (stage 3, curve E), although they fluctuated in capacity. In maturing animals the seminal vesicles did not fill up until a day or two before eggs were laid. On two occasions, populations were brought into the laboratory which had large eggs in their ovaries and seminal vesicles which were no further advanced than stage II (not hollowed out). Within 24-48 hours, most of these animals had laid eggs and had large seminal vesicles containing spermatophores. It must be assumed on the basis of Figs 1 and 5 that these animals were laying eggs for the first time, the sperm and eggs thus becoming available at the same time. If this were the case, it would seem to facilitate self-fertilization in the natural environment. In the laboratory self-fertilizaton must have taken place, since larvae were obtained from eggs which came from individual adults which had been isolated from the juvenile stage. Pairing, presumably to facilitate cross-fertilization, was witnessed by Murakami (1959) in *Sagitta crassa*, although I have not as yet seen it in laboratory populations of *Sagitta hispida*. Alvariño accepted it as the rule. Hyman on the other hand, considered that self-fertilization was normal in the genus *Sagitta*, basing the statement on two earlier observations. At present, there is so little information in the literature on this point that no general conclusions can be made.

Death, in many of the animals which were studied for their egg-production ability, occurred soon after the initiation of a progressive degeneration of the tail (e.g. animal no. 2, Fig. 5). This phenomenon was not in evidence in natural populations. Apart from this, however, there were signs that egg production fell, ovaries tended to become smaller, and shrinkage in overall length occurred before death. The high death rate of mature animals is reflected in Fig. 6, where the mean total eggs produced per animal did not exceed 100. As noted above, it is probable that maximum rather than average figures are more indicative of the natural environment. Furthermore, there was no certainty that animals had not already laid some eggs before capture. So the data of Fig. 6 are perhaps only to be considered in relation to temperature rather than as absolute values. Between temperatures of 16-31°C, survival was inversely related to temperature, whereas over most of this range the survival of growing animals was not affected (see above). Perhaps the survival time is a constant in metabolic terms at the onset of maturity, and higher temperatures accelerate the processes towards death. The total number of eggs produced was directly proportional to temperature except at the highest temperature, when the survival was presumably too short to result in further increase.

The experiments on egg production indicate that there is much to be discovered concerning the reproductive behaviour of chaetognaths before any estimation of their overall production can be derived. This behaviour is far more complex than a description of preserved material can reveal.

Gross growth efficiency was determined in terms of both dry weight and total organic nitrogen. The latter was used by Corner et al. (1965, 1967) in one of the few other studies of gross growth efficiency in marine zooplankton.

It has been noted already that animals tend to shrink during egg-laying. This is confirmed clearly in terms of dry weight in graph A, of Fig. 7. In graph C, where the weight of the eggs produced was taken into account, there was virtually no overall weight change, and the gross growth efficiency was approximately zero. In these, and graphs B and D for nitrogen efficiency without and with eggs, the three populations largest in size were those which produced eggs (i.e. the three points nearest the right-hand side of each graph). There appears, however, to be no overall loss of nitrogen in egg-bearing populations even when the eggs are not taken into consideration (graph B). When egg-laying is expressed as a form of growth (graph D), nitrogen gross growth efficiencies remain as high during this period as for the preceding rapidly-growing but immature population. The immature populations, whether on a basis of dry weight or nitrogen, exhibited gross growth efficiencies over similar ranges with a combined range of 19-50% yielding a mean value of 34·5%. This mean value was calculated from observations to be found in columns 11 and 12 of Table 3. Of three observations for efficiency in terms of dry weight for egg producers, only two have positive values, the highest being about 8%. On the basis of nitrogen, however, the three values for efficiency of egg producers yield a mean figure of 40·9%. Taking the overall mean for efficiency in terms of nitrogen for the seven experiments under consideration (graph D) a figure of 36·0% may be derived over the entire size range studied. Since dry weight and nitrogen figures are in agreement for experiments in which populations were immature, it would seem likely that their divergence in egg-producing populations is not an artifact. It is possible that lipid reserves are accumulated during growth and used extensively during the period of egg-laying, which might explain the overall loss of weight of mature animals. Since a nitrogen supply is necessary for the production of eggs, nitrogen being assimilated from the food during the mature phase is appearing in the eggs rather than being utilized for maintenance metabolism. Thus in terms of nitrogen incorporated into new tissue (gross growth efficiency) values remain high throughout life. Conover (1967) noted that the marine copepod, Calanus hyperboreus, "completes its life cycle using stored lipids as a carbon source [even though] nitrogen is required to form the protein for reproductive products".

Graph F indicates that in considering nitrogen rather than dry weight an interesting temperature relationship is yielded in which gross growth efficiency is inversely related to increasing temperature. This relationship appears little affected by the varying mean size ranges of the populations, and the highest values for each temperature were obtained in the largest (egg-laying) populations. It has been shown above that both growth rate and egg production increase over this temperature range, so that here is a situation where the rate of production of new biomass is inversely related to efficiency. This was also shown (Reeve, 1963) for *Artemia* feeding on algal suspensions, though in respect of feeding level rather than temperature.

Gross growth efficiency determinations may be found for a variety of invertebrates (see Reeve, 1963) but very few of these are marine zooplankton. Conover (1964), working with the copepod *Calanus hyperboreus*, obtained values for dry weight and calories of 4-36 and 5-50% respectively with means of 21 and 30%. Corner *et al.* (1965, 1967) working with the same genus (*C. finmarchicus*), quoted values of 36 and 34% on the basis of total nitrogen. Lasker (1966) on the basis of carbon determinations on a euphausid crustacean, derived a mean value of 30% over the whole lifetime of the animals including eggs and moults. Mullin and Brooks (in this volume) reported values of 30-45% for juvenile marine copepods also in terms of carbon. It is significant that there is a surprisingly close agreement between all these mean values (including those derived in this paper) despite their range of units. All fall between 30 and 45%. My mean value for nitrogen (36%) may be directly compared with those of Corner (36, 34%). It is dangerous and probably unwise to try to interpret these laboratory values in terms of yields from one food chain level to the next. On a relative basis however, it is interesting to note that the typically herbivorous calanoid copepods have closely similar growth efficiencies to the typically carnivorous chaetognaths.

Finally, it may be noted that if my estimation of the biomass of chaetognaths being about 30% of that of copepods in the oceans is accepted, and that this ratio is also proportional to their relative production capacities, chaetognaths must form a very significant intermediate link in the food chain whereby the majority of energy converted into animal material by copepods must be distributed to higher trophic levels via chaetognaths. If copepods are the primary herbivores of the marine environment, chaetognaths may be considered as primary carnivores. Furthermore, if the two groups can approach the conversion efficiencies in natural populations which they demonstrate in the laboratory the planktonic community as a whole must be very conservative of the energy being passed up the food chain.

ACKNOWLEDGEMENTS

The financial assistance of the National Science Foundation (Grant BG-6232, I.M.S. no. N8920) is gratefully acknowledged. This work represents Contribution no. 1113 of the Institute of Marine and Atmospheric Sciences of the University of Miami.

REFERENCES

ALVARIÑO, A. 1965. Chaetognaths. *Oceanogr. mar. Biol.*, **3**, 115-94.

BEERS, J. R. 1964. Ammonia and inorganic phosphorus excretion by the planktonic chaetognath *Sagitta hispida* Conant. *J. Cons. perm. int. Explor. Mer*, **29**, 123-9.

BSHARAH, L. 1957. Plankton of the Florida Current. V. Environmental conditions standing crop, seasonal and diurnal changes at a station forty miles east of Miami *Bull. mar. Sci. Gulf Caribb.*, **7**, 201-51.

BURFIELD, S. T. 1927. *Sagitta. L.M.B.C. Mem. typ. Br. mar. Pl. Anim.*, (28), 104 pp.

CONOVER, R. J. 1964. Food relations and nutrition of zooplankton. *Symp. exp. mar. Ecol. Occ. Publ. Narragansett mar. Lab.*, *R.I.*, (2), 81-91.

CONOVER, R. J. 1967. Reproductive cycle, early development, and fecundity in laboratory populations of the copepod *Calanus hyperboreus. Crustaceana*, **13**, 61-72.

CORNER, E. D. S., Cowey, C. B., and Marshall, S. M. 1965. On the nutrition and metabolism of zooplankton. III. Nitrogen excretion by *Calanus. J. mar. biol. Ass. U.K.* **45**, 429-42.

CORNER, E. D. S., Cowey, C. B., and MARSHALL, S. M. 1967. On the nutrition and metabolism of zooplankton. V. Feeding efficiency of *Calanus finmarchicus. J. mar. biol. Ass. U.K.*, **47**, 259-70.

CURL, H. 1962. Standing crops of carbon, nitrogen and phosphorus and transfer between trophic levels, in continental shelf waters south of New York. *Rapp. P.-v. Réun. Commn int. Explor. scient. Mer. Méditerr.*, **153**, 183-9.

DAVID, P. M. 1955. Distribution of *Sagitta gazellae.* "*Discovery*" *Rep.*, **27**, 235-78.

DUNBAR, M. J. 1941. The breeding cycle in *Sagitta elegans arctica* Aurivillius. *Can. J. Res.*, *Ser. D*, **19**, 258-66.

FURNESTIN, M. 1953. Contribution à l'étude morphologique, biologique et systématique de *Sagitta serrodentata* Frohn des eaux Atlantiques du Maroc. *Bull. Inst. océanogr. Monaco*, (1025), 39 pp.

GHIRARDELLI, E. 1961. Istologia e citologia degli stadi di maturita nei chetognati. *Boll. Pesca Piscic. Idrobiol.*, **15**, 5-19.

GRICE, G. D., and HART, A. D. 1962. The abundance, seasonal occurrence and distribution of the epi-zooplankton between New York and Bermuda. *Ecol. Monogr.*, **32**, 287-307.

HORRIDGE, A. 1966. Some recently discovered underwater vibration receptors in invertebrates. In *Some contemporary studies in marine science*. Edited by H. Barnes. London, Allen and Unwin Ltd, 395-406.

HORRIDGE, G. A., and BOULTON, P. S. 1967. Prey detection by Chaetognatha via a vibration sense. *Proc. R. Soc., Ser. B*, **168**, 413-19.

HYMAN, L. H. 1958. The occurrence of chitin in lophophorate phyla. *Biol. Bull. mar. biol. Lab.*, *Woods Hole*, **114**, 106-12.

HYMAN, L. H. 1959. *The invertebrates*, **5**. New York, McGraw-Hill, Inc., 783 pp.

KUHL, W. 1938. Chaetognatha. *Bronn's Kl. Ordn. Tierreichs*, **4**, (2), Buch 2, T. 1, 226 pp.

LASKER, R. 1966. Feeding, growth, respiration, and carbon utilization of a euphausid crustacean. *J. Fish. Res. Bd Can.*, **23**, 1291-1317.

LEBOUR, M. V. 1922. The food of plankton organisms. *J. mar. biol. Ass. U.K.*, **12**, 644-77.

MURAKAMI, A. 1959. Marine biological study on the planktonic chaetognaths in the Seto Inland Sea. *Bull. Naikai reg. Fish. Res. Lab.*, **12**, 1-186.

MURAKAMI, A. 1966. Rearing experiments of a Chaetognatha, *Sagitta crassa. Inf. Bull. Planktol. Japan*, **13**, 62-65.

NEWELL, B. S. 1967. The determination of ammonia in sea water. *J. mar. biol. Ass. U.K.*, **47**, 271-80.

NICOL, J. A. 1967. *The biology of marine animals*. 2nd edition. New York, John Wiley & Sons, Inc., 707 pp.

OWRE, H. B. 1960. Plankton of the Florida Current. VI. The Chaetognatha. *Bull. mar. Sci. Gulf Caribb.*, **10**, 252-322.

PARRY, D. A. 1944. Structure and function of the gut in *Spadella cephaloptera* and *Sagitta setosa. J. mar. biol. Ass. U.K.*, **26**, 16-36.

PIERCE, E. L. 1953. The Chaetognatha over the continental shelf of North Carolina with attention to their relation to the hydrography of the area. *J. mar. Res.*, **12**, 75-92.

REEVE, M. R. 1963. Growth efficiency in *Artemia* under laboratory conditions. *Biol. Bull. mar. biol. Lab.*, *Woods Hole*, **125**, 133-45.

REEVE, M. R. 1964a. Studies on the seasonal variation of the zooplankton in a marine sub-tropical in-shore environment. *Bull. mar. Sci. Gulf Caribb.*, **14**, 103-22.

REEVE, M. R. 1964b. Feeding of zooplankton, with special reference to some experiments with *Sagitta*. *Nature, Lond.*, **201**, 211-13.

REEVE, M. R. 1966. Observations of the biology of a chaetognath. In *Some contemporary studies in marine science*. Edited by H. Barnes. London, Allen and Unwin Ltd., 613-30.

RUSSELL, F. S. 1932. On the biology of *Sagitta*. II. The breeding and growth of *Sagitta elegans* Verrill in the Plymouth area, 1930-31. *J. mar. biol. Ass. U.K.*, **18**, 131-46.

THOMPSON, J. M. 1947. The Chaetognatha of south-eastern Australia. *Bull. Coun. scient. ind. Res., Melb.*, **222**, 1-43.

TOKIOKA, T. 1955. Notes on some chaetognaths from the Gulf of Mexico. *Bull. mar. Sci. Gulf Caribb.*, **5**, 52-64.

WIMPENNY, R. 1936. Distribution, breeding and feeding of some important plankton organisms of the southwest North Sea. *Fishery Invest., Lond.* Ser. 2, **15**, (3), 56 pp.

Part Three

FEEDING
MECHANISMS

Introduction

C. Barker Jørgensen

A complete appreciation of the transfer of energy through the food chains and food webs of any biotope presupposes a detailed knowledge of the physiology of feeding and digestion in the animals involved. The quantitative analysis of marine food chains is often hampered by incomplete knowledge of the feeding mechanisms, of the kinds and amounts of food eaten, of the factors which control food intake, and of the digestibility and assimilation of ingested food. Examples may be found in the papers of this Symposium.

The papers grouped in the present section have in common that they emphasize ecological aspects of feeding. Studies on feeding are, however, covering a wide range of organisms.

The methods of study of feeding vary with the feeding type, especially whether the animals in question are macrophagous or microphagous feeders.

In macrophagous feeders much information on feeding and food can be obtained from analysis and measurements of the contents of the digestive tract combined with analysis of potential food items in the habitat, and of other environmental factors. Such techniques have been adopted by Allen Keast in his extensive studies on feeding in populations of cohabiting fishes in small lakes (this volume).

Microphagous feeders are often characterized by poor, or even lacking, selection of the material ingested. Analyses of the contents of the digestive tract are therefore often of restricted value in determining the diet of the species. However, some microphagous feeders, especially amongst deposit feeders, do show some selectivity in the ingestion of material. The mullet is an example. It ingests deposited detritus and benthic microalgae, and selects preferentially material of particle size smaller than about 30 μ. William E.

193

Odum has successfully developed the technique of analysis of the content of the various parts of the digestive tract to determine the diet of the mullet in various habitats. He has, too, exposed the difficulties inherent in the method when applied to a microphagous animal (this volume).

In non-selective microphagous suspension feeders indirect methods have been adopted in determining the diets. These methods include measurement of the rate at which the surrounding water is transported through the feeding organs and the efficiency with which these organs retain the various types of suspended matter in the water, including potential food sources as phyto plankton, detritus, bacteria, and organic matter in solution. The suspension feeders are especially suited for such quantitative studies on feeding because of their generally highly automatous and stereotyped mechanisms of feeding. Jürgen Winter has studied rates of filtration and ingestion in the suspension feeding bivalves *Arctica islandica* and *Modiolus modiolus*, in animals of various body size, and at different temperatures and concentrations of sus pended algal cells.

In the sea suspended food is usually present in low concentrations and food appears often to be a limiting factor. Calculations on the amounts of phytoplankton that could be obtained by suspension feeders have often sug gested that these amounts were insufficient to meet the food requirements of the animals. In some instances this discrepancy between needed and cal culated ingested amounts of phytoplankton seems to have arisen from too low estimates of the amount of water transported through, and thus of phytoplankton retained in, the feeding organs. In other instances, however the discrepancy still persists, for instance, between the rates at which zoo plankton crustaceans filter the surrounding water and the phytoplankton concentrations in the water, especially in off shore and oceanic regions. The discrepancy may be due to the fact that no zooplankton crustacean feeds exclusively as a filter feeder. Even calanoid copepods and euphausians with feeding organs that are typically constructed for filter feeding can also feed on larger prey. It seems that filter feeding may be adopted only at higher con centrations of suspended food (see T. R. Parsons and R. J. LeBrasseur this volume).

However, there still appears to be a need for experiments on feeding in un disturbed zooplankton suspension feeders.

Suspension feeding in invertebrates is generally characterized by the highly retensive food collecting structures, which often efficiently retain particles down to about one or a few microns in diameter, in some groups even less than one micron. Suspension feeding vertebrates usually possess more porous filters, especially the larger forms. Thus, in baleen whales suspended material smaller in size than that of the zooplankton freely pass the filters, which thus only retain a well defined fraction of the organic matter suspended in the water. Moreover, this fraction is of well-established food value. Analyses and measurements of stomach contents and of the zooplankton in the sur rounding waters are therefore important in these suspension feeders, and the technique has been used extensively by Takahisha Nemoto in his studies on the types of feeding and on quantitative aspects of feeding in the baleen whales (this volume).

Knowledge of the structure and function of the feeding organs not only helps in defining the place of the animals in the food webs of which they are members. It is also needed in the descriptions of animal communities in terms of feeding types. Our knowledge of feeding mechanisms, though still incomplete in many animals, is, however, sufficiently detailed to allow estimates to be made of the distribution of the animals according to feeding types in certain biotopes. V. I. Zatsepin has estimated the relative importance of a number of feeding types in bottom communities of the northern seas, and he has correlated the feeding pattern of the communities with environmental factors, such as depth, types of bottom sediments, and water currents.

Filter feeding and food utilization in Arctica islandica L. and Modiolus modiolus L. at different food concentrations

JÜRGEN WINTER
Institut für Meeresforschung
Bremerhaven, Germany

ABSTRACT. Filtration rates and the extent of phagocytosed food particles were determined in the offshore lamellibranchs *Arctica islandica* and *Modiolus modiolus*, in relation to body size, temperature, and offered particle concentration. Pure cultures of *Chlamydomonas* and *Dunaliella* were used as food.

The filtration rates were determined photometrically by the decrease of food density. Since the algae taken in by the mussels were continuously replaced, the concentration was kept constant. Therefore, filtration rates could be determined in undisturbed animals.

The phagocytosed food was measured by the biuret method as algal cells ingested minus algal cells in faeces. This method could be applied, because the mucus produced by the mussels gives a negative reaction, the micro-organisms are transported only by ciliary activity passing the gut alive and without any mechanical damage, and because of the poor development of extracellular digestion.

The filtration rates vary continuously. As a rule during a period of 24 hours two phases of high food consumption alternate with two phases of nearly exclusive food digestion.

A significant correlation exists between the body size and the filtration rate. The bigger the individuals the higher is the rate of filtration and the extent of phagocytosed food (on an average: 76% of the algae ingested).

The filtration rates and the extent of phagocytosed food in adapted mussels increase in significant relations at increasing temperatures (4°, 12° and 20°C).

The density of micro-organisms in the range of natural concentrations influences the filtration rate and the grade of food utilization in significant correlations. There are rather definite concentrations above which the density begins to interfere with the feeding process. Only in concentrations being below the threshold of pseudo-faeces production an accurate correlation is to be expected. At different temperatures the range of concentrations being favourable to the filtration activity and to the extent of food utilization was determined.

INTRODUCTION

In suspension feeding bivalves the gill-filter may be compared with a sticky sieve or, better, with a glue-twig mechanism (Tammes and Dral, 1955; Dral, 1967) able to remove very fine particles from the inhalant water. The pumping rate is the amount of water pumped per unit of time (litres/hour). This has been determined by a "direct method" by isolating the exhalant stream of water or by measuring the velocity of the in- or exhalant stream. The filtration rate is the amount of water filtered free of particles per unit of time (litres/hour). This was determined by an "indirect method" as a function of decreasing particle concentration. The decrease of particle concentration was measured photometrically or by the decrease of radioactivity, when labelled (^{14}C, ^{32}P) algae were used. In connection with the indirect method the classical work of Fox, Sverdrup and Cunningham (1937) as well as the works of Jørgensen and his school (1943, 1949, 1952, 1953, 1960), Willemsen (1952), Rao (1953, 1954), Chipman and Hopkins (1954), Ballantine and Morton (1956), Rice and Smith (1958), Kuenzler (1961), Allen (1962), Theede (1963), Durve (1963) and Mikheyev and Sorokin (1966) must be mentioned.

When using the indirect method one has to face the fact that the particle concentration of the experimental medium is high at the beginning and low at the end of the experiment. Furthermore, the experiments cannot be continued over longer periods without any interruptions. I therefore modified the classical indirect method and kept the concentration of the experimental medium constant to $\pm 1\%$. When the bivalves removed particles from the medium, new particles were added and by the amount of particles added per unit of time I calculated the filtration rate in my experiments.*

EXPERIMENTAL ARRANGEMENT

I worked with *Arctica islandica* and *Modiolus modiolus* from various localities around Heligoland from depths between 25 and 50 m. Pure cultures of *Chlamydomonas* sp. and *Dunaliella* sp. in the exponential phase of population growth were used to prepare the experimental medium. Before the algal cultures were used in my experiments they were centrifuged from the culture medium, washed twice with a small amount of sea water by centrifuging and then resuspended in filtered sea water. The concentrations of the concentrated food cultures were determined with a counting chamber.

During the experiments the bivalves were kept in a 5 l glass aquarium (Fig. 1) which was connected by tubes and a pump with a flow colorimeter (Langrohr-Kolorimeter 400, Lange, Berlin; length of the cuvette: 52 cm). According to the concentration of the medium the light source was regulated with a potentiometer; thus measurements were made only within the most

* The research work was made possible by a research grant from Deutsche Forschungsgemeinschaft.

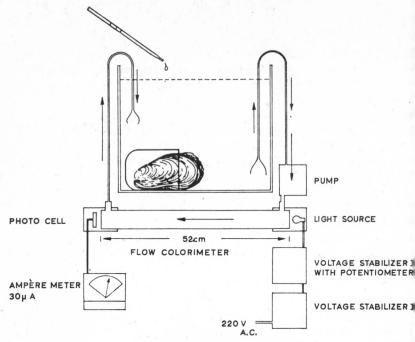

PHOTO CELL

PUMP

LIGHT SOURCE

52cm

FLOW COLORIMETER

VOLTAGE STABILIZER II
WITH POTENTIOMETER

AMPÈRE METER
30µ A

VOLTAGE STABILIZER I

220 V
A.C.

Fig. 1. Experimental arrangement.

sensitive range of the meter. Whenever the meter indicated the slightest
decrease in concentration, new concentrated culture was added with a graded
pipette. Time and quantity of the culture added were recorded for the calculation of the filtration rate. To simulate the natural environment the
animals were kept in darkness. Since living algal cells kept in the light would
multiply, they were kept in darkness, too. By means of counts it was proved
that the concentration was constant over a period of 48 hours. Since the light
source was close to the incoming stream of the flow colorimeter and the algal
cells were not able to counteract the velocity of the current in the cuvette,
there was no accumulation near the light source.

The extent of food utilization was determined by Kuenzler (1961) in
Modiolus demissus, by Allen (1962) in *Mya arenaria* and *Venus striatula* and
by Mikheyev and Sorokin (1966) in *Dreissena polymorpha*, using labelled
(^{14}C, ^{32}P) algae.

In my experiments food utilization was measured by determining the
protein content of the ingested algae minus the protein content of faeces.
The protein content was analysed with the biuret method and expressed in the
form of albumin equivalents. The biuret method was developed by Krey
(Krey, 1951; Krey *et al.*, 1957; Boje, 1965, 1966) to determine the protein
content in small amounts of plankton. For the determination of the percentage of phagocytosed algae this method is usable, because (*i*) the mucus produced by the mussels and contained in the faeces gives a negative biuret reaction, (*ii*) the algae are transported only by ciliary activity (a typical situa-

tion in suspension feeding bivalves), so that they pass through the gut without any mechanical damage, and (*iii*) extracellular digestion is, if at all, poorly efficient. Due to these facts the protein content of a definite number of algal cells in the food culture is equal to that of faeces containing the same number of algal cells. This result was not unexpected, since the biuret method gives a positive reaction only with larger protein molecules like those found in living organisms but not with fragments of digested protein (Krey, 1951). The amount of protein detected in the faeces was therefore assumed to be localized in the algae and not in the waste of the process of phagocytosis. These circumstances made it possible to calculate accurately the numbers of cells that were phagocytosed per unit of time.

In order to prepare the experimental medium at the desired food concentration a certain amount of highly concentrated culture was added to 5 l of filtered sea water. A corresponding amount of the concentrated culture was filtered through a paper filter (Schleicher and Schüll, no. 1575, ϕ 4·5 cm) to determine the dry weight of the algal material suspended in the experimental medium (abscissa in Fig. 3B). A corresponding amount of food culture was centrifuged and subjected to the procedure of the biuret method to determine the protein content. The total amount of algae filtered (regarding the numbers of cells or the dry weight or the protein content) was calculated by means of the filtration rate. In the case of pseudo-faeces production the pseudo-faeces were treated separately. The number or the dry weight of algal cells contained in faeces was calculated by comparing the protein content in faeces with that of the food culture used in the experiments.

During the experimental time a large amount of faeces was siphoned out with a pipette. Since the specific weight of the faeces is a little higher than the surrounding medium, the faeces settle down at the bottom of the aquarium. Consequently, when the experimental time was over, the experimental medium could be reduced to 2 l with the outflow tube of the flow colorimeter; the meter indicated that no faeces were contained in the 3 l of the experimental medium, which was discarded. After the reduction of the experimental medium the mussels were transferred into a second aquarium with filtered sea water, where they remained for a second period of 24 hours to produce their faeces. A large amount of the filtered sea water was afterwards siphoned out by a very narrow tube, so that no faeces were stirred up from the bottom. After this reduction the animals were cleaned and put back into larger vessels. The total amount of the reduced experimental medium (2 l) and of the filtered sea water including the faeces was centrifuged, so that the faeces were collected quantitatively. A great advantage of this method is that the experimental medium was kept constant over the experimental time, so that it was not necessary to wash the faeces and pseudofaeces: the amount of algal cells (suspended in the experimental medium), that were siphoned out together with the faeces, and that were contained in the 2 l of the reduced experimental medium were afterwards subtracted in the calculation. This is possibly the only way to work quantitatively, because washing of faeces is connected with a high loss of algal cells, which may be entangled in the mucous mass only in a very loose way. Since the animals were kept without food for 24 hours before starting the experiment, the postfeeding period of 24 hours was necessary to obtain the total amount of faeces that belonged to the amount of algae ingested during the feeding period.

Though during the experiments all factors (darkness, temperature, oxygen, algal concentration) of the medium were kept constant, the filtration rate varies from hour to hour (Fig. 2). In order to determine an average value, a standard time of 24 hours for all experiments was chosen. In spite of the great variations, the average values for an experimental time of 24 hours were

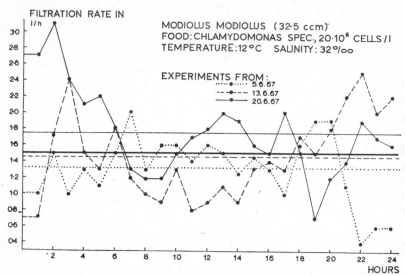

Fig. 2. Parallel experiments to demonstrate the variations in the rate of filtration in *Modiolus modiolus* over a period of 24 hours.

very similar. When there was good agreement between average values, the number of experiments was restricted to three, otherwise it was necessary to carry out up to six experiments. As a rule during such a period of 24 hours two phases of high food consumption alternate with two phases when food digestion prevails.

DISCUSSION

Experiments regarding the relation to body size and temperature are given by Winter (1969). In this paper I shall only discuss some experiments concerning influence of food concentration upon the filtration rate and the degree of food utilization.

The filtration rate with different particle concentrations was investigated by Jørgensen (1952, 1960) in *Mytilus edulis*, by Chipman and Hopkins (1954) in *Pecten irradians*, by Ballantine and Morton (1956) in *Lasaea rubra*, by Rice and Smith (1958, fed with *Nannochloris*) in *Venus mercenaria*, and by Kuenzler (1961) in *Modiolus demissus*, but all these authors could not detect

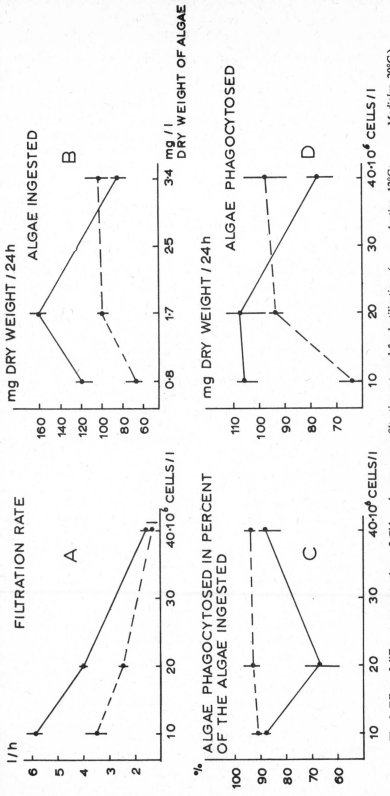

Fig. 3. Effect of different concentrations of *Chlamydomonas* upon filtration rate and food utilization. (——— *Arctica*, 12°C; ------- *Modiolus*, 20°C.).

any correlations between the filtration rate and the offered particle concentration. Only the results by Rice and Smith (1958) in *Venus mercenaria* fed with *Nitzschia* showed a slight correlation. But these experiments are not significant, because during the first 10 minutes after starting an experiment the highest concentration was reduced to about 45%. The method used in all these experiments was not suited to find positive results.

In my experiments the filtration rate and the extent of food utilization were determined in *Arctica islandica* (total volume 74 cm^3) and *Modiolus modiolus* (32·5 cm^3) of medium size with different food concentrations which resemble seston concentration of the North Sea: 10×10^6, 20×10^6 and 40×10^6 cells/1, corresponding to 0·8 mg, 1·7 mg and 3·4 mg/1 algal dry weight. It was not the idea to imitate the phytoplankton concentration of the North Sea; the natural phytoplankton concentration of the North Sea is very small and an imitation of such a small fraction of seston would result in a more artificial situation than is given in my experiments, where the total amount of natural seston was replaced quantitatively by pure cultures of one species of uniform size (*Chlamydomonas* sp. 7·5 $\mu \times 10$ μ and *Dunaliella* sp. 5 $\mu \times 7·5$ μ). As consequence of the relatively small protein content of the algal cells, the protein contents of the offered food concentrations resemble that of natural seston, too. Since the filter feeding mechanism probably does not give significant responses to different qualities of "particles" (excluding particle size) the cultures used in my experiments were not thought to be very unnatural. In this connection an important finding of Rice and Smith (1958) was that the rate of filtration by *Venus mercenaria* of a labelled natural plankton suspension was about the same as the rate for unialgal suspensions. The process of phagocytosis, of course, may depend to a very high degree on the nature of particles involved.

Fig. 3A shows that the filtration rates increase with decreasing concentrations. At the experimental temperature of 12°C *Arctica* filters out the highest amount of algae at medium concentrations (Fig. 3B). Fig. 3C shows the percentage of phagocytosed algae in relation to the total amount of filtered algae. The small amount of algae ingested at low concentrations is utilized to a very high percentage (88%), and therefore the amount of phagocytosed algae in mg dry weight (Fig. 3D) is nearly the same as at medium concentrations. At high concentrations the amount of phagocytosed algae is reduced.

An explanation of these results can be derived from Fig. 4. Only at high concentrations does a production of pseudofaeces take place when 30% of the total amount of algae filtered are accumulated as pseudofaeces. At medium concentrations the largest amount of algae is ingested (= filtered) and it seems that in this case the filter mechanism, regarding its straining and transporting ciliary activity, is close to its maximal capacity. At *low concentrations* it seems that the filtration rate (or better the amount of water pumped by the bivalves) is close to its maximal capacity. Little faeces (12%) are formed, 106 mg algal dry weight/24 hours are utilized. At *medium concentrations* the filtration rate is reduced, because at this concentration even a reduced filtration rate brings so many algae to the filter mechanism, that its

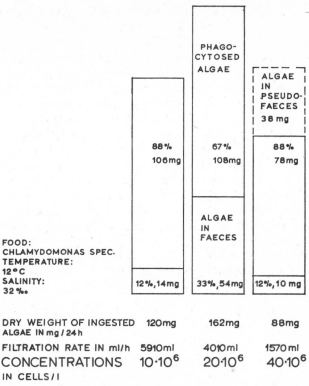

FOOD:
CHLAMYDOMONAS SPEC.
TEMPERATURE:
12°C
SALINITY:
32 ‰

DRY WEIGHT OF INGESTED ALGAE IN mg/24h	120mg	162mg	88mg
FILTRATION RATE IN ml/h	5910ml	4010ml	1570 ml
CONCENTRATIONS IN CELLS/l	$10 \cdot 10^6$	$20 \cdot 10^6$	$40 \cdot 10^6$

Fig. 4. Influence of different concentrations on filter feeding and food utilization in *Arctica islandica* (74 ccm) during 24 hours.

maximal capacity seems to be reached even with medium filtration effort. Regarding the high amount of faeces production (33%) and the relatively low percentage of food utilization (67%), it seems that the maximal capacity of phagocytosis is reached. Since at low and at medium concentrations the same amount of algae is phagocytosed, the assumption of the maximal capacity of the process of phagocytosis is supported.

At different temperatures, filtration rate and food utilization differ. The different temperatures used in my experiments were 4°, 12° and 20°C. The natural temperature range around Heligoland lies between 4° and 16·5°C, so that the experimental temperature of 20°C is very high and does not occur in depths of 25-50 m. The experimental animals were adapted to this high temperature for about one month. Additional experiments at 8° and 16°C need to be done.

Fig. 5 shows that the decrease of temperature from 12°C to 4°C results in a 50% reduction of the filtration rate and of the amount of phagocytosed algae. Filter feeding and phagocytosis are therefore equally reduced. With an increase in temperature from 12°C to 20°C the filtration rate does not change much, but the amount of phagocytosed algae increases from 108 mg at 12°C to 144 mg algal dry weight/24 h at 20°C.

It may be mentioned that *Arctica islandica* at a temperature of 20°C did not produce pseudofaeces at the food concentration of 40×10^6 cells/1, but that pseudofaeces occurred at 60×10^6 cells/1.

The filtration rate and the extent of food utilization in relation to the same range of food concentrations in *Modiolus modiolus* were determined at the experimental temperature of 20°C (Fig. 3). In agreement with *Arctica islandica* at 20°C a production of pseudofaeces takes place at the concentration of 60×10^6 cells/1, at 12°C the same situation is given at a concentration of 40×10^6 cells/1. From Fig. 3B can be seen that with increasing concentrations the amount of algae ingested increases too, but at the concentrations of 20×10^6 and 40×10^6 cells/1 there hardly exists a difference. Since the percentage of food utilization in these experiments is nearly the same, the amount of phagocytosed algae is consequently nearly the same at the two higher concentrations (Fig. 3D). According to the much higher activity of the animals at 20°C the low concentration is too small and as a consequence the amount of phagocytosed algae is relatively small.

Since both species accumulate considerable amounts of algae as pseudofaeces at the concentration of 40×10^6 cells/1 at 12°C and of 60×10^6 cells/1 at

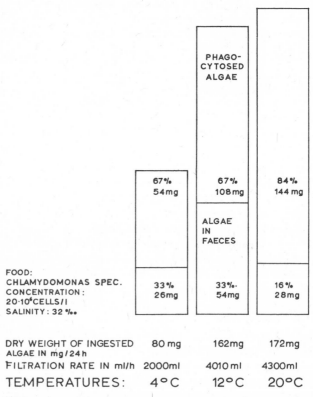

Fig. 5. Influence of different temperatures on filter feeding and food utilization in *Arctica islandica* (74 ccm) during 24 hours.

20°C, and since *Arctica* and *Modiolus* both reach a very high percentage of phagocytosed algae at the concentration of 20×10^6 cells/1 at 20°C (Fig. 5), it can be assumed that both species are quite similar in their reactions. When this assumption is true, the effect of concentration upon the amount of phagocytosed algae is shifted in relation to an increase of temperature, as is obvious from Fig. 3D. It may be said that the optimal food density at 12°C lies between 10×10^6 and 20×10^6 cells/1, whereas at 20°C the optimal range lies between 20×10^6 and 40×10^6 cells/1. As to particle concentration, there exists a threshold above which an increase of particle density is not followed by an increase of particles ingested. Such an increase will correspond to a larger formation of pseudofaeces.

REFERENCES

ALLEN, J. A. 1962. Preliminary experiments on the feeding and excretion of bivalves using Phaeodactylum labelled with ^{32}P. *J. mar. biol. Ass., U.K.*, **42**, 609-23.

BALLANTINE, D., and MORTON, J. E. 1956. Filtering, feeding and digestion in the lamellibranch *Lasaea rubra. J. mar. biol. Ass. U.K.*, **35**, 214-74.

BOJE, R. 1965. Die Bedeutung von Nahrungsfaktoren für das Wachstum von *Mytilus edulis* L. in der Kieler Förde und im Nord-Ostsee-Kanal. *Kieler Meeresforsch.*, **21**, 81-100.

BOJE, R. 1966. Proteine. *Limnologica*, **4**, 383-86.

CHIPMAN, W. A., and HOPKINS, J. G. 1954. Water filtration by the bay scallop, *Pecten irradians*, as observed with the use of radioactive plankton. *Biol. Bull. mar. biol. Lab. Woods Hole*, **107**, 80-91.

DAVIDS, C. 1964. The influence of suspensions of micro-organisms of different concentrations on the pumping and retention of food by the mussel (*Mytilus edulis* L.). *Neth. J. Sea Res.*, **2**, 233-49.

DRAL, A. D. G. 1967. The movements of the latero-frontal cilia and the mechanism of particle retention in the mussel (*Mytilus edulis* L.). *Neth. J. Sea Res.*, **3**, 391-422.

DURVE, V. S. 1963. A study on the rate of filtration of the clam *Meretrix casta* (Chemnitz). *J. mar. biol. Ass. India*, **5**, 221-31.

FOX, D. L., SVERDRUP, H. U., and CUNNINGHAM, J. P. 1937. The rate of water propulsion by the California mussel. *Biol. Bull. mar. biol. Lab. Woods Hole*, **72**, 417-38.

JØRGENSEN, C. B. 1943. On the water transport through the gills of bivalves. *Acta physiol. scand.*, **5**, 297-304.

JØRGENSEN, C. B. 1949. The rate of feeding by *Mytilus* in different kinds of suspension. *J. mar. biol. Ass. U.K.*, **28**, 333-44.

JØRGENSEN, C. B. 1952. On the relation between water transport and food requirements in some marine filter-feeding invertebrates. *Biol. Bull. mar. biol. Lab. Woods Hole*, **103**, 356-63.

JØRGENSEN, C. B. 1960. Efficiency of particle retention and rate of water transport in undisturbed lamellibranchs. *J. Cons. perm. int. Explor. Mer*, **26**, 94-116.

JØRGENSEN, C. B., and GOLDBERG, E. D. 1953. Particle filtration in some ascidians and lamellibranchs. *Biol. Bull. mar. biol. Lab. Woods Hole*, **105**, 477-89.

KREY, J. 1951. Quantitative Bestimmung von Eiweiss im Plankton mittels der Biuretreaktion. *Kieler Meeresforsch.*, **8**, 16-29.

KREY, J., BANSE, K., and HAGMEIER, E. 1957. Über die Bestimmung von Eiweiss im Plankton mittels der Biuretreaktion. *Kieler Meeresforsch.*, **13**, 35-40.

KUENZLER, E. J. 1961. Phosphorus budget of a mussel population. *Limnol. Oceanogr.*, **6**, 400-15.

LOOSANOFF, V. L. 1962. Effects of turbidity on some larval and adult bivalves. *Proc. Gulf Caribb. Fish. Inst., 14th Ann. Sess.*, 80-94.

LOOSANOFF, V., and ENGLE, J. 1942. Effects of different concentrations of plankton forms upon shell movements, rate of water pumping and feeding and fattening of oysters. *Anat. Rec.*, **84**, 86-87.

LOOSANOFF, V. L., and ENGLE, J. B. 1947. Feeding of oysters in relation to density of microorganisms. *Science, N.Y.*, **105**, 260-61.

MIKHEYEV, V. P., and SOROKIN, Yu. J. 1966. Quantitative studies of *Dreissena* feeding habits by the radiocarbon method. *Zh. obshch. Biol.*, **27**, 463-72.

RAO, K. P. 1953. Rate of water propulsion in *Mytilus californianus* as a function of latitude. *Biol. Bull. mar. biol. Lab. Woods Hole*, **104**, 171-81.

RAO, K. P. 1954. Tidal rhythmicity of rate of water propulsion in *Mytilus* and its modificibility by transplantation. *Biol. Bull. mar. biol. Lab. Woods Hole*, **106**, 353-59.

RICE, T. R., and SMITH, R. J. 1958. Filtering rates of the hard clam (*Venus mercenaria*) determined with radioactive plankton. *Fishery Bull. U.S. Fish Wildl. Serv.*, **58**, 73-82.

SMITH, R. J. 1958. Filtering efficiency of hard clams in mixed suspensions of radioactive phytoplankton. *Proc. natn Shellfish Ass.*, **48**, 115-24.

TAMMES, P. M. L., and DRAL, A. D. G. 1955. Observations on the straining of suspensions by mussels. *Archs néerl. Zool.*, **11**, 87-112.

THEEDE, H. 1963. Experimentelle Untersuchungen über die Filtrationsleistung der Miesmuschel *Mytilus edulis* L. *Kieler Meeresforsch.*, **19**, 20-41.

WILLEMSEN, J. 1952. Quantities of water pumped by mussels (*Mytilus edulis*) and cockles (*Cardium edule*). *Arch. néerl. Zool.*, **10**, 153-60.

WINTER, J. E. 1969. Über den Einfluss der Nahrungskonzentration und anderer Factoren auf Filterierleistung und Nahrungsausnutzung der Muscheln *Arctica islandica* und *Modiolus modiolus*. *Mar. Biol.*, **4**. (In press).

On the significance of various ecological groups of animals in the bottom communities of the Greenland, Norwegian and the Barents Seas

V. I. ZATSEPIN
Chair of Hydrobiology
Moscow State University
Moscow, USSR

ABSTRACT. There are definite ecological zones in the distribution of bottom fauna communities in the Barents Sea and in the southern parts of the Greenland and Norwegian seas. These zones are closely connected with depth, the character of bottom sediments, the strength of bottom currents, etc. Different ecological groups of animals, according to their mode of feeding, play different roles in the composition and the size of biomass of these bottom fauna communities.

The ecological groups differ from one another by their mode of feeding and of getting the food: filter-feeders from epifauna and infauna; detritus feeders; mud-eaters; predators, etc. In the bottom communities which inhabit mixed firm grounds in the regions with a constant change of bottom waters, in shelf or upper bathial zones, epifauna communities are widely distributed; the filter-feeders are the leading forms here making up almost 90% of the average biomass of the community.

In the communities inhabiting the soft mud sediments from the sublittoral to the abyssal zones of the Norwegian Sea the detritus feeders and mud-eaters are dominant forms constituting up to 70% of the average biomass of the community.

On sands mixed with mud, in the lower parts of the sublittoral and bathial zones there are bottom communities where detritus feeding forms and mud-eaters or filter feeders from the infauna are predominant. Of a similar ecological structure and distribution are bottom fauna communities in the Barents Sea.

Under similar conditions of depth, bottom sediments and hydrodynamics of the bottom water layers in all the above-mentioned seas (Barents, Norwegian and Greenland), the bottom fauna communities of a definite ecological character were found (with a predominance of filter feeding or detritus feeding forms or mud-eaters). This ecological pattern (by the mode of feeding) of the communities is constant even in quite different seas. The zoogeographic character of the leading forms of the communities distributed in the whole region can change very significantly—from cold-water (arctic) forms in the Barents Sea to the temperate or warm-water (boreal) forms in the northern parts of the North Sea.

207

INTRODUCTION

The purpose of the present study is to elucidate the composition and quantitative distribution of separate groups of invertebrates with different modes of nutrition and methods of obtaining food, and to determine their relative importance in the bottom fauna of the Barents Sea and adjacent parts of the Norwegian Sea. Approximately one thousand quantitative samples were used, taken with Petersen grabs mainly with a sampling area of 0·25 m². Besides these, trawl samples were used for counts of larger forms. The region sampled covered about 85% of the Barents Sea. The material was collected for the greater part by the author in the late 1940s and early 1950s, and also in the 1930s by the ships of the Knipovich Polar Institute of Marine Fisheries and Oceanography. Earlier bottom-sampler and trawl collections were also used (Brotskaya and Zenkevitch, 1939). A comparison of samples taken in the same regions of the sea in different years showed no substantial changes in the quantitative distribution of the bottom fauna. This permitted the author to use all the materials mentioned above.

The bottom fauna of the Barents Sea comprises various forms and groups of invertebrates with diverse food habits. However the differences in the main sources of food are on the whole not very significant. Thus the basic diet of bottom invertebrates (with the exception of predators and, partly, filter-feeders from the epifauna) in the sublittoral and pseudoabyssal consists of detritus, bacteria, minute bottom animals (micro- and meiobenthos, including also juveniles of larger forms) and the organic matter of the ground itself. According to several investigations, the nutritive value of detritus seems to be determined mainly by the bacteria present in them (Rodina, 1963). In the diet of filter-feeders from the epifauna, the bacterial population of the organic-mineral particulate matter is also important. The source of all the organic matter on which the bottom fauna of the open parts of the Barents Sea subsists is plankton produced in the upper water layers, macrophytes (brown and red algae) growing in the coastal and shallow regions of the sea, and detritus brought from land. Many of these nutrients are lost (through disintegration of organisms in the "rain of dead bodies", expenditure on energetic exchange, removal by migratory animals and zonal distribution of zooplankters), so that only some of the organic matter produced in the upper layers reaches the bottom invertebrates inhabiting the intermediate depths of the sublittoral and pseudoabyssal.

ECOLOGICAL GROUPS IN THE BENTHOS

Seven ecological groups are distinguished among the benthic invertebrates of the Barents Sea, according to their feeding habits, namely: (*i*) filter-feeders from the epifauna, (*ii*) filter-feeders from the infauna, (*iii*) detritus-feeders, (*iv*) deposit-feeders, (*v*) predators, (*vi*) omnivores and scavengers, (*vii*) herbivores-scrapers. In such a general outline, many species with a mixed type of feeding had to be assigned to the category considered dominant. Details are given in Zatsepin (1962, 1968).

The biomass of each of these ecological groups was assessed for each station and their relative significance was expressed as a percentage of the total biomass of benthic invertebrates at the given station. Using these indices, charts were compiled, showing the distribution of the total benthic biomass and, of the biomasses of filter-feeders from the epi- and infauna, detritus-feeders and deposit-feeders (Figs 1-5). The weights are of material preserved in alcohol. These include the shells of molluscs and the gut contents of all invertebrates (excluding the larger forms of deposit-feeders, such as *Ctenodiscus crispatus, Trochostoma, Brisaster*, etc.).

On the basis of these charts (taking into account the dimensions of areas between isobenthos), computations were made of the biomass, both for the total macrobenthos as a whole, and for the separate ecological groups. It was found that the biomass of macrobenthos in the open parts of the Barents Sea averages 102 g/m². The total biomass of the benthos in the part of the Barents Sea investigated was found to be of the order of 120 million tons and for the whole sea about 140 million tons.

The most important group (Table 1) in the total biomass of macrobenthos are filter-feeders, feeding on plankton and detritus and composing more than 40% of the whole benthic biomass of the Barents Sea.

Judging from samples (circa 300) taken with the bottom-samplers, the nearshore regions of Murman, show an analogous partition of the total biomass (Zatsepin, 1968).

TABLE 1. Division of biomass into feeding groups

	Total biomass	Epifauna filter feeders	Infauna filter feeders	Detritus feeders	Deposit feeders	Others: (predat., omnivor., scaveng., etc.)
Million tons	120·0	24·6	27·6	31·2	20·3	16·3
% of total biomass	100·0	20·6	22·9	26·0	17·0	13·5

TABLE 2. Distribution of separate ecological groups in the Barents Sea

Biomass g/m²	Percentage of the total area investigated			
	filter-feeders		detritus- feeders	deposit- feeders
	epifauna	infauna		
less than 10	67	68	33	54
10-25	17	10	36	22
25-50	8	8	22	21
50-100	3	8	5	2
over 100	5	6	4	11

The distribution of the total biomass of macrobenthos and the biomasses of its separate ecological groups is far from homogeneous (Figs 1-5). Thus, areas of the Barents Sea where the total benthic biomass is between 25 and 100 g/m² occupy more than half of the investigated part of Barents Sea; areas with a biomass greater than 500 g/m²—less than 1%; areas with a biomass of 300-500 g/m²—5%; with a biomass of 200-300 g/m²—about 9%, of 100-200 g/m²—15%; and about 7% of the sea has a biomass less than 1 g/m². The areas occupied by biomasses of separate ecological groups are unequal too, Table 2.

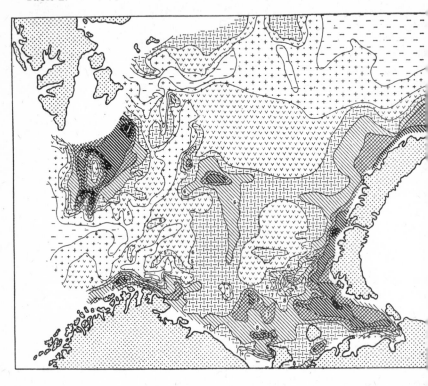

Fig. 1. Distribution of the total biomass of macrobenthos in the Barents Sea (g/m²). 1. up to 10; 2. 10-25; 3. 25-50; 4. 50-100; 5. 100-200; 6. 200-300; 7. 300-500; 8. 500-1000; 9. over 1000.

A trend is observed in the distribution of biomass and in the importance of separate ecological groups. Thus the biomass tends to decrease with depth and distance from shore, as well as from the central regions of the sea toward its north-western part bordering on the Kara Sea (Fig. 1). In shallower water filter-feeders rank first in biomass and relative importance. With increasing depth detritus feeders become more important and at still greater depths deposit feeders. Furthermore there are regions of maximum and minimum development of total bottom fauna and of the separate ecological groups. Naturally the distribution of the total benthic biomass reflects the basic features of the quantitative distribution of these groups. Thus the highest biomass of benthos, reaching several hundred grammes per square metre and in places even 1 kg/m² is found in the south-eastern part of the Barents Sea, on the Kaninsk-Kolguevsk, Pechersk and Novozemelsk Shallows. It is accounted for by an abundant development of infauna filter-feeders and detritus-feeders (Figs 3 and 4) with mass populations of *Astarte borealis* (Chemnitz), *A. montagui* (Dillwyn), *Macoma calcarea* (Chemnitz), *Clinocardium ciliatum* (Fabricius), *Serripes groenlandicus* (Brugiere), *Nuculana pernula* (Müller), *Yoldia hyperborea* (Loven) Torell, *Portlandia arctica* (Gray) and other species characteristic of the bottom communities of these regions. There are also filter-feeders from the epifauna, especially in places rich in communities of various Bryozoans, Hydroids, Ascidians and barnacles. The high biomass of benthos of the Medvezhinsk Bank is likewise determined by a mass development of epifauna filter-feeders; in various communities, characteristic for these regions are Bryozoans, Hydroids, barnacles, with colonies of *Hiatella arctica* (Linné), *Chlamys islandicus* (Müller), *Cucumaria*

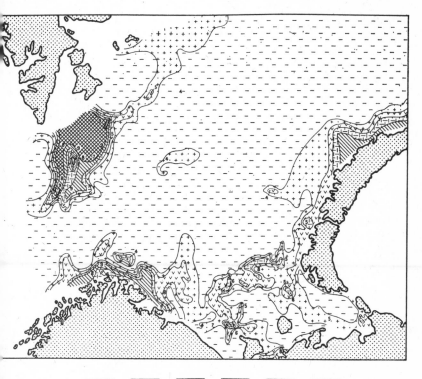

Fig. 2. Distribution of the biomass of filter-feeders from the epifauna. Notation 1-5 see Fig. 1; 6. over 200.

frondosa (Gunnerus). The abundant benthic biomass on the southern and south-eastern slopes of the Medvezhinsk bank is formed mainly by infauna filter-feeders, in communities with mass populations of *Astarte borealis*, *A. montagui*, *A. elliptica* (Brown), *A. arctica* (Gray), *A. crenata* (Gray), *Macoma calcarea* and other bivalves molluscs. On the south-western slopes of the bank, epifauna filter-feeders are predominant (Fig. 2), in communities of sponges, bryozoans and brachiopods. The same pattern of distribution is observed on the Gussinaya bank and its slopes. The high biomass of benthos (in places up to 1 kg/m² and even more) in the south-western part of the sea, in areas directly influenced by the North Cape current is also determined by a mass development of filter-feeders from the epifauna; diverse variants of sponge, bryozoan and brachiopod communities (Fig. 2). Regions and smaller areas with a high benthic biomass up to 200 g/m², on the Central Elevation, Central Plateau, northern slopes of the Murmansk Bank and Kanin Shallows are associated with an intensive development of detritus-feeders and infauna filter-feeders in the respective groupings of the central-Barents Sea community (Figs 3 and 4). Here denser populations are observed first of all of *Astarte crenata*, then of *Bathyarca glacialis* (Gray) or

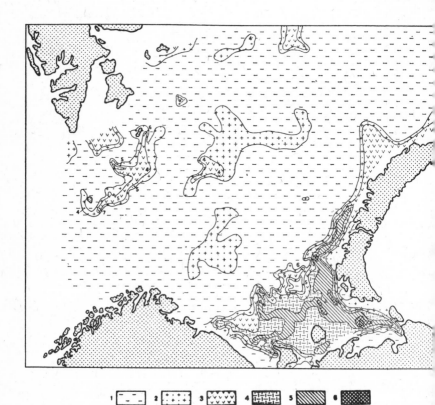

Fig. 3. Distribution of the biomass of filter-feeders from the infauna. Notation see Fig. 2.

other bivalve molluscs, of the polychaetes, *Spiochaetopterus typicus* M. Sars, *Ampharetidae*, *Cystenides hyperborea* Malmgren, ophiuroids, most often *Ophiura sarsi* Lutken, less frequently *Ophiacantha bidentata* (Retzius), *Ophiopholis aculeata* (Linné) or other species characteristic of the respective groupings of this community. In deeper and silted areas of this central region of Barents sea deposit-feeders are of considerable importance too, such as *Ctenodiscus crispatus* (Retzius), *Maldanidae*. Compared with these values, a somewhat higher biomass of benthos up tó 75. g/m², found in separate places of the western parts of the central regions of the Barents Sea, is determined by an intensive development of detritus- and deposit-feeders (Figs 4 and 5); by certain species of the holothurians *Trochostoma* and *Ankyroderma*; and in adjacent and deeper areas in the south-western part of the sea, also by the irregular echinoid *Brisaster fragilis* (Duben et Koren), a leading species of the local community.

The regions of the Barents Sea with the lowest total benthic biomass, delimited by the isobenthos of 10 and 25 g/m², are as a rule confined to deep-sea troughs and pits (as for instance the Medvezhinsk Trough) or to more stagnant zones, such as the stagnant and halistatic areas of the Central depression, and to regions with brown grounds (Northern Plateau, Persey

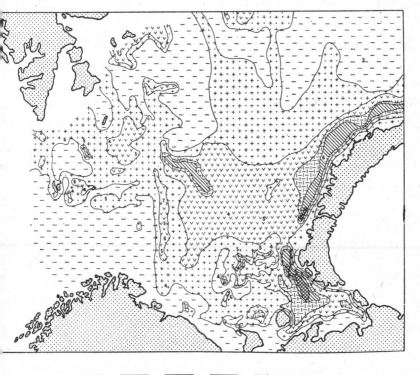

Fig. 4. Distribution of the biomass of detritus-feeders. Notation see Fig. 2.

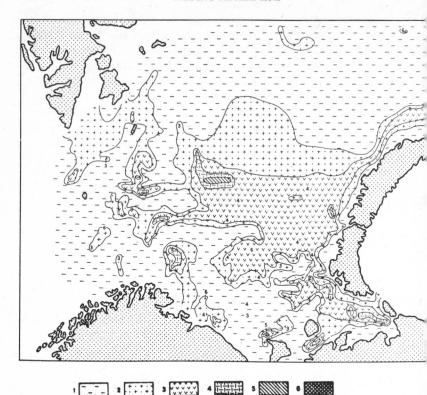

Fig. 5. Distribution of the biomass of deposit-feeders. Notation see Fig. 2.

Elevation). All these regions are characterized by a relative predominance of detritus- and deposit-feeders, and an extremely poor development of filter-feeders from the epifauna. A somewhat higher biomass (up to 50 g/m², seldom more) found in places in the northern part of the sea, is mainly determined by a somewhat greater development of detritus-feeders and some filter-feeders from the infauna, mostly populations of *Astarte acutucostata* (Jeffreys and Friele), less frequently by *A. crenata*, *Ophiacantha bidentata*, *Ophiocten sericeum* (Forbes), *Ophiopleura borealis* Danielssen and Koren, *Ophioscolex glacialis* Müller and Troschel or others, characteristic of the communities of this part of the sea. But occasionally, as for instance on the Persey Elevation, a slight increase in the biomass of benthos is associated with a somewhat greater development of some species of filter-feeders from the epifauna or of deposit-feeders.

DISCUSSION

Some of the basic features in the distribution both of the total biomass of macrobenthos and the biomasses of its separate ecological groups are closely related to the distribution of depths in the Barents Sea, to its hydrodynamic regime (currents, exchange of near-bottom water-masses, etc.) and to the

Figs 6 and 7 show the relation of the total biomass and the biomasses of separate ecological groups to depths, to type of bottom sediments in the central and southern parts of the Barents Sea and in its northern part, to the region of predominance of brown sediments enriched by ferrous and manganese hydroxides. In relating benthic biomass and character of ground, data

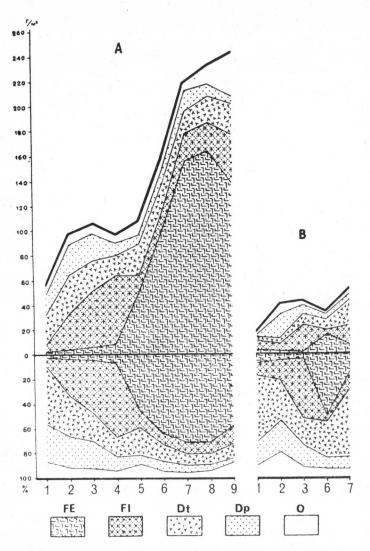

Fig. 7. Changes in the total biomass of macrobenthos and biomasses of separate ecological groups related to type of bottom. 1—silt; 2—sandy silt; 3—silty sand; 4—sand; 5—silt and gravel; 6—sandy silt with gravel; 7—silty sand with gravel; 8—sand with gravel; 9—shell with sand and gravel. Other denotations see Fig. 6.

were used on the percentage of the bottom sediment less than 0·01 mm
(Klenova, 1960; Vinogradova and Litvin, 1960); sand being defined as containing less than 5% of this fraction; silty-sand, from 5 to 10%; sandy-silts,
from 10 to 30%; and silts, 30 to 50%.

It must be noted that the environmental factors considered above (depth,
sediments, hydrodynamic regime) affect the quantitative distribution of the
Barents Sea bottom fauna and its ecological groups, as a rule in their action
and mutual determination. These factors are only to a certain degree indicative of the potential food supply of the bottom fauna, its possible reserves and
their renewal, i.e. of the primary and most important cause determining the
quantitative development of the fottom fauna as a whole and of its separate
ecological groups.

The quantitative distribution of zoobenthos and its ecological groups in its
relation to changing depth, type of bottom sediments and hydro-dynamic
regime depends first and foremost on the food supply, i.e. on the reserves of
organic matters in different regions of the sea.

Almost no data are available as yet on the food supply of the bottom fauna
of the Barents Sea. For example we need to know the quantity of plankton
suitable as food for filter-feeders in the near-bottom water layers, the quantity
of detritus, the assimilable-organic matter content in the layers inhabited by
bottom animals, etc.

The value of reserves of organic matter as food for deposit-feeders and
partly also for detritus-feeders, may be deduced from the content of non-
carbonate carbon and organic nitrogen in the upper layers of sediments.
The quantity of organic matter in the bottom depends first of all on sedimentation conditions and intensity of primary production in the different regions
of the sea. It will also reflect the loss of nutrient organic matter produced in
the upper water layers during its passage through the water column (resulting
from consumption by zooplankton, disintegration and bacterial decomposition) and from consumption by benthic filter-feeders and partly, detritus-
feeders. Not all the organic matter by far can be assimilated by deposit-
feeders. The portion of organic matter, which is highly resistant to decomposition by saprophytic bacteria common in the shelf zone (and consequently, resistant to the action of the digestive ferments of bottom invertebrates), the so-called "marine humus", seems to form about several tens of
per cents of the whole organic-matter content of marine sediments (Skopintsev, 1950). The amount of nitrogenous organic matter readily assimilable to
bacteria (and, most probably, to invertebrates too), contained in the sediments is as a rule rather low, often of the same order as the biomass of
heterotrophic bacteria in the upper layers of the bottom (Kusnezov, 1952,
1959). The biomass of the latter in the topmost bottom layer may reach one
or even several grammes per square metre and several per cent of the total
organic matter content (Kriss, 1959; Jørgensen, 1966). According to some
calculations made for the northern part of the Barents Sea the biomass of
bacteria in the upper 4-5 cm layer may be of the same order as the biomass
of benthos (Butkevich, 1932).

H

In the sediments of the Barents Sea the content of organic carbon varies from 0·15 to 3·12% and that of organic nitrogen from 0·02 to 0·42% depending on the type of sediments (Gorschkova, 1957; Klenova, 1960). The distribution of organic nitrogen follows the same pattern as the distribution of non-carbonate carbon.

The content of the latter increases as a rule with an increase of the fine fraction (less than 0·01 mm), and in silty and silty-sandy grounds with decreasing depth, tending to increase from the northern regions towards the southern (more productive) regions of the sea, as well as shorewards. Brown coloured sediments, prevalent in the northern regions, are poor in organic carbon, the latter being most abundant in the greenish-gray sediments characteristic of the central and southern regions. Thus brown silty-sands, sandy-silts and silts contain on an average 0·6%, 1% and 1·8% respectively, while in sediments of the same type but of a greenish-gray colour the content of organic carbon increases to 0·7%, 1·5% and 2·2% respectively (Gorschkova, 1957; Klenova, 1960). The content of chlorophyll (or, more exactly, of its derivates) in the sediments of the Barents Sea is associated first of all with the oxidative character of the benthic environment but also reflects the intensity of primary production in different regions of the sea. Brown ground with a high oxidative activity determined by the presence of ferrous and manganese hydroxides in a colloidal state, contains no chlorophyll, which occurs in appreciable quantities in greenish-gray sediments only. Shallows and nearshore regions are richest in chlorophyll. In many cases a high chlorophyll content coincides with areas of intensive plankton development in regions of the so-called "polar front" or near it (Klenova, 1960).

A comparison of the distribution of non-carbonate carbon in the sediments of the Barents Sea (Gorschkova, 1957) with the quantitative distribution of the different ecological groups of its bottom fauna shows that regions of maximum development of detritus- and deposit-feeders coincide as a rule with regions of maximum organic-matter content (Figs 4 and 5). It must be noted, however, that in the open parts of the Barents Sea, areas with a maximum content of organic carbon (2-3%) are usually located at great depths, near submarine slopes, at the foot of elevations or in the vicinity of near-shore shallows. Here the accumulation of detritus seems to be determined by the washdown of detritus from regions with a rich bottom fauna, first of all of epifauna filter-feeders (from banks and their slopes, nearshore shallows), or by a greater supply of organic matter from the producing layers of the sea. The vast central part of the Barents Sea with sediments containing from 1 to 2% of non-carbonate carbon including areas with a maximum content of 3%, is a region of highest quantitative development of detritus- and deposit-feeders, which taken together may account for more than half of the whole benthic biomass of this region.

In the marginal zone of all this central part of the sea a marked increase in the importance of detritus-feeders is observed: they alone may form here more than 50% of the total benthic biomass. This zone of prevalent detritus-feeders reaches its greatest width in its southern and western parts. The relatively restricted regions confined to the great depths of the Central Depression and to the areas of the south-western part of the sea and the southern regions of the Medvezhinsk Trough that border on the western side of the Central Plateau, are regions of highest relative importance of deposit-feeders in the formation of the total benthic biomass (constituting not infrequently more than 50%). To the south of this vast region of detritus- and deposit-feeders, the non-carbonate carbon content of the sediments usually decreases (down to 1%) and both these ecological groups lose much of their importance.

Detritus-feeders are also predominant in grounds rich in organic carbon (up to 2-3%) of the deeper zone situated in the southern part of the slopes of Yuznyi Island (Novaya Zemla). The biomass of deposit-feeders too is fairly high, up to 50 g/m² (in places even more). The regions of the Northern Plateau and Persey Elevation with their brown grounds, characterized by a maximum content of manganese and a low organic-matter content, are also a zone of prevalent important of detritus-feeders, which usually contribute more than 50% of the low total benthic biomass characteristic of this region (Fig. 4).

The sandy-bottom regions of Prikaninsk and Kanin-Kolguevsk Shallows with a low non-carbonate carbon content (up to 0·5%) coincide with a region of maximum development of infauna filter-feeders, which here form more than 50% (not infrequently up to 75%) of the total benthic biomass, whose value may reach several hundred grammes per square metre. The region of maximum development of infauna filter-feeders includes also some places in the south-eastern part of the sea with a somewhat higher non-carbonate carbon content (up to 1%).

The regions of maximum development of epifauna filter-feeders (the south-western part of the sea, along the shore of northern Norway and the Rybachyi Peninsula, some areas in the southern part of the "Kola meridian", marginal areas and slopes of the Medvezhinsk and Gussinaya Banks) are associated with coarser and mixed bottom sediments, characterized by a low organic-content (usually less than 1%).

The low content of non-carbonate carbon in areas with a predominant development of epi- and infauna filter-feeders is determined first of all by the high rates of organic matter oxidation in these regions of a markedly "running-water" type with most active near-bottom waters, as well as by the already mentioned wash-down of detritus into adjacent deep-water areas with a maximum content of organic carbon. The low organic matter content is determined by a high consumption of plankton and detritus by bottom filter-feeders, and in regions of commercial fishing, such as Medvezhinskaya and Gussinaya Banks, and the shallow south-eastern part of the sea, also by the removal of organic matter by migratory bottom-feeding fishes which eat out the local bottom fauna.

SUMMARY

The data discussed above are evidence not only of the paramount importance of detritus-feeders in the benthic fauna of the Barents Sea, but also of a

definite sequence in the interchange of its diverse ecological groups in accord
ance with changes in their food supply caused by increasing depth, reduced
hydrodynamic activity of the near-bottom water masses, different type o
bottom and different organic-matter content of sediments.

With increasing distance from shore the zone of predominance of filter
feeders from the epifauna is succeeded by a zone of infauna filter-feeders
which in its turn cedes its place to a zone of maximum development o
detritus-feeders, followed still further off-shore by a zone of deposit-feeders
Each of these zones has its own bottom communities in which most of the
leading and characteristic species are respectively filter-feeders from the epi
and infauna, detritus-feeders or deposit-feeders. Of most value for bottom
feeding fishes are the ecological zones and the benthic communities with
predominance of detritus-feeders and filter-feeders from infauna (Zatsepin
1939, 1962, 1968; Zatsepin and Petrova, 1939). An analogous pattern o
distribution is observed in the southern part of the Norwegian Sea and in the
northern part of the North Sea.

REFERENCES

BROTSKAYA, V. A., and ZENKEVITCH, L. A. 1939. Quantitative evaluation of the bottom
fauna of the Barents Sea (in Russian, English summary). *Trudȳ vses. nauchno-issled
Inst. morsk. rȳb. Khoz. Okeanogr.*, 4 (8), 5-126.
BUTKEVICH, V. S. 1932. Zur Methodik der bakteriologischen Meeresuntersuchungen und
der Bakterien im Wasser und in den Boden des Barents Meeres (in Russian, German
summary). *Trudȳ gos. okeanogr. Inst.*, 2 (2).
FILATOVA, Z. A. 1938. The quantitative evaluation of the bottom fauna of the south
western part of the Barents Sea (in Russian, English summary). *Trudȳ polyar. nauchno
issled. Inst. morsk. rȳb. Khoz. Okeanogr.*, 2, 1-58.
GORSCHKOVA, T. I. 1957. Organic matter and carbonates in the sediments of the Barent
Sea (in Russian). *Trudȳ polyar. nauchno-issled. Inst. morsk. rȳb. Khoz. Okeanogr.*
10, 260-80.
JØRGENSEN, C. B. 1966. *Biology of suspension feeding.* Oxford, Permagon Press. 357 pp
KLENOVA, M. V. 1960. *The geology of the Barents Sea* (in Russian). Moscow, Acad. Sci.
USSR.
KRISS, A. E. 1959. *Marine microbiology (deep sea)* (in Russian). Moscow Acad. Sci.
USSR. *Meeresmikrobiologie.* Jena, Gustav Fischer Verlag. 1961. English trans
1963. Edinburgh, Oliver and Boyd.
KUSNEZOV, S. I. 1952. *The role of micro-organisms in the cycle of matter in lake*
(In Russian), Moscow, Acad. Sci., USSR, *Die Rolle der Mikroorganismen im Stoff
kreislauf der Seen.* Berlin, Dtsch. Verlag Wiss. 1959.
RODINA, A. G. 1963. Microbiology of detritus of lakes. *Limnol. Oceanogr.*, 8, 388-93.
SKOPINTSEV, B. A. 1950. Organic matter in natural waters (Aquatic humus) (in Russian)
Trudȳ gos. okeanogr. Inst., 17 (29).
TOURPAEVA, E. P. 1948. Feeding of several bottom invertebrates of the Barents Sea (i
Russian). *Zool. Zh.*, 27 (6).
VINOGRADOVA, P. S., and LITVIN, V. M. 1960. Studies of bottom relief and sediments in th
Barents and Norwegian Seas (in Russian). *Soviet Fisheries Investigations in Nort
European Seas.* Edited by Yu. Yu. Marty *et al.* Moscow, VNIRO. 101-10.
ZATSEPIN, V. I. 1939. The feeding of the haddock of the Murman Coast in connection wit
the bottom fauna (in Russian, English summary). *Trudȳ polyar. nauchno-issled. Ins
morsk. rȳb. Khoz. Okeanogr.*, 3, 39-95.
ZATSEPIN, V. I. 1962. The communities of bottom invertebrates of the Barents Sea in th
Murman region and their relation to the North Atlantic communities (in Russian)
Trudȳ gidrobiol. Obshch., USSR, Moscow, 12, 245-344.

ZATSEPIN, V. I., and RITTIKH, L. A. 1968. Quantitative distribution of the bottom fauna and their ecological groups with different food habits in the Murman region of the Barents Sea. *Trans. Moscow Soc. Nat.*, 30, 49-80.

ZATSEPIN, V. I., and PETROVA, N. S. 1939. The feeding of cod in the south part of the Barents Sea (in Russian, English summary). *Trudy polyar. nauchno-issled. Inst. morsk. rȳb. Khoz. Okeanogr.*, 5.

ZENKEVITCH, L. A. 1963. *Biology of the seas of the USSR* (in Russian). Moscow Acad. Sci. USSR. *Biology of the seas of the USSR.* London, George Allen and Unwin Ltd. 1963, 955 pp.

ZERNOV, S. A. 1949. *General hydrobiology* (in Russian). Moscow Acad. Sci. USSR.

Utilization of the direct grazing and plant detritus food chains by the striped mullet Mugil cephalus*

W. E. ODUM
Institute of Marine Sciences
University of Miami
Miami, Florida, U.S.A.

ABSTRACT. Most foodchains in shallow estuaries are based upon macro-plant detritus and benthic and epiphytic micro-algae rather than on phytoplankton. In such areas the zooplankton are relatively unimportant in energy transfer and are replaced as the critical herbivore link by benthic invertebrates and phytophagous fishes. This paper is an examination of the ability of one of these fish, *Mugil cephalus*, to successfully exploit the first trophic level and, in effect, "telescope the food chain".

INTRODUCTION

In the classical concept of aquatic food chains zooplankton are considered the first important link in secondary production. What is often ignored is that this concept, like so many other basic principles in marine biology, was derived from observations and research carried out either in the open ocean or at relatively high latitudes. To the ecologist whose work is oriented to estuarine areas, especially in the tropics and subtropics, it becomes apparent that zooplankton are not the most important herbivore link. Williams *et al.* (1968) have demonstrated that in the vicinity of Cape Hatteras, North Carolina, zooplankton become steadily less important with decreasing depth in areas shallower than 100 metres; in shallow inshore areas they were shown to be relatively unimportant in food chains since they grazed only two to nine per cent of available phytoplankton.

Moreover, in shallow estuaries the phytoplankton itself has been found to be of secondary importance (Schelske and E. P. Odum, 1961; Ragotzkie, 1959; Teal, 1962; Pomeroy, 1960); the most important primary producers in such systems are marsh grasses (*Spartina*, etc.), sea grasses (*Zostera, Thalassia*, etc.), attached macroalgae, mangroves (*Rhizophora*, etc.) and the benthic

* Cont. no. 1108, Inst. of Marine and Atmos. Sci., Univ. of Miami.

222

microflora (benthic diatoms, dinoflagellates, filamentous green and blue-green algae). In most cases food chains are based on detritus derived from marsh grasses, sea grasses, macro-algae and mangroves, or directly on the benthic and epiphytic microflora. Animals which are able to utilize such energy sources replace zooplankton as the critical herbivore link.

Two of the most important groups of these animals are: (*i*) the benthic invertebrates, whose role as detritus consumers was emphasized by Jensen and Petersen (1911) and many times since (summarized by Jørgensen, 1966); and (*ii*) a limited number of highly successful fishes which consume both benthic microplant material and macroplant detritus. Although this second group includes few species, their great abundance supports many fisheries. Important in this group are the milkfish, *Chanos chanos*, and the mullets, *Mugil* sp., plus their freshwater counterparts, the carps, *Cyprinus* sp. and the tilapia.

Telescoping of the food chain

The phenomenon of relatively large fishes feeding directly from the first trophic level has been referred to by Hiatt (1944) as "a telescoping of the food chain". A fishery based on a species with this ability to bypass steps in the food chain with their accompanying loss of energy will potentially produce far more than a fishery based on a third or fourth level carnivore. Since it is essential to have a basic understanding of how they are able to utilize such a food source, the first step in any management or culture operation should be the analysis of the trophic ecology of the individual species. In this paper I have made such an analysis for the striped mullet, *Mugil cephalus*. By attacking the problem in a number of different ways it is possible to obtain a picture of how *M. cephalus* exploits the first trophic level. Unfortunately, this type of approach often raises more questions than it resolves.

A widespread and successful fish

It would be futile to attempt to describe all the information which has accumulated concerning *Mugil cephalus*. Thompson, (1963, 1966) has published two complete synopses which summarize the literature.

To describe the distribution of the fish very briefly, it is probably the most widespread and successful inshore teleost species, with a range extending around the world between 40°N and 40°S in waters with salinities from near zero (Hellier, 1957) to 113 parts per thousand (Zenkevitch, 1963). It is found not only in neritic areas and estuaries, but also hundreds of miles up fresh-water rivers such as the Colorado (Dill, 1944). During spawning, mullet move offshore, at least in North America, and may be found beyond the continental shelf (Anderson, 1958) over depths as great as 1500 metres (Arnold and Thompson, 1958). In most areas of the world where *M. cephalus* occurs, its success is expressed by great abundance. Sizeable fisheries are supported in many regions, while in others the potential exists for a valuable industry. In Florida over twenty million kilograms are harvested annually even though low demand limits the fishery.

Feeding and feeding behaviour

Typically, *M. cephalus* feeds either by sucking up the surface layer of the mud or by grazing on submerged rock and plant surfaces. An extensive literature reviewed by Thompson (1954) has established that the major contents of the stomach may be categorized as: (*i*) micro-algae including epiphytic and benthic forms, (*ii*) decaying plant detritus, and (*iii*) inorganic sediment particles. The latter appear to function as a grinding paste in the degradation of plant cell walls in the highly modified gizzard-like pyloric stomach (Thompson, 1966). Juvenile mullet are primarily carnivorous until they reach a standard length of about 30 mm. During this period they feed on mosquito larvae, copepods and other zooplankton.

M. cephalus is uniquely equipped for its trophic feeding niche. Its highly modified stomach has been described by Al-Hussaini (1947) as consisting of two parts: (*i*) the cardiac portion, a thin walled, saccular, blind caecum used for storage, and (*ii*) the pyloric portion, a thick, muscular gizzard-like structure used for pulverizing the food. The intestine is extremely long and coiled, often being more than five times the length of the fish. Finally, there is a pharyngeal filtering device, described in detail by Ebeling (1957). This mechanism aids in selecting very fine particles from the coarser sediments which are expelled. This finer material has been shown to be much richer in adsorbed micro-organisms than the particles which are rejected (Wood, 1964; W. E. Odum, 1968).

More than one method of selecting food is used by *M. cephalus*. For sediment feeding the mullet inclines its body at an angle of 15 to 30 degrees to the surface of the sediment and extends the premaxillaries so that particles are picked up in one of two ways: either by taking up small mouthfuls at random, or by skimming along the bottom with the lips barely in contact with the sediments and sucking up the uppermost layer. The latter method often includes shaking of the head back and forth. By either method the result is a small quantity of sediment in the pharynx which is carefully strained by the pharyngeal filter. Any material found unsuitable (usually large particles) is expelled in a small cloud. The presence of mullet is often characterized by a trail of these clouds behind feeding fish.

A similar means of feeding is used when browsing on micro-algae attached to submerged surfaces. With the body at an angle to the substrate, with the premaxillaries extended and with a back and forth movement of the head, the attached algae are either nibbled or sucked off the surface. Finally, the food material is filtered in the pharynx, the suitable material ingested and the remainder expelled.

Under certain conditions *M. cephalus* will feed at the water surface. The location of the mouth in an almost terminal position allows it to be protracted and the surface film sucked in. This type of feeding is used when thick concentrations of micro-algae are present at the air-water interface.

TABLE 1. Important environmental parameters of the seven systems

	Georgia beach	Spartina marsh pool	Spartina marsh stream	Florida mangrove swamp	Florida Thalassia bed	Florida brackish lake	Florida freshwater canal
Normal salinity range (in %)	28-32	12-24	18-26	27-35	33-35	2-12	0
Yearly temperature range	12-30	10-32	12-30	16-36	18-34	14-30	15-30
Benthic substrate	siliceous sand	sandy mud	sandy mud	organic mud	calcium carbonate	organic mud	limestone rock
Organic content (% dry wt.) of sediment	0·5-1·1	3·6-4·8	1·1-3·1	—	—	5·1-7·2	—
Dominant main producer available to mullet	benthic microflora	benthic microflora	benthic microflora	benthic microflora	benthic microflora and epiphytic diatoms	green and blue-green algae	green and blue-green algae
main source of Macrophyte detritus	Spartina	Spartina	Spartina	mangrove	3 species of sea grasses	terrestrial plant material	terrestrial plant material

An "interface feeder"

In most types of feeding by this mullet there is a dependence on the presence of some type of physical surface. These include the top layer of the sediments, the surface film of the water, the surface of *Thalassia* leaves, the rock walls of canals and the exposed trunks and roots of trees along the water's edge. *Mugil cephalus* might be described as an "interface feeder", a role which differs markedly from filter feeders which utilize mucous nets or gill rakers to concentrate food particles.

FEEDING IN RELATION TO THE ENVIRONMENT

Despite the body of knowledge which has accumulated concerning *Mugil cephalus*, including many descriptions of the food materials which are ingested, there has not been a concerted attempt to analyse its feeding in relation to the food available in the environment and how this varies from one area to the next. For the present study seven very different environmental systems were chocen, and the mullet's diet analysed for each system. Four systems were located along the south-east coast of Florida (25°30'N), while the remaining three were six degrees of latitude to the north in the vicinity of Sapelo Island, Georgia.

The Florida systems were: (*i*) a red mangrove, *Rhizophora mangle*, swamp; (*ii*) a turtle grass, *Thalassia testudinum*, bed in Biscayne Bay; (*iii*) a limestone quarry pit filled with brackish water; and (*iv*) a small, freshwater, agricultural drainage ditch. One system at Sapelo Island, Georgia, consisted of an open siliceous beach, while the remaining two were associated with the extensive cord grass, *Spartina alterniflora*, marshes lying between Sapelo Island and the mainland. One of the latter was a small mud-bottom marsh creek and the other a high marsh tide pool which was connected with the sea only on high spring tides. Important ecological parameters that need to be considered in these systems are given in Table 1.

Methods of study

Samples of planktonic, epiphytic and benthic micro-organisms were collected from the environment in each system. Planktonic forms were removed from the water column at the surface and near the bottom with a Niskin Bottle and spun down in a continuous centrifuge (Kimball and Wood, 1964). Sediment samples came from the upper one-half centimetre of mud or sand and were collected with a glass vial. Epiphytes on *Thalassia* and algal growth on rocks were removed with a razor blade which was then washed in a vial of filtered sea water.

Mugil cephalus were collected with a nylon throw net of 4·5 metres diameter and 2 cm stretched mesh. Samples of food material were taken for both identification and quantitative counts from the cardiac stomach and selected sections of the intestine. They were measured volumetrically in a calibrated tube and then suspended in filtered seawater to make 25 ml of solution. Each sample was shaken thoroughly and a known volume pipetted onto a slide with an etched grid.

Identification of stomach contents was done with conventional light field microscopy. Diatoms were identified to species, but other unicellular algae, filamentous algae, Foraminifera and Protozoa were more broadly categorized. For quantitative comparison of various parts of the alimentary tract, fluorescence microscopy (Wood, 1955; Wood and Oppenheimer, 1962) was used in conjunction with acridine orange stain. The usefulness of this method rests in its ability to detect organisms even when they are adsorbed on or covered by clay or quartz particles.

Pillay (1952) has pointed out the difficulty of estimating the per cent composition of the stomach contents of mullet. For the present study Pillay's method (1953) of eye estimation with a graduated slide was utilized. In addition, an eyepiece micrometer was found invaluable to overcome bias on the observer's part. All materials were placed in three categories: (*i*) fresh plant material (live diatoms, blue-green algae, dinoflagellates, etc.). (*ii*) macroplant detritus, and (*iii*) inorganic particles (quartz, clay, calcium carbonate, etc.).

Obviously, some mistakes are inherent in a procedure which attempts three dimensional answers (volume) using two dimensional methods. Some difficulty was experienced with pennate diatoms and large pieces of detritus because of their asymmetrical shapes. In such cases rough estimates were made. Fortunately, most particles ingested by mullet are small and roughly spherical; thus, ratios from two dimensional measurements should closely approximate true volumes.

Determinations of the per cent of organic matter in sediments, stomach and intestinal contents were derived from loss on ignition at 550°C after 10 hours. Due to the large percentage of quartz, heavy metals and other inorganics, it was important to use small amounts which were burned for the full 10 hours to ensure complete ignition. This method was impractical for the majority of south Florida samples due to the high amount of calcium carbonate in the sediments.

The rate of oxygen consumption by the stomach and intestinal contents was measured with a potentiometer and variable resistor and was read with a linear scale recorder. Silver and platinum electrodes were used and the readings were standardized with the Winkler method.

Samples for plant pigment analysis were taken from the sediments and the cardiac stomach of the mullet. Half of each sample was dried to give the dry weight while the other half was extracted with 90% acetone. Extraction was aided by grinding for 10 minutes with a mortar and pestle. The resulting solution was filtered and analysed with a Beckman DU spectrophotometer. The amount of plant pigment was calculated using the formula of Richards with Thompson reported in Strickland and Parsons (1965). As Vallentyne (1955) has demonstrated, a portion of this value may be due to degradation products of plant pigments (phaeophytine, pheophorbides, etc.); however, for purely comparative purposes the origin of the total pigment value is unimportant.

Caloric analyses were made of mullet stomach contents using a Parr oxygen bomb calorimeter. The high ratio of inorganic to organic matter made it impossible to burn the samples directly; this was overcome by adding known quantity of benzoic acid to the sample and later subtracting the combustion value of the benzoic acid from the total combustion value. Even with this method it was impossible to burn sediment samples.

Utilization of two basic food sources

In the standard energy flow diagram there is a direct flow of energy from the first trophic level (the primary producers) to the second (the primary consumers) and then to the secondary and higher consumers. Classically, the bacteria and other decomposers have been considered as primary consumers along with the herbivores. Since there is a considerable time lag involved in decomposition, it is much more realistic to separate herbivores and decomposers. E. P. Odum (1962, 1963) has suggested a Y-shaped diagram emphasizing two principal pathways, which he refers to as the grazing food chain, in which living plants are the primary energy source for the consumer, and the detritus food chain, in which dead and decaying organic materials are the energy source.

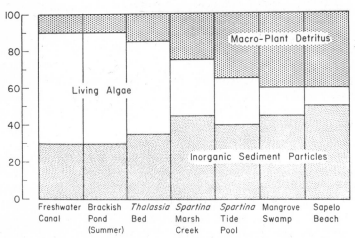

Fig. 1. Estimated percentages of the three important components occurring in the stomach contents of *M. cephalus* in the seven systems. The value for each system represents over 100 fish which were sampled over a period ranging from 3 to 12 months.

The estimated percentages of plant detritus, live plant material and inorganic sediments occurring in the stomach contents of mullet at the seven stations are shown in Fig. 1. It appears that *Mugil cephalus* occupies a position between the two food chains and is able to utilize whichever is the easiest to exploit in a given situation. Thus, mullet at the Sapelo beach station and in the Florida mangrove swamp depend to a great extent upon detritus and its associated micro-organisms, while those from the *Thalassia* bed in Biscayne Bay and from the freshwater canal subsist largely on live plant material. In the other systems investigated, feeding was a more equal combination of items from both food chains. Animal remains (harpacticoid copepods, nematodes, etc.) were found in less than 1% of the stomachs examined.

The preferred food chain

Hartley (1947) has suggested that, "it is . . . possible that the organization of the alimentary system of a particular species, as for example in the relative concentrations of its digestive enzymes, may be such as to obtain maximum advantage for only a limited part of the range of material which the animal is actually capable of ingesting." If this idea is true, *M. cephalus* should show a preference for either plant detritus or live plant material when both are present in abundance.

A situation exists in one of the systems where conditions are suitable to give an indication of which food source is preferred. In the Florida freshwater canal the bottom is covered with large quantities of fine terrestrial plant detritus which has been washed in from surrounding fields. Since the canal was dug through limestone rock the sides present an ideal substrate for the accumulation of a large mat of filamentous green and blue-green algae intertwined about diatoms and desmids. Faced with a choice between these two resources, *M. cephalus* feeds almost exclusively on live plant material (see Fig. 1). A similar situation exists at the *Thalassia* bed where mullet feed primarily on epiphytic diatoms in preference to copious quantities of plant detritus on the bottom. A small amount of detritus is ingested by mullet at both stations when sand grains are sucked from the bottom to triturate the ingested fresh micro-algae.

Parameters of the two diets

The question arises as to why *M. cephalus* would demonstrate a preference for live plant material over plant detritus. In Table 2 a comparison is made of stomach contents which represent three naturally occurring diets; (*i*) a

TABLE 2. Values of stomach contents taken from mullet feeding on three diets. Each value represents the mean for the stomach contents from 25 mullet

	Algal diet (*Thalassia* bed)	Mixed algal and Detritus (*Sapelo marsh*)	Detritus diet (*Sapelo beach*)
Plant pigment (mg/g dry wt.)	0·465	0·315	0·166
Per cent organic matter (dry wt.)	9·81%	6·22%	5·90%
Calorific value (kilocalories per ash-free gram)	5·226	4·552	3·957

predominantly macroplant detritus diet at the Sapelo beach, (*ii*) a mixed detritus and microalgae diet in the Sapelo marshes and (*iii*) a diet made up largely of fresh micro-algae at the Florida *Thalassia* bed. A rough indication of the relative amounts of living plant material in the three diets is given by the milligrams of plant pigment per gram of stomach contents. As a general rule stomach contents from fish feeding on live plant material had values three times as high as those from fish feeding on detritus.

Table 2 also shows that the percentage of organic matter in the stomach contents of mullet feeding on live plant material was almost twice as high as the percentage for those feeding on detritus. It should be remembered that the stomach contents always contain inorganic sediment particles; the weight of these particles is responsible for the low per cent organic matter values.

The most significant difference in the diets is in their caloric content. Clearly, the pure plant diet presents a much greater potential energy source than the plant detritus. According to Odum (1963), plant biomass averages about four kilogram calories per ash-free gram, and animal biomass five. The very high value of 5·226 kilogram calories for the contents of the mullet stomachs from the *Thalassia* bed is probably due to the preponderance of epiphytic diatoms in the diet. Diatoms have been shown to have a high caloric value due to the amounts of light oils which they store. Conover (1964) found a value of 5·342 kilogram calories for *Thalassia fluviatilis*, 5·225 for *Ditylum brightwelli*, and 5·270 for *Rhizosolenia setigera*.

The contribution of phytoplankton to the diet

Very little of the food ingested by mullet originates from the water column; the possible exception would be settled planktonic forms on the benthos. The most common planktonic diatoms occurring over the *Thalassia* bed were *Chaetoceros diversus*, *Ch. affinis*, *Ch. laciniosus*, *Ch. didymus*, *Rhizosolenia fragillissima*, *Nitzchia closterium* and *Skeletonema costatum*. Only the last two were ever noted in mullet stomachs and they are not strictly planktonic diatoms. In the Sapelo marsh the most common planktonic diatoms were *Chaetoceros affinis*, *Ch. curvisetum*, *Hemidiscus hardmanianus*, and *Rizosolenia setigera*. None of these appeared in mullet stomachs with any regularity.

Organic detritus, bacteria and Protozoa as food sources

Organic macroplant detritus was present in virtually all mullet stomachs which were examined (Fig. 1). In the Georgia salt marshes this consisted entirely of particles of decayed *Spartina alterniflora* leaves and stalks. In the Florida habitats it was derived from sea grass leaves, from mangrove leaves, root hairs and bark, and from terrestrial vegetation. Generally, the detritus particles which were ingested were smaller than 200 μ in diameter and often appeared to be organic aggregates of much smaller particles. Fluorescence microscopy with acridine orange stain revealed that these aggregates and particles were covered with large numbers of bacteria and Protozoa. Since the number of these organisms was greatly reduced during passage down the mullet's digestive tract, it is probable that they are utilized as an energy source.

Baier (1935) was one of the first to hypothesize that the nourishment from detritus particles comes from the bacteria involved in decomposition rather than the detritus itself. This idea has been more recently emphasized by Darnell (1967a, 1967b). Zobell and Felthan (1938) demonstrated that certain marine invertebrates could live almost indefinitely on an exclusive diet of bacteria. Zobell (1942) suggested that bacteria were useful as food in two ways: (i) they were important directly as nourishment, and (ii) they assisted the organism's digestion.

Although micro-organisms may not constitute an important fraction of the ingested food mass in a consumer's stomach, they may supply important nutrients (Burkholder and Burkholder, 1956; Burkholder and Burnside, 1957). Teshima and Kashiwada (1967) found that about half of the 198 strains of bacteria which were isolated from the intestinal canal of carp were capable of producing vitamin B_{12}. No vitamin B_{12} decomposing bacteria were found.

Fish (1955) while studying the feeding relationships of tilapia in East Africa concluded that the food of *Tilapia nilotica* and *T. leucosticta* was mainly bacteria and Protozoa adsorbed upon decomposing "debris". He suggested that the micro-organisms were making previously undigestible material available to the fish. Newell (1965) reached much the same conclusion concerning two marine deposit feeders, the prosobranch *Hydrobia ulvae* and the bivalve *Macoma baltica*. Neither organism is able to alter or utilize the basic carbon structure of ingested detritus particles, but both are capable of digesting micro-organisms adsorbed on the particles. These micro-organisms are capable of oxidizing the carbon of the detritus particles and indirectly convert the detritus carbon into mollusc carbon.

From these examples the following recurring cycle emerges. First, the adsorbed bacteria and fungi begin oxidation, hydrolysis and assimilation of the basic carbon structure of the detritus particle. During the process of microbial breakdown, these decomposers are continuously grazed by Protozoa. This creates a rich Protozoa-bacteria-detritus system with great potential food value. At intervals the entire complex may be ingested by a larger organism such as a snail or mullet and most of the bacteria, fungi, and Protozoa digested off the particle. In the case of organisms such as *Mugil cephalus* the detritus particle may be further fragmented by grinding or chewing. Once the particle (or fragments of the original particle) is released as faecal material into the environment, the entire process begins again.

Feeding periodicity

Under normal conditions mullet appear to feed almost continuously: however, the intensity of feeding is not always the same. Since ingested material is continuously transferred from the cardiac stomach to the pyloric stomach, the feeding intensity is reflected by the amount of material present in the cardiac stomach. If feeding is intensive the cardiac stomach will remain filled, but if feeding is sporadic the cardiac stomach will be only partially filled.

In areas which are influenced by tidal exchange there is a definite relationship between the rate at which food is ingested and the state of the tide. In

Fig. 2 the percentage of the cardiac stomach capacity which is occupied by ingested material is plotted against the stage of the tide. There is a marked increase in the amount of food ingested as the tide rises. This is not surprising as many of the optimal feeding areas become accessible to the mullet with the rising tide.

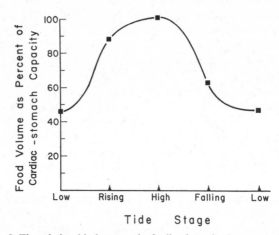

Fig. 2. The relationship between the feeding intensity (represented by the fullness of the cardiac stomach) and the stage of the tide. Each of the four data points represents the mean value for 50 mullet sampled at the Sapelo beach.

A prolonged cessation of feeding was noted only under unusual circumstances. Mullet with empty digestive tracts were found at the Georgia beach station during extreme weather conditions when there was a great deal of turbidity and turbulence. In Biscayne Bay mullet ceased feeding immediately before and during the fall spawning migrations to offshore areas.

Food as a limiting factor

For a "broad spectrum" herbivore such as *Mugil cephalus* food does not appear to be a limiting factor. The ability to ingest and apparently utilize benthic and epiphytic diatoms, dinoflagellates, Protozoa, green and blue green algae along with macroplant detritus and its associated micro-organisms ensures a constant energy source. In all the environmental systems examined at least two of these components were present in copious amounts. In the event of severe depletion or competition for any of these food materials the mullet possesses the ability to switch to a different type of feeding.

Although adult mullet do not appear to be food limited, it has been noted by Suzuki (1965) and others that during the first several weeks of life (the Querimera stage) the young mullet feed chiefly on planktonic micro-crustaceans. It is possible that the survival of Querimera stage mullet may be controlled by the availability of micro-crustaceans in a density-dependent manner.

Selection-rejection rate

By comparing the amount of plant pigment in the contents of the cardiac stomach with the amount occurring in the sediments upon which the fish feed, it is possible to obtain a rough estimate of the amount of mud filtered by a mullet to obtain one gram of ingested material. Such a determination will be valid only if feeding is confined exclusively to the sediments.

In the Sapelo marshes all substances ingested by *M. cephalus* originate from the sediments. The upper one-half centimetre of these sediments was found to contain 2-4 micrograms of plant pigment per dry gram of sediment (range of values for 20 sediment samples). Mullet feeding upon these same sediments contained 350 micrograms of plant pigment per dry gram of stomach contents (mean value of stomach contents of 25 mullet). This indicates that a mullet must filter almost one hundred grams of sediment to obtain one gram of material in its digestive tract.

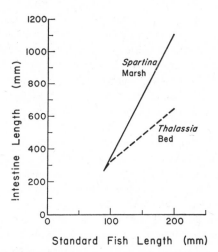

Fig. 3. The relationship between the standard fish length and the intestinal length for two populations of mullet; (1) a population from the *Spartina* marsh creek and (2) a population feeding in the vicinity of the Florida *Thalassia* bed. Each regression line was constructed from measurements made on 50 mullet.

Turnover effect on the sediments

A mullet of 200 mm standard length commonly contains about 3 grams (dry weight) of sediment material in its pyloric and cardiac stomachs and intestine. This would require filtration of 300 dry grams of sediment. Assuming more or less continuous feeding and a turnover rate of five times in 24 hours (discussed later in this paper), such a mullet would filter 1500 dry grams of sediment per day or over 450 kilograms (half a metric ton) a year. This means that a single mullet would be responsible for altering every year an area of sediment measuring 45 square metres and $\frac{1}{2}$ centimetre deep. The effect of a large school of mullet on the environment is not difficult to imagine.

Morphological adjustment to the environment

The ratio of the length of a fish's intestinal tract to the total length of the fish is often a good indication of the type of food commonly ingested. Generally, this ratio is less than unity in carnivores, from one to three in omnivores and even greater in herbivores. Al-Hussaini (1949) has stated that this ratio is nearly constant for each individual species, but evidence suggests that this is only true for carnivores. Klust (1939) found that the length of the gut increased relative to the body length with age in the *Cyprinidae*. Kostomarov (1942) confirmed this for the carp, *Cyprinis*. However, in the tench, *Tinca*, which is normally a carnivore, the proportion of the gut length remained the same at all ages (Kostomarov and Pulankova, 1942).

Hiatt (1944) examined *M. cephalus* growing in Hawaiian fish ponds and found that the intestinal length and total fish length maintained a constant ratio of 3·2:1 in mullet of different sizes. On the other hand, milkfish, *Chanos chanos*, in the same ponds demonstrated heteronomous growth of the intestine by increasing the ratio of the intestinal length to total fish length from 3·5:1 for fish of 90 mm total length to 7·2:1 for fish of 115 mm. For larger fish the ratio tended to decrease slightly and finally reached 6·5:1 for fish of 272 mm. The great increase of intestinal growth for fish between 90 and 115 mm was correlated with a change in diet from unicellular algae to filamentous algae and plant fragments.

The length of intestine versus standard fish length of two populations of *M. cephalus* are plotted in Fig. 3; only fish between 75 and 200 mm were measured. The extremely steep slope for the Sapelo marsh population shows a much faster growth of the intestine—six or seven times greater than the increase of the standard fish length. As a result the ratio of the intestine to the standard length increased from 3·2:1 in 100 mm fish to 5·5:1 in those measuring 200 mm. In contrast, the data from the mullet feeding in the area of the Biscayne Bay *Thalassia* bed displays a slope of slightly greater than one or an almost constant ratio of 3·2:1, a result which is in close agreement with Hiatt's data.

The apparent discrepancy of these data can be explained by noting the most important components of the diet of these two groups of mullet and the population described by Hiatt in Hawaii. The Sapelo marsh mullet included a great deal of plant detritus in their diet while those in Biscayne Bay and the Hawaiian fish ponds fed predominantly on benthic and epiphytic diatoms, a food source which is much more easily broken down by the mullet's pyloric stomach and presumably easier to absorb. It seems that an intestine to fish ratio of about 3·2:1 is adequate to effectively assimilate a diatom diet, while a longer intestine is needed to extract nourishment from plant detritus.

Schuster (1949) has found a similar difference in intestinal length to fish length ratios for *Chanos chanos* feeding on different diets. In one pond where *Cyanophyceae* predominated the ratio was 7·9:1, while in another pond where

Chlorophyceae predominated the ratio was only 5·7:1. He concluded that the widespread utilization of *Chanos* as a pond fish could be partially explained by the ability of the species to adjust morphologically to the environment. The same could be said for *Mugil cephalus*.

Intestinal flagellate populations

Large numbers of very small (10-20 μ) colourless flagellates were usually present in the intestinal contents of *M. cephalus*. These flagellates, which could not be identified, probably function in the breakdown of organic particles, particularly the cellulose walls of plant detritus particles. They were characteristically observed clustered around the broken cell walls of pieces of mangrove and *Thalassia* detritus.

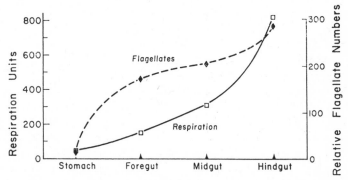

Fig. 4. Flagellate numbers and oxygen consumption rates for cardiac stomach contents and 3 portions of intestinal contents from a single Sapelo marsh mullet. The flagellate numbers were counted from 20 grids of a calibrated counting chamber using fluorescence microscopy. One respiration unit = $1·70 \times 10^{-4}$ milligram atoms/O$_2$/hour.

Since most teleost intestinal fauna are facultative anaerobes (E. J. F. Wood, personal communication), it was possible to gain some idea of the activity of intestinal micro-organisms in general and the flagellates in particular by measuring the respiration rate of the intestinal contents at various points under aerobic conditions. To do this, small portions of intestinal contents from a freshly killed mullet were suspended in 25 ml of filtered and sterilized sea-water and measured in the dark with a respirometer. In Fig. 4 these respiration values for one mullet are plotted along with counts of flagellates made from the same samples using fluorescence microscopy and acridine orange stain. The lowest values for respiration and counts were in the cardiac stomach and were followed by a steady increase down the intestinal tract.

It is very likely that *M. cephalus* has the ability to digest these flagellates. Fish which have been starved for a day or two have nothing in the intestinal tract except quantities of mucopolysaccharide. Even under fluorescence microscopy with acridine orange no micro-organisms could be found.

Both Hunter (1920) and Wood (1940) have suggested that the intestinal tracts of many fish become sterile during periods when no food is ingested.

Wood found it difficult to obtain cultures from the guts of non-feeding, migrating *M. cephalus* in Australia; when organisms were found they were sporulating aerobes. He postulated a bactericidal or at least bacteriostatic influence in the intestinal tract, thus ruling out permanent intestinal flora and fauna.

Maximum power output

It is commonly accepted that the ideal natural system should approach maximum efficiency in converting food. In contrast to this supposition is the maximum power hypothesis of H. T. Odum and R. C. Pinkerton (1955) which is predicated upon the hypothesis that natural systems tend to operate at an efficiency which produces a maximum power output. Thus, a predator which may be feeding upon a scarce and potentially limiting food source will have a low rate of ingestion coupled with a high assimilation efficiency, while an omnivore feeding upon plentiful and easily obtained supplies of micro-algae and plant detritus will achieve the greatest power output from a high rate of ingestion and a relatively low assimilation efficiency.

The digestive tract of *M. cephalus* is efficient at breaking up diatoms; other particles such as plant detritus and blue-green algae are only partially destroyed. Mullet feeding upon a bloom of the dinoflagellate *Kryptoperidinium* have been found to pass from 20 to 30 per cent through the digestive system in a viable condition (W. E. Odum, 1968). The explanation for this lies in the ability of the dinoflagellate to encyst rapidly upon meeting adverse conditions.

It was of interest to see if the apparent low assimilation efficiency and high ingestion rate of *Mugil cephalus* were accompanied by a low retention time. To determine the retention time of food material, several groups of mullet ranging in length from 100 to 200 mm were fed a mud-detritus mixture containing fluorescent "Day-glo" (fireorange AG-14) particles (see Haven and Morales-Alamo, 1966). These particles, which have a mean diameter of 4·5 μ, were readily ingested and subsequently labelled the faecal matter very clearly.

A comparison of retention times in fish may be misleading due to the effects of temperature, the condition of the fish and the rate of feeding. For this reason several groups of fish were tested at three different temperatures: 20°C, 23°C, and 26°C. Retention time of the ingested material was highly variable, but could not be correlated with these temperature changes. Depending upon the intensity of feeding, the retention time ranged from two to six hours. The mean value was between four and five hours indicating a turnover rate of five times in 24 hours.

In summary, the maximum power output for *Mugil cephalus* results from a relatively large and continuous ingestion rate and short retention time coupled with an assimilation efficiency which is probably high for a few components of the diet such as epiphytic and benthic diatoms, but rather low when the entire bulk of ingested organic matter is considered.

SOME CONCLUSIONS AND HOW THEY RELATE TO POND CULTURE

Although *Mugil cephalus* is widely used as a pond culture species, in most cases it either is of secondary importance to other species such as eels (Nakamura, 1949; D'Ancona, 1955) and *Chanos chanos* (Acosta, 1953) or when it is the primary species, techniques have not been refined to the point where optimal yields are obtained. In reviewing the results of the present investigation several ideas emerge which might enhance the production of mullet in ponds.

First, it should be realized that mullet in a culture pond differ from those living in a natural ecosystem in that their growth may be easily limited by the availability of food. A culture pond is an artificial ecosystem with an artificially increased population size and, ideally, a lack of predators. To develop an unnaturally large population size and still retain optimal growth, it is necessary to increase the amount of food available to the mullet. Addition of commercial inorganic fertilizers can increase the production of mullet ponds somewhat, but this by itself may not be sufficient.

Earlier in this paper *M. cephalus* was characterized as an "interface feeder" dependent upon the availability of physical surfaces to concentrate food. If this is true, mullet in ponds are limited ultimately by the surface available for grazing. The way to increase production, then, is to increase the available surface area and at the same time supply adequate nutrients. There are two ways in which this might be done.

The first is to create an artificial grass bed in the pond. Natural sea grasses are difficult to transplant and in unnatural conditions are susceptible to sudden salinity changes and disease. Artificial grass beds, however, can be easily and inexpensively made from strips of polypropylene film. W. L. Rickards at the University of Miami has built such structures from long strips of this positively buoyant film attached at the bottom to panels of synthetic netting. The netting, which has a mesh size sufficiently large to allow fish to feed on the bottom, is weighted down every few square metres with concrete weights. The result is a luxuriant bed of the desired density and with the filaments stretching from the bottom to the surface. Within two weeks the surfaces of the filaments become encrusted with epiphytic micro-algae, increasing by at least 20 times the surface available for mullet grazing.

A second way to increase production in mullet ponds is to increase the surface area of the sediment particles. This can be accomplished by constructing the pond so that the bottom consists of clay-sized inorganic particles. In addition, large quantities of very finely chopped macroplant detritus can be periodically added to the bottom of the pond. The increased surface area of the fine inorganic and plant detritus particles presents an ideal substrate for growth of micro-algae, bacteria, Protozoa and fungi. This in turn should increase mullet production. It is important that both the inorganic sediment particles and the artificially created detritus particles be very fine (less than $200\,\mu$) since most of the particles ingested by mullet are of this size (see W. E. Odum, 1968).

SUMMARY

1. It is suggested that in shallow estuaries most food chains are based not on phytoplankton, but on macroplant detritus and benthic and epiphytic micro-algae. It is the animals capable of utilizing such energy sources which replace zooplankton as the critical herbivore link.

2. To obtain the maximum yield of fish from a shallow estuary it will be necessary to concentrate management and culture operations upon species such as *Mugil cephalus* which obtain their energy directly from the first trophic level.

3. From analyses of the diet in seven different environmental systems it appears that *M. cephalus* is able to utilize either the direct grazing or plant detritus food chains as an energy source depending upon which is the easiest to exploit. When both food sources are present in abundance, the mullet exhibits a preference for living micro-algae. Such an algal diet has a higher caloric content than a macroplant detritus diet.

4. Fine particles of decaying plant detritus (mangrove, *Spartina*, *Thalassia*) were present in virtually all mullet stomachs which were examined. Bacteria and Protozoa adsorbed to these particles may be important in the diet of the striped mullet, either as a source of essential nutrients or by providing assistance in the breakdown of plant material. Unidentified colourless flagellates, often present in the intestinal contents, seem to function in the destruction of cellulose cell walls. Starved fish have a digestive tract which is practically sterile.

5. Although feeding is almost continuous, a marked increase in the rate of ingestion was found during rising and high tides. Feeding ceases during adverse weather and at the time of offshore spawning migrations.

6. For a "broad spectrum" herbivore such as *M. cephalus* food does not seem to be a limiting factor. The ability to ingest and apparently utilize benthic and epiphytic diatoms, dinoflagellates, Protozoa, green and blue-green algae along with macroplant detritus ensures a constant energy source.

7. By comparing the amount of plant pigment in the contents of the cardiac stomach with the amount in the sediments upon which the fish feed, it was found that *M. cephalus* filters almost 100 dry grams of sediment to obtain one dry gram of material in its digestive tract. Assuming more or less continuous feeding and a turnover rate of five times in 24 hours, a 200 mm S.L. mullet would filter 1500 dry grams of sediment per day or over 450 dry kilograms (half a metric ton) a year.

8. Mullet populations adjust morphologically to the type of food available in the environment by lengthening the intestine for coarser diets. An intestine length to fish length ratio of 3·2:1 is adequate to effectively assimilate a diatom diet, while a longer intestine is needed to extract nourishment from plant detritus and blue-green algae.

9. The maximum power output for *Mugil cephalus* results from a large, continuous ingestion rate and short retention time coupled with an assimilation efficiency which is probably high for diatoms, but low when the entire bulk of ingested organic material is considered.

10. To obtain the maximum yield of mullet from a culture pond it is necessary to increase the available surface area for growth of microfauna and
microflora which are subsequently grazed by *M. cephalus*. This can be
accomplished by creating artificial grass beds and constructing the bottom of
the pond so that it is composed of very fine inorganic and plant detritus
particles.

REFERENCES

ACOSTA, P. A. 1953. Inland fisheries of the Philippines. *Philippine fisheries. Handb. 8th Pacif. Sci. Congr.*, 124-31.
AL-HUSSAINI, H. 1947. The feeding habits and the morphology of the alimentary tract of some teleosts living in the neighborhood of the Marine biological Station, Ghardaqa, Red Sea. *Publs. mar. biol. Stn Ghardaqa*, (5), 61 pp.
AL-HUSSAINI, H. 1949. On the functional morphology of the alimentary tract of some fish in relation to differences in their feeding habits: anatomy and histology. *Q. Jl. microsc. Sci.*, **90**, 109-39.
ANDERSON, W. W. 1958. Larval development, growth and spawning of striped mullet (*Mugil cephalus* L.) along the south Atlantic coast of the United States. *Fishery Bull. U.S. Fish Wildl. Serv.*, **58**, (144), 501-25.
ARNOLD, E. L., and THOMSON, J. R. 1958. Offshore spawning of the striped mullet (*Mugil cephalus*) in the Gulf of Mexico. *Copeia*, 1958 (3), 84.
BAIER, C. R. 1935. Studien zur Hydrobakteriologie stehender Bennengewasser. *Arch. Hydrobiol.*, **29**, 183-264.
BURKHOLDER, P. R., and L. M. 1956. Microbiological assay of vitamin B12 in marine solids. *Science*, **123**, 1071-73.
BURKHOLDER, P. R., and BURNSIDE, G. H. 1957. Decomposition of marsh grass by aerobic marine bacteria. *Bull. Torrey bot. Club*, **84**, 366-83.
CONOVER, R. J. 1964. Food relations and nutrition of zooplankton. *Proc. Symp. exp. mar. Ecol.*, 1963, *Occ. Publ. Narragansett mar. Lab. R.I.* (2), 81-91.
D'ANCONA, U. 1955. Fishing and fish culture in brackish lagoons. *Fish. Bull. F.A.O.*, **7**, 147-72.
DARNELL, R. M. 1967a. The organic detritus problem. In *Estuaries*. Edited by G. H Lauff. *AAAS Publ.*, **83**, 374-75.
DARNELL, R. M. 1967b. Organic detritus in relation to the estuarine ecosystem. In *Estuaries*. Edited by G. H. Lauff. *AAAS Publ.*, **83**, 376-82.
DILL, W. A. 1944. The fishery of the lower Colorado River. *Calif. Fish Game*, **30** (3), 109-214.
EBELING, A. W. 1957. The dentition of eastern Pacific mullets, with special reference to adaptation and taxonomy. *Copeia*, 1957 (3), 173-85.
FISH, G. R. 1955. The food of Tilapia in east Africa. *Uganda J.*, **19** (1).
HARTLEY, P. H. T. 1947. The natural history of some British freshwater fishes. *Proc. zool. Soc. Lond.*, **117**, 129-206.
HAVEN, D. S., and MORALES-ALAMO, R. 1966. Use of fluorescent particles to trace oyster biodeposits in marine sediments. *J. Cons. perm. int. Explor. Mer*, **30**, 267-69.
HELLIER, T. R. 1957. *The fishes of the Santa Fe River system*. Unpublished M.S. thesis, University of Florida, 99 pp.
HIATT, R. W. 1944. Food chains and the food cycle in Hawaiian fish ponds. Part I. The food and feeding habits of mullet (*Mugil cephalus*), milkfish (*Chanos chanos*) and the tenpounder (*Elops machnata*). *Trans. Am Fish. Soc.*, **74** (2), 250-61.
HUNTER, A. C. 1920. Bacterial decomposition of salmon. *J. Bact.*, **5**, 345.
JENSEN, P. B., and PETERSEN, C. G. 1911. Valuation of the sea. I. Animal life of the sea-bottom, its food and quantity. *Rep. Danish biol. Stn*, **20**, 1-81.
JØRGENSEN, C. B. 1966. *Biology of suspension feeding*. Oxford, Pergamon Press, 357 pp.
KIMBALL, J. F., and WOOD, E. J. F. 1964. A simple centrifuge for phytoplankton studies. *Bull. mar. Sci. Gulf Caribb.*, **14**, 539-44.

KLUST, G. 1939. Über Entwicklung, Bau and Funktion des Darmes beim Karpfen (*Cyprinus carpio* L.). *Int. Revue ges. Hydrobiol. Hydrogr.*, **39**, 498.

KOSTOMAROV, B., and PULANKOVA, B. 1942. Die Beziehung zwischen der Korpergrosse des Karpfens seiner Darmgrosse. *Z. Fisch.*, **40**, 157-70.

NAKAMURA, N. 1949. Some notes on the pond-culture of mullets. *Bull. Jap. Soc. scient. Fish.*, **14**, 211-15.

NEWELL, R. 1965. The role of detritus in the nutrition of two marine deposit feeders, the prosobranch *Hydrobia ulvae* and the bivalve *Macoma baltica*. *Proc. zool. Soc. Lond.*, **144** (1), 25-45.

ODUM, E. P. 1962. Relation between structure and function. *Jap. J. Ecol.*, **12**, 108-118.

ODUM, E. P. 1963. In *Ecology*. New York, Holt, Rinehart and Winston.

ODUM, H. T., and Pinkerton, R. C. 1955. Time's speed regulator; the optimum efficiency for maximum power output in physical and biological systems. *Am. Scient.*, **43**, 331-43.

ODUM, W. E. 1968. Mullet grazing on a dinoflagellate bloom. *Chesapeake Sci.*, **9**, 202-204.

ODUM, W. E. 1968. The ecological significance of fine particle selection by the striped mullet *Mugil cephalus. Limnol. Oceanogr.*, **13**, 92-97.

PILLAY, T. V. R 1952. A critique of the methods of study of food of fishes. *J. zool. Soc. India*, **4** (2), 185-200.

PILLAY, T. V. R. 1953. Studies on the food, feeding habits and alimentary tract of the grey mullet, *Mugil tade. Proc. natn. Inst. Sci. India*, **19**, 777-827.

POMEROY, L. R. 1960. Primary productivity of Boca Ciega Bay, Florida. *Bull. mar. Sci. Gulf Caribb.*, **10** (1), 1-10.

RAGOTZKIE, R. A. 1959. Plankton productivity in estuarine waters of Georgia. *Publs Inst. mar. Sci. Univ. Tex.* **6**, 146-58.

SCHELSKE, C. L., and ODUM, E. P. 1961. Mechanisms maintaining high productivity in Georgia estuaries. *Proc. Gulf Caribb. Fish. Inst.*, **14**, 75-80.

SCHUSTER, W. H. 1949. On the food of the bandeng, *Chanos chanos. Meded. alg. Proefstn Landb., Buitenz.*, **86**. 20 pp.

STRICKLAND, J. D. H., and PARSONS, T. R. 1965. A manual of sea water analysis. *Bull. Fish. Res. Bd Can.* (125), 815 pp.

SUZUKI, K. 1965. Biology of striped mullet *Mugil cephalus* Linne. I. Food contents of young. *Rep. Fac. Fish. prefect. Univ. Mie*, **5** (2), 295-305.

TEAL, J. M. 1962. Energy flow in the salt marsh ecosystem of Georgia. *Ecology*, **43**, 614-24.

TESHIMA, S., and KASHIWADA, K. 1967. Studies on the production of B vitamins by intestinal bacteria of fish. III. Isolation of vitamin B12 synthesizing bacteria and their bacteriological properties. *Bull. Jap. Soc. scient. Fish.*, **33**, 979-83.

THOMPSON, J. M. 1954. The organs of feeding and the food of some Australian mullet *Aust. J. mar. freshwat. Res.*, **5**, 469-85.

THOMPSON, J. M. 1963. Synopsis of biological data on the grey mullet *Mugil cephalus Fish. Synopsis, C.S.I.R.O.*, (1), 82 pp.

THOMPSON, J. M. 1966. The grey mullets. *Oceanogr. and mar. Biol.*, **4**, 301-35.

VALLENTYNE, J. R. 1955. Sedimentary chlorophyll determination as a paleobotanical method. *Can. J. Bot.*, **33**, 304-13.

WILLIAMS, R. B., MURDOCH, M. B., and THOMAS, L. K. 1968. Standing crop and importance of zooplankton in a system of shallow estuaries. *Chesapeake Sci.*, **9** (1), 42-51.

WOOD, E. J. F. 1940. Studies on the marketing of fresh fish in eastern Australia. Part 2. The bacteriology of spoilage of marine fish. *C.S.I.R.O. Pamphlet* (100), 92 pp.

WOOD, E. J. F. 1955. Fluorescence microscopy in marine microbiology. *J. Cons. perm. int. Explor. Mer*, **21**, 6-7.

WOOD, E. J. F. 1964. Studies in microbial ecology of the Australian region. *Nova Hedwigia* **8**, 461-568.

WOOD, E. J. F., and OPPENHEIMER, C. 1962. Note on fluorescence microscopy in marine microbiology. *Z. allg. Mikrobiol.*, **2**, 164-65.

ZENKEVITCH, L. 1963. *Biology of the seas of the U.S.S.R.* Trans. by S. Botcharskaya London, Allen and Unwin. 955 pp.

ZOBELL, C. E. 1942. The bacterial flora of a marine flat as an ecological factor. *Ecology*, **23**, 69-78.

ZOBELL, C. E., and FELTHAN, C. B. 1938. Bacteria as food for certain marine invertebrates. *J. mar. Res.*, **4**, 312-27.

Feeding pattern of baleen whales in the ocean

TAKAHISA NEMOTO
Ocean Research Institute
University of Tokyo
Japan

ABSTRACT. A study of the main feeding structures of baleen whales–baleen plates, ventral groove extension, tongues and the shape of heads—prove that there are three feeding types, blue whale (blue, fin, humpback, Bryde's), right whale (right, Greenland) and grey whale types. These types are divided into "skimming", "swallowing" and a combination of these two. Food species are described both for the southern hemisphere and northern Pacific, and a clear selection of foods is observed. The swallowing type whales demand heavy patches of food organisms such as euphausiids and gregarious fish, but skimming whales depend on more sparse plankton patches. Feeding activity is high in the morning and becomes low in daytime. Food quantities in stomachs are considered to coincide with the probable caloric intake estimated from the huge weight and big proportion of the blubber tissue in whales. Two main food chains involving whales in the Antarctic and a more complex one in the northern hemisphere are discussed. Seasonal feeding migration to seek heavy patches of zooplankton is suggested chiefly for swallowing type whales. The feeding range and areas are also restricted by the biological strength (number, size, school, time of migration) and interspecific interferences.

INTRODUCTION

The status of baleen whales in marine food chains is very interesting considering their huge body sizes, in conjunction with their consumption of comparatively minute foods. Baleen whales had been considered to feed generally on small zooplankton, but recent studies show that the habits and selection of food vary considerably from species to species according to their feeding apparatus (Tomilin, 1954; Nemoto, 1959; Kulmov, 1961).

For a study of feeding patterns eleven baleen whales belonging to three families and six genera are considered here.

241

TABLE 1

Balaenidae		
Balaena mysticetus	(Greenland right whale)	Northern Polar Seas
Eubalaena glacialis	(Right whale)	Northern and Southern hemisphere
Caperea marginata	(Pygmy right whale)	Southern hemisphere
Eschrichtiidae		
Eschrichtius gibbosus	(Grey whale)	North Pacific
Balaenopteridae		
Balaenoptera musculus	(Blue whale)	Northern and Southern hemisphere
Balaenoptera physalus	(Fin whale)	Northern and Southern hemisphere
Balaenoptera edeni	(Bryde's whale)	Northern and Southern hemisphere
Balaenoptera borealis	(Sei whale)	Northern and Southern hemisphere
Balaenoptera acutorostrata	(Little piked whale, Minke whale)	Northern and Southern hemisphere
Balaenoptera bonaerensis	(New Zealand piked whale)	Southern hemisphere
Megaptera novaeangliae	(Humpback whale)	Northern and Southern hemisphere

Feeding apparatus

The main structures affecting the feeding of baleen whales are baleen plates, head structures including mouth opening and tongue, and ventral grooves in the abdominal part of the body. The characteristics of baleen plates are summarized in Table 2. The shape of plates in right, Greenland and pygmy right whales is slender and elastic, and the fringes along the inner margin of plates are very fine and numerous. The plates of Balaenopteridae whales (blue, fin, humpback, etc.) are short and tough, and have rather rough baleen fringes. Sei whales have somewhat slender plates and sei and minke whales have finer baleen fringes. Grey whales have thick, short plates with coarse, short fringes. The younger whales generally have finer baleen fringes than the adults, and local differences in the character of baleen plates in the same species are also observed in many species if they feed regularly on different kinds of foods.

TABLE 2. Characteristics of baleen plates of whales in the North Pacific

	Blue	Fin	Bryde's	Sei	Minke	Hump-back	Right	Grey
Mean number in one side	360	355	300	340	280	330	245	160
Mean diameter of baleen fringe (mm)	1·1	0·8	0·6	0·2	0·3	0·7	0·2	1·0
Number of baleen fringes per 1 cm	10-30	10-35	15-35	35-60	15-25	10-35	35-70	10-15
Shape and quality	tough	tough	tough	tough	somewhat elastic	tough	slender elastic	tough

The arrangement of baleen plates indicates three types. Balaenidae whales lack the plates or smaller tuft at the tip of the upper palate. Grey whales also lack the plates at the tip but the filtering area of baleen plates is limited as they have short plates. *Balaenoptera* whales have smaller plates or tufts at the tip. The skull is curved in Balaenidae whales and rather straight in *Balaenoptera* whales, although sei whales show a somewhat different curve from other *Balaenoptera* whales. Right and Greenland whales have panels in both sides of the lower jaw, which grows upwards to fit the lower edge of the curved arch of the skull.

Although the tongue of right and grey whales is rather tough, in *Balaenoptera* whales it is very flabby. The former is effective in conducting water which contains planktonic food along the inner plane of the rows of baleen plates in continuous feeding; the latter works in such a way as to make the mouth cavity expand for swallowing large amounts of water with the food.

Balaenidae whales lack the ventral groove. Grey whales have only two to four furrows in the abdominal part instead of grooves. The grooves of humpback whales are wide, and their number (about 18 to 24) is far less than other *Balaenoptera* whales (about 52-82). However, the extension is the same with blue and fin whales ranging from 55 to 58% of the body length. Sei and minke whales have a shorter extension of grooves ranging from 45 to 47% of the body length.

TABLE 3. Feeding apparatus types of baleen whales in the North Pacific*

| Whale species | Baleen plates | | | Head, mouth and tongue | Ventral grooves | Apparatus type |
	Shape	Fringe	Row			
Blue	Blue	Blue	Blue	Blue	Blue	Blue
Fin	Blue	Blue	Blue	Blue	Blue	Blue
Bryde's	Blue	Blue	Blue	Blue	Blue	Blue
Sei	Blue	Right	Blue	Blue (Sei)	Sei	Blue (Sei)
Minke	Blue	(Blue)	Blue	Blue	Sei	Blue (Sei)
Humpback	Blue	Blue	Blue	Blue	Blue	Blue
Grey	Grey	Grey	Grey	Grey	Grey	Grey
Right	Right	Right	Right	Right	Right	Right
Greenland	Right	Right	Right	Right	Right	Right

* Modified from the list by Nemoto (1959)

Those characteristics suggest three main types of feeding apparatus, namely blue, grey and right whale types (Table 3), although some parts of sei whales and sometimes minke whales demonstrate a separate sei whale type.

Food species and selection of food

Antarctic. *Euphausia superba* had been considered as the only important food for baleen whales in the Antarctic; however, recent investigations (Nemoto, 1962; 1968) have described many other planktonic animals as food of baleen whales (Table 4). Blue and fin whales feed mainly on euphausiids and the species they take are considered to form dense swarms in the sea.

TABLE 4. Stomach contents of baleen whales caught by Japanese pelagic catch from 1961 to 1965 in the Antarctic*

Food species	Whale species				
	Blue†	Fin	Sei	Humpback	Minke
Euphausiids	517	16158	5936	7	88
Euphausiids and others	4	18	4	—	—
Copepods	2	—	2472	—	—
Amphipods	6	9	1514	—	—
Munida decapods	—	—	75	—	—
Fish	—	76	31	—	—
Squids	—	—	5	—	—
Vacant	674	18878	16145	2	10
No. of whales examined	1203	35139	26182	9	98

* Sei whales include 1966 season.
† Mainly subspecies *Balaenoptera musculus brevicanda* distributed in the lower Antarctic.

Euphausia crystallorophias is often observed in a heavy patch along the Antarctic continental shelf and is fed upon by blue and minke whales (Marr, 1956). *Euphausia superba, Thysanoessa macrura* and *T. vicina* are found within the Antarctic convergence. There are other euphausiids in the Antarctic waters, such as *Euphausia triacantha, E. longirostris, E. lucens, E. fringida,* (John, 1936; Baker, 1965) which have not been described as the main food of baleen whales owing to their non-swarming habit in the ocean (Baker, 1959, for *E. triacantha*). *Euphausia vallentini* is mainly found along the Antarctic convergence (Nemoto, 1962). The subspecies of blue whales, pigmy blue whales, are confined to those waters of lower latitudes where *E. vallentini* is abundant, where they are separated from the larger normal form of blue whales.

Sei whales in the southern hemisphere feed on various plankton, i.e. animals such as carnivorous amphipods (*Parathemisto gaudichaudi*) and herbivorous copepods (*Drepanopus pectinatus*, etc). These are mostly found along the Antarctic and subtropical convergences, and this is the reason why sei whales feed in the comparatively lower latitudes of the Antarctic and subantarctic region. There are also large stocks of zooplankton, including the planktonic decapod *Munida gregaria* (or lobster krill), in the coastal waters of Patagonia.

Minke and Antarctic minke (*Balaenoptera bonaerensis*) whales feed on *Euphausia superba* or *E. crystallorophias* in the pack ice region and offshore feeding is rarely observed.

North Pacific. The food in the north Pacific (Table 5) varies considerably (Nemoto, 1959). Blue whales feed almost entirely on euphausiids. Fin whales feed on both euphausiids and copepods (*Calanus cristatus* and *Calanus plumchrus*). Shoaling fish, herring (*Clupea pallasii*), Alaskan pollack (*Theragra chalcogramma*), capelin (*Mallotus catervarius*) also form part of their diet. Sei whales on the other hand feed mainly on copepods (*Calanus*

TABLE 5. Stomach contents of baleen whales caught by Japanese pelagic catch from 1952 to 1965 in the North Pacific

Food species	Whale species				
	Blue	Fin	Sei	Humpback	Right
Euphausiids	455	4818	85	238	—
Euphausiids and Copepods	5	321	2	2	—
Euphausiids and others	—	34	—	12	—
Sergestes shrimp	1	—	—	—	—
Copepods	6	1877	1459	2	9
Copepods and others	—	3	12	—	—
Fish	—	469	36	53	—
Fish and others	—	—	1	—	—
Squids	—	51	21	1	—
Vacant	504	8794	3565	150	0
No. of whales observed	971	16367	5181	458	9

plumchrus) and those feeding on euphausiids are rather scarce. Humpback whales appear to feed mainly on euphausiids and gregarious fish, although the number of observations is low. Right whales feed only on copepods such as *Calanus plumchrus* and *C. cristatus*. Bryde's whales feed in comparatively warmer waters both in the northern and southern hemispheres (Nemoto, 1959). In the north Pacific they sometimes feed on oceanic micronekton such as lantern fish, *Yarrella microcephala* and *Myctophum asperum*.

Species of euphausiids important as food of baleen whales in the northern Pacific are *Euphausia pacifica*, *Thysanoessa inermis*, *T. longipes*, *T. spinifera* and *T. raschii*, which form extensive swarms in the sea (Nemoto, 1959). In the southern waters of the north Pacific, *E. pacifica*, *E. similis*, *E. recurva* and *E. nana* are important as food for fin whales. The copepods *Calanus cristatus*, *C. plumchrus* and *Metridia lucens* are important as food. *Eucalanus* and other non-swarming copepods never form part of the food supply of baleen whales.

Thus, the order of selection of the food in baleen whales is as follows (= shows equivalence and > shows the dominance to the left).

Blue whale	Euphausiids
Fin whale	Euphausiids = Copepods (large) = Gregarious fish> Copepods (small) > Squids
Bryde's whale	Euphausiids = Gregarious fish > Copepods (small)
Sei whale	Copepods ⩾ Amphipods ⩾ Euphausiids = Swarming fish = Squids
Humpback whale	Euphausiids = Gregarious fish
Minke whale	Euphausiids = Swarming fish > Copepods
Right whale	Copepods > Euphausiids

Feeding types

Two basic feeding types, defined as swallowing and skimming have been proposed (Nemoto, 1959; 1968). *Balaenoptera* whales (except *B. borealis*,

B. acutorostrata) show swallowing type feeding. Sei whales show both types according to food species present (skimming was observed by Ingebrigtsen, 1929) and *Balaenidae* whales show skimming type feeding. Swallowing type whales (blue, fin, Bryde's and humpback whales) swallow the food found in the patch or swarm, along with water, then discharge the sea water through the baleen plates while the food remains in the mouth cavity. They never take scattered plankton nor carry out feeding by swimming with their mouths open. Skimming type whales (right,* Greenland whales) take their food by swimming with their mouths open and the food retained in the mouth cavity is gulped. They may sometimes take a sparse or smaller patch of copepods and amphipods.

The exact position of grey whales is still not clear. However, they possibly combine these two types since field observations have shown them to feed intensively on swarms of zooplankton and also to skim off mud in shallow water in the Bering Sea to feed on benthic amphipods.

The feeding types can be classified as follows:

> Swallowing type
> > blue whale
> > fin whale
> > Bryde's whale
> > humpback whale
> > minke whale
> > Antarctic minke whale
>
> Skimming type
> > right whale
> > Greenland whale
> > pygmy right whale
>
> Skimming and swallowing type
> > sei whale
> > grey whale

Diurnal and seasonal variations

The feeding activity in the open ocean is high in the early morning, becomes lower later in the day and increases again late in the evening (Nemoto, 1959). Other marine mammals such as fur seals in the north Pacific also show the same tendency in feeding on squid and myctophid fish (Tayler, Fujinaga and Wilke, 1955). The variation in the feeding activity is due to the diurnal migration of food species. In the shallow water region on the continental shelf of Anadyle gulf in the north Pacific, fin whales still feed very actively on euphausiids *Thysanoessa raschii*, and capelin (*Mallotus catervarious*) in the daytime. When fin whales are feeding on copepods, *Calanus cristatus* and

---

* Field observations show two right whales swimming along the current rip with their mouths open in order to take a sparse patch of copepods which were so scattered as not to be recognizable from the airplane above (W. Schevill, personal communication on a film shown at Harvard University in 1962).

C. plumchrus, they do not exhibit such a clear diurnal variation in their feeding activity, the reason for which may be due to less intense migration of copepods in the summer.

The seasonal vertical migration of plankton also affects the feeding of baleen whales. The copepod *Calanus cristatus* in the north Pacific spends spring and summer in the upper layers, especially as copepodite stage V. They feed to store oils in their bodies and at this stage are fed upon by fin whales. After summer, *Calanus cristatus* goes deeper than 500 metres (Heinrich, 1957) and then fin whales cannot take them as the diving range of large fin whales is limited to about 300 metres (Scholander, 1940). Adults of both *C. plumchrus* and *C. cristatus* have not been observed in the stomachs of baleen whales (Nemoto, 1963).

TABLE 6. Examples of weights of foods in the stomachs of baleen whales in the North Pacific

Food and whale species	Food species	Amount of foods (kg)
Euphausiids		
Fin whale (18-19 m)	*Thysanoessa* and *Calanus*	425 (Ponomareva, 1949)
Fin whale (18 m)	*Thysanoessa inermis*	113 (Nemoto, 1959)
Bryde's whale (12 m)	*Euphausia similis*	204 (Nemoto, 1959)
Fin whale (?)	?	364 (Betesheva, 1954)
Fin whale (18-19 m)	*Thysanoessa raschii*	340 (Ponomareva, 1949)
Copepods		
Fin whale (18-19 m)	*Calanus plumchrus*	255 (Ponomareva, 1949)
Fin whale (19 m)	*Calanus cristatus*	107 (Nemoto, 1963)
Fin whale (18 m)	*Calanus plumchrus*	72 (Nemoto, 1963)
Fin whale (18 m)	*Metridia lucens*	80 (Nemoto, 1963)
Sei whale (?)	*Calanus plumchrus*	370 (Betesheva, 1954)
Fish		
Fin whale (17 m)	*Theragra chalcogramma*	759 (Nemoto, 1959)
Fin whale (?)	*Corolabis saira*	464 (Betesheva, 1954)
Squids		
Fin whale (?)	*Todarodes pacificus*	560 (Betesheva, 1954)
Sei whale (?)	*Todarodes pacificus*	600 (Betesheva, 1954)

More than 1000 kg of *Euphausia superba* were found in the stomachs of baleen whales in Antarctic waters (Marr, 1962). For the north Pacific, data on the quantity taken by baleen whales are shown in Table 6. The fresh condition of food suggests it is usually recently taken. There is no information about the weight of amphipods in the north Pacific, but sei whales took 200 kg amphipods in the southern hemisphere (Kawamura, 1968).

Food chain through baleen whales

Namoto (1968) gives two main flows through baleen whales in the Antarctic. One is "phytoplankton→herbivorous euphausiids *Euphausia superba*→blue and fin whales" which is possibly the shortest one in the ocean (Mackintosh, 1965). The other is "small phytoplankton→protozoa, larvae of zooplankton and small copepods→carnivorous amphipods (*Parathemisto gaudichaudi*)→

sei whales". In the northern hemisphere, the relationships are far more complicated as there are many different feeding patterns observed.

The feeding patterns and feeding mechanisms of euphausiids (Nemoto, 1968) show that the gregarious euphausiids feed mainly on phytoplankton or on small zooplankton. True carnivorous euphausiids however are never considered to form heavy swarms in the sea.

The relation between food concentrations and the filtering volumes of the baleen whales may be the key in solving the question of the feeding of these large animals. Large right whales have about 2·70 cm baleen plates in the centre of the baleen plate row. Left and right rows form about 13·5 m² of a

TABLE 7. Approximate filtering area of baleen whales formed by baleen plate row

Species	Body length (m)	Filtering area (m²)	Maximum length of baleen (cm)
Blue	27	4·6	85
Fin	24	4·0	80
Sei	16	2·2	70
Bryde's	15	1·7	60
Humpback	15	2·6	70
Black right	17	13.5	270
Fin*	18-19	2·4-2·8	62-70

* Estimated by Kulmov (1961).

filtering curtain which skims small zooplankton in continuous feeding as a maximum estimate. If the right whales swim with the mouth open, the mouth opening may be about 8·9 m² as a maximum. The filtering areas of other baleen whales by baleen plates are shown in Table 7. If right whales filter the sea water by swimming at a speed of 6 km per hour (Nishiwaki, 1965), then the amount of food taken by filtering is shown in Table 8. These values give the maximum, as the mouth opening may be less than the filtering area, and the values with 8·9 m² are also given. They show us that they can take suitable amounts of Calanoid copepods (*Calanus plumchrus* and *Calanus*

TABLE 8. Amounts of food taken according to different food plankton concentration in the sea by skimming method (kg per hour)

Speed in feeding (km)	Food concentration (mg/m³)				
	100	500	1000	2000	4000
13·5 m² filtering area					
1	1·4	6·8	13·5	27·0	54·0
2	2·7	13·5	27·0	54·0	108·0
4	5·4	27·0	54·0	108·0	216·0
6	8·1	41·1	81·0	162·0	324·0
8·9 m² filtering area					
1	0·9	4·5	8·9	17·8	35·6
2	1·8	8·9	17·8	35·6	71·2
4	3·6	17·8	35·6	71·2	142·4
6	5·3	26·7	53·4	106·8	213·6

cristatus), as the feeding activity of right whales is still high even in the day-time. For swallowing type whales (blue, fin, humpback and others) the volume of the mouth cavity is calculated as 4·5 m³ (Fraser, in Marshall and Orr, 1955) and 6·0 m³ by Kulmov (1966). The stretch of the ventral grooves (Matsuura, 1943) when feeding suggests that the cavity volume may be slightly greater than these values. If we use these volumes the amount of food taken by a swallowing action is shown in Table 9. Tables 8 and 9 show the filtering by baleen plates in skimming food is effective for right and Green-land whales which have broader filtering areas of baleen plates and curved skulls. For the swallowing types it is difficult to explain the quantity of food found in the stomachs from the average zooplankton biomass. In the feeding grounds of baleen whales in the north Pacific, Bogorov and Vinogradov (1956) describe the maximum biomass slightly exceeding 2000 mg/m³. Ponomareva (1966) also described the abundance of euphausiids as slightly more than 1/m³ both in the Antarctic and Arctic. Thus the standing stock is 100 to 200 mg/m³ in most parts of the feeding area of the north Pacific and 0·75 to 1·5 g/m³ in the Antarctic.

TABLE 9.   Amount of food taken according to different food concentration in the sea and mouth cavity volume swallowed by baleen whales (g per swallowing)

Swallowing volume one action (m³)	Food concentration (g/m³)								
	0·1	0·5	1	10	100	500	1000	5000	10000
4·5	0·45	2·25	4·5	45	450	2250	4500	22500	45000
6·0	0·6	3	6	60	600	3000	6000	30000	60000
10·0	1	5	10	100	1000	5000	10000	50000	100000
15·0	1·5	7·5	15	150	1500	7500	15000	75000	150000

It is very clear that food organisms, plankton, fish and squid must swarm in very heavy concentrations for swallowing type whales to be able to feed. The study of the actual concentration of plankton in patches is very important for the study of these food chains in the future. If the concentration of euphausiids is considered according to the maximum values shown in the literature (Boden, 1952; Marr, 1962; Ponomareva, 1966; Nemoto, 1968), then swallowing whales might take sufficient food. The calorimetric balance in baleen whales is discussed by Kulmov (1966). The basal metabolism measured by Benedict (in Nakaya, 1961) and Sherman (1949) gives us insight into the problem. Kulmov (1966) has already drawn attention to the agree-ment of calories in the food with metabolic rate. Data from different sources for fin and sei whales are given in Table 10, which also confirm the consider-able amount of food which must be taken. The basal metabolism in baleen whales as a fraction of their weight must be very low. The high proportion of blubber shown in Table 11, which is very peculiar to baleen whales, also reduces the basal metabolism even further and improves flotation.

I

TABLE 10. Tentative kilo-calories of standing stocks of food of fin (18 m, 35 ton) and sei (16 m, 18 ton) whales in the North Pacific per 1 kg flesh of the body*

FIN WHALE

	Food quantity (kg)					
Food species	300	400	500	600	700	800
Euphausiids	9	13	**16**	19	22	25
Copepods	12	15	**19**	23	27	31
Allaska pollack	7	9	11	13	15	**18**
Squids	7	10	12	**14**	17	19
Saury	14	19	**23**	28	32	37

SEI WHALE

	Food quantity (kg)					
Food species	100	200	300	400	500	600
Mackerel	6	13	19	25	32	38
Saury	9	18	27	**36**	45	54
Anchovy	7	14	21	28	35	42
Squid	5	9	14	19	23	**28**
Copepods	8	15	23	**30**	38	45
Amphipods†	10	20	30	40	50	60

* Field observed food amounts are in bold figures
† Based on Kulomov's calorie value (1961).

TABLE 11. Body weight of baleen whales

Species	Length (m)	Weight (ton)	Blubber weight (ton)	Locality
Blue	29·5	108·3	20·9 (19·3%)	Antarctic
	27·8	96·5	18·7 (19·4%)	Antarctic
	26·2	81·2	15·8 (19·5%)	Antarctic
Fin	26·2	60·8	14·2 (23·4%)	Antarctic
	24·6	54·0	12·4 (23·0%)	Antarctic
	23·0	45·6	10·5 (23·0%)	Antarctic
	19·7	36·3	8·3 (22·9%)	Antarctic
	18·2	30·0	7·0 (22·9%)	Antarctic
Sei	18·0	22·8	3·9 (17·1%)	North Pacific
	16·4	18·2	3·1 (17·0%)	North Pacific
	14·7	14·1	2·5 (17·7%)	North Pacific
Right	16·4	78·5	29·2 (37·2%)	North Pacific
	15·1	55·3	23·0 (41·6%)	North Pacific
	14·1	47·6	20·0 (42·0%)	North Pacific

Figures concerning examples for sei and fin whales in the north Pacific are given, underlined in Table 10. If whales can take this amount of food twice a day in the summer, they may easily store the surplus calories as storage oil in their bodies for their winter migration to the south (Kulmov, 1961).

The distribution of baleen whales and zooplankton standing stocks in the ocean show good agreement. Skimming type whales such as right whales

(*Eubalaena glacialis*) are distributed mainly in the waters where the copepods are dominant. An example of this occurs in the sea off Japan where *Calanus plumchrus* occupy two-thirds of the total zooplankton standing stock in the summer season (Nakai, 1942), and many right whales have been caught there. Right whales were also distributed between the Antarctic and subantarctic convergences in the southern hemisphere, and in the northern part of the north Pacific, and give evidence of a short migration from winter to summer season (Townsend, 1935). Greenland whales are confined to the polar seas. On the other hand, swallowing type baleen whales show a typical seasonal migration from the high latitudes where the swarm of zooplankton (euphausiids and copepods) is available, to poorer waters in winter for breeding (Mackintosh, 1965). The feeding range of baleen whales in the ocean is also restricted by inter-specific competition (Nemoto, 1959; Sergent, 1968). In Antarctic waters, blue whales had occupied the pack ice region in former years, but in recent years fin whales have penetrated more into higher latitudes because of the decrease in the stock of blue whales. The fact that sei whales are also often observed south of the Antarctic convergence in recent years may be attributable to the decrease of fin and blue whales in those areas (Nemoto, 1962). In the north Pacific, the main areas of feeding in blue, fin and sei whales often show discrepancies. Of course few sei whales feed on euphausiids but they compete with fin whales for *Calanus* copepods in the feeding areas.

The number of baleen whales in the Antarctic has been estimated by the International Whaling Commission and the national scientific groups concerned. More than 400 000 blue, fin and humpback whales were estimated as a virgin stock in the Antarctic before 1930 (Mackintosh, 1965) and the total weight of baleen whales would have been 25·76 million tons (Crisp, 1962). At that time they must have consumed more than 772 800 tons of *Euphausia superba* per day, as the caloric value of *E. superba* is 1000 cal/kg and 1 kg flesh of baleen whales needs 30 calories per day. The summer feeding season may be about 100 days in the Antarctic and so the *Euphausia superba* consumed amounted to 77·3 million tons a year, which is somewhat larger than the value given by Marr (1962, about 37·8 million tons). This is only speculation on the food chains through the Antarctic whales. However, it suggests that at present a vast stock of *E. superba* remains unused because of the decrease of baleen whales in the Antarctic. The decrease of an ocean harvester, such as baleen whales, may thus cause very large changes in marine food chains and the consequences of this are still a problem to be solved both in the Antarctic and in the northern hemisphere.

REFERENCES

BAKER, A. de C. 1959. Distribution and life history of *Euphausia triachantha*, Holt and Tattersall. *Discovery Rep.*, **29**, 309-40.
BAKER, A. de C. 1965. The latitudinal distribution of *Euphausia* species in the surface waters of the Indian Ocean. *Discovery Rep.*, **33**, 309-34.

BETESHEVA, E. I. 1954. Data on the feeding of baleen whales in the Kurile region. *Trudÿ inst. Okeanol.*, **2**, 238-45.

BODEN, B. P. 1952. Plankton and sonic scattering. *Rapp. P.-v. Réun. Cons. perm. int. Explor. Mer*, **153**, 171-76.

BOGOROV, B. G., and VINOGRADOV, M. E. 1956. Some essential features of zooplankton distribution in the North-Western Pacific. *Trudÿ Inst. Okeanol.*, **18**, 60-84.

CRISP, D. T. 1962. The tonnages of whales taken by Antarctic pelagic operations during twenty seasons, and an examination of the blue whale unit. *Norsk Hvalfangsttid.*, **51** (10), 389-93.

HEINRICH, A. K. 1957. The propagation and the development of the common copepods in the Bering sea. *Trudÿ vses. gidrobiol. Obshch.*, **8**, 143-62.

INGEBRIGTSEN, A. 1929. Whales caught in the North Atlantic and other seas. *Rapp. P.-v. Réun. Cons. perm. int. Explor. Mer*, **56**, 1-26.

JOHN, D. D. 1936. The southern species of the genus *Euphausia*. *Discovery Rep.*, **16**, 193-324.

KAWAMURA, A. 1968. Quantity of foods in baleen whales in the Antarctic. *Geiken-Tsushin*, (201) (in press).

KULMOV, S. K. 1966. Plankton and the feeding of the whalebone whales (Mystacoceti). *Trudÿ Inst. Okeanol.*, **51**, 142-56.

MACKINTOSH, N. A. 1965. *The stocks of whales.* London, Fishing News (Books) Ltd. 232 pp.

MARR, J. W. S. 1956. *Euphausia superba* and the Antarctic surface currents. *Norsk Hvalfangsttid.*, **45** (3), 127-34.

MARR, J. W. S. 1962. The natural history and geography of the Antarctic Krill (*Euphausia superba* Dana). *Discovery Rep.*, **32**, 33-464.

MARSHALL, S. M., and ORR, A. P. 1955. *The biology of a marine copepod.* Edinburgh, Oliver and Boyd. 388 pp.

MATSUURA, Y. 1943. (Kujira satsuroku) Miscellaneous note on whales, 3, *Dobutsu-Shokubutsu (Animals and plants)*, **11** (12), 1001-3.

NAKAI, J. 1942. The chemical composition, volume, weight and size of the important marine plankton. *J. oceanogr. Soc. Japan*, **1** (1), 45-55.

NAKAYA S. 1961. (Eiyogaku) *Nutrition* 2, Tokyo, Japan Women. University Press.

NEMOTO, T. 1959. Food of baleen whales with reference to whale movements. *Scient. Rep. Whales Res. Inst., Tokyo*, **14**, 149-290.

NEMOTO, T. 1962. Food of baleen whales collected in recent Japanese Antarctic whaling expeditions. *Scient. Rep. Whales Res. Inst., Tokyo*, **16**, 89-103.

NEMOTO, T. 1963. Some aspects of the distribution of *Calanus cristatus* and *C. plumchrus* in the Bering and its neighbouring waters, with reference to the feeding of baleen whales. *Scient. Rep. Whales Res. Inst., Tokyo*, **17**, 157-70.

NEMOTO, T. 1968. *Feeding of baleen whales and krill, and the value of krill as a marine resource in the Antarctic.* Presented at the Symposium on Antarctic Oceanography at Sanchiago in 1966. (In press).

NISHIWAKI, M. 1965. *Whales and pinnepeds.* Tokyo, University of Tokyo Press.

PONOMAREVA, L. A. 1949. On the nourishment of the plankton eating whale in the Bering sea. *Dokl. Akad. Nauk SSSR*, **18** (2).

PONOMAREVA, L. A. 1966. Quantitative distribution of Euphausiids in the Pacific Ocean. *Oceanology*, **6** (4), 690-92.

SCHOLANDER, P. F. 1940. Experimental investigations on the respiratory function in diving mammals. *Hvalråd. Skr.*, **22**, 1-131.

SERGENT, P. F. 1968. *Feeding ecology of marine mammals. Symposium on diseases and husbandry of aquatic mammals at Florida, Feb. 21 and 22, 1968.*

SHERMAN, H. 1949. *Chemistry of food and nutrition.* New York, Macmillan and Co.

TAYLER, R. J. F., FUJINAGA, M., and WILKE, F. 1955. Distribution and food habits of the fur seals of the north Pacific Ocean. *Rep. Coop. Invest. Canada, Japan and U.S.A., Feb-July, 1952.*

TOMILIN, A. G. 1954. Adaptive types in the order Cetacea (The problem of an ecological classification of Cetacea). *Zool. Zh.*, **33** (3), 677-92.

TOWNSEND, C. H. 1935. The distribution of certain whales, shown by logbook records of American whaleships. *Zoologica, N.Y.*, **18** (1), 1-50.

Part Four

FOOD REQUIREMENTS FOR FISH PRODUCTION

Introduction

G. HEMPEL
Institut für Meereskunde
an der Universität Kiel

The food chain efficiency of the large stocks of commercially important fish has been discussed frequently. In fresh water fish, Ivlev (1945) and Winberg (1956) have summarized much of the older studies. While their interest concentrated on food chain efficiency in ponds, later authors (Gerking, Le Cren, Mann) put more emphasis on natural habitats such as rivers and lakes. In marine fish some of the most detailed studies have been carried out on plaice by Danish workers, particularly C. G. J. Petersen and Blegvad.

Petersen (1918) described the results of transplantations of 445 000 plaice = 15 000 kg in a restricted area of Limfjord, Denmark. 118 000 fish = 68 000 kg were caught after one year (10-20 kg/ha). Annual production of at least 53 000 kg plaice required 20 g food/m^2; a similar amount was presumably taken by eels. In the same area annual production of food organisms (polychaetes, crustacea, some bivalves and gastropods) fluctuated between 42 and 77 g/m^2, as estimated on rough assumptions about the ratio of production to biomass in macrobenthos.

While obviously in Limfjord most of the available food was taken by commercially important fish, no close relationship between food supply and yield of the fisheries has been found in the open sea (Blegvad, 1928; Jensen, 1928).

Petersen (1918) estimated the benthos biomass in the Kattegatt as being 5 . 10^6 tons starfish and other "useless" invertebrates, 1 . 10^6 tons plaice food but only 0·012 . 10^6 tons cod and plaice. From those figures and from more recent data on growth efficiency Hardy (1959) stated that only 1-2% of the suitable food is actually taken by commercially important fish.

Birkett (this volume) has studied the energy budgets of major predator and prey organisms in the food chain of plaice in the Dogger area of the North

Sea. He describes the sharing of the major food species, *Mactra*, amongst fish, asteroids and the drilling snail *Natica*. During the first summer of its life 70% of *Mactra* are taken by predators, mostly by fish. With increasing thickness of the shells, the importance of fish as predators decreases while predation by asteroids and *Natica* increases. In order to make good use of the molluscs, the fish population should be rather dense on the grounds of young *Mactra*.

Selective feeding in juvenile cod was demonstrated by Arntz (personal communication) in his comparative study of macrobenthos and of stomach contents of cod and dab in Kiel Bight. Cod leaves untouched some of the dominant bivalves as well as most of the smaller crustacea. Dabs, being the only other important fish in the area, only partly fill the niche left by cod. Those studies indicate that macrobenthos is only partly used by food fish. Furthermore it is evident that fish are often fixed in their preference of particular food items. This preference changes in the course of life. There is increasing evidence that in fish and fish larvae, preference for food depends on early experience. The "printing" to particular food has a long lasting effect on the individual fish as shown by Beukema (1968) in sticklebacks. The ecological importance of the behavioural side of feeding and food selection in marine fish is not yet studied in detail.

Gulland (this volume) assumes that fish rarely "overfish" the populations of food organisms. Haddock stocks normally underexploit the food supply of an area. Highest overall production of fish biomass is achieved by dense fish stocks even if individual growth in haddock is retarded in very dense fish populations. Hempel (1958) and Gulland (this volume) have shown that retardation of growth in large stocks of plaice as built up during the war is not as severe as previously suggested. The limited (or non-existent) importance of stock density on growth in adult fish facilitates predictions of the effect of changes in fishing intensity on fish production. Two other factors, however, have to be considered. Food chain efficiency depends largely on the age composition of the stock, decreasing with increasing average age of the population. Highest overall production would be achieved by keeping the stock at a high density but low average age.

Particular attention has been paid to the nursery grounds of plaice where individual growth is adversely affected by very high abundance of O-group plaice and I-group plaice (e.g. Bückmann, 1932). Young fish of several commercially important demersal species, particularly flat fish, live in the littoral and uppermost sublittoral. After the pelagic phase of the larvae and before moving to the deeper feeding grounds of the maturing and adult fish those young fish are members of an ecosystem of a rather particular structure. Several papers and contributions to the discussion of the Symposium dealt with the food chain in very shallow waters. One of the outstanding features of their ecosystem is the high importance of benthic primary producers and of detritus compared with the relative scarcity of phyto- and zooplankton. Fish are either bound to browse on benthic algae and filter mud or to feed on filter feeders and consumers of meiofauna. Herbivorous fish depend on a continuous supply of primary producers. Detritus feeding requires very high concentrations of particulate organic matter of suitable composition. Those conditions are to be found in tropical waters as described at the Symposium by Odum and by Qasim. Odum concluded that in adult *Mugil*, population size is controlled by predators and parasites rather than by shortage of food.

Little is known about the feeding of the early stages of tropical lagoon fish. The very high food requirement in young and fast growing fish might prohibit the various kinds of filter feeding and browsing described for the adults. Odum refers to Suzuki (1965) who found young striped mullets mainly feeding on planktonic crustacea.

In shallow waters of the temperate zone seasonal variations in abundance of primary producers prevent the establishment of permanent populations of herbivorous fish. Those grounds, however, are occupied by high concentrations of young fish during summer. The zone of less than 2 m depth is populated mainly by early stages of plaice which have passed metamorphosis in spring. In slightly deeper water 1-2-year-old fish are found. Food of the young flat fish consists partly of annelids and crustaceans feeding on meiofauna and partly on filter feeding bivalves. Daily ration in plaice (*Pleuronectes platessa*) and other flat fish during their first summer, studied by Müller (1968, 1969) in their natural habitat, amounts to about 25% of body weight. The percentage decreases rapidly with increasing body size in very young fish but levels off at older stages as shown for common dab (*Limanda limanda*) by Pandian (contribution to the discussion of the Symposium) and earlier authors (e.g. Bückmann, 1952). Trevallion *et al.* (this volume) demonstrated the striking increase in energy requirements of a population of juvenile plaice during the first summer. Later in the year energy requirements of the fish population decrease due to mortality, reduced growth, and emigration.

Populations of flat fish are suitable for studies of growth efficiency because of the stationary life of both the fish and the food organisms. From analysis of gut contents, Petersen (1918) estimated a daily ration of 30 g wet weight in a plaice of 300 g, the annual requirements being 7200 g (720 g dry matter) which were converted into 400 g (100 g dry matter) of plaice flesh, i.e. a growth efficiency of 14% in terms of dry matter. Müller (1969) estimated growth efficiencies of about 20% based on field observations of growth and daily ration in O-group plaice and flounder, turbot feeding on fish being slightly higher. Under optimal experimental conditions Edwards *et al.* (1968) found a growth efficiency of 36% in young plaice and dab. Conversion of *Enchytrea* in fish was measured in calories, while previous authors had based their figures on live weight of fish and food. Edwards and Steele (1968) studied feeding in O-group plaice and dab in the sea. High growth rates and high growth efficiency can only be achieved by a high supply of first class food which has to be immediately available in spring and might be grazed down in a few months. While O-group plaice and dab differ to some extent in their distribution (dab occupying somewhat deeper habitats) and selection of food, there is an increasing overlap and competition for food between the two species with increasing age. Competition for food and hence retardation in growth might become serious in years and areas where bivalves—being the primary food of plaice—are relatively scarce. Those effects complicate the models of density dependent growth in juvenile fish.

A rather special case of a predator/prey relationship in young plaice has been described by Trevallion *et al.* (this volume). Instead of causing high mortality by cracking the young bivalves, plaice can also crop the bivalves by feeding on the inhalant siphons of *Tellina* and other bivalves. The bivalve survives the amputation and regenerates the siphon. In tanks rapid growth of O-group plaice is accompanied by a reduction in siphon weight of the *Tellina* population. At a low level of siphon weight plaice reduce their feeding on the siphons so that regeneration can overcompensate the cropping. At that stage plaice do not grow. In the littoral zone of Loch Ewe *Tellina* siphons are the dominant food for newly metamorphosed plaice in their first summer. Total daily food requirements of the plaice population reached almost the level of daily regeneration of the *Tellina* population. Obviously competition for food is highest during May to August. A strong brood of plaice may find a shortage of adequate food. That is in conformity with the older finding of reduced growth in strong year-classes of O-group plaice.

There is no consensus about the effect of food abundance in giving an inverse relationship between growth efficiency and food supply in plaice. Bückmann (1952) got higher figures of growth efficiency than Dawes (1930/31) but no indication of a food dependent efficiency. In their experiments on feeding and respiration rate of O-group plaice and dab Edwards *et al.* (1968) have shown that basic metabolism increases with the $\frac{2}{3}$ power (or slightly higher) of body weight. It is not a constant value but increasing with amount of food provided. Therefore it follows that growth efficiency seems rather independent of food supply above a certain minimum level.

Food requirements of large pelagic populations of marine fish are far more difficult to assess. It has to be based on rather precise estimates of population size, growth, and reproduction in fish and on food production in the area occupied by the fish population. Large-scale migrations and mixing of populations as found in fish like herring and tuna prohibit reliable estimates. In other species, e.g. the Pacific sardine (*Sardinops caerulea*), better opportunities for those studies are provided. The vast research effort put into the California Cooperative Oceanic Fisheries Investigations resulted in a detailed knowledge of the population dynamics of the Pacific sardine and of the abundance and production of phyto- and zooplankton in Californian waters. Lasker (this volume)—for the first time—used the available information for an overall energy budget of the population over a period of 30 years. Estimates of total energy consumption by the sardine population are based on total biomass figures decreasing from $2\cdot8 \times 10^{12}$ g in 1932 to $0\cdot2 \times 10^{12}$ g in 1956. In the early years the sardine population—according to those figures—might well have utilized the major part of the zooplankton produced in the area.

It is interesting to compare Lasker's figures on metabolism in sardines with data on plaice as published by Edwards *et al.* (1968). The percentage of food energy which is incorporated in the fish amounts to 40% in O-group plaice (and 20% in I-group plaice—Bückmann, 1952; Hemple, unpublished). In sardines O-group fish convert only 18% into growth. That figure drops sharply during the second and following years of life, leaving only 1% for growth in 6-year-old fish. The requirements for reproduction remain almost constant over the mature life at a level of about 1% of total energy intake. In plaice, as in other fish, respiration rate per unit body weight decreases with increasing size of the fish. Median values of respiration per body weight in sardines, however, increase considerably with size (about $0\cdot25$ ml O_2/gramme wet weight per hour in 38 g fish compared with $0\cdot44$ ml in fish of 156 g—in plaice of the same size $0\cdot03$ ml O_2/g per hour—Hempel, unpublished). Lasker points to the very high energy demand required for continuous swimming in sardines. The increasing energy requirements relative to body weight in sardines together with the low growth rate of mature fish are presumably causally linked.

Taking the population as a whole the effect of reduction of average age by increasing exploitation on total food requirement is quite different in plaice and sardines. While in plaice energy demands per unit of population weight increases with total mortality (absence of density dependent effects provided), there is a decrease in sardines.

CONCLUSIONS

Reviewing some of the older literature in the light of the papers presented in the Symposium we might conclude that in the sea—in contrast to many fresh water habitats as studied by several authors—plankton or benthos biomass in general is normally not the limiting factor to size and productivity of populations of fish.

While a large part of potential food in benthos and plankton remains unutilized by fish some food elements are heavily grazed during early stages in the life history of fish. A well founded knowledge of the particular food requirements of the various developmental stages of fish in relation to abundance and productivity, rather than overall biomass estimates of plankton or benthos are needed for a proper understanding of the connection between the dynamics of fish population and the food supply. In this context the growing number of combined laboratory and field studies in metabolism, feeding and growth of fish in relation to food abundance, size, and age of fish, was of particular interest.

REFERENCES

BEUKEMA, J. 1968. Predation by the three-spined stickleback (*Gasterosteus aculeatus* L.): The influence of hunger and experience. *Behaviour*, **31**, 1-126.
BLEGVAD, H. 1928. Quantitative investigations of bottom invertebrates in the Limfjord with special reference for the plaice food. *Rep. Dan. biol. Stn*, **34**, 33-52.

BÜCKMANN, A. 1932. Die Frage nach der Zweckmässigkeit des Schutzes untermassiger Fische und die Voraussetzungen für ihre Beantwortung. *Rapp. P-v. Réun. Cons. perm int. Explor. Mer*, **80**, VII, 1-16.

BÜCKMANN, A. 1952. Vorläufige Mitteilung über Fütterungs- und Wachstumsversuche mit Schollen im Aquarium. *Kurze Mitt. fischbiol. Abt. Max-Planck-Inst. Meeresbiol. Wilhelmsh.*, **1**, 8-21.

DAWES, B. 1930-1. Growth and maintenance in the plaice (*Pleuronectes platessa* L.) Parts I-III. *J. mar. biol. Ass. U.K.*, **17**, 103-74; 877-947; 949-75.

EDWARDS, R., and STEELE, J. H. 1968. The ecology of O-group plaice and common dabs at Loch Ewe. I. Population and food. *J. exp. mar. Biol. Ecol.*, **2**, 215-38.

EDWARDS, R. R. C., FINLAYSON, D. M., and STEELE, J. H. 1969. The ecology of O-group plaice and dabs in Loch Ewe. II. Experimental studies of metabolism. *J. exp. mar. Biol. Ecol.*, **3**, 1-17.

HARDY, A. C. 1959. *The open sea*, Vol. 2. London, Collins, 322 pp.

HEMPEL, G. 1968. Zur Beziehung zwischen Bestandsdichte und Wachstum in der Schollen-bevölkerung der Deutschen Bucht. *Ber. dt. wiss. Kommn. Meeresforsch.*, **15**, 132-44.

IVLEV, V. S. 1945. The biological productivity of waters. *Usp. sovrem. Biol.*, **19**, 98-120.

JENSEN, A. J. C. 1928. The relation between the size of the plaice stock and the quantity of "first-class plaice-food" in certain parts of the Limfjord. *Rep. Dan. biol Stn*, **34**, 87-98.

MÜLLER, A. 1968. Die Nahrung junger Plattfische in Nord- und Ostsee. *Kieler Meeres-forsch.*, **24**, 124-43.

MÜLLER, A. 1969. Körpergewicht und Gewichtszunahme junger Platt-fische in Nord- und Ostee. *Ber. dt. wiss. Kommn Meeresforsch.* **20**. (In press.)

PETERSEN, C. G. J. 1918. The sea bottom and its production of fish food. *Rep. Dan. biol. Stn*, **25**, 1-62.

SUZUKI, K. 1965. Biology of striped mullet, *Mugil cephalus* Linné. I. Food content of young. *Rep. Fav. Fish. prefect. Univ. Mie*, **5**, 295-305.

WINBERG, G. G. 1956. Rate of metabolism and food requirements of fish. *Nauch. Trud. Beloruss. gosud. Univ. Minsk*, 253 pp. *Fish. Res. Bd Can. Trans. Ser. no. 194.*

Experimental determination of food conversion and its application to ecology

L. BIRKETT
Fisheries Laboratory
Lowestoft, U.K

Introduction

The transfer of energy and organic matter via the benthos is a vital component of secondary marine production, since a large proportion of the waste matter formed in the water-column reaches the sea-bed; there both remineralization and transfer to higher trophic levels occur, the latter being of more direct importance to man.

The number of marine benthic situations in which a trophic sequence has been studied still appears rather limited, no doubt due to the general complexity of trophic relationships. The food chain centred around *Mactra stultorum* on the Dogger Bank is not only comparatively clear but also is the main channel of transfer in this area, leading to higher production potentials of the leading demersal fishes, especially the plaice.

Life-history of Mactra

Eggs and larvae are planktonic and become distributed during their approximately two months long tour of the Dogger Bank, in dense patches covering as much as 2000 square kilometres. Densities may rise to 200 000/m² at spatfall. On settling, the *Mactra* are immediately pounced on by waiting predators including especially fishes, asteroids and prosobranchs. Some of them have travelled, as larvae, in the same plankton. Yet others, especially fishes, arrive later, to stay and engage in the business of killing *Mactra*.

Growth

By the first winter, *Mactra* are approximately 20 mm long. The maximum size, after 5-11 years, is 45-50 mm. Growth in weight is apparently marked by a succession of exponential growth seasons, interrupted by winter checks.

Thus, the mean growth and production could be calculated from the formula

$$\frac{dw}{dt} = k.w$$

during the summer seasons.

Survival patterns

Two distinct patterns of survival occur. Many year-classes die away fairly rapidly in more or less exponential fashion. The others, in which high early mortality rates are succeeded by a slowing down of mortality, may survive for as long as eleven years. In such cases, the survival curve is hyperbolic and it appears that the numbers surviving are inversely proportional to time, or in other words that the mean area occupied per individual increases linearly with time. The exponential decrease is

$$\frac{dN}{dt} = -i.N$$

while the hyperbolic curve is

$$\frac{1}{N} = a(t-t_0).$$

Production and mortality estimates

Production and mortality were estimated by interpolating at monthly intervals on the growth and survival curves, and combining the estimate of mean weight and mean density. Production of dry matter, and consumption, assumed to account for total mortality, were calculated from the ratio between the exponents of the growth and mortality curves:

$$\text{Production} \quad = \text{net change of biomass} \times \frac{k}{k-i}$$

$$\text{Consumption} \quad = \text{net change of biomass} \times \frac{i}{k-i}$$

and a curve showing the seasonal fluctuations of monthly production, and the non-seasonal changes in consumption, was calculated in this way for the first four years of life of the average year-class.

Partition of mortality between predators

Very fortunately, the proportional roles of three main predators—fish asteroids and *Natica*—can be estimated from the dead shells which accumulate in the patches as the *Mactra* are consumed. It is assumed that, among the whole shells depositing:

mortality due to *Natica*   = number of drilled shells deposited,
mortality due to asteroids = number of undrilled shells deposited,
mortality due to fish   = residual between total mortality and that due to asteroids and *Natica*.

The length distribution of the different categories of dead shells shows that
n years subsequent to the first year of life of the *Mactra*, the total mortality
s due to *Natica* and asteroids, in the proportion of 1:2; during the first year,
ishes caused up to 94% of the total mortality, but they could only eat the
iphons of older *Mactra*, thus not adding to the mortality.

Experimental determination of energy budgets of plaice and starfish

Small plaice, and starfish *Asterias rubens*, were fed on *Arenicola* and *Mytilus*
espectively, to determine the following items of their energy budgets:

  (*i*) efficiency of energy absorption;
  (*ii*) efficiency of conversion of absorbed energy to either growth or heat;
  (*iii*) the maintenance energy requirements.

Individual animals were kept for two to three weeks on maximum rations,
and the following calculations were made from the observed growth, feeding
ates and faecal output:

Energy intake rate, I	gcal/g live wt/day
Faecal energy, F	gcal/g live wt/day
Energy absorption rate, A	gcal/g live wt/day
Chemical energy excretion rate, E_c	gcal/g live wt/day
Metabolizable energy absorption rate, metA	gcal/g live wt/day
Energy of growth, R	gcal/g live wt/day
Heat output, E_h	gcal/g live wt/day

Chemical energy is the energy content of metabolites, estimated as 7 gcal per
ng N excreted;
Metabolizable energy is thus metA $= A - E_c$, and the
Heat output is $E_h = $ metA $- R$.

Simple percentage energy budgets for these two predators were then con-
tructed for different levels of intake.
Maintenance energy is the energy intake equivalent to $R = 0$.

Energy budget of the food chain

To compile the energy budget of the whole food chain, it was assumed that
he minor predator, *Natica*, and also the *Mactra*, had similar energy budgets
o *Asterias rubens*. It was further assumed that, at the peak of the production
eason, all these species including the fish would feed at three times the
naintenance rate. The energy content of *Mactra* was further assumed to be
qual to that of the *Mytilus edulis* used in the starfish experiment, i.e. 0·8
cal/mg live weight. In this way, the energy consumption by each main class
f predators per m² per day, during the peak production season of the *Mactra*
n their first and second years of life, was partitioned into energy becoming
vailable for re-cycling by micro-organisms (as faecal energy, and also as
nergy of metabolites), energy transfer to growth, and heat output.

From the maintenance-energy of each predator, the amounts they consumed per day provided estimates of the biomass of each predator supported by the *Mactra* population.

Table 1 summarizes the rates obtained, and the total energy flow through the part of this ecosystem based on *Mactra* can be determined by summation of the various forms of energy output (i.e. growth, chemical energy and heat output) by total predators, compared with that of the *Mactra*. The least certain segment in this budget is the energy utilization by *Mactra*, except in so far as growth was actually determined from the surveys. The data for *Natica* are also not directly determined, but the proportion of the total energy flow represented by this species is small in any case.

TABLE 1. Energy budgets expressed as per cent of total intake of energy, at different rates of intake (gcal g body/day), assuming energy intake is at *three times* the maintenance level

Total intake	*Plaice* 48 gcal/g day	*Starfish* 36 gcal/g day
Faecal	8%	5%
Total absorbed	92%	95%
Chemical	8%	5%
Metabolizable	84%	90%
Heat	65%	35%
Growth	16%	55%

TABLE 2. Energy budgets in the food-chain based on *Mactra stultorum* on the Dogger Bank. High summer peak rates (energy rates as gcal/m²/day)

	Mactra	Fish	Asteroids	Natica	Total Predators
A. During 1st summer life of *Mactra*					
Biomass—g/m²	37	13·75	0·78	0·39	14·92
Energy consumed	993	660	27·8	14·4	702
Energy faecal	50	53	1·4	0·7	55·1
Energy chemical metabolites	50	53	1·4	0·7	55·1
Heat output	348	429	10	5	444
Growth	546	125	15	8	148
B. During 2nd summer					
Biomass—g/m²	50	0·10	1·47	0·74	2·31
Energy consumed	608	4·9	53·0	26·7	84·6
Energy faecal	30	0·4	2·7	1·4	4·5
Energy chemical metabolites	30	0·4	2·7	1·4	4·5
Heat output	213	3·2	18·5	9·3	31·0
Growth	335	0·9	29·1	14·6	44·6

Utilization of zooplankton energy by a Pacific sardine population in the California current

Reuben Lasker
United States Department of the Interior
U.S. Fish and Wildlife Service
Bureau of Commercial Fisheries
Fishery-Oceanography Center
La Jolla, California 92037

ABSTRACT. The caloric requirements for a single year class of Pacific sardines (*Sardinops caerulea* Girard) was obtained by analysis of growth (tissue addition), fat deposition, respiration, and reproduction. Approximately $1 \cdot 2 \times 10^6$ calories are required by the age 3 sardine for one year. Of the assimilated energy 91% is accounted for by respiration, 2% as final fat deposition, 2% for reproduction and 5% as growth. Information on the efficiency of energy utilization by other year classes and by the population of sardines as a whole from 1939 to the present is also presented. An assessment of the amount of zooplankton necessary to obtain the population is also given.

INTRODUCTION

One aim of studies in marine food chain dynamics is the quantitative evaluation of the amount of energy which ultimately appears in fishes. The trophodynamics of single important species of pelagic fishes is often neglected, chiefly because of the difficulty in relating laboratory findings to what is happening in the ocean. The population usually is not sufficiently understood in terms of biomass, temporal fluctuations or areal distribution. However, unlike many pelagic fish species, and as a result of long standing and intensive investigations, a vast amount of information is now available describing the growth, population dynamics and temporal biomass fluctuations of the Pacific sardine, *Sardinops caerulea* Girard (= *Sardinops sagax*). For this report I combined much of this information with my own laboratory and field data in order to describe an energy budget for the sardine, and I attempt to deduce from these findings the quantity of zooplankton needed to sustain the population observed at all biomass levels since 1932. Surveys of zooplankton biomass have been made in the California Current, where the sardine resides, as part of the continuing California Cooperative Oceanic Fisheries

265

Investigations (CalCOFI) since its inception in 1949. The surveys provide the basis for comparing the zooplankton standing stock with the amount of zooplankton energy used by the sardine population. Information on seasonal biochemical changes in the sardine population is also included where it is pertinent to the description of the sardine's energy budget.

METHODS

Adult sardines were obtained from baitboats from October to December 1960 and throughout 1961. In early 1962, samples were obtained from the commercial catch. All samples were frozen. Each month during the sampling period 14 to 20 fish from a single school were dissected and analysed: in some months two or more samples were obtained. There are no data for the three-month period February to April 1961, when no sardines were caught off San Diego and San Pedro.

The following information was taken routinely for each sardine: sex, standard length, total weight, age, weight of gonads, weight of mesenteric fat, and percentage of fat in the dorsal muscle.

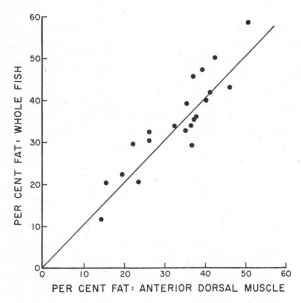

Fig. 1. Percent fat of dry whole individual sardines against percent fat found in the dried anterior dorsal muscle: The line, $y = 1·05x$, was determined by the method of least squares assuming the line passes through the origin.

Analyses of sardine tissues for fat were made by Soxhlet extraction with chloroform-methanol (2:1, V:V) in specially constructed tubes as suggested by Kvarić and Mŭzinić (1950). Before extraction, fish tissue was dried to a constant weight in a vacuum oven at 70°C. Because of differences in fat content in different parts of the fish body, muscle sections for fat analysis were removed, free of scales, exclusively from the dorsal section of the sardine body between the dorsal fin and head and the relation between total fat and dorsal muscle fat was established (Fig. 1).

Standard length and weight were measured by the methods described by Sette (1926).

Caloric measurements were made with a Parr semi-micro bomb calorimeter (Parr Instrument Co., Moline, Illinois, U.S.A.).* Fish tissue was dried or lyophilized before combustion.

Subpopulations (races) were determined for most of the samples taken from July 1960 through August 1961 by Mr Andrew M. Vrooman, to whom I am indebted for such services. The immunological techniques used to determine the population grouping of each sample were those described by Sprague and Vrooman (1962). Two subpopulations off California and Baja California, Mexico are now recognized as "northern" and "southern". The two-thirds of the samples that represented the northern subpopulation were the only fish used in this study. Vrooman (1964) has demonstrated that all the sardines taken north of San Diego during the 1961-62 season (September through February) were from the northern subpopulation.

Most of the fish obtained by periodic sampling were 2, 3, 4, and 5 years old. The bait fishery from which most of my samples were taken did not set on fishes younger than 2 years. Mainly mature Pacific sardines varying in size from 152 to 242 mm standard length, age 3 to 4, were represented.

Respiration measurements of single sardines were made after acclimation to laboratory sea water temperatures for 1 week or more and to a circular plexiglass chamber containing 7·84 l of sea water for 5 hours or more. A Beckman model No. 777 laboratory oxygen analyser was used to monitor percentage oxygen saturation. Water was continually recirculated past the oxygen sensor electrode in a closed system. Oxygen uptake by each sardine was recorded as a drop in saturation and calculated as millilitres of oxygen consumed per hour per gram wet weight by reference to a standard table of oxygen saturation values (Carpenter, 1966). Uptake was always linear over the experimental periods which lasted up to 4 hours depending on the size of the fish. Sea water with no fish showed negligible oxygen uptake. Temperature was maintained in any single experiment to $\pm$ 0·1°C by submergence of the experimental chamber in running sea water, and for different experiments ranged between 17 and 21°C. Each fish was observed during the entire measurement period. The results of respiration experiments are given for sardines at "normal" cruising speed, 0·3 km per hour (after a period of tank acclimation, usually overnight) and for rapid movement ($> 0·5$ km per hour).

Digestion

Two Pacific sardines, 85 mm standard length, acclimatized to a closed sea water aquarium at 18°C, were starved for 3 days, then fed successively 100 to 200 *Artemia* per day for 18 days. The *Artemia* were stained red by feeding

* The use of a trade name does not imply endorsement by the Bureau of Commercial Fisheries.

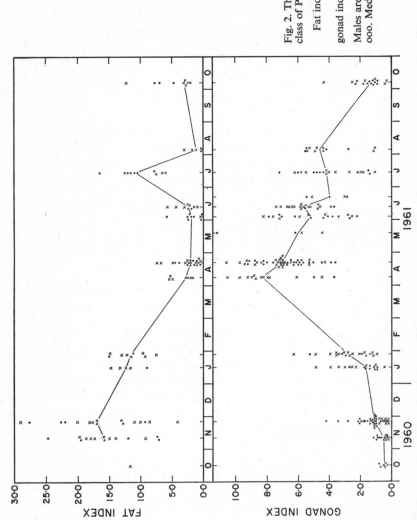

Fig. 2. The fat and gonad cycle of the 1957-class of Pacific sardine.

Fat index = $\dfrac{\text{weight of body fat}}{\text{wet weight of fish}} \times 100$;

gonad index = $\dfrac{\text{weight of gonads}}{\text{wet weight of fish}} \times 100$.

Males are indicated by xxx and females by ooo. Median values are connected by the line.

them in a mixture of carmine particles and carmoisine. Sardine faecal pellets resulting from digested *Artemia* were therefore easy to see and retrieve. Over the period of the experiment 1883 *Artemia* with a dry weight of 954·7 mg were fed to the two sardines. A sardine empties its gut within 12 hours after food ingestion. Faecal pellets recovered weighed 162·2 mg and digestion was 83% of dry weight. The average percentage of ash, 49·7 in dried faecal pellets as compared with 11·4% ash in dry *Artemia*, implies that digestion of organic matter was greater than that calculated on a dry weight basis. If these digestion figures are assumed to apply to all ages of the Pacific sardine, about 20% more calories should probably be added to the respiratory calories given later in this paper to obtain a total caloric intake requirement.

The reproductive cycle

Spawning off southern California in years earlier than this study (1950-56) was confined to the spring (April-June), but with the higher water temperature in 1957-59 spawning occurred earlier in the year and continued for a longer period, from January through July (Ahlstrom, 1960). In 1961, spawning in the southern California region was once again limited to the spring. Fig. 2 shows the gonad index (defined as the wet weight of the gonad divided by the fat-free wet weight of the fish × 100) of individuals of the 1957-year-class in the northern subpopulation during the 1961 spawning season. Males and females showed the same reproductive periodicity.

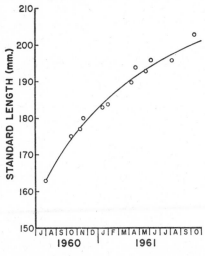

Fig. 3. The growth rate of the 1957-class Pacific sardine. Each dot is the mean length of a sample of this year class taken with other fish ranging from 6 to 40 fish of this year class per sample. The line is drawn in by eye.

The gonads of Pacific sardines attain a large size before spawning. The accumulation of gametes represents an amount of stored energy which is almost entirely lost to the fish at spawning. Individual gonad indices at the peak of the reproductive season are usually at their maximum at 10% of the body weight. At the end of spawning, females always retain some eggs and have indices of about 1%, whereas males spawn out almost completely. Each school of sardines appears to respond as a unit during the spawning

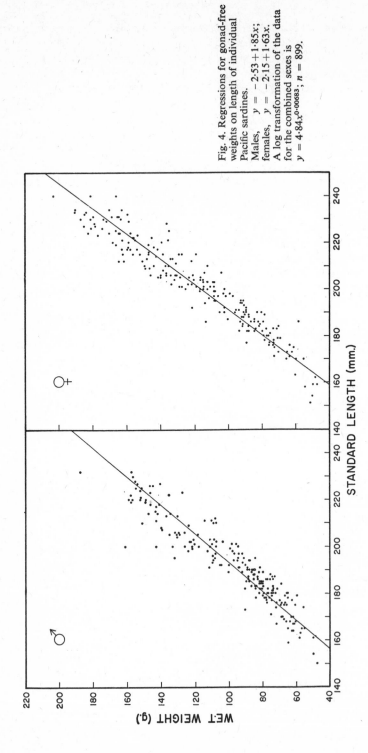

Fig. 4. Regressions for gonad-free weights on length of individual Pacific sardines.
Males, $y = -2.53 + 1.85x$; females, $y = -2.15 + 1.63x$.
A log transformation of the data for the combined sexes is $y = 4.84x^{0.00683}$, $n = 899$.

eason; some schools spawn and are depleted of their gametes by the middle
of the season.

A calculation of the energy used for reproductive products is made in this
paper in terms of estimates of population biomass assuming that half of all
the 2-year-old fish spawn and that the gonads of males and females weigh
10% of the fat-free, wet weight. Ovaries contain 5·43 kilocalories per gram
dry weight and testes 4·85 kilocalories per gram dry weight.

Growth in length

Any attempt to estimate the accumulation of energy by the sardine with time
must take into consideration the sardine's growth rate. The mean lengths of
the 1957-year class of sardines captured for this study are given in Table 1 and
are compared with growth data for earlier years given by Phillips (1948).
Fig. 3 is a graphic presentation of the same 1957-year class data.

The average length of fish of the 1957-year class did not differ from that of
sardines of different year classes caught along the United States Pacific coast
for the 7-year period 1941-47. For descriptive purposes and calculation of
energy budgets for younger and older fish not represented in my sampling, I
rely on the age-length data of Phillips (1948). Judging from Phillips' data for
several earlier years, it seems unlikely that the 1957-class sardine differed
significantly from 121 mm in length at 1 year of age. Thus, the average sardine
grows 121 mm in the first year, 46 mm in the second, 30 mm in the third, 20
mm in the fourth, 14 mm in the fifth, and 9 mm in the sixth year.

TABLE 1. Mean standard lengths, Pacific sardines

Group of sardines	Approximate age* (years)					
	1	2	3	4	5	6
Length in mm of the 1957-year class, this study	—	163	196	215	—	—
Phillips (1948) all years, 1941-47 all Pacific coast U.S.A. (mm)	121	167	197	217	230	240

* Assumes fish were spawned in July.

Increase in weight

In an entire year, weight varies widely in sardines of the same length because
changes in weight are a result of gain by addition of fat or other tissue, or loss
by spawning and utilization of fat reserves. Fig. 4 is a plot of standard length
versus gonad-free fish weight with a regression line describing the relation for
each sex; males $y = -2.53 + 1.85x$; females $y = -2.15 + 1.63x$. The re-
gression line for fat-free, gonad-free fish (combined sexes), $y = -2.35 + 1.73x$, is shown in Fig. 5, and the 50 and 90% confidence intervals for length
on weight are given in Table 2. These limits are applicable to all calculations
based on length in this paper. Fig. 6 is a summary of the values taken from
the regression line in Fig. 5 and added to length-weight values given by Clark
(1928) for smaller fish which were not sampled in this study. Fish smaller

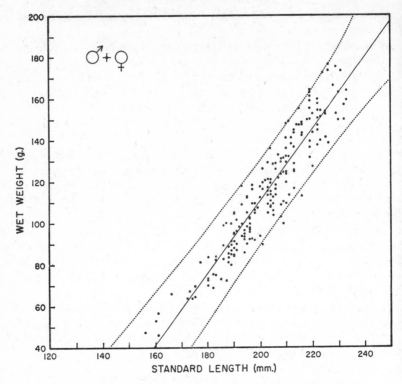

Fig. 5. Regression for gonad-free, fat-free weight on length of combined sexes of Pacific sardines. $y = -2 \cdot 35 + 1 \cdot 73x$. The dashed lines are the 90% confidence limits. The log transformation of y on x is $y = 5 \cdot 80x^{0 \cdot 00637}$; $n = 256$.

than 150 mm are immature and have relatively little fat so that the continuity of the plot is very good. I have also plotted caloric values on the right ordinate which can be used to describe growth in terms of caloric increase (fat-free sardine tissue is 1·4 kilocalories per gram wet weight).

TABLE 2. Confidence intervals for length-weight regression of $y = -2 \cdot 35 + 1 \cdot 75x$; male and females of fat-free, gonad-free Pacific sardines

Standard length (mm)	Weight (g) from $y = -2 \cdot 35 + 1 \cdot 73x$	50% limits (g)	90% limits (g)
167	54	47·0- 61·0	35·8- 71·1
197	106	99·1-113·0	89·0-123·0
217	140	133·0-147·0	123·0-157·0
231	165	158·0-172·0	148·0-182·0
240	180	173·0-187·0	163·0-197·0

Calories used in respiration

The uncertainties of obtaining an accurate estimate of the metabolic utilization of calories are due to the variability of oxygen consumption measurements of

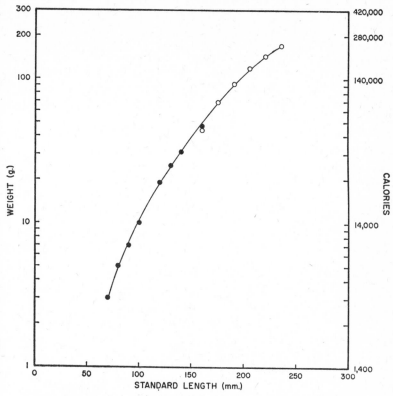

Fig. 6. Weight versus length of the Pacific sardine and the energy, in calories, represented at each weight. Closed circles represent data of Clark (1928), open circles from the regression line shown in Fig. 5.

individual fish, and the difficulty in relating these measurements to the conditions prevailing in nature. Mann (1965) described some reasons for the large variations found with individual fish in the laboratory. They include the need to acclimatize the fish to the respiratory chamber, their starvation, and fluctuations in their swimming related to the day-night cycle. Winberg (1956) has pointed out the necessity of relating metabolic rates to activity, but we lack information on the average swimming activity of Pacific sardine in nature.

The Pacific sardine has no true "basal" rate of oxygen consumption because it swims continually with constant tail beats, rarely resting, in contrast to other clupeids, e.g. the anchovy, which spends much of its time gliding, after a beat of its tail. Bursts of swimming activity increase oxygen consumption by as much as two or three times the rate when the fish is "calm" and cruising slowly. This relation is shown in Fig. 7, where an acclimatized sardine cruising at 0·3 to 0·5 km per hour had a steady and low rate of oxygen consumption. When 200 Artemia were added to the chamber, feeding began immediately, oxygen consumption increased, swimming was extremely rapid and the fish was agitated. The food was consumed in a few minutes but the low rate of oxygen consumption prior to feeding was not achieved for 2-3 hours.

Changes in rate of oxygen consumption due to changes in temperature were masked by the relatively large changes in uptake due to muscular activity. A total of 80 observations were made of the respiration of 10 sardines ranging in length from 121 to 246 mm long (Table 3) whose weights closely approximated the fat-free, gonad-free weights shown in Fig. 5. My data are expressed

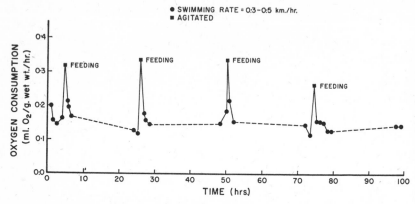

Fig. 7. Rate of oxygen consumption by a single Pacific sardine at normal cruising speed and while feeding.

TABLE 3. Respiration of Pacific sardines. $QO_2 = $ ml O_2 consumed per gram wet weight per hour. Temperature range $= 16\cdot5$ to $22°$ C

Standard length (mm)	Wet weight (g)	QO_2 "calm"			QO_2 "active"		
		Number of observations	Median	Range	Number of observations	Median	Range
122	25·0	—	—	—	3	0·28	0·28
132	27·6	9	0·28	0·20-0·32	4	0·36	0·30-0·44
139	42·6	5	0·15	0·14-0·16	6	0·24	0·19-0·26
146	31·7	13	0·20	0·17-0·25	8	0·35	0·26-0·42
162	46·1	3	0·16	0·15-0·16	—	—	—
179	50·8	—	—	—	4	0·20	0·16-0·29
203	114·1	4	0·17	0·16-0·21	4	0·42	0·39-0·47
220	150·3	5	0·33	0·27-0·35	—	—	—
226	172·6	4	0·56	0·46-0·62	3	0·77	0·70-0·77
246	189·4	5	0·38	0·35-0·38	—	—	—
		Median = 0·24			Median = 0·35		

in median values, since the median is not unduly influenced by extremes. The median oxygen consumption for "calm" fish is 0·24 ml O_2 per gram wet weight per hour and for "active" fish it is 0·35 ml O_2 per gram wet weight per hour. Assuming that the sardine in nature spends half its time in an excited feeding or actively swimming state and the other half cruising slowly, the general median, 0·3 ml O_2 per gram wet weight per hour, is the most applicable value. The sardine chiefly metabolizes fat utilizing 4700 calories per litre of oxygen consumed (Hawk et al., 1954). As I stated earlier, weight may

TABLE 4. Energy budget for the Pacific sardine, *Sardinops caerulea* Girard. Calculations of respiration and growth are based on gonad-free fat-free fish. Sardines have 1·4 kilocalories per gram (wet weight) of tissue; weight of gonads was calculated on the basis of 10% wet weight at the end of each time interval with 66% water in ovaries, 80% water in testes, 5·43 kilocalories per gram dry weight ovaries and 4·85 kilocalories per gram dry weight testes. Only half of the 2 year olds are believed to be capable of spawning

Time interval (years)	Standard length (mm)	Weight (g)	Kilocalories				Per cent assimilated calories		
			Respiration	Growth	Reproduction	Total required	Respiration	Growth	Reproduction
0-1	121	19	117·3	26·6	0·0	143·9	81·5	18·5	0·0
1-2	167	56	475·0	51·8	4·0	530·8	89·5	9·8	0·7
2-3	197	106	1000·5	70·0	11·0	1081·5	92·5	6·5	1·0
3-4	217	140	1519·3	47·6	19·9	1586·8	95·8	3·0	1·2
4-5	231	165	1883·6	35·0	23·5	1942·1	97·0	1·8	1·2
5-6	240	180	2130·7	21·0	25·8	2177·5	97·9	1·0	1·1

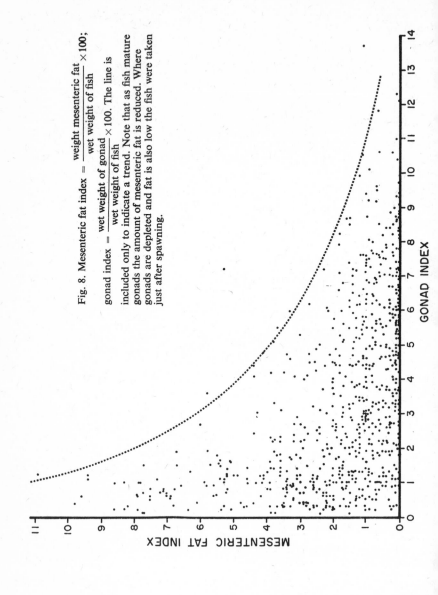

Fig. 8. Mesenteric fat index = $\dfrac{\text{weight mesenteric fat}}{\text{wet weight of fish}} \times 100$;

gonad index = $\dfrac{\text{wet weight of gonad}}{\text{wet weight of fish}} \times 100$. The line is

included only to indicate a trend. Note that as fish mature gonads the amount of mesenteric fat is reduced. Where gonads are depleted and fat is also low the fish were taken just after spawning.

vary widely at a given length but the ergression line for fish taken at random at all seasons indicates that, on the average, the values obtained for the population for weight against length are not affected by the fluctuations in fat. Therefore, I have used the average values for weight against length for fat-free, gonad-free fish to determine the oxygen consumption and caloric utilization on a weight specific basis.

Energy needed by an "average" sardine

Table 4 is a summary of the results of the changes in energy requirements for an average sardine through 6 years of life. Respiration continually dominates the requirements of the fish increasing steadily through the years from 81·5% of the assimilated energy in its first year to over 97% in its sixth. Assimilated energy invested in growth decreases from a high of 18·5% in the first year to 1·0% in the sixth year; reproduction always requires less than 2% of the assimilated energy and therefore is a negligible drain on the resources of the sardine.

Fat as an energy store for the Pacific sardine

I have shown in Fig. 2 that the Pacific sardine follows a seasonal cycle in fat deposition, a characteristic of other sardines and for most marine fishes (Shul'man, 1960). Fat in the Pacific sardine occurs in two main depots, in muscle and discrete mesenteric fat. In Fig. 8 I have plotted the relationship between the percentage wet weight of mesenteric fat in a fish against the percentage wet weight of the gonad. This graph shows clearly that no appreciable quantity of mesenteric fat is ever present when the gonads have matured, another illustration of the negative relationship between fat deposition and gonad maturation. The fat of the sardine yields 8970 calories per gram and in November through December the tissues contain an appreciable quantity. For example, in November 1960 at the peak of the fat cycle, the 1957-year class sardine had an average fat content of 16% (wet weight). In Table 5, I

TABLE 5. Calorific value of stored fat at the peak of the sardine fat cycle

Time interval (years)	Total energy required (kilocalories)	Deposited fat energy at peak of fat cycle (kilocalories)	Fat calories as percentage of total required energy
1-2	530·8	80·5	15·2
2-3	1081·5	152·6	14·1
3-4	1586·8	202·0	12·7
4-5	1942·1	238·0	12·3
5-6	2177·5	259·0	11·9

have compared the amount of stored fat energy found within a fish of a given weight to the amount of total energy required by that fish for a year. These data show that about 12-15% of the energy required by a fish throughout a year can be found stored within its tissues at the peak of the fat cycle. This implies that mature fish must continue to feed throughout the spawning season but may do so at a reduced level.

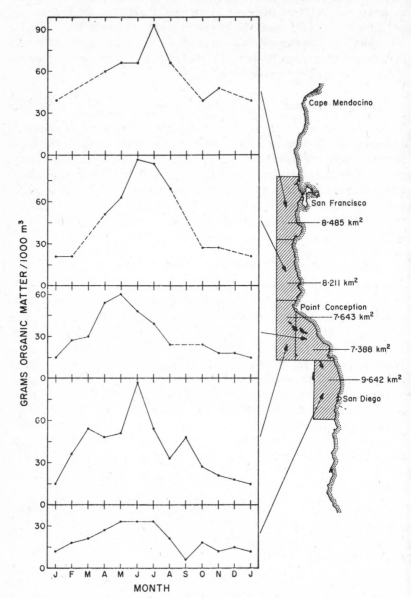

Fig. 9. Areal distribution of the northern subpopulation of the Pacific sardine with 10-year median values of grams of organic matter per 100 m³ plotted for each month and segregated by area.

Food of the Pacific sardine

A considerable amount of information is available on the zooplankton stand-ing crop in the California Current system from about 20 000 plankton net tow collections over 10 years; the data were published in a series of reports (Thrailkill, 1959, 1961, 1963, MS). Dr Paul E. Smith of the Fishery-Oceano-graphy Center, Bureau of Commercial Fisheries, has kindly supplied me with his analysis of data on zooplankton volume which, with the proper conversion factors, can give an estimate of the usual zooplankton standing crop in the California Current. Standard tows (oblique, from 140 m deep to the surface with a net of 1 m mouth diameter and a mesh size of 0·505 mm) were made in the CalCOFI area from 1951 through 1960. The volumes of preserved zooplankton were measured after a period of time (usually 2 years) and con-verted to grams of organic matter per 1000 m^3 by the conversion factor of 0·082 times the volume which was derived from Ahlstrom and Thrailkill (1963; their Table 8).

Recent work with a plankton pump in the California Current (Beers and Stewart, 1967) and work with fine meshed nets (0·202 mm mesh size) by Smith (unpublished) suggest that zooplankton volumes (ml per 1000 m^3)

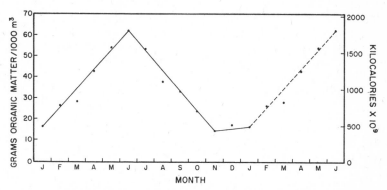

Fig. 10. Grams of organic matter and calories plotted as median values for the entire area shown in Fig. 9. Medians were obtained from 10 years of collections covering the entire area. The least number of plankton tows was in September, $n = 60$, all other months had at least 120; a maximum of 266 tows in March. To show the October to February trough in this graph, the January to June points on the right hand side of the graph are repeated. The line was drawn by eye through the points. Low points in March and August appear to reflect the lack of sampling in the north during those months (see Fig. 9).

obtained with the CalCOFI net must be multiplied by three to include the smaller zooplankters lost through the mesh and which make up the bulk of the food of the Pacific sardine (Hand and Berner, 1959), and by four if even smaller organisms are to be accounted for. I have used the factor of three in this study, since Smith's work was done in April and a multiplication factor of three is probably a good estimate during the spring increase in zooplankton standing stock.

Records of Pacific sardine egg and larval distribution, spawning intensity, and area (Ahlstrom, 1954, 1959), the areal extent of the sardine catch in good and bad fishing years (Clark, 1937; Pinkas, 1951), and recent studies indicating that the southern distributional limit of the northern subpopulation is at about the latitude of San Diego, California (Vrooman, 1964), lead me to conclude that the northern subpopulation of sardines, which is the chief concern of this investigation, has an areal distribution of approximately $41 \cdot 4 \times 10^9$ m^2 (shown graphically in Fig. 9).

The median values for 10 years of zooplankton volumes (converted to grams of organic matter per 1000 m^3) for specific regions of the California Current, are also plotted in Fig. 9 by month. Fig. 10 presents the median values by month for the entire area after the data were pooled and the medians reselected. I have converted grams of organic matter to calories by using the simple expedient of multiplying by 5 kilocalories per gram and these estimates are included on the ordinate of the graph.

The data in Fig. 10 show a strong seasonality of zooplankton standing crop—an increase during the spring with the peak in June. Dr William G. Pearcy of Oregon State University (personal communication) suggested that the synchronous peaking in June argues against advection from the adjacent northern areas off Washington and Oregon as a major factor in producing plankton seasonality in the sardine's area of the California Current because of the time interval needed to transport zooplankton into the California Current off California. Thus, production is probably the result of local up-

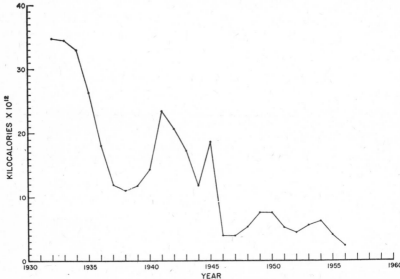

Fig. 11. Energy used in respiration by the calculated biomass of sardines for each year from 1932 to 1956.

welling within the area circumscribed by the California Current sampling grid. Murphy (1966; his Table 14) in his study of the population biology of the Pacific sardine for each year from 1932 to 1956 deduced the magnitude of the sardine population from total catches and other data. Using my regression line values for gonad-free wet weight (Fig. 5), I have converted his numbers of fish to weight in grams and in Table 6 I present the summary of sardine biomass for each year. From these values of total sardine biomass and the median respiration value based on fat-free, gonad-free wet weight for sardines I have plotted the yearly utilization of energy by the northern subpopulation of sardines due to respiration (Fig. 11). Sardine abundance in 1932-34 was essentially stable as to biomass and Murphy (1966) suggested that this species was an important, if not the dominant, consumer of zooplankton in the California Current for those years.

TABLE 6. Biomass of the Pacific sardine and energy used in respiration for each year, 1932-56

Year	Biomass $g \times 10^{12}$	Kilocalories $\times 10^{12}$ used in respiration	Year	Biomass $g \times 10^{12}$	Kilocalories $\times 10^{12}$ used in respiration
1932	2·82	34·8	1945	1·50	18·5
1933	2·80	34·5	1946	0·31	3·9
1934	2·68	33·1	1947	0·33	4·0
1935	2·14	26·4	1948	0·43	5·3
1936	1·45	18·0	1949	0·62	7·6
1937	0·96	11·9	1950	0·60	7·4
1938	0·89	11·0	1951	0·41	5·1
1939	0·87	10·7	1952	0·36	4·4
1940	1·16	14·3	1953	0·44	5·5
1941	1·89	23·4	1954	0·49	6·1
1942	1·67	20·6	1955	0·33	4·0
1943	1·39	17·2	1956	0·19	2·3
1944	0·95	11·8			

About 34×10^{12} kilocalories were used for respiration by the Pacific sardine for each year during 1932-34, or $2·8 \times 10^{12}$ kilocalories per month. This figure is contrasted with the years 1951-56 where an average of only $4·6 \times 10^{12}$ kilocalories were used per year or $0·38 \times 10^{12}$ kilocalories per month. Documentation is ample, summarized by Murphy (1966), that the severe reduction in the sardine population was due to successive years when the sardine spawn did not survive sufficiently, coupled with severe fishing on the spawning stock which crippled the future spawning potential of the population.

The zooplankton standing stock in the California Current is at its lowest November through January, about $0·4$-$0·5 \times 10^{12}$ kilocalories; the rate of increase, i.e. the net increase above the losses due to grazing and mortality is $0·26 \times 10^{12}$ kilocalories per month from January to June and the standing stock at its peak in June is $1·8 \times 10^{12}$ kilocalories. (Although the monthly median values for zooplankton calories given in Fig. 10 are for 1951-60, they are taken here to be representative of all previous years as well.)

The respiratory caloric requirement of the 1932-34 sardine population was $2·8 \times 10^{12}$ kilocalories per month or 10 times the rate of increase of the observed zooplankton standing stock. In 1951-56 the sardine population, at its lowest level, consumed 0.38×10^{12} kilocalories per month or very little more than the observed increase in zooplankton standing stock.

To put these figures into their proper perspective it is desirable to compare them with primary productivity in the spring and summer. Very few data are available but Dr J. D. H. Strickland (personal communication) has kindly shown me recent figures for spring-summer primary production in the California Current taken within 13 km of La Jolla, California. The net increase of photosynthetic carbon is about 1 gram m² per day, which, if extrapolated to the entire area of the California Current delineated in this paper, is 1.24×10^{12} g C per month or, in terms of the calorific equivalent of glucose, 12.7×10^{12} kilocalories per month. Table 7 summarizes the food chain information for 1932-34 by assuming that secondary productivity is 10% of the primary and that the Pacific sardine was the only major zooplankton consumer in the California Current during those years.

It seems obvious that the maintenance requirements for the 1932-34 sardine population (three times·the probable energy production of the zooplankton) are either too high or the estimate of both primary and secondary production for those years is too low. Alternatively, the sardine may have taken advantage of phytoplankton directly, since Hart and Wailes (1931) showed that sardines feed on phytoplankton in the Pacific north-west. Despite the discrepancies, it is clear that the virtual disappearance of the sardine has made available a prodigious quantity of zooplankton food energy to organisms using the same food. In the California Current many possible

TABLE 7. Energy transfer through the food chain (January-June) in the California Current during 1932-34, see text

	Kilocalories $\times 10^{12}$ *per month*
Primary production	12·7
Secondary production	1·27
Sardine grazing	2·8

successors to the sardine were available, among which are the Pacific mackerel (*Scomber japonicus*), jack mackerel (*Trachurus symmetricus*), the northern anchovy (*Engraulis mordax*), and carnivorous zooplankters. The anchovy is believed to have benefited greatly showing an increase in biomass roughly equivalent to the decline in the biomass of sardines (Murphy, 1966).

SUMMARY

An energy budget for the Pacific sardine, *Sardinops caerulea* Girard, was constructed from information acquired in laboratory experiments and from statistics accumulated over a 24-year period by the California Department of Fish and Game and the U.S. Bureau of Commercial Fisheries. It was found that growth accounts for 18·5% of the sardine's assimilated energy during its first year of life. Energy needs for growth decline in succeeding years to 9·8, 6·5, 3·0, 1·8 and in the sixth year, 1·0%. Respiration is the dominant energy-consuming process throughout the sardine's life (82-99% of assimilated calories). Reproduction requires only abut 1·0% of the energy.

An annual fat cycle alternates with the reproductive cycle. The fat accumulated at the peak of the fat cycle can provide enough energy for only about 1 month. Fat is used up almost completely just before maturation of the gonads which, at their maximum, weigh 10% of the fat-free weight of the fish.

The biomass of sardines and the amount of energy, in calories, needed by the standing stock of sardines in the California Current were calculated for each year from 1932 to 1956. The maximum population in 1932-34 had a stable biomass and the energy required per month for its respiration, $2·8 \times 10^{12}$ kilocalories, was 10 times the monthly increase in the usual zooplankton standing stock between January and June (data from 1951-60) but only 3 times the probable secondary production.

With the decline of the sardine biomass to the 1956 level, a major fraction of zooplankton food energy became available for exploitation by other predators and may be reflected, as others have postulated, in the resurgence of the numbers and biomass of the northern anchovy, now the major planktotrophic fish in the California Current.

ACKNOWLEDGEMENTS

Thanks to Mrs Gail H. Theilacker, Mr Jack D. Metoyer and Mr Gerald H. Sanger for their able assistance throughout this investigation.

REFERENCES

AHLSTROM, E. H. 1954. Distribution and abundance of egg and larval populations of the Pacific sardine. *Fishery Bull. Fish Wildl. Serv. U.S.*, **56**, 83-140.

AHLSTROM, E. H. 1959. Distribution and abundance of eggs of the Pacific sardine, 1952-56. *Fishery Bull. Fish Wildl. Serv. U.S.*, **60**, 185-213.

AHLSTROM, E. H. 1960. Synopsis on the biology of the Pacific sardine (*Sardinops caerulea*). In *Proceedings of the world scientific meeting on the biology of sardines and related species*, **2**, U.N.F.A.O., 415-51.

AHLSTROM, E. H., and THRAILKILL, J. R. 1963. Plankton volume loss with time of preservation. *Calif. coop. oceanic Fish. Invest. Rep.*, **9**, 57-73.

BEERS, J. R., and STEWART, G. L. 1967. Micro-zooplankton in the euphotic zone at five locations across the California Current. *J. Fish. Res. Bd Can.*, **24**, 2053-68.

CARPENTER, J. H. 1966. New measurements of oxygen solubility in pure and natural water. *Limnol. Oceanogr.*, **11**, 264-77.

CLARK, F. N. 1928. The weight-length relationship of the California sardine (*Sardina caerulea*) at San Pedro. *Fish Bull. Calif.*, **12**, 58 pp.

CLARK, F. N. 1937. Fishing localities for the California sardine, *Sardinops caerulea*, 1928-1936. *Fish Bull. Calif.*, **48**, 11 pp.

HAND, C. H., and BERNER, L. D. 1959. Food of the Pacific sardine (*Sardinops caerulea*). *Fishery Bull. Fish Wildl. Serv. U.S.*, **60**, 175-84.

HART, J. L., and WAILES, G. H. 1931. The food of the pilchard, *Sardinops caerulea* (Girard), off the coast of British Colmbia. *Contr. Can. Biol. Fish.*, N.S., **7**, 247-54.

HAWK, P. B., OSER, B. L., and SUMMERSON, W. H. 1954. *Practical physiological chemistry*. New York, McGraw Hill, 1439 pp.

KVARIĆ, M., and MŮZINIĆ, R. 1950. Investigation into the fat content in the sardine tissues (*Clupea pilchardus* Walb.). *Acta adriat.*, **4**, 291-314.

MANN, K. H. 1965. Energy transformations by a population of fish in the River Thames. *J. Anim. Ecol.*, **34**, 253-75.

MURPHY, G. I. 1966. Population biology of the Pacific sardine (*Sardinops caerulea*). *Proc. Calif. Acad. Sci.*, **34**, 84 pp.

PHILLIPS, J. B. 1948. Growth of the sardine, *Sardinops caerulea*, 1941-42 through 1946-47. *Fish Bull. Calif.*, **71**, 33 pp.

PINKAS, L. 1951. Yield per area of the California sardine fishing grounds 1937-1949. *Fish Bull. Calif.*, **80**, 9-14.

SETTE, O. E. 1926. Sampling the California sardine: A study of the adequacy of the various systems at Monterey. *Fish Bull. Calif.*, **11**, 67-123.

SHUL'MAN, G. E. 1960. Dynamics of the fat content of the body of fish. *Russ. Rev. Biol.*, **49**, 209-22.

SPRAGUE, L. M., and VROOMAN, A. M. 1962. A racial analysis of the Pacific sardine (*Sardinops caerulea*) based on studies of erythrocyte antigens. *Ann. N. Y. Acad. Sci.*, **97**, 131-38.

THRAILKILL, J. R. 1959. Zooplankton volumes off the Pacific coast, 1957. *Spec. scient. Rep. Fish. U.S. Fish Wildl. Serv.*, (326), 57 pp.

THRAILKILL, J. R. 1961. Zooplankton volumes off the Pacific coast, 1958. *Spec. scient. Rep. Fish. U.S. Fish Wildl. Serv.*, (374), 70 pp.

THRAILKILL, J. R. 1963. Zooplankton volumes off the Pacific coast, 1959. *Spec. scient. Rep. Fish. U.S. Fish Wildl. Serv.*, (414), 77 pp.

THRAILKILL, J. R. (Manuscript) Zooplankton volumes off the Pacific coast, 1960. *Spec. scient. Rep. Fish. U.S. Fish Wildl. Serv.*

VROOMAN, A. M. 1964. Serologically differentiated subpopulations of the Pacific sardine *Sardinops caerulea. J. Fish. Res. Bd Can.*, **21**, 691-701.

WINBERG, G. G. 1956. Rate of metabolism and food requirements of fishes. *Nauch. Trudȳ Belorussk. gos. Univ. V. I. Lenina Minsk*, 253 pp. (Fish. Res. Bd Can. Trans. Ser. No. 194).

Dynamics of a benthic bivalve

ANN TREVALLION, R. R. C. EDWARDS and J. H. STEELE
Marine Laboratory
Aberdeen
Scotland (U.K.)

ABSTRACT. The bivalve *Tellina tenuis* is a filter feeder living near low water on sandy beaches. In the area studied, Loch Ewe on the west of Scotland, the siphons are a major source of food for plaice in the first few months after metamorphosis. There is also a mortality of whole *Tellina* of about 30 per cent per year. A study of the energetics of *Tellina* has shown that their growth and, especially, their reproduction can be limited by food concentration. The regeneration of the siphons as a result of predation is an important energy requirement during the period when reproduction can occur and this may affect the probability of successful stock recruitment. However, it appears that predation on the siphons may be density dependent with a threshold density of *Tellina* below which predation does not occur. This should increase the probability of recruitment at low densities and so provide long term stability. This form of population regulation is similar to that proposed for some terrestrial systems.

INTRODUCTION

For the three years 1965-67 a detailed study has been made of the ecology and energy flow of the littoral and sublittoral regions of Firemore, a sandy bay in Loch Ewe on the west coast of Scotland (Fig. 1). Some of the work is already published (Ansell and Trevallion, 1967; Trevallion and Ansell, 1967; McIntyre and Eleftheriou, 1968; Steele and Baird, 1968; Edwards and Steele, 1968; Edwards, Finlayson and Steele, 1968) and much of the information used here is taken from these papers. For this reason few descriptive details are given. This paper examines some of the factors which affect the dynamics of two species, the bivalve *Tellina tenuis* da Costa which is common from mid-tidal level to a few metres below low water, and O-group plaice, *Pleuronectes platessa*, L. which live just below the tide mark. Each year the plaice population is a completely new one, depending on recruitment, but the *Tellina*

12

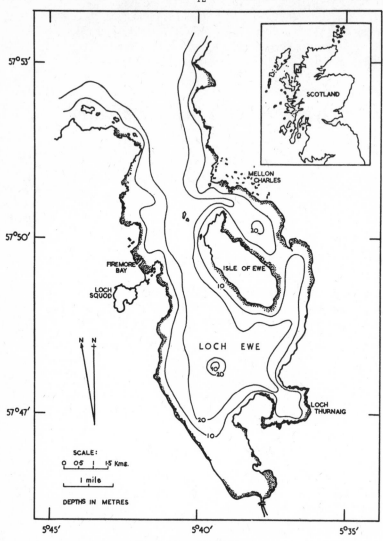

Fig. 1. Map of Loch Ewe.

population consists of many of the same individuals, which live for five to six years, changed only by growth, recruitment and mortality. During the first few weeks after settlement the young plaice may feed on the inhalant siphons of *Tellina* which extend up to or beyond the sand surface. O-group plaice and dabs were the only observed predators on *Tellina* siphons out of 12 species of fish examined in Firemore.

The density of the natural plaice population depends on the recruitment of larvae from a breeding stock of adults offshore. Thus one will expect random fluctuations in this initial settlement which are not dependent on the density

f food in the bay. At the position of low water mark, the major utilization
f organic matter available from primary production in the water is by the
nterstitial ecosystem of bacteria and meiobenthos (McIntyre, Munro and
Steele, this volume). The macrobenthos as a whole, and *Tellina* in particular,
assimilate a relatively small fraction, so that there is unlikely to be any control
xerted by the feeding rate of the population on the concentration of their
ood. Thus, the amount of food available to the *Tellina* is effectively another
andom variable. These two factors make it appear likely that control of the
Tellina population may be mainly dependent on a functional response, i.e.
n increase in the number of prey consumed per predator, as prey density
ises.

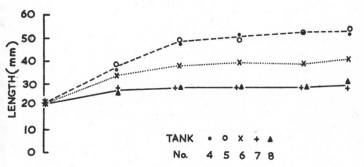

Fig. 2a. Growth of 0+plaice in tanks.

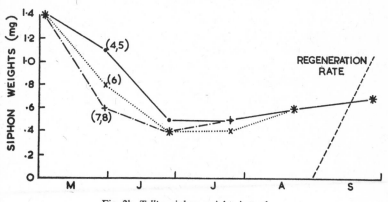

Fig. 2b. *Tellina* siphon weights in tanks.

EXPERIMENTS ON SIPHON CROPPING

he growth of *Tellina* and plaice with different predator/prey ratios, has been
tudied. Large, shore-based tanks were set up with sand in the bottom, void
f macrofauna. The water supply was pumped up directly from the sea, with
daily exchange of 25%. The tanks were stocked with local *Tellina tenuis*
nd newly metamorphosed plaice in varying densities and ratios as in Table 1.

TABLE 1. Tank experiments

Tank	No. of Tellina	No. of Plaice	Ratio
4	9000	30	300:1
5	3000	10	300:1
6	3000	30	100:1
7	3000	100	30:1
8	900	30	30:1

This experiment began in May and continued until September. The fish were counted and measured monthly. Samples of *Tellina* were also taken monthly for biochemical analysis. Shell length was measured on marked individuals, and the inhalant siphons were measured by weighing the dry siphon in relation to the dry body weight. The base of the siphon was distinct and easily separated from the mantle. The increase in length of plaice in the tanks is shown in Fig. 2a together with simultaneous changes in the siphon weights of *Tellina* (Fig. 2b). Both the growth of the fish and the reduction in siphon weights appear to be related to the predator/prey ratios, rather than the absolute densities of either predator or prey. During the first month the siphons were cropped intensively so that the weights were rapidly reduced, roughly in proportion to the predator/prey ratios, while the fish grew rapidly. During the second month the siphon weights continued to drop to the minimum. The drop was rapid in tanks 4 and 5 and successively slower in tanks 6, 7 and 8. Fish growth was correspondingly greater in tanks 4 and 5, less in tank 6 while in tanks 7 and 8 growth had almost ceased by the end of the first month. After the siphon weights had fallen to the minimum of 0·4-0·5 mg the fish growth ceased in all tanks. The weights of the siphons then increased gradually with no significant differences between tanks, but the rate was much slower than regeneration rates measured in laboratory experiments. This suggests that a limited amount of predation continued, supplying enough food for maintenance of the fish but not for growth.

Estimates of the energy requirements of the different plaice populations in the tanks, based on experimental data (Edwards, Finlayson and Steele, 1968) agree reasonably with the energy calculated from the regeneration rate of

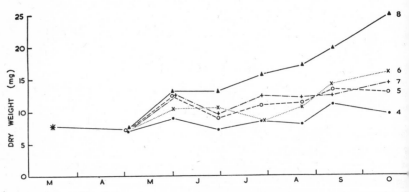

Fig. 3. Growth of *Tellina* of 4 + year group.

siphons observed in laboratory experiments (Edwards, Steele and Trevallion, in press). Thus it is concluded that the regeneration rate is maintained in all the tanks to provide the energy required by the plaice populations. On the other hand the growth rate of the *Tellina* is different in different tanks varying inversely with *Tellina* density. Food limitation is expected to occur at the higher densities due to the limited water exchange and thus it appears that this limitation affects growth rather than siphon regeneration (Fig. 3).

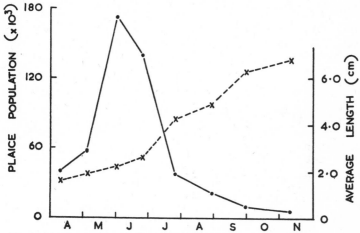

Fig. 4a. Numbers and mean sizes of individuals in O-group plaice population.

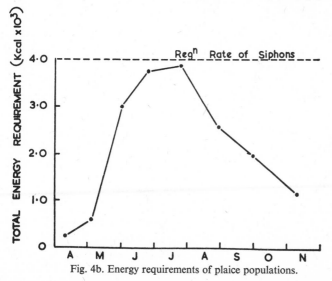

Fig. 4b. Energy requirements of plaice populations.

Preliminary calculations have been made for the beach populations in 1965 (Fig. 4a). As a result of settlement of newly metamorphosed fish, the population increases until the end of May and then decreases rapidly as a consequence of predation. During the initial phase of this decrease there is rapid

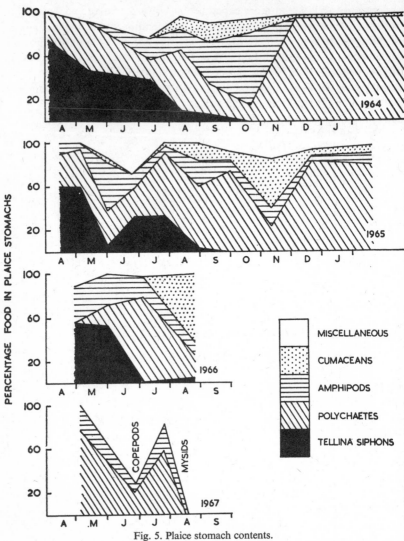

Fig. 5. Plaice stomach contents.

growth of fish and so (Fig. 4b) the maximum energy requirement of the population occurs after the peak in population numbers. Fig. 4b also shows that for this year, if siphons had been the sole food of the plaice, then growth would have been limited. As expected, other foods are taken but Fig. 4b is an indication of the potential limitation on siphon predation in the natural environment. The actual food intake is indicated by stomach contents (Fig. 5*) which displays the importance of *Tellina* in the first two years but also

---

* The May-30th sample in 1965 is considered anomolous due to very severe wave action on that day which probably caused the very low stomach contents, particularly of siphons, observed on this occasion (Edwards and Steele 1968).

the decline in 1966 and complete absence of *Tellina* in the food in 1967.

Several other parameters of the two populations have also changed during the years 1965-67. In 1965 the maximum plaice population density was about 170 000 while in 1966 and 1967 the maximum densities were around 70 000. The *Tellina* population has undergone a steady decline in size, since there is a mortality of around 30% partly due to predation of whole *Tellina* by two and three-year old plaice, and there has been no recruitment in the bay for three years Thus it appears that there may be a threshold in *Tellina* density below which the plaice do not prey on the population. This cessation of predation may be expected to have effects on the growth of individuals.

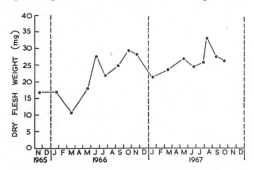

Fig. 6. Weight of standard *Tellina*, 1965-67.

The condition of *Tellina* as expressed by the dry flesh weight of an animal of standard size (15 mm) has shown a progressive change since measurements began in the winter of 1965-66 (Fig. 6). At this time the weights were very low. They increased in summer and autumn as part of the normal seasonal cycle due to build up of gonad and storage material, and although gonads were developed, most of this seemed to be resorbed and used as overwinter storage material. The minimum for the winter of 1966-67 was not nearly as low as in the previous winter and the maximum weight in August 1967 was higher than in the previous summer. The minimum weights for the 1967-68 winter were similar to those of the previous winter, and this pattern of varia-tion suggests that the population has been improving in condition from a very low level in 1965, and would support the hypothesis that lack of siphon predation permits increased somatic growth. To consider these questions quantitatively details are needed on the energy expended on each aspect of

TABLE 2. Budget for *Tellina* (m²/year)

	g Carbon	kcal
Growth { seasonal	0·6	5·37
{ annual	0·1	0·89
Spawning	0·1	0·89
Siphon regeneration	0·2	1·79
Respiration	0·9-1·8	8·05-16·10
Production { annual growth } { spawning } { siphons }	0·4	3·58

metabolism. In Table 2 a budget calculated for standard *Tellina* and expressed as g carbon and kcal/m² year is presented. A standard animal, 15 mm in length, is taken as 20 mg dry weight, and the average biomass of *Tellina* in 1966, the year on which calculations are based, was 1·3 g/m². Growth has been divided into seasonal growth, mainly gonad, and annual somatic growth (Ansell and Trevallion, 1967). The spawning value was calculated from the sudden drop in weight which occurred in July and corresponded with observations on the gonad. (In Table 2 the spawning value has been subtracted from the value for seasonal growth; the balance is probably gonad which is resorbed, and overwinter storage material.) However, this drop in weight could also be explained if the gonad was resorbed to provide energy for increased respiration due to high temperatures at this time. This year, when the population was in a poor condition, spawning was probably less than normal. The continued rise in body weight after July in 1966 also suggests that much of the gonad was resorbed.

The value for siphon regeneration was based on the weight of siphons found in plaice stomachs and the density of the plaice. The respiration value was based on laboratory experiments, taking the oxygen consumption at 10°C as an average for the whole year. This respiration value is probably low, as even when doubled, the efficiency Production/Production + Respiration works out as 20% which is rather high. However, it is clear that the major energy demand is for respiration and that the rest is fairly equally divided between somatic growth, storage products and reproduction, and siphon regeneration.

If the rate of siphon regeneration is constant up to a normal siphon weight, then heavy siphon predation will result in most of the energy available after the respiration requirements are satisfied, being channelled into the siphons, and the rest being used for somatic growth and storage products. A certain amount of storage material is essential for the overwintering period when little food is available in the environment. Possibly only when there is excess from these demands can there be a successful maturation of gametes and spatfall. This idea was tested in tank experiments.

EXPERIMENTS ON *Tellina* IN ENRICHED SEA WATER

These experiments were carried out in a tank, similar to those used in the siphon cropping experiments, but divided into four compartments. Each of these was stocked with 1000 *Tellina*. Daily water exchange was 10%. One compartment was supplied with natural sea water while the others had three levels of nutrient enrichment. Nutrients were added daily, and nitrate, phosphate, organic carbon and chlorophyll were monitored almost continuously. *Tellina* were sampled monthly. There were no fish in this tank.

The results are summarized in Fig. 7. Tank A was the compartment with natural sea water, and tank B received the lowest level of enrichment (around 3 × the natural level of nutrients). In C and D the higher enrichment gave rise to pollution effects, and conditions were not so desirable as in B, so those results are not shown.

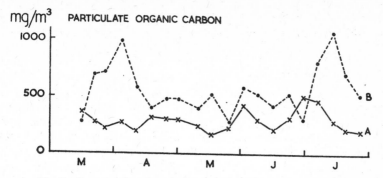

Fig. 7a. Concentration of particulate organic carbon in enriched tanks.

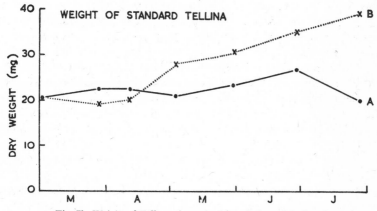

Fig. 7b. Weight of *Tellina* of standard length in enriched tanks.

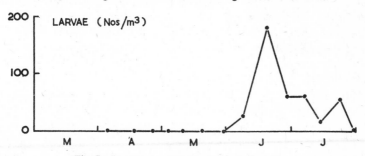

Fig. 7c. Occurrence of *Tellina* larvae in tank B.

The nutrients in B resulted in a twofold increase in the levels of organic carbon for most of the time, with occasional higher peaks (Fig. 7a). The weights of a standard animal in tank B corresponded to the increase in food density so that by the end of the second month, they were markedly greater than in A. The difference was even more marked by the end of July (Fig. 7b). Plankton samples were taken daily in the tanks, and only in tank B did any bivalve larvae appear (Fig. 7c). The numbers were small, but evidence enough to suggest that an increase in available food and lack of predation may contribute to successful spawning.

DISCUSSION

From these experiments and observations it would appear that conditions in Firemore Bay over the last three years have inhibited successful reproduction by the *Tellina* population. One of these adverse conditions is probably the predation on the siphons by O-group plaice. Regeneration of cropped siphon may take place at the expense of energy which might otherwise be channelled into growth and reproduction. There is also the behavioural aspect to consider. A newly cropped inhalant siphon may undergo a latent period in which it does not function. Certainly, once contracted after a stimulus such as a bite or attempted bite by plaice, there is a time lag of a few minutes before the siphon is re-extended, and then it may not function as efficiently as a full grown siphon. The tank experiments suggest that once the availability of siphons falls below a certain threshold, the plaice no longer feed so intensively on them. Under natural conditions with alternative food available, the threshold to start feeding again may be higher than that to stop feeding. Such a threshold, however, would also depend on the densities of alternative foods. The fact that during 1967 the plaice in Firemore Bay did not feed on *Tellina* siphons suggests that a threshold had been reached where the density of siphons relative to other sources of food was below that which stimulated predation by the plaice. With the reduction in the amount of predation over the last three years, the condition of the *Tellina* population has improved so that we might expect successful reproduction and recommencement of siphon predation in subsequent years.*

On this basis it is suggested that although concentrations of food and predators are obviously important to the *Tellina* population the fluctuations in population size may depend mainly on the functional response of the predators to prey density. In a series of observations of a *Tellina* population covering twelve years, Stephen (1938) found peaks in recruitment every three or four years. He attempted to relate these to environmental factors and observed that years of successful reproduction corresponded to some but not all of the years with above average temperatures. This would agree with the concept that favourable environmental conditions could be a necessary but not a sufficient condition for successful reproduction.

Failure of recruitment may result not only from unsuccessful spawning but through mass mortality in the planktonic larval stage due to such factors as heavy predation, disease, insufficient food or unfavourable climatic conditions, or through failure to settle on the bottom. Sampling of *Tellina* has recently begun in other bays of Loch Ewe, and in two of them, at Mellon

---

* Preliminary analysis of 1968 data shows slight predation on siphons but no sign of spatfall, although this had occurred in adjacent bays in Loch Ewe.

Charles and Inverasdale, there is evidence of recruitment in 1967. No data have been collected on predation in these areas, but the observations suggest two possible situations. One is that spawning occurred in the Firemore Bay population in 1967 as in the other bays, but Firemore being more exposed did not offer suitable conditions for settlement of spat. Alternatively, the functional relation between predator and prey was in a different phase in Firemore compared with other areas. If this were the case, then it would indicate the importance of such relations over any climatic control.

The details of the prey/predator relation described here are highly specialized but the essential feature is the concept of a density of prey below which predation is negligible. Similar relations have been described for vertebrate predators on insects.

Tinbergen (1960) defined a threshold density below which predation by birds on insects ceased. Holling (1959) described the general form for the predation of small mammals on insects as an "S-shaped" curve. Both authors point out the importance of this type of relation for providing stability to the prey populations. Holling also suggests that this pattern may be a specific feature of vertebrate carnivorous predation and the results given here would support this idea.

REFERENCES

ANSELL, A. D., and TREVALLION, A. 1967. Studies on *Tellina tenuis* Da Costa I. Seasonal growth and biochemical cycle. *J. exp. mar. Biol. Ecol.*, 1, 220-35.
EDWARDS, R., and STEELE, J. H. 1968. The ecology of O-group plaice and common dabs at Loch Ewe. I. Population and food. *J. exp. mar. Biol. Ecol.*, 2, 215-38.
EDWARDS, R., FINLAYSON, D. M., and STEELE, J. H. 1968. The ecology of O-group plaice and common dabs at Loch Ewe. II. Experimental studies of metabolism. *J. exp. mar. Biol. Ecol.*, 3, 1-17.
EDWARDS, R., STEELE, J. H., and TREVALLION, A. 1969. The ecology of O-group plaice and common dabs at Loch Ewe. III. Prey-predator experiments with plaice. *J. exp. mar. Biol. Ecol.* (in press).
HOLLING, C. S. 1959. The components of predation as revealed by a study of small mammal predation of the European pine sawfly. *Can. Ent.*, 91, 293-320.
McINTYRE, A. D., and ELEFTHERIOU, A. 1968. The bottom fauna of a flatfish nursery ground. *J. mar. biol. Ass. U.K.*, 48, 113-42.
McINTYRE, A. D., MUNRO, A. L. S., and STEELE, J. H. 1970. Energy flow in a sand eco-system. This volume, pp. 19-31.
STEELE, J. H., and BAIRD, I. E. 1968. Production ecology of a sandy beach. *Limnol. Oceanogr.*, 13, 14-25.
STEPHEN, A. C. 1938. Production of large broods in certain marine lamellibranchs with a possible relation to weather conditions. *J. Anim. Ecol.*, 7, 130-43.
TINBERGEN, L. 1960. The natural control of insects in pinewoods. I. Factors influencing the intensity of predation by songbirds. *Archs néerl. Zool.*, 13, 265-535.
TREVALLION, A., and ANSELL, A. D. 1967. Studies on *Tellina tenuis* Da Costa II. Preliminary experiments in enriched sea water. *J. exp. mar. Biol. Ecol.*, 1, 257-70.

Food chain studies and some problems in world fisheries

J. A. GULLAND
FAO
Rome

ABSTRACT. The world's catch of marine fish, now 40 million tons per year, is expanding at around 7% per year. The maintenance of this expansion requires advice from marine scientists on the management and, where necessary, regulation of stocks already exploited, and on the quantitative evaluation of other potentially valuable stocks so far not exploited.

Knowing the primary production and the quantitative transfer between trophic levels an estimate can be made of the potential production of fish in an area, both first stage carnivores (zooplankton eaters) and predators. Preliminary analyses for certain areas give reasonable agreement between such estimates and estimates from direct studies of the fish population—but suggest that the efficiency may be greater than the 10-12% often used and that this efficiency may be greater in the discontinuous production of high latitudes than in the continuous, tropical system.

In many areas, fishery management is increasingly dependent on knowing the quantitative interaction between several species, at different positions in the food chain, e.g. how much herring is eaten in producing one ton of dogfish. Examples using slightly different values for the efficiency of utilization by the predators show how critically management decisions depend on this efficiency.

INTRODUCTION

The world's catch of marine fish in 1966 was some 44 million tons (FAO, 1967); in addition some 5 million tons of crustaceans, molluscs, etc. were landed. While this is only a small proportion of the total food production, fish supplies an important fraction of the animal protein, especially in some areas, e.g. eastern Asia. Also, unlike most other foods, fish production has been increasing at a rate (around 7% per year) which is rather faster than the increase in the human population. Since shortage of food, and especially shortage of protein, is possibly the potentially most explosive problem in the

296

world today, it is most desirable that the fish catches should continue to expand as least as fast as they have in the past decade. To do this, marine science, and those aspects of marine science discussed in the present symposium, can contribute in two main fields, first in the conservation and management of the presently exploited resources, and secondly in the identification and quantitative evaluation of potential new resources.

The present catch, equivalent to about 5×10^6 tons of carbon, may seem very small compared to the total annual primary production in the oceans of about 20×10^9 tons of carbon synthesized into living matter. Two factors should, however, be taken into account that show that production of the presently harvested types of fish are not so greatly in excess of the catches as the 4000-fold difference between catch and primary production would suggest. First, the fish being harvested are several stages removed from the primary production, and a 90% reduction at each stage would give the following estimates of production, in terms of carbon, at each trophic level:

phytoplankton	$20\ 000 \times 10^6$ tons
zooplankton	2000×10^6 tons
zooplankton eaters (e.g. herring)	200×10^6 tons
predatory fish (e.g. cod)	20×10^6 tons

Secondly, these figures are of production in all parts of the ocean. Present fisheries, with the exception of whaling and tuna fishing, are generally limited to the waters over or close to the continental shelf. Though these regions have, on the average, a higher primary production per unit area than the oceanic regions, the latter have much the greater extent, and most production takes place over the deep oceans. The area of the oceans with depths less than 1000 m is only about 12% of the total ocean (Menard and Smith, 1966). Allowing for the greater production over the shelves, and the catches of pelagic fish over deep water (though still close to the continental shelf), the estimates of production given above should be reduced by a factor of perhaps 5 to give quantities potentially available in the areas at present fished. This brings the production of predatory fish, and also of zooplankton eaters within an order of magnitude of the present catches, thus giving some confirmation of the supposition, on other grounds, that the catches of several of the larger and more valuable species are approaching their upper limit, and that the potential for great expansion of catches are among species lower in the ecological pyramid.

ESTIMATES OF POTENTIAL

The rough calculations above suggest a fair agreement between direct assessment of fish resources, and estimates from primary production, through the food chain. There are as yet few detailed studies of particular areas to determine how good this agreement is quantitatively, and thus the extent to which measurement of primary production together with estimates of ecological efficiency can be used to produce estimates of the potential catches of fish at various trophic levels.

One problem is that the production of, say, herring in the North Sea, will be considerably greater than the potential catch of herring. Man is only one among several predators, and production is also lost due to other deaths, e.g. from disease. In the usual notation of fish population dynamics, the catch C is given by:

$$C = \frac{F}{F+M} \times P$$

$$= \frac{F}{F+M} \times (P_R + P_G)$$

where P = total production of commercial size fish, which may be expressed as the sum of P_R, the net production before recruitment, i.e. the weight of fish reaching a fishable size, and P_G the production in the form of growth after recruitment. An increased share of the production can be caught by increased fishing, but excessive fishing can result in a falling catch—in the present terms due to a reduction in production of fish, implying that the reduced fish stock cannot efficiently harvest the production of its food. The best that can be achieved in a well managed fishery is for the catch to be around half the total production, including the production of fish smaller than the sizes being caught.

The area for which there are the best data on both primary production and fish production is probably the North Sea. This has been discussed by Steele (1965). Since the time his paper was prepared there have been big changes in the fisheries, and as a result additions to the knowledge of the fish stocks, though these events (and an error in taking the area of the North Sea as 3·3 rather than $4\cdot4 \times 10^5$ km²) do not affect his argument—that the ecological efficiency must be high.

Steele divides the fish catches into two groups; the zooplankton feeders (herring, mackerel, etc.) conveniently labelled "pelagic", though including sandeels which spend most of their life on, or indeed in, the bottom, and predatory and benthos-eating fish (cod, haddock, flatfish, etc.), conveniently labelled "demersal". He estimated the total production of each group of fish by multiplying the average recent catches by the ratio $(F+M)/F$, giving estimates of $4/3 \times 0\cdot43 = 0\cdot57$ and $2 \times 1\cdot0 = 2\cdot0$ million tons respectively. Since 1960 the catches from the North Sea have been rapidly increasing, reaching in 1966 1·01 million tons of demersal fish and 1·87 million tons of pelagic fish (Table 1). The demersal catches in particular are well in excess of the total production estimated by Steele. At least three explanations are possible; the catches are indeed greater than the production, through a rapid increase in fishing removing an accumulated stock; the estimates of production were too low; or the production has in fact increased.

For the herring and mackerel fisheries the situation is certainly not steady.

TABLE 1. Catches (thousands of metric tons) of fish from the North Sea

	1938	1948	1953	1960	1962	1963	1964	1965	1966
"Demersal" species									
Lemon sole	4·4	1·8	5·4	4·2	4·6	3·8	4·8	5·7	6·1
Plaice	46·0	84·0	78·9	86·3	87·4	107·1	110·4	96·9	100·1
Sole	5·2	4·7	18·8	18·5	26·8	26·1	11·3	17·0	31·8
Cod	58·3	70·4	81·1	104·3	89·6	105·9	121·6	179·5	219·7
Haddock	71·5	71·1	60·4	66·4	52·4	59·4	198·7	221·7	269·0
Saithe	26·6	28·1	21·3	29·0	22·3	27·6	55·1	68·9	86·9
Whiting	36·6	59·1	62·5	53·1	69·0	98·7	91·5	106·7	155·2
Dogfishes	6·5	10·9	19·6	31·8	30·4	34·4	21·6	26·1	23·4
Skates and rays	7·9	14·8	11·5	11·0	9·6	10·7	11·0	10·0	8·2
All demersal species*	343·3	429·5	487·4	517·0	590·9	690·3	773·9	858·8	1010·7
"Pelagic" species									
Norway pout	—	—	—	40·2	157·0	166·8	121·6	59·3	52·7
Sandeels	—	—	5·0	112·9	110·0	162·1	128·5	130·8	161·1
Herring†	1034·5	1290·5	1211·4	787·0	678·5	805·3	932·0	1230·3	1038·9
Sprat	46·2	9·9	21·5	16·5	31·3	67·7	70·8	76·2	106·6
Mackerel	25·4	31·0	41·3	72·9	66·3	55·4	79·4	151·7	505·1
All pelagic species*	1106·1	1336·3	1286·4	1030·5	919·6	1145·8	1367·6	1733·2	1871·5

* Includes species not individually specified.
† Includes some Norwegian catches of Atlantic-Scandian herring, not strictly part of North Sea production.

Following the development by the Norwegians of advanced techniques of fishing using purse-seines with power blocks, and sonar, Norwegian North Sea catches of herring, and later mackerel have expanded very rapidly. There are clear signs that the recent high catches of herring cannot be maintained; after building up to a peak of 1 230 000 tons in 1965 catches fell to 1 038 000 tons in 1966, and the Norwegian catches were only maintained by increased fishing on mackerel. Mackerel catches increased from 79 000 tons in 1964 to 505 000 tons in 1966, almost all the increased catches being used for reduction to meal and oil. These catches comprised fish of several age groups up to about 10 years old, and therefore elements of the production from several years. Also mackerel appear to be highly migratory, fish tagged in the English Channel and Irish Sea having been caught in the North Sea (U.K M.A.F.F., 1967). Thus the mackerel catches in the North Sea contain an element of the production from outside the North Sea, though because the mackerel come into the North Sea in the summer (Postuma and Zijlstra, 1964), which is probably the main feeding season, a large proportion of the mackerel production is based ultimately on the primary production within the North Sea.

The amount of fishing for demersal fish has not changed greatly. The most dramatic increase is in the abundance of haddock and is due principally to the outstanding 1962 year-class which was clearly detectable as being outstanding among research vessel catches in the autumn of 1962 (Jones, 1964). The cod stock and cod catches have also increased due to good year-classes; there has not been one quite so outstanding year-class as the 1962 haddock year-class, but in one or more of the different stocks of cod in the North Sea the 1961, 1962, 1963 and 1964 year-classes have all been good (Raitt and Symons, 1967), while the increased plaice catches since around 1955 have been ascribed by Gulland (1968a) mainly to more rational fishing (lower mortality and protection of the small fish), though there also seems to have been better growth, at least among the small plaice.

Production of demersal fish in the North Sea has therefore definitely increased. For the main exploited stocks the total production may be estimated, following Steele, from the average of the 1966-67 catches as $4/3 \times 0.93$ = 1.24 million tons. Since not all demersal stocks are as heavily exploited as the haddock, this estimate should be increased somewhat, particularly for any stocks that are very lightly exploited. Gulland (1967a) reviewed the information on incidental catches by commercial and research vessels, distribution of eggs and larvae, etc., and concluded that the potential catches (and by implication the production) of demersal species of minor demersal importance (gurnard, dab, long rough dab) was not large. The potential catches of all demersal species, under average conditions of year-class strength, etc. was estimated as 800 000 tons, equivalent to a total production of over 1 million tons of which cod, haddock, whiting, coalfish and plaice accounted for 650 000 tons.

The production of pelagic fish includes a bigger proportion from stocks that are still lightly exploited. Gulland (1967a) estimated potential catches of 3·2 million tons; this estimate may be rather too high, but it seems unlikely that the total production of pelagic fish is less than 3 million tons.

Ignoring for the present the question of changes in fish production, the figures of what seems to be the more usual level of North Sea fish may be compared as in Steele's Fig. 2 with the primary production. Dividing by the area of the North Sea,* 0·55 × 10⁶ km², and assuming a 10% carbon content, gives estimates of demersal and pelagic fish production per square metre of 0·2 and 0·6 g C per year respectively, almost exactly equal to those estimated by Steele. Comparing these figures with the annual primary production of around 100 g C/m², and recognizing that no major fish species in the North Sea eats phytoplankton, and several (e.g. cod and whiting) eat few herbivorous animals, two important conclusions are, as Steele points out, inevitable— that the ecological efficiency, at one or more stages, must be appreciably greater than 10%. Also, unless even higher values of the ecological efficiency are possible, the commercial fish yield is more closely limited by the primary production than is commonly realized.

Similar conclusions may be deduced for other areas, though generally the data are not so good either for primary or fish production. Thus in parts of the Gulf of Thailand the recently developed Thai trawling fishery has been obtaining yields of about 300 000 tons (Tiews, 1965). Adding the production from pelagic fishes, which are producing annual catches of about 70 000 tons from stocks which are probably not fully exploited (Shomura and Gulland, 1968), the total fish production is probably at least 300 000 × 3/2 + 70 000 × 2 = 600 000 tons, from an area of about 80 000 km², i.e. a production of 8 g/m² per year. Not much is known in detail concerning the food of the large number of species concerned, but most are at two stages removed from the primary production, which is not believed to be exceptionally high.

Again the vast catches of anchoveta from off Peru come from a relatively small area, about 1500 km long, extending perhaps 200 km offshore. Recent catches have been about 9 million tons per year, and fishing accounts for only about half the total deaths (Schaefer, 1967; Gulland, 1968b), so that the production of fish of commercial size is about 18 million tons, or about 6 g C/m². This figure would not seem at all large compared with the primary production, which is probably high, if anchoveta did, as is sometimes suggested, feed entirely on phytoplankton. Data at the Instituto del Mar at Callao, shows, however, that while sometimes anchoveta stomachs contain almost only phytoplankton, at other times, especially in May (the southern winter) the contents can be almost entirely zooplankton—mostly copepods but including sometimes large euphausids. The juvenile fish appear to be very largely zooplankton feeders. The proportion of zooplankton in the total food consumption of anchoveta is not known, but even quite a moderate proportion can reduce the overall efficiency of transfer of energy from phytoplankton to fish close to that for fish that are wholly zooplankton feeders. Since

* The area of the North Sea, as defined by ICES for statistical purposes, includes areas west of Shetland and is therefore larger than the North Sea in the strict sense.

this effect is quite general, it is worth setting out in detail. Mathematically, denoting:

C_3 = total consumption by fish (third trophic level)
where aC_3 = consumption of phytoplankton by fish
and $(1-a)C_3$ = consumption of zooplankton by fish

and also writing $_1C_2$ = consumption of phytoplankton by that part of the zooplankton population which is ultimately eaten by fish, then, defining $_1E_2$ as the efficiency of utilization of phytoplankton by zooplankton and similarly for $_2E_3$ and $_1E_3$

$$(1-a)C_3 = {}_1E_2 \, {}_1C_2$$

and total production of fish $P_3 = a_1E_3C_3 + (1-a)_2E_3C_3$
and consumption of phytoplankton $= aC_3 + {}_1C_2$
$$= aC_3 + (1-a)C_3/{}_1E_2 = {}_1C$$

Therefore the efficiency $= P_3/{}_1C$ (phytoplankton into fish)

$$= \frac{[a_1E_3 + (1-a)_2E_3] \, {}_1E_2}{a_1E_2 + (1-a)}.$$

(Note that this efficiency will be greater than the ecological efficiency, which is generally calculated from the production of one trophic level to production of the next level, since this equation does not take into account that part of the phytoplankton production which is not consumed by fish or zooplankton.)

This equation may be simplified by assuming that the Es are equal, i.e.

$$\text{efficiency} = \frac{E}{1 - a(1-E)}.$$

The values of this efficiency, for values of $E = 0.1$, 0.2 and 0.3 and various values of a are tabulated below:

a	0	0·2	0·4	0·5	0·8	0·9	1·0
E							
0·1		0·01	0·012	0·016	0·022	0·036	0·052	0·10
0·2		0·04	0·05	0·06	0·08	0·11	0·14	0·20
0·3		0·09	0·10	0·12	0·17	0·20	0·24	0·30

Though too much significance should not be attached to the precise numerical values in this table, it does show clearly how the effective efficiency approaches the efficiency corresponding to a three-stage food chain (phytoplankton-zooplankton-fish) even when the food of the fish is as much as 60-80% phytoplankton, especially at the lower (and perhaps more probable) values of E.

CHANGES IN FISH PRODUCTION

Recent events in the North Sea show that the fish production can change appreciably. There is no clear evidence that the primary production has changed appreciably; most probably these changes in fish production are due less to changes in the total primary production than to changes in the efficiency with which the primary production is converted into fish. Though Steele (1965) suggests that so far as herring is concerned the efficiency may have changed due to qualitative changes in the zooplankton—more large *Calanus*, on which the herring can feed more effectively than on the small *Temora* and *Pseudocalanus*—apparently independent changes in the fish stock, such as outstanding 1962 haddock year-class, and increased plaice and cod stocks appear to have had a greater affect on the demersal production.

Though the 1962 year-class of haddock is exceptional, the year-classes of this stock have always been highly variable; various hypotheses have been put forward concerning the causes of good or bad year-classes, none of which have proved satisfactory over a period (Carruthers *et al.*, 1951; Saville, 1959). Certainly the strength is determined in the pelagic state, before the young fish settle in the bottom in the autumn of their first year of life (haddock spawn in the spring), since fair measures of the year-class strength can be obtained by trawl surveys with small-meshed nets at that time. Less certainly food supplies, possibly of a particular type at a particular time—e.g. when the yolk-sac supplies are finished—may be critical, though there seems no relation to the total food supplies to the larval haddock.

Detailed data on the growth of haddock have been collected over a long period by the Aberdeen Laboratory (e.g. Raitt, 1939), and from these a definite reduction in growth can be detected during periods of high stock abundance (Beverton and Holt, 1957; Gulland, 1962). Using Raitt's data on the mean length at the end of each year and assuming that weight is proportional to the cube of the length, a measure of the growth in weight during the year can be calculated. This can be related to the abundance; there seems little relation between growth and abundance in the first year of life and the clearest relation is in the second year. As suggested by Raitt, since the 1- and 2-year group haddock live together, but rather separate from the older haddock, the growth in the second year has been related to the abundance of 1- and 2-year-old fish, as measured by the mean abundance of the two-year-classes concerned. The relation for Raitt's Western and Central areas are given in Fig. 1. The points are reasonably fitted by a straight line; using these lines, an index of the total haddock production can be easily calculated as the product of growth and numbers. This gives the parabolas shown in Fig. 2. These curves may be compared with the similar curves relating the yield of a fishery (production) to the fishing intensity (year-class abundance), especially with the parabolas deduced by Schaefer (1954). Strictly a better comparison would be the curves (also a parabola for Schaefer's model) relating net economic yield to fishing intensity, i.e. value of yield less the costs, which can be considered comparable to the maintence requirements in the terms of this symposium. In the fishery case economic yield can be lost by "over-fishing" involving excess costs, or "under-fishing" when much less than all the available fish are harvested. Similarly it appears that the haddock can "over-

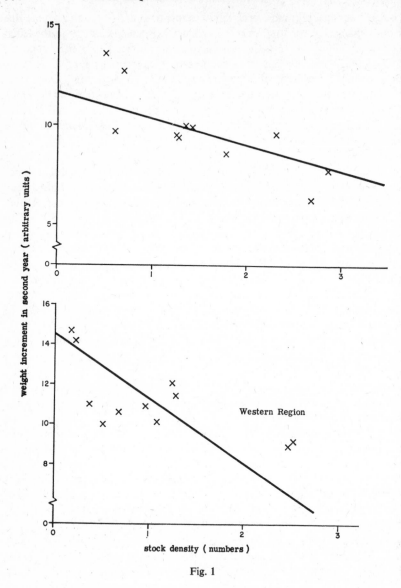

Fig. 1

fish" or "under-fish" their food supply, and that the maximum production occurs at levels of year-class strength well above the average. For average or less than average year-class strengths, the haddock stock greatly under-exploits its resource. The detailed implications of this supposition regarding the population dynamics of the food of the haddock have not been examined, but by analogy with lightly exploited fish stock there would be a high proportion of large and old individuals which are growing little, and have a very low ecological efficiency.

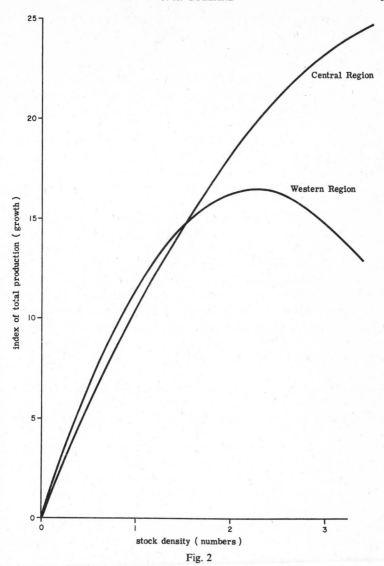

Fig. 2

THE EFFECT OF FISHING

Fishing affects both the abundance of fish stocks and their composition. Increased fishing reduces the stocks, which unless the stocks are "overfishing" their food supplies (as may be the case in some stunted fish populations in lakes) will tend to reduce the production; however, the increased proportion of younger and faster growing fish means that a more efficient use is being made of the food that is being consumed. For this reason it has been suggested that the proper fishery management should keep the proportion of large fish as low as possible, and that some regulatory measures often recommended, such as the use of larger mesh sizes in trawls, have less beneficial effect than predicted from the simpler models of the population dynamics of the stocks concerned.

As regards the preferential capture of small fish such a suggestion is clearly erroneous; if, say, cod less than 50 cm are more efficient users of the food consumed than cod greater than 50 cm, the rational harvesting procedure (ignoring other factors) would be to avoid catching cod less than 50 cm, but to catch them once they reached that size. The catches would therefore consist of fish which, at the size when caught, are inefficient consumers of food, but which for nearly all their lives were efficient consumers.

The more serious question is the extent to which the greater efficiency with which a small population of young and fast-growing fish use the food they consume is counteracted by the great consumption of a larger population including many old fish. The greatest fish production would be achieved by maintaining a *large* population of small fish. The numbers of young (the year-class strength) naturally occurring is usually too low to produce a sufficiently large stock of small fish, especially if heavy fishing should reduce the number of young produced. It is therefore sometimes suggested that the natural recruitment should be increased by the hatching and rearing of additional numbers of young fish for liberation in suitable areas of the sea. Though some of the technical problems of doing this may be solved (Shelbourne, 1964), it is very doubtful whether such operations are at all close to being economically justifiable, except perhaps for very highly priced fish such as salmon.

Exploited fish populations therefore consume less food and have a lower production than when unexploited. These reductions will depend partly on the effect fishing has on the average number of recruits, and on the changes, if any, in the growth rate of the individual fish. If growth does not change, the reduction can be very large; production can fall to half or less (Gulland, 1967b, Fig. 2); Beverton and Holt (1957, Table 18.9) estimated that the food consumption of North Sea plaice stock in the unfished state would have been some 10 times that in the inter-war period if the growth and numbers of recruits were unchanged.

Evidence of the quantitative effects on growth of fishery-induced changes in abundance are scarce. Beverton and Holt found that there was no significant difference in the growth of plaice in the North Sea between the inter-war period and 1946-47, when the stock abundance was at least twice as high. They found, however, that in the 1946 samples the 10-13-year-old fish were appreciably smaller than the pre-war average, and estimated that this change could be represented by a reduction in the growth parameter W_∞ from an interwar value of 2867 g to a mean wartime value of 1982 g. This is probably an over-estimate of the change in growth of the population as a whole, since Margetts and Holt (1948) showed that there was very little difference on the weight-at-age of fish sampled in 1938 and 1946 for fish less than 10 years old. Since 1946 there have been other changes in the apparent growth of plaice; young fish are growing faster, but old fish more slowly than between the wars, and the latter at least may be correlated with the increased abundance of large fish (Gulland, 1968a), though the analysis is complicated by the appreciable difference in size at age between different areas. Also tending to over-estimate the real decrease in growth is the increased proportion of male fish in the post-war catches of the older fish (Beverton, 1964), since among the older fish male plaice are smaller than females, the increased proportion of males will reduce the average size at a given age in combined samples.

Accepting Beverton and Holt's figures as being an over-estimate of the reduction in growth, their calculations can be used to give a lower limit to the food consumption of the plaice stocks at the high wartime abundance. This could not be given in absolute units, but Beverton and Holt express the food consumption, with reference to Dawes (1931) experiments as Mytilus-gram-equivalents per recruit. In these units the consumption in the 1930s was 1122 units, and that during the war was 2800 units, i.e. 250% of the inter-war value. The production can be estimated from the values of yield per recruit given by Beverton and Holt (hypothesis f_1 in Fig. 18.13) by multiplying by $(F+M)/F$, and is

between wars, $\dfrac{0.83}{0.73} \times 194 = 221$

during war (equivalent to a steady $F = 0.27$) $\dfrac{0.37}{0.27} \times 215 = 295$, or 133% of the interwar period.

These figures are equal to the total production of fish above the age at recruitment, plus the weight of recruits. The production due to growth after the age at recruitment may be estimated by subtracting the weight at recruitment, 123 g from the above estimates, giving 98 and 172 g respectively. The wartime production, on this basis, is then 176% of the interwar value. The age of recruitment, if not the clear distinct and rapid process that it is, for ease of calculation, assumed to be in simple stock assessment computations, does correspond to a fairly well defined movement from the predominantly nursery grounds along the coast to the main offshore fisheries. The food consumed in the pre- and post-recruitment phase differs in its detailed species composition—as it does between different inshore or offshore grounds—but in both phases the food is almost entirely bottom-living invertebrates, especially molluscs and worms, which probably fit to a large extent into the same position in the food web. The more meaningful of the above estimates of production is therefore that for total production, including the net production during the pre-recruitment phase, which indicates an increase of fish production per recruit during the war of a little over 30%.

The year-class strength of North Sea plaice appears to be independent of the adult stock (Beverton, 1962). Specifically those present during the war were of about average strength, though immediately followed by strong year-classes born in 1946, 1947 and 1948. The above figures of food consumption or production per recruit may therefore also be used on measures of total consumption or total production.

The increase in adult stock brought about by the temporary cessation of fishing, therefore increased the total food consumption by 150%, fish production by 33%, and if the same average density had been achieved by a reduced fishing effort constant over a period, this would have increased fish catches by 11%. Expressed another way, the efficiency with which the plaice harvest their food supply increased by 150%, though the efficiency with which they used the food consumed decreased by $100 - 133/250 = 47\%$, and the efficiency of fishing (in the sense of the proportion of the fish production harvested by man) decreased by $100 - 111/133 = 17\%$. (In economic terms the efficiency of the fishing operation has greatly increased since a slightly greater catch could be taken with a much reduced amount of fishing.)

One problem in the above analysis is the degree to which the wartime events may be used as measures of what would occur in a steady state, with reduced fishing, and an equivalent stock density of plaice. The calculation of the plaice stock, its food consumption and production have been based on steady state conditions using the parameters of growth and mortality estimated from the transitional wartime situation; the remaining question is whether the relation between stock density and growth derived from a transitional state holds good for a steady state. The steady state plaice population will differ slightly from the transitional population of the same mean density in its age composition, and hence (because one fish of 800 g will not eat precisely the same quantity of food as two fish of 400 g) have a rather different food consumption, but this difference is not large. However the food population will not react instantaneously to changes in the fish population, but there will be a delay whose extent will depend on the population dynamics of the food organisms, especially their average lifespan, and the relation between the adult stock and average number of young surviving to a size at which fish can eat them. This situation is similar to the problem of analysing the effect on fish stocks of a fishery whose intensity is changing. The density may be related to the mean fishing intensity over a period equal to the average lifespan of fish in the fishery (which will be considerably less than the potential maximum lifespan of the fish) (Gulland, 1961), though the possible stock/recruit relation must also be investigated. The food organisms are all small, with short lifespans, so that the average lifespan is probably a fraction of a year. Nothing is directly known about their possible stock/recruit relations, but most of the animals concerned have as high or higher fecundities as most fish; for fish it is argued that only a small change in the high mortality rate among the young would be required to compensate for big changes in adult stock, thus resulting in a constant average number of recruits. The same may well be true of the bottom invertebrates. Certainly the settlement of young molluscs is extremely variable in space and from year to year, and the major influence on the number of young is some, as yet undetermined, environmental factor, and the number of adults seems, above a certain level, to be unimportant. Lacking further evidence it is not unreasonable to assume that the recruitment is independent of adult stock, and therefore that there is only a short time-lag between changes in the stock of plaice, and changes in the food population, and that the calculations above may be taken as valid for steady state conditions.

Plaice is only one demersal species. During the war all the demersal stocks of major commercial importance increased, and in fact the percentage of plaice in the catches in 1946 was, on several grounds, less than in 1938 (Margetts and Holt, 1948).

Calculations of the changes in food consumption and production could in principle be made for each major species (cod, haddock, sole, etc.) though even less of the necessary data are available. The most critical information is on the changes, if any, in the growth-rate. Parrish (1948) gives data on the average lengths of haddock in different areas; these, with the sampling data, and also the pre-war mean lengths used by Beverton and Holt, and the post-war lengths given by Parrish and Jones (1953) are given in Table 2.

TABLE 2.   Average lengths of North Sea haddock

Age	Pre-war	1945 North (November)	1945 Central (July)	1945 South (September)	1946-50 North	1946-50 South
0+	17·0	13·5	—	—	15·5	17·8
1+	24·5	28·4	24·9	29·2	25·0	28·0
2+	29·5	33·7	34·7	38·4	30·5	34·5
3+	33·5	38·0	40·7	43·5	34·5	40·0
4+	37·0	39·4	42·8	46·5	38·0	43·0
5+	40·0	40·6	45·2	49·5	41·0	48·5
6+		41·8	47·1	51·4		

In interpreting this table it should be noted that for all except the 1945 data the lengths refer to the lengths achieved at the end of the growth period. In 1945 the fish sampled in November would have probably finished growing for the year, but those sampled in July and September 1945 would be expected to grow somewhat by the end of the year. Table 2 shows that while there has been an increase of size since before the war, there is little difference between the mean lengths observed in 1945 and 1946-50, the former fish being if anything larger. There is therefore no evidence of even the small decrease in growth estimated for the plaice. This conclusion is somewhat at variance with the previous observation that there is a negative correlation between year-class strength and growth, which may be explained by the fact that this latter correlation concerns the younger fish (which did not increase in abundance during the war), and that the year-class fluctuations are more extreme than the changes in total stock density due to the wartime reduction in fishing. That is, there may have been a wartime reduction in growth but not sufficient to be apparent in the available data.

For the other species (cod, whiting, sole, etc.) there is no direct observation on the wartime growth rate, but the existence of unusually large quantities of very big cod (e.g. Margetts and Holt, 1948), suggest that the growth rate was little, if at all, below normal. Most of these species (including haddock) are shorter lived than plaice—plaice up to 24 years old were present in the 1945 catches, while few haddock over 10 years old were present (Parrish, 1948)—so that for these species the transitional wartime conditions would approach even closer to steady state conditions with the same mean density. The previous conclusions regarding the plaice stocks can therefore be applied to the total demersal stocks, that is, during the wartime period of high abundance the food consumption of the fish stocks was greatly increased, probably by between two and three times (i.e. by about the increase in stock density) that the fish production also increased, but not so greatly, and that, in a steady state condition of similar density, the catch would also be increased, but not so much as the total production.

This conclusion, that during the periods of intense fishing, which cover most of the last seventy years, fish production, and the food consumption by the fish stocks have been less than they might have been, is surprising in view of the fact, noted earlier, that fish production during the same period has been near the theoretical limit set by the primary production and the generally accepted 10-15% efficiency at each trophic level. The earlier conclusion, that the efficiency must, for some stages, be greater than 10-15%, is reinforced, and the potential efficiency, reached under wartime conditions, is even higher than the high efficiencies during periods of heavy fishing. It does not require much increase in the efficiency, if occurring at each of several stages, to double the potential fish production. For instance, if the latter is three stages from the phytoplankton, a difference in efficiency between 11% and 14% can make more than a twofold difference.

Another tentative conclusion made earlier, that the fish production and potential fish catches could be directly estimated from the primary production and efficiency factors receives less support. Even with little fishing the haddock data suggest that there is some "under-fishing" of the food stocks except when there is an exceptionally strong year-class present. If man's use of the fish production is made more efficient by moderately intense fishing, the reduced fish stock makes progressively less efficient use of the production of the organisms on which they feed. Thus, for at least the last two stages (fish-food to fish, fish to man) maximum efficiency in one stage can only be achieved at the cost of reducing efficiency in the other stage. The loss in efficiency appears to be of the order of a factor of two or three, which means that the estimates of potential fish catch made from food-chain considerations may well be correct within an order of magnitude—or even closer if the figures for efficiency less than the maximum possible are used—and this degree of precision can be of definite assistance.

Though these studies show that the potential of species of fish several stages along the food-chain is larger than predicted using some commonly used values for the ecological efficiency, catches of these species are now, in some areas, e.g. the North Sea, reaching a level beyond which they cannot be increased much further. Greatly increased catches from these areas can only come from organisms earlier in the food-chain. Thus the main recent increases in the North Sea catches which have been due to changes in the fishery, are the increased catches of herring and the new fisheries for sandeel and Norway pout (*T. esmarkii*). In the North Sea it seems that the catches even of the latter species cannot be very greatly expanded—perhaps by a factor of two or three—and the unexploited resources (mainly the food of the presently exploited fish) are the zooplankton and bottom invertebrates. Further exploitation of the latter seems possible. In most other areas, even where fishing is intense, there are underexploited stocks of smaller fish. For instance, just outside the North Sea, on the edge of the shelf to the west of Scotland, where the hake stocks are heavily fished, one of the main foods of the hake, the blue whiting (*M. poutassou*), forms an unexploited resource which has been estimated to be of the order of hundreds of thousands of tons.

MANAGEMENT OF MULTI-SPECIES FISHERIES

Exploitation of several stocks on different trophic levels can raise difficult problems of management which can only be resolved by good knowledge of the quantitative relationships between predator and prey, including the weight of prey consumed to produce unit weight of predator. It is in the nature of fishermen to complain about the activities of any other fishermen fishing in the same sea. Two such complaints recently made in the North Sea of relevance to food-chain dynamics are complaints by demersal fishermen against those fishing for sandeels, and of herring fishermen against dogfish, and hence against those wishing to conserve the stocks of dogfish.

Regarding the sandeels, the complaints take two forms, first that fishing for sandeels reduces the food available to the demersal stocks, and hence the growth of the individual fish, and the weight caught. Second, when the local concentrations of sandeels are reduced, the normal patterns of aggregation of the demersal fish, which at times are locally feeding intensively on sandeels are disturbed. Thus the abundance of demersal fish is unchanged, but catching becomes more difficult. The latter point is one that is difficult to prove or disprove, though it seems somewhat unlikely. Earlier discussion in this paper suggested that the demersal stocks in the North Sea were, at least in periods of full exploitation by man, not using all the available food supplies. Also, as Macer (1966) points out, in the southern North Sea, sandeels are only a minor part of the total food supplies of even those demersal species which feed most heavily on them (cod and whiting) and the density of sandeels at which commercial fishing becomes uneconomic still provides a good food supply. (However to the north of Scotland sandeels form much the most important food item of cod (Rae, 1967a), though these stocks of sandeels are at present unexploited.) Therefore, the available, not very detailed, information on the feeding of demersal fish is sufficient to determine that the complaints against the sandeel fishery are not justified, though similar complaints might become justified if there was intense exploitation of all or most of the stocks of food of demersal fish. Also, if sandeel fisheries had been established before those for, say, cod, other complaints concerning the effect of cod on the sandeel stocks would probably have occurred.

Better quantitative data are required to resolve the dogfish-herring controversy. Ignoring the problem of direct damage to herring drift nets and other gear, and complaints of swarms of dogfish driving away the herring (Rae, 1967b), the essential question is whether a larger or more valuable catch is taken by catching dogfish which have eaten herring or by harvesting herring directly—politically the question is complicated by the fact that different groups of fishermen are involved in harvesting the two different species.

Holden (1965) calculated that the Scottish-Norwegian stock of dogfish consumed annually some 227 000 tons of food, of which 29 500 was herring. The quantity eaten is not the same as the loss of catch to the herring fishermen and should be adjusted to allow for possible growth of herring before capture, and for the proportion of herring that are not caught, but die of natural, non-fishing, causes. Also the contribution to the catch of dogfish will, due to losses from natural mortality, be less than the total increment in weight of the dogfish stock, which may from Holden's data be estimated as $120 \times 1 \cdot 25 \times 10^8 g$ or 15 000 tons. (This figure is less than recent dogfish catches, and the stock is being reduced by too heavy fishing.) Allowing for a rather greater value per unit weight for dogfish, and a greater exploitation rate (the ratio of fishing to total mortality) for dogfish than for herring, it appears that the value of the dogfish catch may be quite close to the reduction in value of the herring catch. In addition, if the overall effect on the economic success of the fisheries is to be examined, some assumptions have to be made, as were done by Andersen and Knudsen (1965), on the effects of costs of fishing. To harvest the dogfish would entail a specialized fishery, while the reduction in herring stocks would result in lower catches for the same fishing effort. These considerations will not be analysed further here.

Generalizing the dogfish/herring analysis, in mathematical terms, we can write:

Nw = weight of food (herring) eaten where w is the mean weight of an individual herring eaten.

k = conversion ratio, or the mean weight increment of the predator from consuming unit weight of prey. It will be greater than the ecological efficiency to the extent that not all the production of prey is consumed by the predator, and will depend on the age composition of the predator population.

$\bar{w}$ = mean weight of individual prey species (herring) in the catches.

E = exploitation rate of prey; strictly E = proportion of prey of size w which ultimately would be caught.

E' = exploitation rate of predators.

V, V' = value per unit weight of prey and predators respectively.

Then, if prey are eaten, the ultimate value of the resulting catch, as predators (dogfish) is

$$V' \, E' \, k \, Nw$$

while, if prey are not eaten, the value of the catch of prey (herring) is

$$V \, E \, N \, \bar{w}$$

A greater value of catch is therefore obtained by harvesting the predators if

$$V' \, E' \, k \, w > V \, E \, \bar{w} \tag{1}$$

which may also be written as $k > V/V' \times E/E' \times \bar{w}/w$
i.e. the efficiency of utilization should be greater than the product of the relative values, the relative exploitation, and the probable growth of the prey.

Since both the conversion efficiency and the type of food consumed will depend on the size of the predator, both sides of this expression should be expressed as functions of the size, so that the conditions as far as predators, of length l, are concerned becomes:

$$V' \, E'_l \, k_l \, w_l > V \, E \, \bar{w}$$

and, integrating over the predator population as a whole, in a similar manner to Andersen and Knudsen (1965), the condition becomes

$$\int V' \, E'_l \, k_l \, N_l \, w_l \, dl > \int V \, E \, \bar{w} \, N_l \, dl \tag{2}$$

where N_l is the number of prey consumed per unit time by predators of length between l and $l+dl$.

Also more than one species of prey is likely to be concerned. For instance dogfish eat in addition to herring, such commercially exploited fish as cod, haddock and sandeels, as well as species at present unexploited, including fish such as blue whiting (*M. poutassou*) and dragonet (*Callionymus lyra*) as well as various crustaceans (e.g. hermit crabs) and molluscs. Equation (1) should therefore be written as the sum for the various food organisms, i.e.

$$\sum_1^n V' \, E' \, k_i \, w_i > \sum_1^n V_i \, E_i \, \bar{w}_i, \tag{3}$$

using the obvious notation and assuming there are different food organisms with different efficiencies of utilization by the predator.

In these equations the various parameters can be estimated with greatly varying degrees of precision. The relative value can be calculated from normal past statistical data with great precision, though it can vary from year to year, often depending on the use to which the catch is put. The average size of prey fish in the catches is easily established, as may be the average size eaten, by examination of stomach contents—using measurements of otoliths if the actual fish are well digested. Measurement of exploitation rates is more difficult, but increasingly accurate estimates are now being obtained for the major species; even where there are important doubts, e.g. for the herring of the northern North Sea, the range of uncertainty for the larger fish (for which $E = F/(F+M)$ for constant mortalities) may extend over no more than a two-fold range, e.g. from $E = 0\cdot4$ to $E = 0\cdot8$ for the herring. There is more uncertainty for smaller fish, below the sizes commercially exploited, since the value of E for these sizes is less than $F/(F+M)$ in proportion to the natural deaths occurring before the fish reach a commercial size. The natural mortality among these small fish is not easy to estimate. However, the greatest uncertainty concerns the value of k, the efficiency of utilization of the food consumed. This uncertainty relates to both the maintenance requirements of fish in their natural conditions (which may well be greater than that measured in the restricted environment of a laboratory) and the efficiency of conversion of that part of the food intake above maintenance requirements.

Where there appears to be a close balance, at least in dogfish and herring, between the desirability of catching predators or catching prey fish, proper management will require close estimates of the various parameters involved. Since the parameter which appears to be least accurately known is the conversion factor, better knowledge of this factor, and its variation with size of predator and of prey, with species of prey, etc. will become increasingly important as the range of species exploited increases.

REFERENCES

ANDERSON, K. P., and KNUDSEN, H. 1965. The yield from a fishery on a fish-eating stock as compared with the value of the quantity of fish consumed by this stock. *I.C.E.S.*, *C.M.*, *1965*, *Near Northern Sea Comm.*, Doc. No. 75.

BEVERTON, R. J. H. 1962. Long-term dynamics of certain North Sea fish populations. In *The exploitation of natural animal populations*, edited by E. D. Le Cren and M. W. Holdgate. Oxford, Blackwell Scient. Publs., 242-49.

BEVERTON, R. J. H. 1964. Differential catchability of male and female plaice in the North Sea, and its effect on estimates of stock abundance. *Rapp. R-v. Réun. Cons. perm int. Explor. Mer*, **155**, 103-12.

BEVERTON, R. J. H. and HOLT, S. J. 1957. On the dynamics of exploited fish populations. *Fishery Invest., Lond. ser. 2*, **19**, 533 pp.

CARRUTHERS, J. N., LAWFORD, A. L. and VELLEY, V. F. C. 1951. Fishery hydrography: brood strength fluctuations in various North Sea fish with suggested methods of prediction. *Kieler Meeresforsch.*, **8**, 5-15.

DAWES, B. 1931. Growth and maintenance in the plaice (*Pleuronectes platessa*). Part 2. *J. mar. biol. Ass. U.K.*, **17**, 877-947.

FAO YEARBOOK OF FISHERY STATISTICS, 1967. *Catches and landings, 1966*, **22**.

GULLAND, J. A. 1961. Fishing and the stocks of fish at Iceland. *Fishery Invest., Lond., ser. 2*, **23** (4), 52 pp.

GULLAND, J. A. 1962. The application of mathematical models to fish populations. In *The exploitation of natural animal populations*, edited by E. D. Le Cren and M. W. Holdgate. Oxford, Blackwell Scient. Publs., 214-17.

GULLAND, J. A. 1967a. Area reviews on living resources of the world's oceans: Northeast Atlantic. *FAO Fish. Circ.*, **109** (6) Rev. 1, 40 pp.

GULLAND, J. A. 1967b. The effect of fishing on the production and catches of fish. In *The biological basis of freshwater fish production*, edited by S. D. Gerking. Oxford, Blackwell Scient. Publs., 399-415.

GULLAND, J. A. 1968a. Recent changes in the North Sea plaice fishery. *J. Cons. perm. int. Explor. Mer*, **31**, 305-22.

GULLAND, J. A. 1968b. Report on the population dynamics of the Peruvian anchoveta. *FAO Fish. tech. Pap.* (72), 29 pp.

HOLDEN, M. J. 1965. The food intake of the spurdog *Squalus acanthias*. *I.C.E.S.*, *C.M. 1965*, *Near Northern Seas Comm.*, Doc. No. 88.

JONES, R. 1964. Haddock. North Sea stock, Scottish investigations. *Annls biol., Copenh.*, **19**, 103-5.

MACER, C. T. 1966. Sandeels (Ammodytidae) in the southwestern North Sea; their biology and fishery. *Fishery Invest., Lond., ser. 2*, **24** (6) 55 pp.

MARGETTS, A. R., and HOLT, S. J. 1948. The effect of the 1939-45 war on the English North Sea trawl fisheries. *Rapp. P.-v. Réun. Cons. perm. int. Explor. Mer*, **122**, 26-46.

MENARD, H. W., and SMITH, S. M. 1966. Hypsometry of ocean basin provinces. *J. geophys. Res.*, **71**, 4305-25.

PARRISH, B. B. 1948. The haddock stocks in the North Sea during the second half of 1945. *Rapp. R-v. Réun. Cons. perm. int. Explor. Mer*, **122**, 47-54.

PARRISH, B. B., and JONES, R. 1953. Haddock bionomics 1. The state of the haddock stocks in the North Sea 1946-50 and at Faroes 1914-50. *Mar. Res.*, 1952 (4), 27 pp.

POSTUMA, K. H., and ZIJLSTRA, J. J. 1964. Some remarks on the estimation of the abundance of herring and mackerel from data on the catches of Netherland trawling fleet. *Rapp. R-v. Réun. Cons. perm. int. Explor. Mer*, **155**, 117-21.

RAE, B. B. 1967a. The food of cod in the North Sea and on West of Scotland grounds. *Mar. Res.*, 1967 (1), 68 pp.

RAE, B. B. 1967b. The food of the dogfish *Squalus acanthias*. *Mar. Res.*, 1967 (4), 19 pp.

RAITT, D. F. S., and SYMONDS, D. J. 1967. The Scottish cod fishery in the North Sea. *Mar. Res.*, 1967 (5), 24 pp.

RAITT, D. S. 1939. The rate of mortality of the haddock of the North Sea stock 1919-38. *Rapp. P.-v. Réun. Cons. perm. int. Explor. Mer*, **110**, 65-80.

SAVILLE, A. 1959. The planktonic stages of haddock in Scottish waters. *Mar. Res.*, 1959 (3), 23 pp.

SCHAEFER, W. B. 1954. Some aspects of the dynamics of populations important to the management of commercial marine fisheries. *Bull. inter-Am. trop. Tuna Commn*, **1** (2), 26-56.

SCHAEFER, M. B. 1967. Dynamics of the fishery for the anchoveta *Engraulis vingenus* off Peru. *Boln. Inst. Mar. Peru*, **1** (5), 191-303.

SHELBOURNE, J. E. 1964. The artificial propagation of marine fish. *Adv. mar. Biol*, **2**, 1-83.

SHOMURA, R. S., and GULLAND, J. A. 1968. Areas review on living resources of the world's oceans: West Central Pacific. *FAO Fish. Circ.*, **109** (9), 18 pp.

STEELE, J. 1965. Some problems in the study of marine resources. *Spec. Publs int. Comm. NW. Atlant. Fish.*, **6**, 463-76.

TIEWS, K. 1965. Bottom fish resources investigation in the Gulf of Thailand and an outlook on further possibilities to develop marine fisheries in South East Asia. *Arch. FischWiss.*, (**16** (1), 67-108.

U.K. M.A.F.F. 1967. *Rep. Dir. Fish. Res.*, Lowestoft, 1966, 122 pp.

Part Five

FOOD ABUNDANCE AND AVAILABILITY IN RELATION TO PRODUCTION

Introduction

L. M. Dickie
Fisheries Research Board of Canada
Marine Ecology Laboratory
Bedford Institute
Dartmouth, N.S.

In principle it is already possible to calculate the minimum food intake needed
to support a specified amount of production at various levels in marine food
chains. However, to estimate potential fisheries yield requires that this
process be reversed and that the production or that part of it realized as yield
be predicted from an observed food abundance. This latter process raises
problems which are as yet unsolved but whose importance is indicated in a
simple food-chain model constructed by Riley (1963). He showed that a
change in the amount of food available at lower and intermediate links of the
chains may be magnified several times in the abundance of a terminal predator, which he called "potential tuna" but which might well have been any of
the fish which form the basis of commercial fisheries.

Initial approaches to a solution of the problem have consisted of attempts
to relate yield from a system to its gross ecological efficiency, calculated as
the ratio of yield to net primary production. The results show that there are
major differences among areas, which seem to be related to differing ecological efficiencies at various levels in the food chains. Thus, for example, while
Slobodkin (1962) found that ecological efficiencies in several simple terrestrial-
and aquatic predator-prey systems differed little from values of 10-15%,
Steele (1965) found evidence for significantly higher values in the North Sea,
and Mann (1965) has suggested lower values among benthic fishes in a section
of the River Thames. From such examples there remains little doubt that

gross ecological efficiency and yields will differ from one situation to another, depending on the internal structuring of the food-chain system and the nature of the pathways which the energy follows.

Further progress towards an understanding of mechanisms which would permit predictions depends on two developments. The first is a fuller description of natural systems. The second is an interpretation of the structuring of food-chains in a way which relates the abundance of a group of potential food organisms to their availability to their predators. Taken in the context of current ecological research, the papers presented here show that both these developments are possible with present methodology. The papers also indicate the likely shape of future researches in this field.

Holling (1966) has pointed out that in an analysis of the individual predator-prey systems which make up food chains, it is frequently an advantage to sacrifice simplicity and mathematical convenience for reality, even where this may cost much in terms of research effort. The paper by Keast, describing the feeding habits of a single natural fish community, indicates the extent of the effort involved. The amount of detail and complexity which can be revealed seems limited only by the willingness and capacity of the investigator to make observations and report them. However, Keast's work in unravelling complexity of diurnal, and seasonal cycles among species and sizes of fish food and fishes, already indicates the possibility of useful generalization. Behavioural and physiological specializations underlying particular food selections by the predators are shown to be strongly related to their size and anatomical form, potentially conferring a degree of structural stability on the system. This will permit a more quantative classification of energy pathways than has been possible by resort to such vague definitions as benthic or pelagic feeders. Further work of the same sort, if followed through experimental or natural alterations in the balance of numbers would make it possible as a minimum to test the relation between the degree of body differentiation among predators and the selection of a particular food, as has been suggested by Nikolski (1962). It is clear from the review of this one study that it is only by such detailed observations of electivity of food types and division of resources among associated predators that we will be in a position to determine what constitutes an adequate classification of a feeding type and of food potentially available to it. If the classifications prove to be stable or at least predictable, the objectives of future descriptions are clarified and the task made considerably simpler.

The question of electivity or selectivity of a given type of food by a predator is one important aspect of availability. The importance of the related aspect of aggregation of food particles was recognized by Ivlev (1961). He demonstrated that at a given level of abundance, feeding success per unit time was as strongly dependent on the variance of the number of food particles per unit area as it was on major changes in average abundance of food. Parsons and LeBrasseur approach this problem for various parts of the food chain, in an ingeniously designed set of laboratory and field experiments which directly relate availability indices to the structure of natural food chains. They summarize earlier work on zooplankton grazing and report new data on the feeding of juvenile fish. In both cases the rates of food intake by the predators are dependent on abundance or density of their prey in the same manner as described by Ivlev (1961) and Holling (1965). As a second step they analyse the observed array of potential food organisms in terms of size classes, and compare the rate of feeding at various densities for several of the modal classes. Size of food particle may in a sense be considered an index of the variance of distribution of food particles. To this extent their finding that at the lower sizes an increase in size resulted in higher feeding success at given abundance agrees with Ivlev's conclusions with respect to effects of aggregation. However, above an apparently optimal size of particle (zooplankton), feeding success decreased, possibly because individual particles become too large for successful catching or handling (cf. Holling 1964). Finally, they compare the concentrations of food required for a given food intake with the concentrations found over various sample distances in nature. They demonstrate that while average abundance of food organisms over long distances is low, compared with requirements, local plankton aggregations which are displayed by detailed sampling provide conditions necessary for successful grazing by juvenile fishes. Their set of experiments, carried out with the plankton associations and fishes observed in their field studies permitted them to identify the actual pathways of energy transfer in the changing seasonal structure of the ecosystem, and to judge on the basis of the Ivlev "feeding-success" coefficients the relative importance of particular energy pathways to the production at higher trophic levels. Their clear demonstration of the dependence of the energy flows on the sizes and distributions of food organisms lays the foundation for a general methodology of designing sampling procedures which will permit the calculation of indices of abundance of food organisms weighted according to their availability in natural systems. There is a further implication in their work that the method may be directly useful for study of the relation of larval fish survival to food abundance. This would be heartening news for the fisheries biologist who has learned to regard direct studies of stock-recruit relationships as one of the most intractable of his problems. The lead given by Parsons and LeBrasseur will be eagerly followed by many other marine biologists.

The paper by Rosenthal and Hempel provides a welcome complement to the findings of Parsons and LeBrasseur. They report a functional analysis of the development of feeding behaviour in herring larvae. It demonstrates that not only does a "critical" stage in development exist, but that survival through this stage must be importantly related to the availability of food. This is reflected in several ways. For example, the period of resorption of the yolk is accompanied by lowering of specific gravity and development of vertical swimming movements which would ensure concentration of the pelagic larvae at the surface and tend to hold them in convergence zones of Langmuir circulation where the concentration of food organisms is highest. The development of horizontal movement and search swimming coincides with disappearance of yolk material. The subsequent degree of activity is strongly correlated with feeding success, an energy conservation device which would appear essential in view of their finding that increased activity of the larva at a particular size is unlikely to lead to a commensurate increase in the probability of encountering food. Finally, the rapid increase in searching capacity of the larvae with increase in their size ensures a positive feedback for given degree of successful feeding and growth. It is difficult to imagine a set of reactions more closely adapted to overcoming problems of feeding and growth, offering strong evidence that this is a major developmental hurdle. Such adaptation clearly implies that fluctuations in year-class success must be importantly linked with the availability of food to the larvae, an implication which provides added assurance that the methods developed by Parsons and LeBrasseur are appropriate to direct studies of recruitment and production at the higher trophic levels.

The significance of food-chain structure, sizes of organisms, and availability are brought into fuller relief by the paper of Kerr and Martin, based on earlier observations of fish production in some 20 small northern Canadian Lakes by Martin (1966). Martin had reported that differences in food-chain length were reflected in the size composition of the top carnivore but that their total production appeared unaffected. Kerr and Martin point out that the basic production in the most regularly exploited lakes is predictable from their edaphic and morphometric characteristics, indicating that the explanation of the failure of food-chain length to affect the production must be sought within the food-chain systems. They then demonstrate, on the basis of present knowledge of the relations of metabolism and growth efficiency of fishes to feeding on different types of food, that the interaction of abundance and availability related to size of food organisms is a necessary and sufficient explanation of the similarity of production levels. They suggest additional experimental and field observations which are appropriate if all reasonable doubt of their hypothesis is to be removed. But even their demonstration that changes in the availability of different sizes of food can have an effect on ecological efficiency commensurate with a major change in effective food-chain length, places the whole problem in a new perspective. A further inference which one is tempted to draw from this work: that there may exist some kind of ecosystem homeostasis which will be reflected in food-chain structure, is probably too startling to be an acceptable generalization without comparable data from other systems. However, the implications of any such generalization with respect to the ultimate dependence of ecosystem production on the physical structure of the environment, and the implications that such features may affect the outcome of various management procedures on fisheries yields are so great, that even the possibility demands closer examination. Data required for such study may already exist in the remarkable time-series of observations following the experimental manipulation of fish populations in Lake Windermere (LeCren, 1958), or Heming Lake (Fish. Res. Bd Canada, 1961-62, 62-63), or in the observations of fish yields in various small lakes and ponds in many parts of the world.

It is clear from the foregoing that simple integral models of food-chains are an inadequate basis for predicting predator yield from its food abundance. The papers presented show instead that for a given predator-prey link in the chain there are likely to be important differences between ecological efficiency (yield/food intake) and food-chain efficiency (yield/food abundance). They also indicate that it is highly probable that the abundance of potential food energy in particular pathways can be adequately understood in terms of its availability when measured by parameters describing particle type, size and distribution. This has been in need of demonstration for some time. What is perhaps even more important however is, first, the evidence that these descriptive parameters are related to intrinsic morphometric, physiological and behavioural properties of the component animals; hence the mathe-matical-statistical description of the system must be approaching a specifica-tion in terms of the real mechanisms responsible for regulation. The second important feature is that to a large extent the papers have specified methods and sampling procedures appropriate to precise measurement. This brings the study of food chains out of the realm of speculation into that of serious quantitative experimental and observational study.

A contribution of this sort is unexpected justification of a small group of papers in one symposium. It comes as extra premium that these food-chain studies should also demonstrate the essential similarity of the natural and man-made links. The parameters of predator searching times and perception distances, together with the indices of prey aggregations: "school" size and distribution, and within school distribution densities and particle sizes, which they show are required for description of natural predator-prey production systems, are the same as have been proposed for the study of the relation of fish abundance to the success of fishing vessels (cf. Paloheimo and Dickie, 1964). The understanding of one will contribute to an understanding of the other, firmly linking the study of yield to man with the overall ecological study. There is indeed a firm foundation for approaching the problem of predicting yields from natural systems, and assessing the effects of natural and man-made changes on it.

REFERENCES

HOLLING, C. S. 1964. The analysis of complex population processes. *Can. Ent.*, 96, 335-47.
HOLLING, C. S. 1965. The functional response of predators to prey density and its role in mimicry and population regulation. *Mem. ent. Soc. Can.*, (45), 5-60.
HOLLING, C. S. 1966. The functional response of invertebrate predators to prey density. *Mem. ent. Soc. Can.*, (48), 86 pp.
IVLEV, V. S. 1961. *Experimental ecology of the feeding of fish*. New Haven, Yale Univ. Press, 302 pp.
LeCREN, E. D. 1958. Observations on the growth of perch (*Perca fluviatilis*) over twenty-two years with special reference to the effects of temperature and changes in population density. *J. Anim. Ecol.*, 27, 287-334.
MANN, K. H. 1965. Energy transformation by a population of fish in the River Thames. *J. Anim. Ecol.*, 34, 253-75.
MARTIN, N. V. 1966. The significance of food habits in the biology, exploitation, and management of Algonquin Park, Ontario, lake trout. *Trans. Am. Fish. Soc.*, 95, 415-22.

NIKOLSKI, G. V. 1962. On some adaptations to the regulation of population density in fish species with different types of stock structure. In *The exploitation of natural animal populations*. Edited by E. D. LeCren and M. W. Holdgate. New York, John Wiley and Sons, 399 pp.

PALOHEIMO, J. E., and DICKIE, L. M. 1964. Abundance and fishing success. *Rapp. P.-v. Réun. Cons. perm. int. Explor. Mer*, **155**, 152-63.

Rep. Fish. Res. Bd Can., 1961-62, 205 pp.

Rep. Fish. Res. Bd Can., 1962-63, 172 pp.

RILEY, G. A. 1963. Theory of food-chain relations in the ocean. In *The Sea*, **2**. Edited by M. N. Hill. New York, Inter-sci. Publ., 438-63.

SLOBODKIN, L. B. 1962. Energy in animal ecology. In *Advances in ecological research*, **1**, Edited by J. B. Cragg. London, Acad. Press, 69-101.

STEELE, J. H. 1965. Some problems in the study of marine resources. *Spec. Publs int. Commn N.W. Atlant. Fish.*, (6), 463-76.

The availability of food to different trophic levels in the marine food chain

T. R. Parsons and R. J. LeBrasseur
Fisheries Research Board of Canada
Pacific Oceanographic Group
Nanaimo, B.C.

ABSTRACT. By expressing the concentration of food available to secondary and tertiary producers as a continuous spectrum based on particle size, the relative importance of different foods can be observed as peaks of maximum biomass in the food spectrum. The relationship between the quantity of food represented by a peak, and the quantity grazed by a predator, has been studied and results indicate that the quantity of food consumed is concentration-dependent and can be best explained by a relationship similar to that proposed by Ivlev (1945) for planktivorous fish.

From studies in the field it has been shown that juvenile salmon feed best off *Calanus plumchrus* and less well off *Pseudocalanus minutus* or adult euphausiids. In nature the important position of *C. plumchrus* as food for juvenile salmon may often be filled by a prey of similar size. The feeding of zooplankton on a certain size fraction of phytoplankton has been shown to determine the growth and reproduction of particular zooplankters. In particular the feeding of microzooplankton on nanoplankton appears to be an important pathway for the survival of larval fish.

Food chains summarizing some of the above events in the waters of the Strait of Georgia, British Columbia, have been described and illustrations of different food relationships for species living in an oceanic environment have been given for comparison.

INTRODUCTION

Attempts to estimate fish production on the basis of primary or secondary production data have not been markedly successful outside of controlled environments. In reviewing the difficulties involved in such estimates, Steele (1965) concluded that the important element in energy transfer was not the quantity of basic production but the differing paths the energy followed through the food chain. The essential step forward in production estimates appears to require, therefore, a change in the type of measurements being

325

made so that (*i*) pathways of food exchange can be identified, and (*ii*) quantitative relationships can be established to determine the amount of food actually being transferred by any one pathway. This is expressed in Fig. 1 which illustrates feeding relationships between particular phytoplankters (P.), zooplankters (Z.), and fish (F.). What is required are data on the existence of these pathways and the relative amount of energy transferred by a particular route.

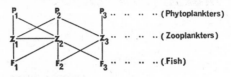

Fig. 1. Diagrammatic exchange of food sources indicated by connecting lines between and within three trophic levels.

In order to investigate the first of these problems we have considered the food available in aquatic systems as a continuous particle size spectrum in which the biomass of material is recorded as the number of particles (or food items) times the volume of an individual particle in a particular size category. Justification for this approach was based on earlier work (e.g. Lebour, 1920; Wiborg, 1948) as well as more recent studies (e.g. Brooks and Dodson, 1965), which tended to show size relationships between predators and prey. While these studies dealt with numbers and species of food organisms, Isaacs (1966), in considering the food of larval fish, introduced the idea of a con-

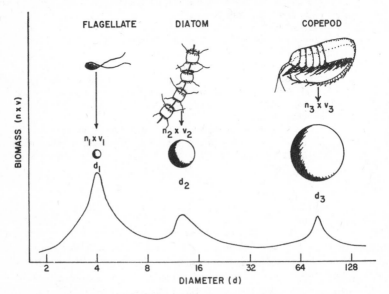

Fig. 2. Particle spectrum representing biomass of material in different size categories determined by the diameter of a sphere equivalent in volume to the original particle.

tinuous size spectrum of particulate food. In an extension of this concept we have used a grade-scale of size distribution described by Sheldon and Parsons (1967). The essential features of this distribution are shown diagrammatically in Fig. 2.

In approaching the problem of a size spectrum we considered that volume was the only measurable property which was common to the very diverse shapes of potential food particles in the oceans. Volume can be measured by a variety of techniques, some of which lend themselves to automation, e.g. electrolyte displacement (Sheldon and Parsons, 1967); laser and sound holography (Knox, 1966; Metherell et al., 1967). The exact choice of methods may to some extent depend on the size of the particles being investigated. In our work to date we have only explored the use of the Coulter Counter for this purpose. Once a measure of particle volume has been obtained, however, the representation of this volume as a sphere of equal volume allows for all data on particle volumes to be normalized into a continuous size spectrum based on a logarithmic progression of particle diameters, such as is shown in Fig. 2.

This approach has been used to identify the principal components of the biomass in any environment by the height and position of a peak in the spectrum, a procedure which readily lends itself to computerization. Our second problem has been to study the association between these peaks.

In order to examine the second problem, we have assumed a predator/prey density-dependent relationship, as described by Ivlev (1945), for planktivorous fish.

$$r = R(1 - e^{-kp}) \qquad [1]$$

where r is the ration at a prey density, p, and R is the maximum ration. The curve represented by this equation passes through the origin which implies that animals continue feeding down to zero prey density. This may be true for fish, which tend to search out their prey, but it appeared from our data on zooplankton/phytoplankton relationships that zooplankton continued to feed down to some low prey density, p_0, and then ceased (Parsons et al., 1967). In order to accommodate this observation, the above equation was modified to the form:

$$r = R(1 - e^{k(p_0 - p)}). \qquad [2]$$

The use of the above equation for examining food relationships between zooplankton and phytoplankton appears to be justified, not only on the basis of our own results, but also on the basis of work done by others (Mullin, 1963). The equation does not, however, explain inhibition of feeding at high prey concentrations, but since this observation has generally been made with laboratory cultures in which the prey density has been unnaturally high, it has not been considered necessary to further modify the equation to accommodate this effect. It is apparent, however, that use of equation [2] discards the concept of zooplankton filtering in barren water (Cushing, 1964) and introduces instead some prey density at which feeding starts. Independent justification for this appears in the data of Adams and Steele (1966), where it may be seen from their Fig. 4 that grazing ceased at a prey concentration of approximately 70 μg C/l. It is also worth noting that data on the volume of water swept clear by a zooplankter has not been considered in this approach but only the amount of food actually eaten.

While the effect of prey size and concentration have been the principal factors in feeding studied here, it is also apparent from our data that other factors (e.g. prey shape, patchiness, etc.) may at times be important. These factors have been noted as causing apparent inconsistencies in some of our data and an attempt has been made to study them further using quantitative indices as suggested by Ivlev (1961) in studies on benthic-feeding fish.

MATERIALS AND METHODS

Size spectra

The number and volume of particles in the size range 2-100 μ were determined using a Coulter Counter. The operation of this instrument for sizing and counting natural distributions of particles in sea water has been described by Sheldon and Parsons (1967). A regression equation was used to convert plankton volumes, as measured with the Coulter Counter, to carbon as measured by total combustion. The size spectrum of zooplankton was determined manually by counting the number of organisms and estimating the volume of individuals from length/weight regressions determined for different groups of animals. The results of these measurements, together with those carried out on smaller particles, were then both expressed on the same grade-scale, a description of which has been given previously (Sheldon and Parsons, 1967). From the particle spectrum obtained for any environment, peaks in the biomass were identified and the relationship between organisms causing maxima in the biomass have been reported in Tables 2 and 3.

Grazing

The amount of plant grazed by a particular zooplankter was determined for different plant concentrations from the size spectrum before and after the addition of the animals. The exact period of a grazing experiment was decided on the assumption that zooplankton which migrated into the euphotic zone at night fed only during the hours of local darkness, while those that were located in the euphotic zone during the day, fed continuously. Particulate material was kept suspended by rotating the incubation vessels on a large wheel enclosed within a water bath. The equipment was readily portable and had a variable speed control from 1 to 0·2 revolutions per minute. Animals were kept at *in situ* surface temperatures by pumping sea water through the water bath in which the incubation vessels were slowly rotated.

Juvenile fish feeding experiments were carried out in eight $\frac{1}{4}$ m³ plastic tanks. These were placed on board a research vessel so that the experiments could be performed in the area from which live fish and zooplankton were collected. Filtered salt water was provided to each tank by pumping water directly from over the side of the ship. Animals were prevented from escaping through the salt water overflow by a piece of fine nylon netting. The tanks were half covered with a black plastic so that lighting conditions varied from daylight to a low light intensity in each tank.

Fish were introduced into each tank and allowed to acclimatize for a minimum period of 24 hours. In a few cases fish were employed from cultures and in these instances the period of acclimatization included feeding the animals with live zooplankton. All fish were starved for 24 hours prior to the start of an experiment.

Preliminary tests showed that with juvenile fish, < 50 mm, food began to pass from the stomach to intestines after 2 hours. Experiments were therefore usually carried out for less than 2 hours during which time the same number of fish in each tank were allowed to graze off different concentrations of size-sorted zooplankton. All fish were preserved with formalin at the end of an experiment and the stomach contents examined for the number of zooplankton eaten. Attempts to perform similar experiments with larval fish were generally unsuccessful.

Selectivity experiments were carried out in $\frac{1}{4}$ m³ plastic tanks. An assortment of plankton was offered to 24-hour starved pink salmon and the animals killed after a 2-hour feeding period. The original concentration of organisms in the water and the number in the stomachs of 10 fish were then determined as a measure of prey selection.

Cultures

In a number of zooplankton grazing experiments, the chrysophyte, *Monochrysis lutheri*, was used as an example of a particularly small phytoplankter. The cultures were grown on a natural sea water medium (27%) enriched with 50 μg at. N/l and 5 μg at. P/l as sodium nitrate and potassium dihydrogen phosphate respectively.

In a few experiments, pink salmon, *Oncorhynchus gorbuscha*, raised as a culture from eggs were employed in fish feeding experiments. These animals were raised at the Biological Station, Nanaimo, and had been fed a synthetic diet up to the time of the feeding experiments described here.

Collection of samples

Samples of sea water containing phytoplankton for use in zooplankton grazing experiments were collected in 7-litre plastic Van Dorn bottles. Live zooplankton samples for the same experiments were collected by towing a fine mesh (200 μ) net at less than 1 knot and at a depth at which a particular zooplankter was known to occur. Zooplankton were further isolated by gently netting or pipetting out unwanted organisms. This procedure led to the selection of zooplankton according to size and in some cases a mixture of two species of approximately the same size were used in grazing experiments.

Samples of larval and juvenile fish for the examination of *in vivo* stomach contents were collected with an Isaacs-Kidd net towed at 6 knots. The mouth opening of the net was 2 m and the mesh size of the codend was 2 mm.

TABLE 1. Theoretical rations for pink salmon

Day	Date	length (cm)	Observed fish* wet wt. (g)	no. days	Daily increment (%)	Av. wt.	Metabolism† at 20° (%/day)	Total requirement (%/day)	Total ration (%/day)	Total ration (mg wet wt./day)
0	April 20	3·5	·245							
6	26	4·0	·395	12	7·7	·39	9·7	17·4	21·8	85
12	May 2	4·5	·60							
18	8	5·0	·84	13	5·3	·84	7·8	13·1	16·3	137
25	15	5·5	1·15							
31	21	6·0	1·55	13	4·6	1·55	6·8	11·4	14·3	222
38	28	6·5	2·06							
44	June 3	7·0	2·65	13	4·2	2·65	6·5	10·7	12·9	342
50	9	7·5	3·35							
57	16	8·0	4·15	13	3·4	4·15	6·1	9·6	12·0	497
63	22	8·5	5·15							
69	28	9·0	6·2	12	3·2	6·2	5·6	8·8	11·0	683
75	July 4	9·5	7·5							
81	10	10.0	9·0	15	3·2	9·0	4·8	8·0	10·0	900
90	19	10·5	11·5							

* Data from LeBrasseur and Parker (1964)
† Data from Winberg (1956), Table 30

TABLE 2. Summary of zooplankton feeding

Animal	Modal food size (dia. μ)	Predominant species	Food density Min.* (μg C/l)	Food density Max.† (μg C/l)	Max. ration obtained (mg C/animal/day)	Mean zooplankton body wt. (mg C)	Ration (% body wt./day)
C. plumchrus nauplii stage V+	4	M. lutheri	75	>300	0·00018	0·0015	12
Oithona sp. and P. minutus‡	4	M. lutheri	40	400	0·0005	0·0014	36
P. minutus‡	32	C. debilis and C. socialis	58	315	0·000052	0·0067	0·8
P. minutus‡ (Oithona sp.)||	14	S. costatum and μ-flagellates	190	760	0·0045	0·010	45
P. minutus	8 and 14	S. costatum and μ-flagellates	81	>163	0·0004	0·010	4·0
C. plumchrus III and IV	5 and 9	μ-flagellates	39	>108	0·0017	0·03	5·7
C. plumchrus III and IV	16	S. costatum and μ-flagellates	119	>520	0·018	0·03	60
Euphausiid furcilia and C. pacificus	8	μ-flagellates	74	285	0·0031	0·078	4·1
C. pacificus and Euphausiid furcilia	32	C. debilis and C. socialis	↘79	285	0·0015	0·078	2·0
C. pacificus	90	Thalassiosira spp.	142	305	0·0168	0·100	16·8
C. pacificus	57	Thalassiosira spp.	85	380	0·0184	0·100	18·4
C. plumchrus V	14	S. costatum and μ-flagellates	62	285	0·026	0·175	15
E. pacifica	32	C. debilis and C. socialis	131	1180	0·302	2·02	15

* Value quoted is p_0, prey density at which grazing started.
† Value quoted is plankton density at 90% R or plankton density above which 90% R would be found.
‡ Size difference in *P. minutus* (mg C) at different times of the year reflect changes in the measured length of organisms.
|| Brackets around animals indicate that the organism was also present in the size fraction as a minor component.

Larval fish were also sorted from zooplankton collections. Live samples of juvenile fish used in feeding experiments were collected with a 100 ft seine net having a mesh diameter of 1 cm.

Growth rates

In order to obtain a comparison of the rations taken by juvenile salmon, as measured by prey density experiments (Equation 1), with rations required by these animals for metabolism and growth, a theoretical calculation was made on the basis of growth data given by LeBrasseur and Parker (1964). From these data the daily per cent increase in weight was calculated using Winberg's formula (1956).

$$\% \Delta W = [10^{(\log W_2 - \log W_1)/t} - 1] \, 100, \qquad [3]$$

where $\% \Delta W$ is the per cent increase in weight per day over the time interval t during which the animals changed in weight from W_1 to W_2. The amount of food required for metabolism by different sized fish was taken from Winberg's Table 30. The sum of the growth and metabolism requirements times 1·25 (to allow for an 80% assimilation efficiency) was taken as the daily ration required by different sized salmon. These estimates, reported in Table 1, are maximum requirements since they assume that the fish was actively swimming at 20°C and that it was also maintaining a growth rate found for healthy fish in nature.

RESULTS AND DISCUSSION

Zooplankton grazing experiments

Zooplankton grazing experiments have been carried out over the past two years using the technique described in the previous section. The results of these experiments have been reported (Parsons et al., 1969) and a summary of the findings, together with some more recent data, are presented in Table 2. With the exception of two feeding experiments using Monochrysis lutheri as food, all results in Table 2 represent the grazing of natural populations of zooplankton on natural blooms of phytoplankton. The results of these experiments show that in many cases the intake of food as a percentage of the body weight is below that required for growth or reproduction. This is due to either the concentration of phytoplankton being too low, (e.g. P. minutus feeding at crop densities of 163 μg C/l obtained only 4% of its body weight per day), or to a phytoplankton being an awkward size or shape (e.g. P. minutus feeding on a Chaetoceros bloom at 315 μg C/l could only obtain 0·8% of its body weight per day). With regard to the concentration of phytoplankton it appears, in fact, that no animal reported in Table 2 could graze phytoplankton below a crop density of ca. 40 μg C/l and that maximum rations were generally not reached until phytoplankton concentrations were 400 μg C/l or more. These concentrations of phytoplankton carbon correspond to chlorophyll a concentrations from ca. 1 to 10 mg/m³. The lower of these two values is of the same order as is found in the Strait of Georgia during the winter, while the higher is characteristic of concentrations found in the area during spring blooms and at times during the summer.

To some extent, therefore, the apparent need for these high concentrations of phytoplankton may be a characteristic of coastal species. However, the same species of zooplankton also exist at Station P (50°N, 145°W), which is a truly oceanic environment where chlorophyll *a* concentrations are seldom greater than 1 mg/m^3. Under these circumstances either the same zooplankers must be adapted to feeding at much lower phytoplankton concentrations, or alternative feeding mechanisms must play a more active part. Among the latter is the possible accumulation of high particle concentrations, such as result from Langmuir circulation or discontinuity layers. Alternatively, an animal may change its feeding habits to take a small number of large prey rather than a large number of small prey. Such is the case with *E. pacifica* which in Saanich Inlet, B.C. is largely herbivore, feeding on diatoms (Parsons et al., 1967), while in an ocean environment it is a carnivore feeding on copepods (Lasker, 1966).

Some indication of the importance of food size and shape is also given in Table 2. In the case of the mixture of *C. pacificus* and euphausiid furcilia, it is apparent, for example, that these animals could obtain only 2% of their body weight per day from a *Chaetoceros* bloom, which on our scale had a modal size of 32 μ diameter (representing a chain). However, when feeding on a *Thalassiosira* bloom of larger unit size (57 to 90 μ diameter), at about the same prey density (*ca.* 300 μg C/l), *C. pacificus* could obtain from 16 to 18% of its body weight per day. In these two examples, the shape of the *Chaetoceros* chains would appear to have been the critical factor in determining the lack of suitability of the food compared with *Thalassiosira* sp. On the other hand we could not feed *E. pacifica* on *M. lutheri* (diameter 4 μ); and even on 8 μ diameter prey, adult euphausiids obtained far less ration than when they were fed off *Chaetoceros* chains (Parsons et al., 1967). In this example, the lack of feeding must be attributed to the relative size of the plants compared with the filtering apparatus of euphausiids. With respect to the very small phytoplankter *M. lutheri*, it appeared that this was a good food source for the very smallest animals tested (nauplii V and a mixture of *Oithona* sp., and *P. minutus*) while for larger foods, such as *Thalassiosira* sp., we could not obtain data on grazing by the smallest animals (although *Oithona* sp. which is reported to eat relatively large prey (Mullin, 1963), was not isolated and tested separately in these experiments).

Thus while nanoplankters may serve as food for both small, and to some degree larger organisms (e.g. see also results of Mullin, 1963), our general experience to date is that they form an essential food for nauplii and adults of the very smallest zooplankton species. Furthermore, it also appears generally true that the larger organisms require food of larger size, or a food which, because of its shape, can be readily concentrated (e.g. chain forming species). These results are in disagreement with the generalized conclusion of Brooks and Dodson (1965), who state that among freshwater herbivorous plankton, "food apportioning according to body size is a path to stable

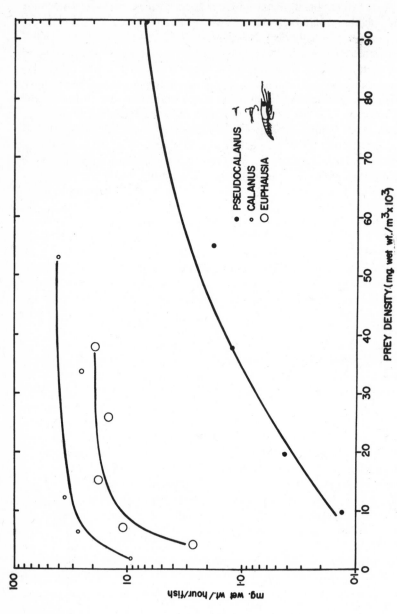

Fig. 3. Juvenile pink salmon rations at different prey concentrations. (●) *Pseudocalanus*;
(○) *Euphausia pacifica*; (○) *C. plumchrus*.

coexistence seldom available to planktonic food collectors". The possibility exists that this is a real difference between marine and freshwater herbivorous zooplankton although such a conclusion could only be reached after a more careful examination of predator/prey size relationships in both freshwater and marine environments.

Juvenile fish feeding

Fig. 3 shows the results of feeding three different sized prey to 90 mm pink salmon. Prey densities and rations have been recorded as equivalent wet weights of plankton in order to compare the relative value of the three food sources. From the data shown, it is apparent that at a prey density of 20 g/m³ juvenile pink salmon can obtain *ca.* 30 mg wet weight of plankton per hour from a *C. plumchrus* crop, less than 20 mg of plankton per hour from an adult euphausiid crop and only about 0·3 mg/hour from a *Pseudocalanus* crop. From Table 1, it may be seen that 90 mm pink salmon require about 583 mg of food per day. Data given in Fig. 3 indicates that this requirement could be obtained from a *C. plumchrus* crop assuming that feeding was continuous. A 24-hr feeding period is consistent with the field observations of Parker and Vanstone (1966), who examined stomach contents at various times of the day and showed that young salmon had a diurnal cycle of feeding intensity, but that feeding was continuous. For an equivalent prey density of euphausiids, the required ration of 683 mg/day would have been barely obtainable, while for a *Pseudocalanus* crop, the young salmon could not have obtained sufficient food. An increase of *Pseudocalanus* prey density up to 90 g/m³ still would not have met the food requirement of 90 mm salmon.

Our first conclusion in this study, therefore, is that zooplankton biomass estimates alone do not give a realistic measure of food available to juvenile fish but that the type of food, as illustrated in Fig. 3, is of fundamental importance. It is not the only factor, however, and the situation with different sized fish, natural prey concentrations and further variations in prey size and type, becomes far more complex. Examples of this complexity can be given as follows:

Regarding prey concentration, it may have been noted already from Fig. 3 that the value of 20 g wet weight/m³, which is a relatively low prey density in our experimental data, is actually a phenomenally high prey density when considered from the point of view of natural prey concentration. This value corresponds to approximately 8 copepods (*C. plumchrus* V)/litre. Prey concentrations of this order have been encountered by us in the Strait of Georgia, particularly in certain inlets, and also in the vicinity of the Fraser River estuary. These concentrations are unusual, however, and have generally resulted from advective effects or local diurnal zooplankton migrations. In general, the concentration of prey as measured with a plankton net towed over a distance, has been found to be very much less than the required prey densities suggested by our data in Fig. 3. Thus it is apparent that in order to obtain these prey densities in nature, the effect of prey patchiness also must be considered.

TABLE 3. Summary of juvenile fish feeding experiments

Fish	Fork length (mm)	Zooplankter	Wet wt. of individual zooplankter (mg)	Food density for max. ration* (mg/m³)	Food density for max. ration* (No./m³)	Max. ration obtained mg/animal/hr	Daily food requirement† (mg/day)	Remarks
Pink salmon	34	C. plumchrus	2·5	10000	4000	20	85	Fish obtained from laboratory culture
Pink salmon	69	C. plumchrus	2·5	20000	8000	25	342	Fish collected in the field
Pink salmon	90	P. minutus	0·134	>90000	>670000	10	683	Fish collected in the field
Pink salmon	90	C. plumchrus	2·5	20000	8000	30	683	Fish collected in the field
Pink salmon	90	E. pacifica	45·0	20000	444	18	683	Fish collected in the field
Chum salmon	41	Anomura juv.	1·2	1900	1600	17·5	ca. 100	Fish collected in the field
Chum salmon	44	Microcalanus	0·06	560	9300	7·0	ca. 130	Fish collected in the field
Pink salmon	34	Microcalanus	0·06	1400	23000	7·6	85	Fish obtained from laboratory culture
Chum salmon	41	C. plumchrus	1·1	3200	2900	25	ca. 100	Fish collected in the field

* Value quoted is plankton concentration at 90% of maximum ration.
† Data from Table 1.

Data on plankton concentrations obtained with a conventional plankton net only provide information on the average amount of prey for the distance over which the net is towed. More important in recent advances in plankton net design is the Longhurst/Hardy recorder (Longhurst et al., 1966) which gives a measure of plankton patchiness over short distances and which has shown (Longhurst, 1967; see also Cassie, 1959) that plankton is highly aggregated in situ. From our own plankton collections made with a Longhurst/ Hardy recorder towed horizontally we have found that there may be at least a 7-fold variation in the number of copepods per unit volume over a sampling distance of 90 metres. The introduction of this information into predator/ prey relationships has been considered by Ivlev (1961), who modified his equation (I in the Introduction) to include an index of aggregation. Thus our second conclusion is that in order to satisfy the need for data on plankton concentrations in situ, collecting apparatus for the measurement of plankton patchiness, such as has been described by Longhurst et al. (1966), is absolutely essential to the further understanding of plankton/fish feeding relationships.

In Table 3 we have reported on other juvenile fish feeding experiments, as well as giving a summary of data already shown in Fig. 3. From these results it is apparent that there are considerable differences in the feeding of different sizes and species of juvenile fish on different prey. For example, 34 mm pink salmon require a much lower concentration of Microcalanus in terms of biomass than of C. plumchrus, in order to obtain a maximum ration. On the other hand, from Tables 1 and 3, the approximate amount of food required by 34 mm pink salmon (85 mg wet weight/day) can be obtained more rapidly (ca. 4 hr feeding) if the young salmon fed off a high concentration (90% asymptotic value of 10 g/m^3) of C. plumchrus, than off a high concentration (90% asymptotic value of 1·4 g/m^3) of Microcalanus (ca. 11 hr feeding). The interpretation of this type of result must at present be speculative, but it is probably related to the ability of the young salmon to capture the two different sized prey; a small prey requiring more time and effort than a larger prey, providing the latter is sufficiently concentrated. The really interesting question arising from this type of result is to discover (over a longer period) what effect the feeding of the two species of copepod would have on the growth of the juvenile fish.

While the data in Tables 1 and 3 give some indication of how efficiently one prey can meet the requirements of a particular size of juvenile salmon, the actual situation in nature, where there are several choices of prey, also requires examination since it is apparent from Table 3 that there are other factors, besides concentration, which may decide the in situ food spectrum of young salmon. Ivlev (1961) grouped all these other factors under a term which he called electivity and to which he assigned an index (E), defined as

$$E = \frac{r_i - p_i}{r_i + p_i} \qquad [4]$$

where r_i is a relative count of different organisms eaten and p_i is a relative count of different organisms in the surrounding water. In our initial studies, attempts to determine electivity coefficients in nature have been made difficult by the fact that as well as the effect of salmon selecting prey, one has also the almost unavoidable effect of net selection (i.e. any plankton net will tend to catch and retain some species better than others). In approaching a measurement of electivity based on *in situ* prey concentrations, to date we have been able only to simulate conditions and report on results obtained when a mixture of different prey is offered to young salmon in a tank. An example of this type of experiment is reported in Table 4. From these results,

TABLE 4. Electivity of 48 mm juvenile pink salmon toward different zooplankters

| Prey | Relative concentration | | | | Electivity Index | |
| | % in water | | % in stomachs | | | |
	Number	Biomass	Number	Biomass	Number	Biomass
Cypris	6 }	1	0 }	0	−1·0 }	−1·0
Limacina	8 }		0 }		−1·0 }	
Copepods total length:						
<1 mm	19	2	9	1	−0·35	−0·33
1-2 mm	45	12	55	13	0·1	0·04
2-4 mm	19	52	35	78	0·3	0·20
>4 mm	1	8	1	6	0·0	−0·14
Amphipods	1	9	trace	1	negative	−0·8
Euphausiids	1	16	trace	1	negative	−0·9
Total number of prey	6461/0·25 m³ tank				2822/10 fish	

it may be seen that on the basis of prey numbers, the young salmon had a high positive electivity towards copepods 2-4 mm in length; were less selective in their choice of 1-2 mm copepods; and were negatively selective toward all other prey except large copepods which were consumed in proportion to their concentration in the surrounding water. If electivity is calculated on the basis of biomass, the relative importance of copepods, 2-4 mm, is further emphasized by a disproportionate decrease in the electivity index of 1-2 mm copepods and a change of electivity for large (>4 mm) copepods from 0 to −0·14. Such an experiment is far from an absolute measure of prey selectivity since the role of training, i.e., the previous experience of the fish, together with such factors as lighting and temperature, may all be expected to modify the results in Table 4. For the purpose of this discussion, however, it is evident that feeding of young salmon is influenced by the animals' selection of certain prey, as well as by prey abundance.

In conclusion to this section, it is apparent that the feeding of pelagic juvenile fish, such as young salmon, is very much more complex than the feeding of zooplankton on phytoplankton, as described in the previous section. Some of the factors contributing to this complexity have been reported on here. The resolution of these factors lends itself to a theoretical treatment similar to that accorded to benthic feeders by Ivlev. No attempt has been made at this point, however, to compare, or in some cases separate, the extent of various effects. In Table 3, for example, where we have considered ration at maximum prey densities, an electivity factor will have been included in these results, since no independent measure of electivity was made at the time of the experiment. For future work, it is apparent, however, that greatly improved oceanographic methodology will be required in order to supply the type of information in nature obtained in our fish feeding experiments. The information required may be summarized as follows:

1. Concentration and size spectrum of predator/prey associations *in situ.*
2. Distributional patchiness of food organisms.
3. Electivity indices of predators and preys obtained from associations found in (1) above, together with food spectra of predator stomach contents.
4. Food requirement for growth and maintenance of different sizes and species of predators, together with the period per 24-hour day available for feeding.

While item 4, in the above summary, is largely a task for the physiologist, items 1, 2, and to a lesser extent 3, are problems in methodology for the biological oceanographer.

Larval fish feeding

Our attempts to perform grazing experiments with larval fish, similar to those reported for juvenile fish, have not been successful to date. Some appreciation of the size fraction of zooplankton grazed by larvae can be obtained, however, from Table 5, which has been summarized from LeBrasseur *et al.* (1969). From these data it is apparent that the size fraction of prey taken

TABLE 5. Size distribution of stomach contents of young fish (from LeBrasseur *et al.*)

Larval species	Stomach contents (% zooplankton in mm size fractions)				
	<0·5	0·6-1·0	1·1-1·5	1·6-2·5	>2·5
Ammodytes hexapterus (<40 mm)	27	33	34	3	3
Hexagrammos decagrammus (<30 mm)	40	39	17	3	1

by two species of larval fish is predominantly in the size range 0·5 to 1·5 mm. Representative species of this size fraction are *Microcalanus, Oithona* and *Pseudocalanus,* as well as the nauplear stages of the larger copepods. The grazing of some of these zooplankters has been shown (Table 2) to be most successful when nanoplankton (2-20 μ) were offered as a food source. Thus it appears from our results that an important step in the food chain for larval fish survival is the association, nanophytoplankton→microzooplankton→larval fish. This is in contrast to the feeding of juvenile fish reported in the previous section, where the most suitable food was shown to be large copepods.

Our lack of success in performing prey density feeding experiments with larval fish was partly due to difficulties in detecting situations where prey had been eaten, but more particularly to a high larval mortality in the experimental vessels. If the latter effect was real, it would appear that for larval fish feeding, a consideration of the amount of energy available to a newly hatched fish and the path it had to travel in order to capture an organism, would be the important factors in survival. A theoretical approach to this type of relationship was given by Ivlev (1944).

RELATIVE TRANSFER OF ENERGY THROUGH FOOD CHAINS

In the two previous sections, we have considered the feeding of zooplankton and of fish as two separate processes. On a number of occasions, data on these two trophic levels were collected simultaneously so that some idea could be obtained of the *in situ* relationship between phytoplankton, zooplankton and fish, as postulated in Fig. 1. An example of such a relationship can now be presented in Fig. 4, where we have summarized part of a food chain based on quantitative data obtained from feeding experiments reported previously (Parsons *et al.*, 1967), as well as in Tables 2 and 3. The heaviness of arrows indicates the most probable path of energy flow; dashed lines indicating poor feeding relationships. The figure shows that the event which led to a suitable food source for 90 mm salmon during the latter part of July was the occurrence of a phytoplankton bloom in June which could be effectively grazed only by a large herbivore (Table 2, *E. pacifica*). In the absence of any competitive grazing by other herbivores and with the maintenance of very high concentrations of phytoplankton over a period of several weeks, large numbers of euphausiid eggs were laid from which euphausiid furcilia developed in sufficient numbers to provide a food source for young salmon. Furthermore, the size of the furcilia (3·5 mm length) was comparable to that of *C. plumchrus*, which has been shown (Fig. 3) to be a good food size for 90 mm juvenile pink salmon. At this time, the stomach contents of juvenile pink salmon caught *in situ* consisted largely of euphausiid furcilia (Barraclough, 1968).

One of the interesting aspects of this small segment of a food chain from Saanich Inlet is the lack of success with which *P. minutus* and *Calanus* were able to graze the *Chaetoceros* bloom. At the start of the time series study there were approximately 1000 *P. minutus*, 400 *Calanus* and 10 euphausiids per m³ in the euphotic zone at night. In terms of both biomass and numbers the amount of adult copepods present was, therefore, considerably greater, than the amount of euphausiid. If a bloom of phytoplankton favourable to grazing by either size group of copepods had become available (e.g. see data in Table 2) it would have undoubtedly altered the food chain shown in Fig. 4, to the possible deprivation of the salmon at this stage of their growth.

In coastal waters, the size spectrum of phytoplankton generally available includes a large quantity of microphytoplankton so that a suitable size fraction of zooplankton for juvenile fish can be produced, either as large copepods (e.g. *Calanus plumchrus*, see Parsons *et al.*, 1969) or through a mechanism such as is shown in Fig. 4. However, in oceanic waters of the subarctic Pacific, microphytoplankton are replaced by nanoplankton as the predominant photosynthetic organisms (McAllister *et al.*, 1960; Anderson, 1965). This must lead to a different food chain for the same species of animal in these waters. The change in feeding of *E. pacifica* from herbivore in coastal waters, to carnivore in ocean waters, has already been mentioned (loc. cit.). In the case of *C. plumchrus*, however, no similar evidence exists to show that the later stages of this animal are carnivorous, although the ability of *C.*

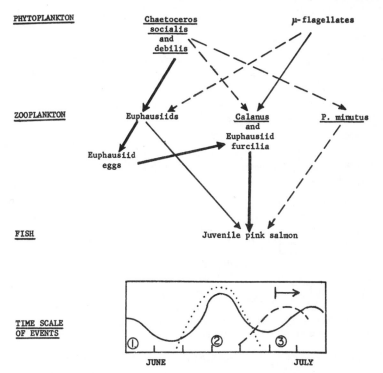

Fig. 4. Tentative food chain in Saanich Inlet, June-July, 1966. (Time scale of events: — relative *Chl. a* concentration; . . . relative numbers of euphausiid eggs (data from Parsons *et al.* 1967); - - - relative numbered euphausiid furcilia (data from Stephens *et al.*, MS 1967); ⊢→ Start of juvenile salmon feeding on furcilia (data from Barraclough, MS 1968). Major phytoplankton species causing bloom: (1) *Distephanus* sp., (2) *Chaetoceros socialis* and *debilis*, (3) μ-flagellates.

finmarchicus in the Atlantic to consume small copepods is known (Gauld, 1966, and references therein). In this respect, the role of such animals as radiolarians may be a significant link in the oceanic food chain. The numbers of these animals at Ocean Weather Station P are at times appreciable (5500/m³), and since their feeding is specifically adapted to the removal of very small particles by mechanisms not available to the crustacean zooplankton (e.g. see Nicol, 1967, for description of radiolarian feeding), they may play an intermediary role in the transfer of food from nanoplankton to large zooplankton.

In summary therefore it is suggested that two basic types of food chains exist at the lower trophic levels in the oceans. These are, firstly, nanophytoplankton → microzooplankton → macrozooplankton → fish; and secondly, microphytoplankton → macrozooplankton → fish. The first of these is probably characteristic of subarctic Pacific ocean waters and the second is generally characteristic of coastal waters, upwelling areas and Antarctic waters (e.g. Marr, 1962, describes the food of *E. superba* in the Antarctic as consisting primarily of chain-forming diatoms). Thus, apart from any differences in the total level of primary production in coastal and oceanic waters, or between two oceanic areas, the efficiency of energy transfer up to larger organisms would be lower in an ocean environment in which the predominant photosynthesizing organisms were nanoplankton, than in an environment which was predominated by microphytoplankton. However, in the interesting paper by Kerr and Martin (this volume), it is suggested that the ultimate yield of fish may be greater on a longer food chain due to lower metabolic expenditures required when animals are feeding on larger prey. This might tend to modify our conclusion suggested above on the efficiency of marine food chains, based on the size of the primary producer.

REFERENCES

ADAMS, J. A., and STEELE, J. H. 1966. Shipboard experiments on the feeding of *Calanus finmarchicus* (Gunnerus). In *Some contemporary studies in marine sciences*. Edited by H. Barnes. London, George Allen and Unwin Ltd., 19-35.

ANDERSON, G. C. 1965. Fractionation of phytoplankton communities off the Washington and Oregon coasts. *Limnol. Oceanogr.*, **10**, 477-80.

BARRACLOUGH, W. E. 1967. Data Record. Number, size composition and food of larval and juvenile fish caught with a two-boat surface trawl in the Strait of Georgia, June 6-8, 1966. *Fish. Res. Bd Can., Ms Rep. Ser.*, (*Biol.*), (928), 58 pp.

BARRACLOUGH, W. E. 1968. Food of larval and juvenile fish caught with a surface trawl in Saanich Inlet during June and July 1966. *Fish. Res. Bd Can., Ms. Rep. Ser.* (*Biol.*) (In press).

BROOKS, J. L., and DODSON, S. I. 1965. Predation, body size, and composition of plankton. *Science, N.Y.*, **150**, 28-35.

CASSIE, R. M. 1959. An experimental study of factors inducing aggregation in marine plankton. *N.Z. Jl Sci.*, **2**, 339-65.

CUSHING, D. H. 1964. The work of grazing in the sea. In *Grazing in terrestrial and marine environments. Symp. Br. Ecol. Soc., no. 4, Bangor 1962*. Oxford, Blackwells Scient. Publ., 209-25.

GAULD, D. J. 1966. The swimming and feeding of planktonic copepods. In *Some contemporary studies in marine science*. Edited by Harold Barnes. London, George Allen and Unwin Ltd., 313-34.

ISAACS, J. D. 1966. Larval sardine and anchovy interrelationships. *Rep. Calif. coop. oceanic Fish. Invest.*, (10), 102-113.

IVLEV, V. S. 1944. The time of hunting and the path followed by the predator in relation to the density of the prey population. (Translation by L. Birkett, M.A.F.F., Lowestoft.) *Zool. Zh.*, **23** (4), 139-45.

IVLEV, V. S. 1945. The biological productivity of waters. *Usp. sovrem. Biol.*, **19**, 98-120.

IVLEV, V. S. 1961. *Experimental ecology of the feeding of fishes*. New Haven. (Trans. by D. Scott.) Yale University Press, 302 pp.

KERR, S. R., and MARTIN, N. V. 1970. Trophic-dynamics of lake trout production systems. This volume, pp. 365-376.

KNOX, C. 1966. Holographic microscopy as a technique for recording dynamic microscopic subjects. *Science., N.Y.*, **153**, 989-90.

LASKER, R. 1966. Feeding growth, respiration and carbon utilization of a euphausiid crustacean. *J. Fish. Res. Bd Can.*, **23**, 1291-1317.

LEBOUR, M. 1920. The food of young fish. No. III. *J. mar. biol. Ass.*, U.K., N.S., **11**, 433-69.

LEBRASSEUR, R. J., and PARKER, R. R. 1964. Growth rate of central British Columbia pink salmon (*Oncorhynchus gorbuscha*). *J. Fish. Res. Bd Can.*, **21**, 1101-28.

LEBRASSEUR, R. J., *et al.* 1969. Production studies in the Strait of Georgia, III. Larval and juvenile fish production under the Fraser River plume, February to May, 1967. *J. exp. mar. Biol. Ecol.*, **3**, 51-61.

LONGHURST, A. R. 1967. Vertical distribution of zooplankton in relation to the eastern Pacific oxygen minimum. *Deep Sea Res.*, **14**, 51-63.

LONGHURST, A. R., *at al.* 1966. A new system for the collection of multiple serial plankton samples. *Deep Sea Res.*, **13**, 213-22.

MCALLISTER, C. D., PARSONS, T. R., and STICKLAND, J. D. H. 1960. Primary productivity at station "P" in the north-east Pacific Ocean. *J. Cons. perm. int. Explor. Mer*, **25**, 240-59.

MADDUX, W. S., and KANWISHER, J. W. 1965. An *in situ* particle counter. *Limnol. Oceanogr.*, **10** (Suppl.), 162-68.

MARR, J. W. S. 1962. The natural history and geography of the Antarctic krill (*Euphausia superba* Dana), *Discovery Rep.*, **32**, 33-464.

METHERELL, A. F., *et al.* 1967. Introduction to acoustical holography. *J. acoust. Soc. Am.*, **42**, 733-42.

MULLIN, M. M. 1963. Some factors affecting the feeding of marine copepods of the genus *Calanus*. *Limnol. Oceanogr.*, **8**, 239-50.

NICOL, J. A. C. 1967. *The biology of marine animals.* Second edition. London, Sir Isaac Pitman & Sons, Ltd., 699 pp.

PARKER, R. R., and VANSTONE, W. E. 1966. Changes in chemical composition of central British Columbia pink salmon during early sea life. *J. Fish. Res. Bd Can.*, **23**, 1353-84.

PARSONS, T. R., LEBRASSEUR, R. J., and FULTON, J. D. 1967. Some observations on the dependence of zooplankton grazing on the cell size and concentration of phytoplankton blooms. *J. oceanogr. Soc. Japan*, **23**, 10-17.

PARSONS, T. R., *et al.* 1969. Production studies in the Strait of Georgia, II. Secondary production under the Fraser River plume, February to May, 1967. *J. exp. mar. Biol. Ecol.*, **3**, 39-50.

PARSONS, T. R., STEPHENS, K., and LEBRASSEUR, R. J. 1969. Production studies in the Strait of Georgia, I. Primary production under the Fraser River plume, February to May, 1967. *J. exp. mar. Biol. Ecol.*, **3**, 27-38.

SHELDON, R. W., and PARSONS, T. R. 1967. A continuous size spectrum for particulate matter in the sea. *J. Fish. Res. Bd Can.*, **24**, 909-15.

SHELDON, R. W., and PARSONS, T. R. 1967. *A practical manual on the use of the Coulter Counter in marine science.* Canada, Coulter Electronics Sales Co., 66 pp.

STEELE, J. H. 1965. Some problems in the study of marine resources. *Spec. Publs. int. Commn. N.W. Atlant. Fish.*, (6), 463-76.

WIBORG, K. F. 1948. Investigations on cod larvae in the coastal waters of northern Norway. *FiskDir. Skr.*, ser. *Fisk.*, **9**, 5-27.

WINBERG, G. G. 1956. *Rate of metabolism and food requirements of fishes.* Minsk, Nauche Trudy Belorusskovo Gosudarstvennova Universiteta imeni V. I. Lenina, 253 pp. [Fisheries Research Board of Canada, Translation Ser. No. 194, 1960.]

Experimental studies in feeding and food requirements of herring larvae (Clupea harengus L.)

H. Rosenthal and G. Hempel

Institut für Hydrobiologie und Fischereiwissenschaft
der Universität Hamburg

ABSTRACT. Larvae of herring of the western Baltic and southern North Sea were hatched and reared in aquaria and small glass vessels for direct observation and filming of their swimming and feeding behaviour. Food consisted of nauplii and metanauplii of *Artemia* as well as of wild plankton.

The volume of water searched for food by a herring larva depends mainly on three factors: (*i*) daily path of the larva; (*ii*) cross-section of the field of perception, and (*iii*) number of hours per day during which light is sufficient for perception of food organisms.

In order to estimate the searching capacity of larvae of different age and size, the swimming pattern of the larvae was analysed in detail. Locomotion in herring larvae is composed of various types of active swimming interrupted by short periods of rest and slow sinking. Each swimming phase is very short. As swimming is basically serpentine (eel-like) the oscillating path of the head is considerably longer than the net progress (straight line distance) of the larva.

Within 5 minutes (i.e. the normal period of observation) the average length of the path including the movements of the head ranges from 1-3 m in yolk sac larvae, but only 0·8-1·5 m in larvae just at the end of yolk sac stage. Larvae of 18-20 mm cover 3-6 m per five minutes.

There is no perception of particles which are underneath the larva nor of those high above. The boundary of the field of perception is not sharp-edged. The 50% perception range increases very much with length and age of the larva. The percentage of successful seizing of food organisms observed is only 3-10% in yolk sac larvae but up to 90% in well developed larvae.

The radius of the field of vision is highest in resting or slowly undulating larvae. Average figures of the radius are 8 mm in yolk sac larvae and up to 40 mm in larvae of 15-20 mm. Estimates of the volume of water searched per day are affected by the fact that pattern of swimming, range of perception and success rate of preying action depend largely on presence of plankton and on hunger of the larvae.

Some caution is therefore necessary if figures of total volume of water searched per five minutes (Table 4) are converted into daily searching capacity. Taking ten hours daylight in Downs larvae, the daily volume searched for food increases from 15-20 litres in one-week-old larvae (near the end of yolk sac stage) to about 60-80 litres in three-week-old larvae of about 14 mm length.

From rate of digestion, optimum feeding rate has been estimated. Larvae of 10-11 mm soon after the end of yolk sac stage can digest about 35-40 *Artemia* nauplii per day, older larvae of 13-14 mm may take 50 nauplii. Evidently digestion of *Artemia* is very incomplete in young larvae, while copepodites are more fully utilized.

There are differences in food preference amongst larvae of similar length and age living together in the same tank. Larvae having started with a particular kind of food tend to stick to it. Threshold values of illumination for feeding as given by Blaxter have been confirmed.

Taking the above figures of optimum daily food consumption and allowing for the increasing rate of successful seizing of food organisms crude estimates of minimum food abundance for optimum feeding can be derived. Two sets of estimates based on different assumptions with regard to catching efficiency are given in Table 5. The most important result is the greater need for high food density in young larvae than in larger ones.

Total need for food increases with size of larva, these higher needs, however, are more than compensated by the increase in volume of water searched per day and by higher catching efficiency. As larger larvae can prey on larger food organisms, the number of food organisms might even decrease with increasing size of the larvae. For very young larvae which have just started feeding, food density should be highest to compensate for low catching efficiency. Furthermore small larvae have a much narrower spectrum of suitable food. All findings point to a short critical period in the early life history of herring larvae.

The figures do not indicate competition for food amongst larval and juvenile fish.

Finally we did not dare to compare the experimental estimates on food consumption per day and on minimum density of food organisms with field observations, as those comparisons would be too crude and might be misleading.

INTRODUCTION

In many stocks of marine fish there is a shortage of recruits so that available food and space cannot be fully used and occupied by the adult population. The causes of inadequate recruitment and of fluctuations in year-class strength are to be found during the first year of life. In this context, shortage of eggs can be disregarded as a major cause in most marine fish stocks. Changes in size or quality of eggs are normally of limited importance for survival of larvae and juveniles of North Sea herring (Blaxter and Hempel, 1963, 1966; Hempel and Blaxter, 1967). Environmental factors, however, affecting the larvae and early O-group fish are considered as dominant determinators of year class strength. For recent discussions of the topic reference is made to Ricker (1954), Gulland (1965) and Hempel (1965).

Many field observations have been published correlating such factors as temperature, currents, food and predator abundance with year-class strength. Tank experiments for a thorough analysis of the basic requirements for growth and survival of larval sea fish have rarely been carried out until recently, when the growing interest in cultivation of marine fish has stimulated those studies. A detailed study on swimming and feeding of larval fresh water fish has been published by Braum (1964).

Since Hjort's (1926) investigations of the fluctuations of Norwegian herring, shortage of planktonic food during an early critical stage just after the resorption of the yolk sac has been regarded as a rather common reason for poor recruitment in herring. It was, however, not until a few years ago that in the course of successful rearing experiments some detailed information on feeding of larval herring was obtained by Blaxter (summarized 1965).

The present study, based on two years' experiments by the first author, was planned to analyse swimming and feeding behaviour of herring larvae at different developmental stages to arrive at an estimate of the minimum abundance of food required by the larva for survival and growth. Bearing in mind the patchiness of plankton distribution, no attempt is made to correlate those results with the abundance figures of food and larvae in the sea. Those figures are only meaningful if they refer to the same order of magnitude of water volume as the body of water searched by a larva per day.

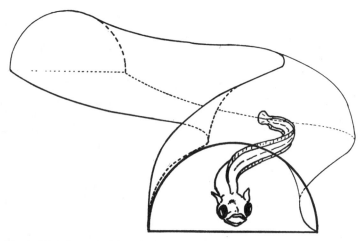

Fig. 1. "Tube" searched by a fish larva. Only those food items which stay in the tube are available to the larva.

As discussed by Ivlev (1944, 1945, 1960) the estimate should be based on the following parameters:

(*i*) Daily ration for maintenance and growth.

(*ii*) The body of water searched per day, determined by the path of the larva during daylight and by the field of perception (Fig. 1).

(*iii*) Percentage of success of feeding actions.

Each of the three parameters is rather complex and might be determined both by external factors such as temperature and light, and by size, age and condition of the larvae.

Preliminary experiments were carried out with the small larvae of herring of the western Baltic and of the Dogger. For the main series, eggs of Downs herring were used. Those winter spawning herring produce relatively large eggs, the larvae hatching at a total length of 7-8 mm, with large yolk sacs which last at 8°C for 15-20 days during which the larvae attain a total length of 10-11 mm (Blaxter and Hempel, 1963). Artificially fertilized eggs of Downs herring were incubated at about 10°C. Most of the larvae were kept at the same temperature in stoneware tanks of 140 l volume. The seawater was circulated through an internal filter (modified after Flüchter, 1964). Artificial illumination by neon tubes was kept at a level of about 1000 lux (max. 1500 lux) at the surface of the tanks for 12 hours per day.

Food consisted of nauplii obtained from natural plankton caught every day near Helgoland and of nauplii of *Artemia salina*. Feeding was done intermittently once to three times a day, the larvae taking the food mainly within the first half hour after provision.

The basic problem in all studies of the swimming path of herring larvae lies in their intermittent movements, short periods of various kinds of loco-motion being followed by periods of rest and slow sinking. This is parti-cularly pronounced in yolk-sac and early post yolk-sac larvae while older stages swim more continously. For the observation of their swimming behaviour single larvae of the yolk-sac and early post yolk-sac stage were transferred into narrow glass vessels. Over periods of five minutes, the observer recorded verbally on tape the positions and movements of the larva relative to a narrow grid. For older stages with more extended periods of faster swimming this method of observation was not practicable and the larvae were kept in larger tanks. Over periods of five minutes the time intervals spent in various types of swimming behaviour were recorded by the observer pressing the keys of a specially designed recording drum. Subse-quently the swimming pattern of similar larvae was filmed in the large tanks using the normal illumination by neon tubes and a microscope lamp in addition. Most larvae did not show any reactions to the beam of brighter light as long as the lamp was at a distance of more than one metre. For each of the three major types of locomotion estimates of average swimming speed and hence of distance covered per second were derived from the analysis of the film. By this indirect method it was possible to transfer the recordings of the behaviour into figures of the distance covered within the five minutes under observation. Although the split of the swimming behaviour into the three types of swimming was somewhat subjective, comparison of both methods applied to young larvae indicated only a small difference in the estimate of distance per 5 minutes, direct observations giving slightly higher values.

Data on the field of vision were difficult to obtain. Direct observations on the reactions of the larvae to moving particles gave rather inaccurate esti-mates. Only filming of single larvae or small groups of larvae to which food organisms were offered gave some approximate figures which, however, need confirmation.

In the course of the work it was possible to rear a fair number of larvae beyond metamorphosis and observe the first steps of shoaling during the early stages of metamorphosis. The present paper, however, will be devoted to larvae at or soon after the critical stage of first feeding, e.g. from yolk-sac stage (~8 mm) to about 15 mm total length, covering the first month of free life.

Acknowledgements

The present study was started under a grant from the Deutsche Forschungsgemeinschaft, at the Institut für Hydrobiologie und Fischereiwissenschaft der Universität Hamburg, where also a considerable part of the analysis of observations was carried out. Collection of eggs of Downs herring was assisted by Rijksinstituut voor Visserijonderzoek, Ijmuiden, Holland. Incubation of eggs and experiments with larvae took place at the Biologische Anstalt Helgoland in 1965 and 1966. The Institut für Wissenschaftlichen Film, Göttingen, put its facilities for evaluating the cinefilm at the disposal of the first author. The kind help of scientists and technicians of the four institutes is gratefully acknowledged.

RESULTS

Swimming

Yolk-sac larvae tend to sink rather rapidly. Swimming at this early stage consists therefore mainly of short periods of an upward wriggling movement followed by sinking. Rate of sinking decreases from about 0·5 cm/sec to 0·2 cm/sec with decreasing size of the yolk sac. At the end of the yolk-sac stage the sinking rate is at its minimum and increases later with increasing size of the larvae. The change in sinking rate might be due to changes in the ratio of weight to surface area which is lowest at the end of yolk-sac stage.

During the yolk-sac stage the vertical movements become gradually transformed into horizontal locomotion. Vertical and horizontal swimming is serpentine (eel-like) in form with a meandering path. Generally three types of swimming can be distinguished.

(a) "*Abrupt swimming*"; very short periods (0·2-0·5 sec) of fast swimming. Normally each dart is connected with a change in the direction.

(b) "*Normal swimming*"; swimming with steady movements for several seconds resulting in a winding path; its "wave length" is about 3 mm in young post-yolk-sac larvae. Each period lasts for 0·5 to about 8 sec in larvae at the stage of first feeding, i.e. towards the end of the yolk-sac stage, 4-7 days after hatching. In older larvae most periods of normal swimming do not exceed 3 sec. Only in the absence of plankton were longer periods of up to 12 sec observed in larvae of one month old.

(c) "*Slow meandering*" ("search swimming"); a slowly winding locomotion with a large amplitude of each meander but with very little net progression of the larvae. This movement is typical of larvae searching for food. With increasing age of the larvae a single period of this type of swimming lasts normally for 0·5 to 2·5 sec in one-week-old larvae and 0·5-4·5 sec in larvae of two weeks or more. There are always a few much longer periods up to about 14 sec, particularly in older larvae.

The locomotory behaviour of post yolk-sac larvae is composed of all three types of locomotion with periods of rest inserted, during which the larvae sink down. In the course of development and growth the slow meandering gains importance, the share of resting periods of the total time amounting to 60% in the first days after hatching but decreasing to about one third of the total time (31-35%) after 10 days of life, e.g. after resorption of the yolk sac. At that stage 30-40% of the time is spent in slow meandering (Table 1).

TABLE 1. Locomotive activity in herring larvae of different age. Duration of the various types of locomotion in percent of total time of observation

Age (days)	Observations (n)	Abrupt+normal swimming (%)	Resting (%)	Slow meandering (%)	Aiming (%)	Snapping (%)
0- 3	33	40·2	59·8	—	—	—
4- 7	51	53·5	39·0	4·2	2·5	0·9
8-11	50	40·0	33·5	22·8	2·9	0·8
12-16	22	28·0	34·9	33·9	2·7	0·5
17-19	15	22·5	31·0	42·8	2·9	0·8

Frequency and duration of the periods of slow meandering, however, are positively related to food supply. On the other hand larvae swim faster if not much food is available. This might be considered as a useful mechanism for survival in the sea. Active swimming might take the larvae into a new patch of food in which they tend to stay for longer by increasing the share of slow meandering. For 7-10 days old larvae, Table 2 gives a percentage split of the observation period with regard to time for normal swimming, slow meandering and rest in relation to plankton supply.

Distance covered per five minutes varies largely with the relative importance of the various types of swimming and with the velocity of the slow meandering. The wide undulating movements of the head in the course of serpentine swimming, particularly during search, add to the direct distance between the starting and ending point of a swimming phase. According to film recordings the head of a 15 mm larva showed a swimming velocity (mainly lateral oscillations) of 1-2 cm/sec while the tail moved 4-5 cm/sec. Taking the straight distance between start and end of a single swimming phase for measuring swimming velocity a considerable increase with length of the larvae can be observed. Mean figures are as follows

total length (mm)	velocity (cm/sec)
8-11	0·7-0·8
11-15	1·0-1·1
19-24	2·1-2·5
32-40	4·0-5·0

Maximum swimming velocities reported by Blaxter (1962) increase from 3

TABLE 2. Swimming activity of 7-10 days old larvae at different levels of food supply. Duration of various types of locomotion in percent of total time of observation. (— = no plankton; (+) = poor plankton; + = moderate plankton; ++ = rich plankton)

Date	Time	Food abundance	Observation time (sec)	Normal swimming (%)	Slow meandering (%)	Resting (%)	Aiming (%)	Snapping (%)
12.12	8.30	—	320	37.2	57.4	5.4	—	—
	10.45	—	302	44.7	54.8	0.5	—	—
	11.15	++	342	42.3	32.1	23.3	2.3	0.5
	1.30	++	338	47.3	27.8	22.0	2.3	—
	12.50	(+)	342	48.4	16.7	32.3	2.7	—
	13.00	+++	658	43.3	23.4	29.7	3.2	0.3
	13.15	+++	291	27.9	35.1	33.1	3.2	0.6
	2100		358	45.5	21.9	31.8	0.8	—
13.12	9.15	—	159	32.2	66.3	1.4	—	—
	8.45	—	333	24.4	65.3	9.0	1.3	—
	9.40	—	302	19.5	76.0	5.7	0.9	—
	9.45	—	317	29.5	68.4	1.6	0.5	—
	11.30	+	333	34.5	15.2	39.8	6.5	4.1
	11.35	+++	234	9.2	8.3	71.9	8.5	2.1
	14.30	+++	352	28.5	18.8	37.2	13.8	2.0
14.12	15.00		399	23.9	12.2	59.8	3.4	0.7
	15.15	++	340	31.9	30.1	35.4	1.3	1.3
	16.50	+++	184	20.4	27.3	44.7	5.0	2.6
	17.00	+	309	27.5	19.0	48.9	3.0	1.6

cm/sec in 8 mm larvae to 30 cm/sec in 20 mm larvae. Those velocities have been maintained by the larvae over short distances only.

Swimming velocity is independent of the duration of the swimming phase. That holds both for the velocity of the movements of the head and for the straight line distance (Fig. 2). More important than swimming velocity (Fig. 3) is the total path covered per unit of time. i.e. 5 minutes observation. This path was calculated as the product of the sums of time spent on the various types of swimming and the average velocity measured by the film recordings.

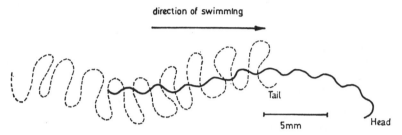

Fig. 2. Path of head and tail (hatched line) of "normal swimming" of a 10 mm larva.

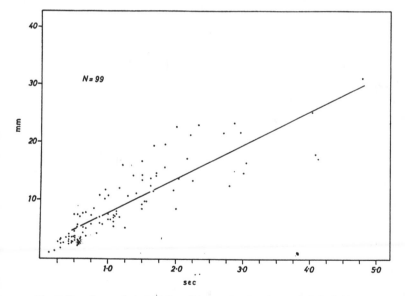

Fig. 3. Duration and straight line distance of swimming periods in 8-11 mm larvae.

The total distance travelled by the head of a yolk-sac larva is 1 to 3 m per five minutes. About 8 days after hatching sinking rate is low and "search swimming" prevails. At this stage the path covered by the head within five minutes amounts to 0·8-1·5 m. Within the following weeks of steady growth of the larvae the length of that path increases to 3 to 6 m per 5 minutes in larvae of 18-20 mm total length. At this length, difference between straight line distance and path covered by the head is less than in early post larvae.

When Downs larvae were 25-30 mm long, i.e. at the end of their second month, signs of shoaling were observed. Larvae came together in small groups for short intervals (Rosenthal, 1968). From this stage onwards registration of the various types of swimming activity and estimates of distance covered became highly variable and any calculations of the path travelled per day are very unreliable. Calculations and conclusions presented in this paper are therefore confined to the first four to six weeks of larval life.

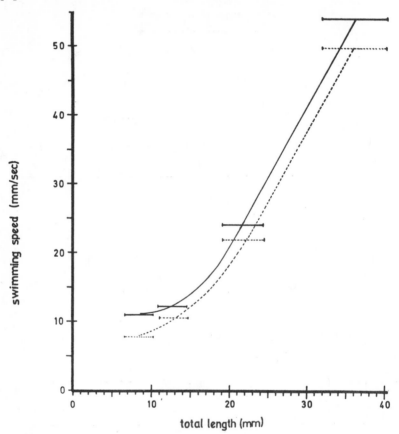

Fig. 4. Swimming velocity (- - - straight line distance, — path covered by the undulating head) in relation to body length.

Perception of food organisms

Perception is determined by the visual abilities of the larvae compared to size and distance of the food item and its visibility (contrast to background, light conditions). Detailed studies on vision in herring larvae have been published (Blaxter 1965, 1966, 1968; Blaxter and Jones, 1967). Perception depends also on the "duration of presentation", i.e. the span of time during which the

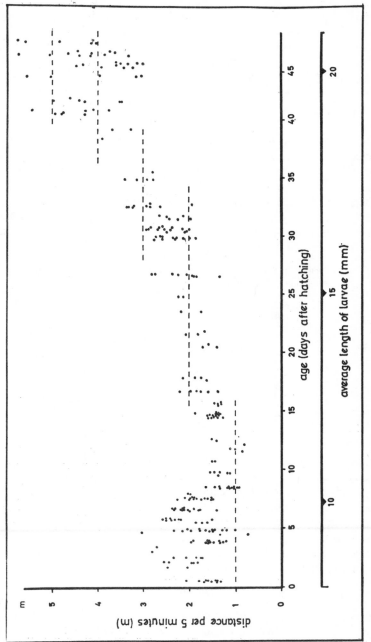

Fig. 5. Path of head per five minutes in relation to age and size of the larva.

image of the food item is projected on the retina. That again is subject to the frequency and amplitude of the undulating movements of the head while swimming forward. During typical search swimming, time of presentation is very long for some sectors but short for others (Fig. 5).

Food items are most frequently perceived at very short distances between the eye and the food item. Percentage perception decreases steadily with increasing distance of the food observed.

Yolk-sac larvae react to food items at a distance of 2 to 8 mm. Larvae of 15-20 mm are able to react to any food organism which is at a distance of 3 to 40 mm. While larvae up to 15 mm took mainly copepod nauplii, larvae of 15 mm aim preferably for copepodites which might be easier to see. Distance of perception is wider when viewing horizontally than when viewing upwards. Larvae do not perceive organisms which are underneath the plane of the horizontal axis of the body.

There is no well defined boundary of the field of perception. For convenience, however, calculations of mean maximum distance of perception have been estimated. It seems that this distance is linearly proportional to the size of the larvae (Table 3).

TABLE 3. Estimated distance of perception (mm) in Downs larvae during periods of rest and of various types of swimming

Age in days	Total length (mm)	Resting	Abrupt or normal swimming	Slow meandering
3- 7	10	8·0	7·0	10·0
8-10	11	10·0	9·0	11·8
11-14	12	11·8	10·8	13·6
15-17	13	13·2	12·2	15·4
18-21	14	14·8	13·8	17·2
22-27	15	16·2	15·2	19·0
28-31	16	17·8	16·8	20·8
33-37	17	19·2	18·2	22·5
38-43	18	20·8	19·8	24·4
44-49	19	22·2	21·2	26·0

Distance of perception differs with the activity of the larva. The slow undulating movements of "search swimming" and the slow sinking during the rest periods ensure a longer presentation of food items to the retina than during normal or abrupt swimming. Field of perception and frequency of reactions to food is therefore larger at "search" and rest than during periods of faster locomotion. Some refinement to the simple concept of the "tube" searched by the larva is required as the "tube" is wider where net progress of locomotion is low.

Volume of water searched per day

From the estimated length of swimming path per five minutes one can calculate an average daily path, taking 12 hours as an average daily period of daylight for the perception of food. In the tanks the path per five minutes was not affected by changes in light intensity between 10 and 1000 lux nor did the length of the path show any difference (diurnal rhythm) between

0600 and 1800 hours GMT. As described by previous authors larvae do not feed in darkness. Bearing in mind the high variance in the estimates of distance per 5 minutes, extrapolation from those short periods of observation to a day's path should be taken with caution. The total distance covered by the yolk-sac larvae is relatively high due to their more rapid sinking and to the low share of slow meandering. On the other hand data given in Fig. 4 have to be corrected for the longer path covered by the oscillating head. The total length of the path covered per day may be of the order of 400 m in yolk-sac larvae and a little less in early post yolk-sac larvae, increasing to about 500-700 m in larvae older than 35 days after hatching.

TABLE 4. Volume of water searched by herring larvae of different age. Rough estimates based on means of several observations

Age in days	Number of obser- vations	Volume searched per 5 minutes (ml)		Volume searched per hour (litres)	
		*	†	*	†
0	6	136	136	1·6	1·6
0	4	118	118	1·4	1·4
2	10	121	132	1·4	1·6
3	7	160	175	1·9	2·1
4	5	150	158	1·8	1·9
5	13	116	126	1·4	1·5
6	17	127	137	1·5	1·6
7	12	165	173	1·9	2·1
8	14	258	278	3·1	3·3
9	16	230	263	2·8	3·2
11	6	247	318	2·9	3·8
13	4	268	335	3·2	4·0
16	18	343	433	4·1	5·2
17	5	325	430	3·9	5·2
19	6	548	658	6·6	7·9
26	2	(790)	(900)	(9·5)	(10·9)
28	9	(940)	(1000)	(11·0)	(12·0)

* Under the assumption that the field of vision is the same for all types of locomotion.
† Under the assumption that the radius of the field of vision is higher by 20-30% during slow meandering.
Brackets indicate considerable inaccuracy of the estimates due to high variability in swimming speed and to low number of motion pictures of larvae in focus.

At the same time the field of vision across the path of the larvae increases with increasing size of the larvae from 1·0 cm² to about 3 cm² at 14-16 mm length. Those figures refer to larvae which swim "normally". At "search swimming" the fields might be 1·6 cm² to 4·6 cm² respectively (Table 4). Blaxter (1966) calculated an average of 1·5 l/h as the volume of water searched by larvae of 12-14 mm. Our figures—as derived from the above estimates— are much higher (3-8 l/h). Estimates of the radius of the field of vision depend largely on the experimental conditions, they are also subjective and are in- fluenced by the training of the observer. Any error in these estimates has a considerable effect on the figures of the volume searched per day as the field of vision is determined by the square of the radius. Further but minor reasons for the difference in the estimates may be found in the fact that we have allowed for a wider field of vision at slow meandering and that we did not correct our figures for the elliptical rather than circular cross-section of the "tube". Blaxter did not allow for the undulating movements of the head, by which the path increases considerably.

It should be emphasized that slow meandering is particularly important for the larvae. Presumably it has low energy expenditure per distance travelled by the head. It results nevertheless in a particularly wide field of vision. Any increase in the percentage of such a "search swimming" will ensure a better exploitation of the plankton. This is a particular advantage for the larger larvae.

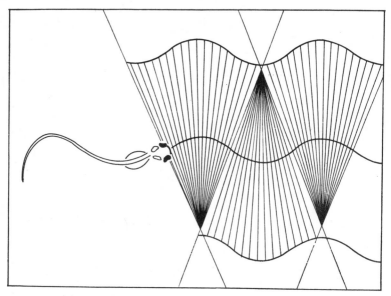

Fig. 6. Areas of different time of presentation within the field of vision of searching herring larvae. Mean radius of field of vision about 6-7 mm.

Feeding behaviour

The process of capture after optical perception of the food organism starts with an S-shaped bending of the body (Fig. 8). The bending is opposite to the direction of swimming of the food organism. The larva adjusts the eyes to achieve optimum binocular fixation of the food and may remain in this S-shape for one to three seconds correcting its position with respect to the moving prey by movements of the tail, pectoral and dorsal fins (Fig. 9). Then the larva darts forward. Even the darting is controlled by the caudal and dorsal fin and this control improves during the first weeks of life as the fins develop. The velocity of darting increases considerably with increasing length of the larva from about 6 cm/sec in yolk-sac larvae to 25 cm/sec in larvae of 15 mm.

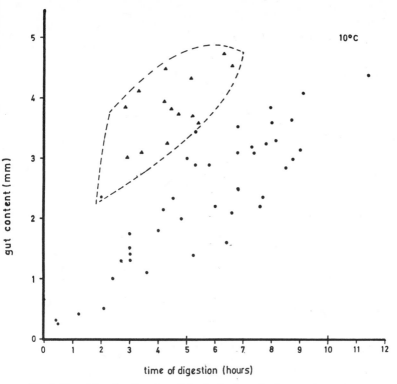

Fig. 7. Time taken for digestion of the food column in the gut after a short feeding period. Larvae of 11-14 mm total length. Abnormal defaecation presumably as a reaction to catch for photographing.

As stated by various authors, herring larvae, like larval coregonids (Braum, 1964), aim for food and perform darting reactions even during the yolk-sac stage. At this stage most of the reactions are not successful because of low velocity and poor aiming. Frequent bending and rapid snapping are favourable to successful feeding as otherwise prey organisms have a fair chance to escape.

In yolk-sac larvae of the western Baltic less than 1 % of all reactions ended with the intake of food. In most cases the S-shaped aiming position was not followed by darting and snapping. After three to four weeks 20 % of the aiming was followed by snapping and about half of the snapping reactions were successful. By this about 10 % of all food items which caused aiming were finally taken.

In Downs-larvae the frequency of darting in per cent of S-shaped positions was always higher, starting with 13 % and rising to about 25 % within four weeks. Later it reached approximately 50 %. The rate of success of darting actions increased quickly from 10 to 60-100 % depending on the kind of plankton encountered by the larvae.

It is suggested that the increase in the success rate with increasing age of

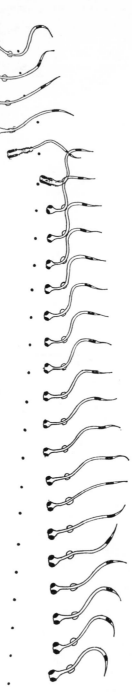

Fig. 8. Sequence of stages during an unsuccessful snapping action of a yolk-sac larva (after moving pictures).

he larvae is caused not only by the process of learning but also by the develop-
1ent of fins, by faster darting in larger larvae and by the growth of the lower
1w. The experiments provided some evidence that larvae get specialized to
articular kinds of food. Even within a single tank and amongst larvae of the
ame origin and size some "preferred" nauplii, others copepodites whereas a
hird group took both (Fig. 10). A specialized larva reacts to other kinds
f plankton organisms within the field of vision but, in general, does not snap
t them. The imprinting of larvae to particular kinds of food might be more
ronounced in the aquaria where during the first weeks food was more
tereotyped than in the sea.

In young larvae success was better in plankton of pure *Artemia* nauplii
han in a mixture of *Artemia* with wild plankton caught with 75 μ mesh-size.
Coarse plankton caught by a 150 μ resulted initially in a low rate of success.

Blaxter in earlier experiments has shown that *Artemia* nauplii have a con-
iderably higher visibility than balanid nauplii. Obviously they are also easy
o catch because of their slow swimming speed. In the sea with a great variety
f different plankton organisms, rate of success will be initially lower than in
he experiments, but will then increase rather rapidly as larvae have an early
pportunity of learning to snap for larger plankton organisms.

Larvae do not follow a prey if the first attempt at snapping fails. There-
ore estimates of minimum abundance of food should allow for the rate of
ailures. Not all organisms perceived by the larva will be taken. In Downs
arvae the rate of successful snapping in per cent of all food items visually
erceived increases within the first month of life from about 1% to 25%; in
he second month it might reach almost 50%, if optimum food is provided.

It is therefore not sufficient for the body of water searched by the larvae to
ontain just as many food items as required by the larva to sustain its main-
enance and growth. Abundance must be much higher to allow for the high
nefficiency of the preying behaviour in young larvae as stated for fresh water
ish by Elster (1944) and Braun (1964). At the stage of first feeding food
abundance has to be 20-100 times higher than for the daily ration of those
arvae. After a month the abundance should be still at least four times the
daily ration.

*Maximum filling of the gut, rate of digestion and estimates of daily
ration*

Figures of the daily amount of food taken by a larva might be based on
estimates of the rate of digestion. With sufficient food provided, herring
arvae can fill almost the total length of their gut. We found in our experi-
nents up to 15 nauplii in 10 mm Downs larvae and up to 20 nauplii in larvae
of 13-14 mm. Figures published by Blaxter (1962) are 10 nauplii in 10-12 mm
arvae but up to 40 food items in larger larvae of 14-15 mm. Blaxter and
Hempel (1961) found 5 to 10 nauplii in Baltic larvae of 8-10 mm. As the guts
are transparent and food items visible as a straight chain, estimates of diges-
tion-rate are easy to make by photographing individual larvae at two-hour
intervals. In this way it was possible to follow the progress of food items from

the fore part of the gut to the anus. During those experiments further food was provided after the initial heavy feeding which filled most of the gut. Fig. 6 shows that time for clearing the gut after the end of feeding is directly proportional to the initial amount of food in the gut. The rate of digestion measured as shrinkage of the food column in the gut is relatively independent of the total amount of food taken. Digestion of nauplii takes slightly longer than if copepodites are the major food item. Larger food items take longer than small organisms. They may block the anus. It was obvious from the photographs of the gut that young larvae do not digest *Artemia* nauplii completely. While copepod nauplii become hyaline in the posterior part of the gut, *Artemia* nauplii remained non-transparent. With an average filling of the gut of 2-4 m = 4-9 food items, i.e. to $\frac{1}{3}$ of its length, food proceeded by 0·4-0·5 mm/h resulting in an average time of passage of 4-10 hours.

In several cases it was observed that larvae suddenly extrude all or a major part of the gut content as soon as they are disturbed. The sudden defaecation might be the reason why in sea-caught larvae the percentage of empty or poorly filled guts is generally very high as observed by all authors dealing with feeding of herring larvae at sea.

If one assumes that herring larvae tend to keep their gut filled during the entire day, a rough estimate of the optimum daily ration is possible. Downs larvae of 10 mm, i.e. at the end of yolk-sac stage, may be taken as an example. These larvae live at 10°C with a period of 8-10 hours of light per day in winter in the southern North Sea. Taking an average clearing rate of 0·5 mm/h and an average size of nauplii of 0·2-0·4 mm, about 2 nauplii are "processed" per hour. During a 10 hours' day of feeding 20 nauplii will be digested. As feeding in the open sea can be more or less continuous as long as light conditions and food supply is sufficient, the gut will be still well filled at dusk. Therefore another 20 nauplii might be digested during the night. The total number of nauplii digested within a 24 hours' period is then about 40 nauplii in 10-11 mm larvae. In larvae of 13-14 mm length the daily ration computed in the same way amounts to about 50 nauplii per 24 hours. Bearing in mind the high variation in digestion rate and the different size and digestibility of the various kinds of nauplii (*Artemia* nauplii are larger and remain longer in the gut than copepod or *Balanus* nauplii), those figures can only be taken as rough estimates.

In larger larvae the amount taken just after the provision of food is very high, making estimates of daily ration very uncertain. In our experiments larvae of 20 mm took up to 50 nauplii within five minutes, but food supply was never sufficient to arrive at good estimates of how much food a larvae of this size may take within a day. Assuming that rate of passage in mm/h is the same as in younger larvae but that the lumen of the gut is considerably wider a digestion rate of five nauplii/hour seems not unlikely. From this we would conclude two fillings of the gut per day (with a light period of 10 hours).

Figures of 40 to 50 nauplii per day at about 10°C for Downs larvae in the first month of larval life and of about 100 nauplii or more in the second month are optimum data arrived under the assumption of continuous food supply.

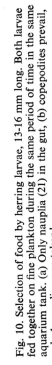

Fig. 9. S-shaped bend of the body, binocular fixation of the prey, and position of fins during a snapping action. Total length of the larva: 18 mm.

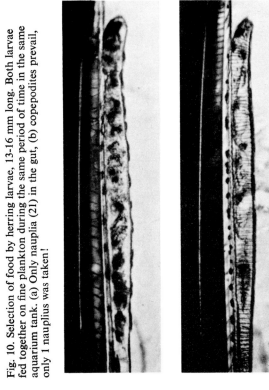

(a)

(b)

Fig. 10. Selection of food by herring larvae, 13-16 mm long. Both larvae fed together on fine plankton during the same period of time in the same aquarium tank. (a) Only nauplia (21) in the gut, (b) copepodites prevail, only 1 nauplius was taken!

TABLE 5. Estimates of density of food required for optimum feeding of herring larvae

Length (mm)	Age (days)	Volume of water searched per day (litres)	Rate of successful feeding actions (%)		Daily ration (number of nauplii etc.)	Number of encounters per day		Required number of nauplii/litre	
			(a)	(b)		(a)	(b)	(a)	(b)
10-11	6- 9	20- 30	6	40	40	750	135	42-21	8-4
13-14	13-30	40-100	10	60	50	1000	170	25-13	4-2

Two estimates of food density are given, under the alternative assumptions (a) and (b) about the best figure for the rate of successful feeding actions:
(a) that only part of the plankters perceived visually cause snapping actions with a low rate of success;
(b) that all plankters visually perceived cause snapping actions at the same rate of success.
As in dense food larvae react to more food than they "intend" to eat, estimate (a) results in an overestimate of required number of food items. Estimate (b) is a minimum estimate, normally too low as it assumes that in hungry larvae all food items in the field of vision will cause snapping reactions.

Under those conditions growth will be at its optimum, provided the food taken is easily digestible and has a high nutritive value. It cannot be decided from the data at hand how far those figures are above the minimum level for maintenance.

Minimum abundance of food for optimum feeding

From the data available conclusions may be drawn about the minimum abundance of food to ensure continuous filling of the gut during daylight hours. Data are complete only for Downs larvae at the end of yolk-sac stage (10-11 mm) and at the end of the first month (13-14 mm). In Table 5 estimates are given differing by an order of magnitude. They are based on different assumptions about the accuracy of predation. The correct figures will presumably be between the two estimates of which (a) refers to the case when food is scarce while (b) might be more correct under continuous food supply.

In spite of the great uncertainty with regard to the accuracy of the data used for the computations we might conclude from the table that abundance of food items must be higher for young larvae than for older ones. There are no reasons to doubt that a similar ratio might be found in the open sea. As the figures found for minimum abundance for optimum feeding lie in the range of plankton abundance recorded on nursery grounds of herring larvae, one might conclude that the very young larvae are more prone to malnutrition than somewhat older or larger larvae. That is in accord with the hypothesis by Hempel and Blaxter (1963) on the better stamina of the larger larvae of autumn-winter spawning herring compared with the smaller larvae of spring-summer spawners.

In addition to their lower requirements of overall food abundance and their energy-saving search swimming, larger larvae have also the benefit of a wider spectrum of food organisms as they can cope with larger (and more nutritive) food items and seem to digest them more fully. It was obvious from our experiments that from the third week onwards metanauplii of *Artemia* and plankton taken by a net of 150 μ mesh-size provided adequate food while in younger larvae only nauplii of *Artemia* and the fine plankton of a 75 μ net were suitable. A homogenous composition of plankton over larger areas on at least some days is favourable for the larvae as any change in composition causes an initial drop in the accuracy of preying. No futher conclusions should be drawn from the experiments as any comparison of the actual figures of estimated food abundance in the experiments with those met in the sea might be very misleading in view of the abnormal discontinuous feeding and swimming conditions in the aquaria.

Computations on the energetics of maintenance and growth leading to estimates of daily food requirements will also be very difficult to compare with our experimental figures as efficiency of digestion might be rather low in young larvae feeding on *Artemia* and figures on conversion factors are not available. Calorimetric studies along these lines have been started.

The results of the experiments might be taken as a basis for discussing competition amongst herring larvae in the sea. This is of particular interest in view of mechanisms leading to a stock-recruitment relationship in herring (Hempel, 1965).

Early post yolk-sac larvae of Banks and Downs herring are normally found in rather low abundance. Dispersion of larvae seems to occur already during the yolk-sac stage. Sampling in different strata of the water column does not indicate concentration of larvae in particular strata or patches (Hempel and Schnack, 1969). Maximum density of larvae of 10 to 11 mm normally does not exceed 20 larvae per m³. Those larvae will search about 20-30 l/day, e.g. the 20 larvae will daily "filter" about half the water mass at their disposal. If food is plentiful, rate of success will be low, about 6% of all encounters. Rate of grazing per day is then about 3% of the food available and will presumably be less than reproduction and growth of the food population.

In slightly larger larvae searching capacity and accuracy of feeding is higher but those larvae are far less abundant. If two larvae of 13-14 mm are found per m³ then grazing rate will be about 2%.

In case of low food abundance grazing rate will be slightly higher as higher estimates of feeding accuracy have to be applied (alternative (b) in Table 2), on the other hand searching capacity might be somewhat lower with restricted food because of a lower proportion of search swimming. No figures of grazing rate under those conditions can be produced with any degree of reliability.

From the information available and under the assumption that larval distribution is not extremely patchy we might conclude that competitive grazing amongst herring larvae is not a major factor controlling larval mortality and causing a stock-recruitment relationship.

REFERENCES

BLAXTER, J. H. S. 1962. Herring rearing—IV. Rearing beyond the yolk-sac stage. *Mar. Res.*, 1962 (1), 18 pp.
BLAXTER, J. H. S. 1965. The feeding of herring larvae and their ecology in relation to feeding. *Rep. Calif. coop. oceanic Fish. Invest.*, **10**, 79-88.
BLAXTER, J. H. S. 1966. The effect of light intensity on the feeding ecology of herring. *Symp. Br. Ecol. Soc.*, **6**, 393-409.
BLAXTER, J. H. S. 1968. Visual threshold and spectral sensitivity of herring larvae. *J. exp. Biol.*, **48**, 39-53.
BLAXTER, J. H S. and HEMPEL, G. 1961. Biologische Beobachtungen bei der Aufzucht von Heringsbrut. *Helgoländer wiss. Meeresunters.*, **7**, 260-83.
BLAXTER, J. H. S., and HEMPEL, G. 1963. The influence of egg size on herring larvae (*Clupea harengus* L.) *J. Cons. perm. int. Explor. Mer*, **28**, 211-40.
BLAXTER, J. H. S., and HEMPEL, G. 1966. Utilization of yolk by herring larvae, *J. mar. biol. Ass. U.K.*, **46**, 219-34.
BLAXTER, J. H. S., and JONES, M. P. 1967. The development of the retina and retina-motor responses in the herring. *J. mar. biol. Ass. U.K.*, **47**, 677-97.
BRAUN, E. 1964. Experimentelle Untersuchungen zur ersten Nahrungsaufnahme und Biologie an Jungfischen von von Blaufelchen (*Coregonus wartmanni* Bloch), Weibfelchen (*Coregonus fera* Jurine) und Hechten (*Esox lucins* L.) *Arch. Hydrobiol.*, **28**, 183-244.
ELSTER, H. J. 1944. Über das Verhaltnis von Produktion, Bestand, Befischung und Ertrag sowie über die Moglichkeiten einer Steigerung der Erträge, untersucht am Beispiel der Felchen des Bodensees. *Z. Fisch.*, **42**, 169-357.

FLÜCHTER, J. 1964. Eine besonders wirksame Aquarienfilterung und die Messung ihrer Leistung. *Helgoländer wiss. Meeresunters.*, **11**, 168-70.

GULLAND, J. A. 1965. Survival of the youngest stages of fish and its relation to year class strength. *Spec. Publs. int. Commn NW. Atlant. Fish.*, **6**, 363-72.

HEMPEL, G. 1965. On the importance of larval survival for the population dynamics of marine food fish. *Rep. Calif. coop. oceanic Fish. Invest.*, **10**, 13-23.

HEMPEL, G., and BLAXTER, J. H. S. 1963. On the condition of herring larvae. *Rapp. P.-v. Réun. Cons. perm. int. Explor. Mer*, **154**, 35-40.

HEMPEL, G., and BLAXTER, J. H. S. 1967. Egg weight in Atlantic herring (*Clupea harengus* L.) *J. Cons. perm. int. Explor. Mer*, **31**, 170-95.

HEMPEL, G., and SCHNACK, D. 1969. Estimates of abundance of larvae on spawning grounds of Banks and Downs herring. *Rapp. P.-v. Réun. Cons. perm. int. Explor. Mer*. (in press).

HJORT, J. 1926. Fluctuations in the year classes of important food fishes. *J. Cons. perm. int. Explor. Mer*, **1**, 5-38.

IVLEV, V. S. 1944. Hunting time and path travelled by predators in relation to the density of prey population. *Zool. Zh.*, **23**, 139-47 (in Russian).

IVLEV, V. S. 1945. Density and distribution of food as factors determining the rations of fishes. *Zool. Zh.*, **24**, 112-25 (in Russian).

IVLEV, V. S. 1960. On the utilization of food by planktophage fishes. *Bull. math. Biophys.*, **22**, 371-89.

RICKER, W. E. 1954. Stock and recruitment. *J. Fish. Res. Bd Can.*, **11**, 559-623.

ROSENTHAL, H. 1968. Beobachtungen über die Entwicklung des Schwarmverhaltens bei den Larven des Herings, *Clupea harengus*. *Mar. Biol.*, **2**, 73-76.

Trophic-dynamics of lake trout production systems

S. R. KERR
Biology Department
Dalhousie University
Halifax, N.S., Canada
N. V. MARTIN
Ontario Department of Lands and Forests
Maple, Ontario, Canada

ABSTRACT. Published information on four Ontario fisheries indicated that production of lake trout (*Salvelinus namaycush*), was proportional to an index of primary productivity, irrespective of the trophic level at which production occurred, although mean size, rate of growth, and age at maturity increased with longer food chains. Trophic-dynamic theory predicts, *ceteris paribus*, that terminal production will decrease with increasing length of food chain. We examine ways in which the trout observations might be reconciled with theory, and conclude that lower metabolic expenditures characterized those trout populations subsisting upon longer food chains, in part because the trout were larger, but also because greater foraging efficiency was attained at a higher trophic level. Thus the larger food resource available to trout from shorter food chains was used less efficiently, with the result that production of trout was relatively independent of trophic level.

INTRODUCTION

The commonly accepted view of trophic-dynamics holds that energy transfers between trophic levels are inevitably accompanied by major losses of energy. Slobodkin (1960) has suggested that ecological efficiencies will ordinarily be in the order of 10%, a prediction which finds support in fish populations (Mann, 1965) indicating that insertion of an extra trophic level into a food chain should decrease the energy resource of the terminal predator by approximately an order of magnitude. Thus trophic-dynamic theory predicts, *ceteris paribus*, that terminal production will decrease substantially with increasing length of food-chain.

Although the quantitative dependence of higher production levels upon primary production has not been explicitly demonstrated for any aquatic system, Ryder (1965) has shown that commercial fish production in 23 North American lakes is predictable from their morphometric and edaphic characteristics which are themselves recognized as determinants of primary productivity. His work implies that long-term production of fish is proportional to primary productivity, largely irrespective of such factors as species composition or food-chain length.

Martin (1966) has reported in detail on nearly two decades of observations of the angling fisheries for lake trout (*Salvelinus namaycush*), of four oligotrophic lakes in Algonquin Park, Ontario. His particular examples were selected from a larger array of data because of evidence that the fisheries were sufficiently intensive for production values to approximate maximum equilibrium yields, in the sense of Ricker (1958). Long-term average production rates for these lakes are in accordance with the trend observed by Ryder (1964) for other angling fisheries, indicating that the relatively small variation in average trout production among the lakes studied by Martin may be accounted for on the basis of morphometric and edaphic characteristics, without recourse to the nature of the food chains concerned. Martin reported considerable differences in diet type among lakes, ranging from 0·9% to 55·6% direct plankton grazing (chiefly planktonic Crustacea, and in particular Cladocera (Martin, 1952)), with forage fishes dominating the residual diet. But although such population parameters as average size, abundance, and age at maturity of the trout varied strikingly with diet, trout yields did not appear to be affected. Martin (1966) concluded ". . . it appears the final production of lake trout flesh in these lakes is not influenced by different utilization of the food chain by the trout."

Because trout feeding on forage fishes are clearly existing on a higher trophic level than are planktivorous trout, and yet yield is unaffected by diet composition, it is of great interest to seek ways in which this might be accomplished without violating trophic-dynamic theory. The problem is complex, because it is necessary to devise a system in which the population yields are independent of food-chain length, yet remain proportional to primary productivity.

A MODEL OF A TROUT PRODUCTION SYSTEM

In viewing the energy budgets of single populations, it is useful to adapt an expression used by Winberg (1956)

$$cF = T + \Delta W / \Delta t \qquad [1]$$

where F indicates potentially available forage corrected by c for non-assimilation and non-exploitation losses, T indicates metabolic expenditure, and ΔW the production increment, all over a common period of time Δt. Thus individual trophic levels can be represented as

$$c_1 h v = T_1 + \Delta W_1 / \Delta t \qquad [2]$$

$$\Delta W_1 / \Delta t = (T_2 + \Delta W_2 / \Delta t) / c_2 \qquad [3]$$

$$\Delta W_2 / \Delta t = (T_3 + \Delta W_3 / \Delta t) / c_3 \qquad [4]$$

etc.

where subscripts denote trophic levels, and where c_1hv indicates the amount of incident solar energy assimilated for primary production. Equations [3] and [4] can be successively substituted, each into its predecessor, to yield

$$c_1hv = T_1 + T_2/c_2 + T_3/c_2c_3 + (\Delta W_3/\Delta t)/c_2c_3 \qquad [5]$$

representing a simple expansion of equation [1] to provide an integral model of the several trophic levels in a generalized food chain. Although [5] is too unwieldy for further use here, it is presented to indicate the general view of food-chains which is further developed below.

Special cases of [5] can be written to describe terminal production at any level in the food chain. Thus where subscript F indicates forage fishes and T indicates trout,

$$\Delta W_{3F}/\Delta t = c_{3F}(\Delta W_2 \Delta t) - T_{3P} \qquad [6]$$

$$\Delta W_{3T}/\Delta t = c_{3T}(\Delta W_2/\Delta t) - T_{3T} \qquad [7]$$

$$\Delta W_{4T}/\Delta t = c_{4T}(\Delta W_{3F}/\Delta t) - T_{4T} \qquad [8]$$

$$= c_{3F}c_{4T}(\Delta W_2/\Delta t) - c_{4T}T_{3F} - T_{4T}.$$

[6], [7], and [8] describe production of forage fishes, planktivorous trout, and piscivorous trout respectively, in terms of their energy resources. In this and the subsequent discussion, we will ignore minor diet items; trout diets will be considered to consist of varying proportions of plankton and forage fishes.

Martin's observation of equivalent production on different food-chain lengths may now be rephrased more specifically. Our problem, in terms of the equation systems, is to seek ways in which [7] could equal [8]. Clearly, from [7] and [8], compensation through either c or T, or less obviously ΔW is possible, provided a mechanism can be found to maintain relatively constant yield with variations in trophic level.

Parameters affecting the production system

The parameter c which includes measurements of non-assimilation and non-exploitation, could vary enough among trophic systems to level potential production differences. For example, energy losses of the required magnitude could conceivably result if planktivorous trout were unable to exploit their energy resources to the same degree as piscivores. The evidence does not support this possibility. Recent work has indicated that planktivorous fish, including alewives, *Alosa pseudoharengus* and *A. aestivalis* (Brooks and Dodson, 1965), yellow perch, *Perca flavescens*, and rainbow trout, *Salmo gairdneri* (Galbraith, 1967), are capable of intensive and selective exploitation of planktonic Crustacea. Galbraith reports that rainbow trout were responsible in his studies for the virtual elimination of particular size-groups of *Daphnia* from the zooplankton. Martin (1952) concluded on the basis of stomach analyses that there was no sign of significant cannibalism in any of the lakes and that where plankton formed an important part of the diet the lake trout were also highly selective planktivores. Martin and Sandercock (1967),

and Qadri (1967) also observed that planktivorous lake trout possess significantly larger numbers and greater development of gill rakers and accessory gill
rakers than do trout from piscivorous populations. It is also appropriate to
note that no differences in condition factor (length-weight ratio) were
observed among Martin's trout populations (Martin, unpublished data).
All of these observations suggest that the trout in all the lakes are capable of a
high degree of food resource exploitation, that they are closely adapted to
making use of the differing natures of their food resources, and individuals in
the populations do not show signs of food deprivation.

We conclude that variation in degree of resource exploitation is an inadequate explanation of Martin's observations.

Because Martin's observations derive from fishery yields rather than from
production values, ΔW in the formulation used here [5] does not take explicit
account of natural (non-fishing) mortality of the terminal predator populations. Actual total production in the populations is greater than the yield
component by the amount of production lost through natural mortality.
Natural mortality differences would therefore help to account for the lower
relative yield from the planktivorous populations if the production lost
through natural mortality of planktivores were significantly higher.

The available data do not sustain this possibility. The yield from the
planktivorous populations is made up of more numerous individuals of smaller
size and younger average age than from piscivorous populations. That is,
production losses through natural mortality, if different at all, would appear
likely to be greater in the piscivorous populations. Providing that a significant
amount of the total trout production occurs among the fishable sizes and
ages, we can only conclude that differences in natural mortality rate do not
provide an adequate hypothesis for the observed independence of trout
production from length of food-chain.

In further support of this conclusion, Martin (1966) has reported on the
lake trout catch and size composition for some 30 additional lakes in the
same geographic region which are fished at various intensities thought to be
somewhat less than required for maximum equilibrium yield. Despite
variable fishing effort there is a strong inverse correlation ($r = -0.83$,
$P < 0.01$) between mean size of fish caught and percentage occurrence of
plankton in the trout diets. On these grounds it appears that mortality rates
are not the prime determinant of either the size structure or productivity of
the trout populations.

Having tentatively eliminated non-exploitation of food resources or
natural mortality of the trout as possible ways to account for the dissipation
of the larger amount of food energy available to planktivores, metabolism
(T) remains a possibility. That is, it remains to inquire whether a higher
metabolic expenditure per unit of growth on the part of planktivorous fish
could account for the fact that production of planktivorous trout does not
exceed production of piscivorous trout despite the fact that the latter are
feeding on a longer food-chain. Formally, the situation can be described as
follows:

By observation, [7] = [8]

i.e., $c_{3T}(\Delta W_2/\Delta t) - T_{3T} = c_{3F}c_{4T}(\Delta W_2/\Delta t) - (c_{4T}T_{3F} + T_{4T})$

but if $c_{3T} = c_{3F} = c_{4T} < 1$ (from the preceding discussion)

then $c_{3F}c_{4T} < c_{3T} < 1$

and $T_{3T} > c_{4T}T_{3F} + T_{4T}$.

That is, we are examining the hypothesis that in the trout lakes being compared the larger basic ration available to planktivores must appear as metabolism since it is not realized as yield.

Effect of changes in metabolic level on trout production

Essentially, the problem becomes one of determining whether differences in growth efficiency on different diets are an adequate explanation of Martin's observations. For this purpose it is useful to express growth efficiency K K_2 of Ivlev, 1955) as:

$$K = \Delta W / cF\Delta t$$

and by substitution from [1],

$$K = \Delta W/(T+\Delta W). \qquad [9]$$

That is, growth efficiency is a function of production and metabolism.

The relation of total metabolism to body size of an organism is adequately described by the familiar

$$T = \alpha W^\gamma$$

where W represents body weight and α and γ are fitted parameters. Winberg (1956) has thoroughly examined the applicability of this expression to a variety of fish species, and concluded that γ usually varies little from 0·8, while α ordinarily can increase to 4 or 5 times its basal (standard) level. Although there are no empirical determinations of α for the lake trout available in the literature, we have taken standard and active levels at 10°C from the laboratory experiments of Job (1955) for the congeneric brook trout (*S. fontinalis*) and adjusted these slightly to correspond with estimates at 10°C made for two sizes of lake trout from Fig. 1 of Gibson and Fry (1954). Thus for standard metabolism we have a value for $\alpha = 0·08$ and for active metabolism, $\alpha = 0·34$. Temperature was standardized at 10°C because this is close to the preferred temperature during the growing season (Martin, 1952). Winberg (1956) has examined at length the advisability of selecting a level of α at twice the standard rate as a routine level of metabolism, and Mann (1965) has further discussed the choice. For comparative purposes we have adopted $\alpha = 0·16$ as a routine level. To calculate the metabolism of the yield component of the several populations, we have assumed that seasonal growth is linear, occurring during a period equivalent to 180 days metabolic expenditure. Respired oxygen may be converted to energy units with the convention (Winberg, 1956) that 1 ml of $O_2 = 5$ cal. Metabolism is then calculated for each population from the mean weight of individual fish appearing in the yield, multiplied by the number of fish per hectare contributed to the yield. Metabolism is therefore calculated as kilocalories per hectare per year expended by the yield component of each population. Estimates of routine and active metabolism for each population are given in Table 1.

TABLE 1. Parameters of lake trout populations in Algonquin Park, Ontario. Magnitude of metabolism and production efficiency (K) values are shown relative to a piscivorous population, from Lake Opeongo. For explanations see text. Original data from Martin (1966)

		Lake name		
	Opeongo	Redrock	Merchants	Happy Isle
% plankton feeding	0·9	22·6	33·2	55·6
Mean weight of individual (kg)	1·58	1·20	0·93	0·62
Routine metabolism of yield component (kcal/hectare/year)	37·3	39·5	41·5	45·0
Active metabolism of yield component (kcal/hectare/year)	—	83·9	88·3	95·6
$\dfrac{\text{Opeongo routine metabolism}}{\text{routine metabolism}}$	—	0·95	0·90	0·83
$\dfrac{\text{Opeongo routine metabolism}}{\text{active metabolism}}$	—	0·45	0·42	0·39
Routine K_T/K_{4T}	—	0·95	0·91	0·85
Active K_T/K_{4T}	—	0·50	0·48	0·45

Several sources of possible error are attendant upon these calculations of metabolic rate. However, it is reasonable to assume that errors are of the same order in each of the four lakes considered, hence if we use the ratio of estimates for individual lakes to the estimate of routine metabolism for Opeongo trout taken as standard much of this error will be obviated. In each case the metabolic rates must underestimate the true metabolic expenditure of each population because we have not included the metabolism expended by the non-yield portion of the populations. Since the average growth increment of a planktivorous trout is less than that of a piscivore, a larger standing crop appears necessary to a planktivorous population if it is to achieve production equivalent to a population of piscivores. Therefore, it seems likely that the use of the piscivorous population of Opeongo as a standard, in which natural mortality of the fished stock is likely to be highest, results in metabolism ratios (T_{4T}/T in Table 1) which are conservative estimates of the true values. That is, the true metabolic differences between populations are likely to be even greater than we assign in subsequent considerations.

Using Happy Isle trout (56% planktivorous) as an example, we can examine the effect on the efficiency of production (K_H) of varying α from the routine to active level.

i.e., from [9], for Happy Isle trout (subscript H),

$$K_H = \Delta W_H/(T_H + \Delta W_H).$$

Similarly for piscivorous Opeongo trout (used as a standard for growth efficiency)

$$K_{4T} = \Delta W_{4T}/(T_{4T}+\Delta W_{4T}).$$

But $\Delta W_H = \Delta W_{4T} = 47 \cdot 18$ kcal/hectare/year (when production is adjusted to the norm for Opeongo), and then,

$$\frac{K_H T_H}{K_{4T} T_{4T}} = \frac{1-K_H}{1-K_{4T}} \qquad [10]$$

From Table 1, $T_{4T}/T_H = 0 \cdot 83$ when both populations are at a routine metabolic level,

therefore $\qquad 0 \cdot 83 K_{4T}/K_H = (1-K_{4T})/(1-K_H)$

and, $\qquad K_H/K_{4T} = 0 \cdot 17 K_H + 0 \cdot 83.$

Since the metabolic expenditure required to obtain rations in an aquarium can be expected to be relatively small, laboratory determinations of K are likely to be in excess of field values. Thus it seems unlikely that K for a natural population of active predators is greater than of the order of $0 \cdot 10$. In this event $0 \cdot 17 K_H$ becomes very small and

$$K_H = 0 \cdot 83 K_{4T}. \qquad [11]$$

Equation [11] indicates that relative production efficiency of the trout populations must be increasing with increasing mean size. Should the planktivorous Happy Isle trout in their search for food exhibit an active metabolic rate ($\alpha = 0 \cdot 34$) while the piscivorous Opeongo trout remain at a routine level, then by the same reasoning as for [11] we have

$$K_H/K_{4T} = 0 \cdot 39 + 0 \cdot 61 K_H. \qquad [12]$$

The term $0 \cdot 61 K_H$ is significant if the value of K_H is as great as $0 \cdot 10$, resulting in the ratio $K_H/K_{4T} < 0 \cdot 45$, although under field conditions it is probable that $K_H \approx 0 \cdot 39 K_{4T}$. But in either case it is clear that the production efficiency of piscivorous trout appears likely to exceed that of 56% planktivorous trout by a factor of two or more in what appear to be reasonably possible situations. That is, the changes in metabolic level, acting in concert with differences in mean size, would mean that planktivorous trout such as those of Lake Happy Isle could be utilizing as forage more than twice the total amount of ration as is taken by piscivores, without surpassing piscivore production. These estimates, and similar values calculated for the trout populations of Merchants and Redrock lakes are set out in Table 1.

If compensation through K is an adequate mechanism to explain equivalent yields on the various diets, then from [12], for the yield component of each population, the energy resource utilized by Happy Isle trout must not exceed two to three times the food resource of Opeongo trout. The available information is inadequate to test this prediction, as it is not possible to determine the relative caloric contributions of diet items from values for the percentage occurrence of the items in trout stomachs. However, data from two lakes for which diet analyses have also been reported as percentages by weight (Martin, 1952) support the supposition that the percentage frequency of occurrence of plankton overestimates its contribution by weight to the lake trout diets. As a consequence, we suggest that the relative trophic levels of the trout populations observed by Martin differ to a sufficiently small degree that the food resource of a moderately planktivorous population, such as Happy Isle trout, is not likely to be more than several times as large as the food resource of piscivores. Useful estimates of relative caloric value of the diet components should be obtainable without great difficulty, and will provide a direct test of the hypothesis.

Growth efficiency and trout production

Paloheimo and Dickie (1966a, b) have recently reviewed literature pertinent to food and growth of fishes. They suggested that equation [1] may constitute a circular causal system in which the metabolic level, α, of the parameter T is established in a complex interaction with ration input, cF. They further concluded that fish fed *ad libitum* on a single diet item exhibit a linear decrease in the logarithm of growth efficiency with increasing body weight (or rations). This relationship they termed the "K-line".

They found evidence for different K-lines (differing in slope and intercept) with various diet items, an effect which they posited may have been due to the size of the food particles offered, since K-line slopes decreased with increasing size of diet item. Their calculations indicated that brook trout fed on a hatchery mash exhibited a relatively steep negative regression of log K against rations, while trout fed on *Gammarus* showed a less steep negative K-line, and a diet of small fish elicited no detectable slope. It should be noted that *ad libitum* rations correlate with body size, as might be expected. Also, with the exception of the initial portion of the K-line determined for hatchery mash, their analysis showed that the trout in these experimental situations all sustained an active metabolic rate, implying that the changes in growth efficiency were a direct result of differences in foraging efficiency, the caloric reward of ration per unit metabolism expended in foraging increasing with increasing size of food particle.

But regardless of the actual mechanism responsible for individual K-line slopes, should the lake trout exhibit a response of growth efficiency to ration type similar to that of the congeneric brook trout, it would appear that we have a possible dietary mechanism for control of production, through the coupling of K with the nature of the food resource.

The few data obviate a check of this hypothesis through direct estimates of K for the various populations. However, limiting values of the ratio of K to a standard, taken as the K value of a piscivorous population at routine metabolism, were given earlier, based on equation [10]. These limits do appear to provide the required assurance of the adequacy of the hypothesis. The percentages of plankton feeding observed for the various trout populations are not a direct measure of their caloric contribution to the diet, but they may be taken as a relative index of the importance of plankton. We may then estimate a K-line for plankton-feeding relative to the standard piscivore K-line. The value of the resulting ratio K/K_{4T} for each population can be considered to arise from two sources. First there is a contribution from the planktivore K-line proportional to the frequency of plankton feeding. Second, there is a contribution from the piscivore K-line proportional to (100—the percentage plankton-feeding). For comparative values we take the slope of the piscivore K-line as zero, as was the case for piscivorous brook trout, and the level of K_{4T} is taken as 1·0.

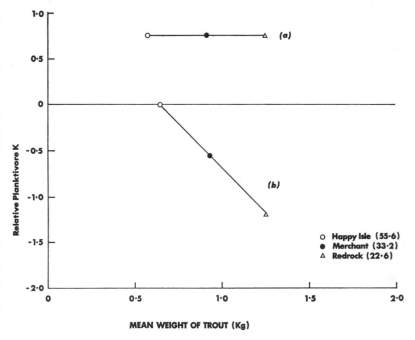

Fig. 1. Planktivore production efficiency K, relative to piscivore production efficiency $K_{4T} = 1·0$. Values are based on the ratios K/K_{4T} set out in Table 1. Line (a) represents routine metabolism line; line (b) represents active metabolism for planktivores and routine metabolism for piscivores. Bracketed numbers under name of lake indicate percentage plankton-feeding by that population. For explanation see text.

Estimates of K/K_{4T} for each population are plotted on Fig. 1, where line (a) results from an assumed routine metabolic level and line (b) an active metabolic level. Line (a), because it deals with the routine level, represents the relative planktivore K-line when the decrease of efficiency due to increased plankton in the diet is least. It therefore represents an upper limit for the relative planktivore K-line. Line (b), dealing with the active metabolic level, represents a condition where there is maximum effect of plankton feeding on K. It thus defines a lower limit of the relative planktivore K-line. There are no data to show whether or not there is an upper limit to the percentage of plankton-feeding at which compensation through K will maintain production on different food chain lengths; however, Fig. 4 of Martin (1966) indicates that at levels of plankton-feeding above 60%, further decrease of mean size no longer takes place, suggesting that for individual organisms the limit of compensation through K has been reached. It may therefore not be fortuitous that the zero intercept on line (b) occurs at about 60% plankton feeding.

It will be recognized that the accuracy of these relative K-lines for planktivorous trout depends upon the accuracy of our estimates of the metabolism of the populations. While no good direct methods are available for field measures of metabolism, many laboratory measurements have been made with related species. Their regularity gives reason for confidence that limits can be set in this fashion. Perhaps the importance of this method of presentation, however, lies not so much in establishing limits for this particular situation as illustrating that in general it is possible to frame the problem in terms which can be experimentally tested. A method for measuring metabolism of planktivorous lake trout in the field would appear to provide an immediate and simple check of the reasoning leading to the construction of Fig. 1.

DISCUSSION

The posited differences in K-line slope for planktivorous and piscivorous lake trout provide mechanisms capable of internal regulation of yield in the face of diet variation. From trophic-dynamic theory, it would be expected that the total energy available to the trout as ration will decrease with increasing proportions of forage fish in their diet. However, the diet change results in changes in the relative contribution from the two K-lines to overall growth, resulting in a growth efficiency increase. From our calculations it appears that the two antagonistic processes are of an order of magnitude which could result in maintenance of a relatively constant trout yield. Such a simple equilibrium would constitute an adequate explanation of Martin's observations.

It is of interest to assess the relative importance of the several ways in which metabolic energy may have been expended by the trout populations. We have shown here that changes in metabolic level and average body weight are potentially of the right magnitude to account for the observed equivalence of yield at different trophic levels, when standing crops are assumed to be equivalent. However, as previously noted, it appears that standing crops can be expected to increase with increasing occurrence of plankton in the diet. If this should prove to be the case, then population metabolism would also increase with shortening of the food chain. In this event the level of metabolism of individuals may remain at a relatively constant level independent of the nature of the food resource, but the population metabolism would increase with standing crop. The equivalence of trout production can then be thought of as primarily a population response to the nature of the food resource.

Alternatively, if the individual level of α is established by the nature of the ration, then the apparent independence of trout production from trophic level is primarily a result of individual, rather than population, response to the nature of the food resource. A test of whether the independence of trout yield from trophic level is primarily an individual or a population response to diet requires field measures of the level of metabolism, in addition to the usual estimates of food-chain parameters, measures which we consider to be of great potential interest.

However, whatever the specific mechanisms underlying the production systems described here, the present analysis suggests two matters of considerable importance to trophic ecology. First, food-chain shortening, at least in oligotrophic environments, does not *necessarily* imply an increase of terminal production. It appears that significant variation in metabolic requirements is permitted the populations which comprise a food-chain, endowing the system with considerable plasticity in its energy requirements.

Second, it appears that the nature of the food resource, aside from its abundance, may provide a causal mechanism regulating the production process. Ivlev (1955) has indicated the effects of varying either the size or the degree of aggregation of food particles upon the foraging efficiency of fishes. In the present paper we have suggested that the metabolic demands imposed upon a trout population in foraging for relatively small zooplankton organisms offsets the potential advantage of the larger energy resource available from exploiting primary consumers rather than other fishes. The result of feeding upon small or dispersed food particles may be to increase the expenditure of metabolism per unit of revenue accruing as rations. The effect upon production efficiency of increasing the ratio of metabolism to rations can readily be seen by dividing equation [1] by cF which yields $\Delta W/cF = K = 1 - T/cF$. In so far as they may affect production efficiency, the consequences of such ration characteristics as particle size would therefore appear to warrant serious consideration in food-chain analyses.

Apart from ecological consideration, the present analysis indicates a general consideration with respect to the maximization of fish yields. From the numerical dynamic models of Ricker (1958), Beverton and Holt (1957) and others, it was held that maximum yields were to be obtained through exploitation at the size of maximum biomass. Further consideration of the dynamics of growth, taking into account the abundance and availability of food resources, suggests that harvesting should rather be done at the smallest practicable size (Paloheimo and Dickie, 1965). To this we would add the further provision that where food type and growth efficiency change with size, and all other things are equal, the growth stanza selected for optimal exploitation should be the best compromise of small size and high growth efficiency obtaining at various stages in the life cycle of the species.

ACKNOWLEDGEMENTS

It is a pleasure to acknowledge the stimulating discussion, encouragement, and support generously provided by Dr L. M. Dickie, Marine Ecology Laboratory, Fisheries Research Board of Canada. Sincere thanks are also due to Profs E. L. Mills, I. A. McLaren, and P. J. Wangersky, all of Dalhousie University, and Prof. F. E. J. Fry, University of Toronto, for critical and constructive review of the manuscript. One of us (S.R.K.) is grateful for financial support through a National Research Council of Canada Postgraduate Scholarship.

REFERENCES

BEVERTON, R. J. H., and HOLT, S. J. 1957. On the dynamics of exploited fish populations. *Fish. Invest., Lond.*, Ser. 2, **19**, 533 pp.

BROOKS, J. L., and DODSON, S. I. 1965. Predation, body size, and composition of plankton. *Science N.Y.*, **150**, 28-35.

GALBRAITH, M. G. Jr. 1967. Size-selective predation on *Daphnia* by rainbow trout and yellow perch. *Trans. Am. Fish. Soc.*, **96**, 1-10.

GIBSON, E. S., and FRY, F. E. J. 1954. The performance of the lake trout, *Salvelinus namaycush*, at various levels of temperature and oxygen pressure. *Can. J. Zool.*, **32**, 252-60.

IVLEV, V. S. 1955. *Experimental ecology of the feeding of fishes.* Trans. by D. Scott, New Haven, Yale University Press, 1961.

JOB, S. V. 1955. The oxygen consumption of *Salvelinus fontinalis*. *Univ. Toronto Stud. biol. Ser.*, (61). *Publs Ont. Fish. Res. Lab.*, (73), 39 pp.

MANN, K. H. 1965. Energy transformations by a population of fish in the River Thames. *J. Anim. Ecol.*, **34**, 253-75.

MARTIN, N. V. 1952. A study of the lake trout, *Salvelinus namaycush*, in two Algonquin Park, Ontario, Lakes. *Trans. Am. Fish. Soc.*, **81**, 111-37.

MARTIN, N. V. 1966. The significance of food habits in the biology, exploitation, and management of Algonquin Park, Ontario, lake trout. *Trans. Am. Fish. Soc.*, **95**, 415-22.

MARTIN, N. V., and SANDERCOCK, F. K. 1967. Pyloric caeca and gill raker development in lake trout, *Salvelinus namaycush*, in Algonquin Park, Ontario. *J. Fish. Res. Bd Can.*, **24**, 965-74.

PALOHEIMO, J. E., and DICKIE, L. M. 1965. Food and growth of fishes. I. A growth curve derived from experimental data. *J. Fish. Res. Bd Can.*, **22**, 521-42.

PALOHEIMO, J. E., and DICKIE, L. M. 1966. Food and growth of fishes. II. Effects of food and temperature on the relation between metabolism and body weight. *J. Fish. Res. Bd Can.*, **23**, 869-908.

PALOHEIMO, J. E. and DICKIE, L. M. 1966. Food and growth of fishes. III. Relations among food, body size, and growth efficiency. *J. Fish. Res. Bd Can.*, **23**, 1209-48.

QADRI, S. U. 1967. Morphological comparisons of three populations of the lake char, *Cristivomer namaycush*, from Ontario and Manitoba. *J. Fish. Res. Bd Can.*, **24**, 1407-11.

RICKER, W. E. 1958. Handbook of computations for biological statistics of fish populations. *Bull. Fish. Res. Bd Can.*, (119), 300 pp.

RYDER, R. A. 1964. Chemical characteristics of Ontario lakes with reference to a method for estimating fish production. *Rep. Ont. Dep. Lands Forests Sect. (Fish)*, (48), 75 pp.

RYDER, R. A. 1965. A method for estimating the potential fish production of north-temperate lakes. *Trans. Am. Fish. Soc.*, **94**, 214-18.

SLOBODKIN, L. B. 1960. Ecological energy relationships at the population level. *Am. Nat.*, **94**, 213-36.

WINBERG, G. G. 1956. Rate of metabolism and food requirements of fishes. *Nauch. Trudy Belorussk. gos. Univ. V.I. Lenina. Minsk*, 253 pp. (Transl. *Fish. Res. Bd Can.*, No. 194).

Food specializations and bioenergetic interrelations in the fish faunas of some small Ontario waterways

ALLEN KEAST
Department of Biology
Queens University
Kingston, Ontario

ABSTRACT. The present paper reviews a series of investigations over the years 1960-68 on the freshwater fish faunas of a series of small self-contained fresh waterways in eastern Ontario. The objective is an understanding of the food and energy interrelationships of the component species with the objective of developing broader concepts that will be generally applicable.

The major findings are as follows:

1. Cohabiting species show a considerable level of food specialization. Many of the apparent food overlaps occur when a particular resource becomes superabundant. Not uncommonly, too, species taking a common item prove, on investigation, to have different habitat preferences so that they are in actuality, taking the items from different places. Sometimes fish species that are apparently taking the same food item are concentrating on larger, and smaller-sized individuals, respectively. Finally, some apparent cases of food overlap prove, on more detailed study, to involve different sub-orders, families, or genera. Accordingly, notwithstanding the finding of Hartley (1948), relative to the impoverished freshwater fauna of the Cam River in England, interspecific competition for food is only moderate in these Ontario faunas. The greatest food overlaps occur in the case of chironomid larvae and Cladocera, both of which are very important fish foods (particularly in Lake Opinicon). The assemblages studied were a blend of specialized and generalized feeders. The fish species also are somewhat differentiated in the levels in the water at which they obtain their prey.

2. Cohabiting species differ widely in their numerical abundance and biomass.

3. The different species are by no means uniformly distributed but show habitat preferences, some preferring sandy, others weedy, and others rocky areas.

4. The cohabiting species differ widely in body size and have a wide range of different body and mouth morphologies: they are structurally specialized for different roles. Whilst some general tendencies are retained predictions about ecology can be made from morphology.

5. The species have different feeding periods, some being day feeders others

377

night feeders, others having both a diurnal and a nocturnal component to their feeding.

6. Optimum temperatures for feeding and growth vary between the species: this is explicable in terms of different origins, the Great Lakes fish fauna being built up, subsequent to glaciation, from the Mississippi-Missouri drainage ("warm water" fishes) and the glacial front lakes ("cold water" fishes).

7. Year groups of species may differ considerably in their diets. They also may differ in their centres of concentration, i.e. they may show some differences in habitat preferences. The adult diet might be attained in a few months, or not for some years. The existence of these ecological differences between year classes greatly complicates the study of the bioenergetic interrelationships between species.

Individually and in combination the specializations and adaptations of the cohabiting species within fish faunas combine to permit, or favour a diversity of species living together. Interspecific food combination is restricted by them.

The studies emphasize the highly complex and sophisticated patterns of interrelationship which exist between fish species in nature. A simplified "food web", or "food chain" cannot completely express these interrelationships. In part, the natural environment is a mosaic of different combinations of species and age forms of species, interrelating with their food organisms, and each other, according to their relative abundances in different areas.

INTRODUCTION

An understanding of the bioenergetics of fish faunas requires an appreciation of the ecological specializations and interrelationships of the species, the living conditions required by each, their individual demands on the resources of the environment, diurnal and seasonal movement patterns, mechanisms for avoiding interspecific competition, and the responses of each to environmental variations and fluctuations. It is difficult to develop such comparative data for the marine environment. Isolated, or semi-isolated bodies of fresh water lend themselves more readily to analysis.

The present paper, in which studies carried out in small lakes and streams in eastern Ontario in the years 1960-68 are reviewed, is an attempt to understand how the component species of fresh water fish faunas interrelate ecologically and to bring out the complexities involved. The synthesis is based partly on published, and partly on unpublished, data. The former has appeared as follows: summer-autumn food studies of cohabiting species in a lake and a stream (Keast, 1965; 1966; 1968a); feeding dynamics at low temperatures (Keast, 1968b); diurnal feeding periodicity and daily ration studies (Keast and Welsh, 1968); the relationships between body and mouth morphology and the ecological niche (Keast and Webb, 1966).

The waterways studied lie in the Lake Ontario basin and have been, or still are, connected intermittently with the Great Lakes proper so that all have had the opportunity to build up balanced and well integrated faunas. Most of the present work has been carried out on the fauna of Lake Opinicon, a small lake with dimensions of $10 \times 1 \cdot 5$ km and a maximum depth of 10 m, which is part of the Rideau Canal system connecting the Ottawa River and Lake Ontario. This fauna numbers 17 species (Figs 1 and 2) as follows: Cyprini-

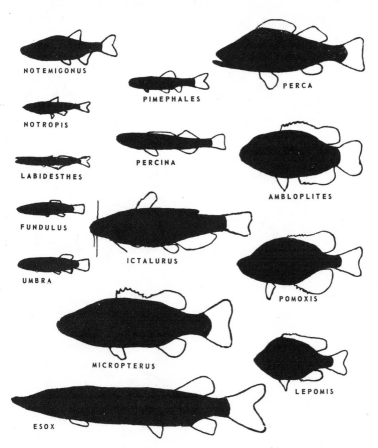

Fig. 1. Relative body sizes and shapes of adult members of the major species of fishes in Lake Opinicon, Ontario. The species vary in total length from 65 mm (*Notropis heterodon*) to over 600 mm (*Esox lucius*). Although diversity of body form is fairly great there are few of the more extreme types found in the African rift lake or Amazonian faunas. This applies to the Great Lakes fish fauna as a whole.

dae: *Notemigonus crysoleucas* (golden shiner), *Notropis heterodon* (blackchin shiner), *Pimephales notatus* (bluntnose minnow); Ictaluridae: *Ictalurus nebulosus* (brown bullhead) and *I. natalis* (yellow bullhead); Umbridae: *Umbra limi* (mud minnow); Esocidae: *Esox lucius* (northern pike); Cyprino-dontidae: *Fundulus diaphanus* (banded killifish); Percidae: *Perca flavescens* (yellow perch) and *Percina caprodes* (log perch); Centrarchidae: *Lepomis macrochirus*, (bluegill sunfish), *L. gibbosus* (pumpkinseed sunfish), *Amblop-lites rupestris* (rock bass), *Pomoxis nigromaculatus* (black crappie), *Microp-terus salmoides* (largemouth bass); Atherinidae: *Labidesthes sicculus* (brook silverside); Clupeidae: *Pomolobus pseudoharengus* (alewife). Of these the yellow-bullhead, log perch, and alewife are rare. The fauna is fairly typical of small lakes of equivalent depth and temperature range in southern Ontario and northern New York, although richer than many.

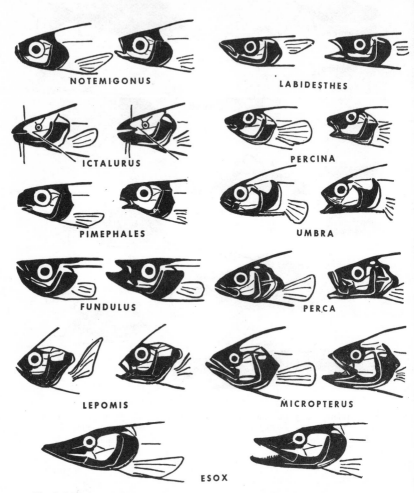

Fig. 2. Mouth morphology (closed on left, open on right) of the major genera of Lake Opinion fishes. The species differ greatly in relative mouth size as well as in mouth shape and position (vide terminal, dorso-terminal, or ventro-terminal). Note barbels but small eyes of the chemo-sensory feeder *Ictalurus*.

The paper discusses the following facets:
1. Foods and feeding interrelationships of cohabiting fish species.
2. The relationship between the relative abundance of major food organisms in the environment and occurrence in the diet of fish.
3. Ecological significance of numerical and biomass differences between cohabiting fish species.
4. Trophic interrelationships of the different age and size classes of species.

5. Body and mouth morphology relative to food and feeding.
6. Distributional patterns within a lake: significance of different habitat preferences in limiting interspecific interactions.
7. Different feeding periodicities of species and their ecological significance.
8. Food uptake rates and the daily ration.
9. Seasonal differences in food uptake rates, a comparison of summer and winter.
10. Experimental work on food uptake rates at various temperatures.
11. Energy requirements for growth at different temperatures, experimental work on the bullhead, *Ictalurus nebulosus*.

FOODS AND FEEDING INTERRELATIONSHIPS OF COHABITING FISH SPECIES

Despite 50 years of research on the foods of fishes few comparative studies exist on the feeding interrelationships of cohabiting species and of dietary relationships within species at different stages of growth. Certain work has suggested a much greater food overlap between cohabiting fish species than in birds or mammals—note the study of Hartley (1948) on the Cam River fauna in England, and the review of Larkin (1956). This possibility needs fuller investigation since it would seem to violate accepted concepts that cohabiting animal species show a high level of "ecological exclusion".

The Birch (Curtis) Bay, Lake Opinicon, assemblage of fishes was chosen for close study as the bay maintains a large and diverse fauna from May to October. The bay is a shallow indentation, about 100 m across and with gradually sloping bottom of sand and mud.

Standardized procedures were followed in the netting of fish and potential food invertebrates in the environment so that the various collections would be comparable. In the former a quarter-inch nylon seine, 2 m deep and 18 m long was used, sweeps of 70 m being made. For bottom samples a square push net, 0·5 m wide, designed to remove roughly the top quarter inch of sediments, was pushed over measured distances of 33 m, the contents being lifted into a bucket every 7 m. The circular 15 cm wide plankton net was towed through the water on a long rope for distances of 33 m. Sampling was carried out between 18th and 25th of each month on four different days, on each day netting being carried out in the early morning, late afternoon, and after dark, to adequately cover the feeding periods of the fish species (see later). The fish were divided into size classes that, in most cases, corresponded to year classes. At any one time only 5-10 fish of each size class of each species was retained to avoid imperilling the status of the less common species. In the initial study (Keast, 1965) about 3000 stomachs were examined: this number has subsequently been extended to 8000.

Identification of the food organisms in the stomachs was routinely taken down to the Order level, to Families in the case of Diptera larvae, and to genera with Cladocera. Both numerical counts and volumetric determinations were made. Only the latter, which are the more important in that they bring out the relative dietary importance of the different organisms, are considered here. These are expressed as "pie diagrams", the size of the segment representing the volumetric proportion of that class of organism consumed during the month.

The purpose of the bottom and plankton samples was to define fluctuations in seasonal abundance of the major food types in the environment and to interpret monthly changes in the diets of fish, and cases of major food overlaps. Numerical counts were used, those for the bottom organisms being calculated on the basis of numbers in a 0.5×3.3 m strip of substrate, and for the plankton organisms in a 15×15 cm column of water 33 m long (Fig. 6). Each is based on 12 day collections. The counts of invertebrates in the fish stomachs represent the totals of all fish examined. Data on the proportion of large-bodied to small-bodied invertebrates was obtained by washing the samples through screens of 5 mm, 2 mm and 1 mm diameter, and counting the numbers of individuals of each Order filtered off by each. The filtrate was analysed by the sub-sampling method of Allanson and Kerrich (1961). For simplicity of expression the invertebrates are grouped into individuals greater than, and less than, 1 mm in diameter in Table 1.

TABLE 1. Invertebrate food organisms in substratum, Birch Bay, Lake Opinicon, as percentage occurrence of newly-hatched or small individuals that passed through a 1-mm square mesh. Based on counts in Keast (1965)

	May %	June %	July %	August %	September %
Amphipoda	20	80	10	75	22
Isopoda	20	20	50	66	0
Chironomid L.	95	73	45	74	84
Chironomid P.	66	64	73	52	0
Ephemeroptera N.	25	14	0	38	0
Trichoptera L.	59	15	14	69	46
Gastropoda	83	61	28	50	0
Lamellibranchia	0	64	60	90	0

The following findings emerged from the Birch Bay food study:

1. The nine species inhabiting the bay differed fundamentally in their diets and there was obviously a marked degree of "ecological exclusion" (Figs 3-5). Major foods were *Notropis heterodon*—Cladocera and small adult Diptera (picked up at the surface); *Pimephales notatus*—Cladocera, organic detritus, and chironomid larvae; *Fundulus diaphanus*—a diverse range of small-bodied aquatic organisms; *Perca flavescens*—Odonata and Ephemeroptera nymphs, small crayfish and fish; *Pomoxis nigromaculatus*—*Chaoborus* larvae, chironomid larvae, and Cladocera; *Lepomis macrochirus*—very wide range of small to medium-sized invertebrates; *Labidesthes sicculus*—*Chaoborus* larvae, Cladocera, and small adult Diptera (caught at the surface); *Micropterus salmoides* young—small aquatic invertebrates; and adult—fish and crayfish; *Ambloplites rupestris* 1st year—chironomids and other small invertebrates; 2nd year—larger hardbodied types such as Odonata and Ephemeroptera nymphs, and Trichoptera larvae; fish of 3 years and older—crayfish.

2. There were marked dietary overlaps between species in some months, thus chironomid larvae were eaten by four to five species. When identifica-

tions were extended below the Order level, however, such "competitive over-laps" proved to be much less than appeared at first sight. Thus, the Odonata nymphs taken by *Perca flavescens* were largely Zygoptera, those eaten by *Ambloplites rupestris*, Anisoptera. The chironomid larvae eaten by the filter-feeding *Pomoxis nigromaculatus* were almost entirely of the free-swimming predatory genus *Procladius*. No other fish species took these in quantity. *Ictalurus nebulosus* took only tube-dwelling genera. Small-bodied fish like *Pimephales notatus* and *Fundulus diaphanus* took only small-bodied genera and/or newly hatched individuals, whilst larger-bodied fish contained mainly medium-sized and larger individuals. This dietary separation between small and large fishes also occurred with Amphipoda, Decapoda, Mollusca, Ephe-meroptera nymphs, Odonata nymphs, and Trichoptera larvae.

3. All fish species showed dietary changes from month to month. In more specialized feeders this was limited to shifts in proportions of different kinds of foods. In the generalized feeders, items prominent one month were some-times insignificant, or absent, the next.

4. Species separated out into specialized feeders, which took only two or three food types, and plastic or generalized feeders which took many. *Lepomis macrochirus* is certainly the most opportunistic feeder in the lake. Ten or eleven different food types occurred in excess of the 3% volume level in some months.

5. Species were divisible into predominantly bottom or bottom and vege-tation feeders, e.g. *Pimephales notatus* and *Perca flavescens*, species with a marked surface component to the feeding (*Notropis heterodon* and *Labidesthes sicculus*), and a mid-water feeder (*Pomoxis nigromaculatus*). The sunfish, *Lepomis macrochirus*, fed at all levels.

THE RELATIONSHIP BETWEEN THE RELATIVE ABUNDANCE OF MAJOR FOOD ORGANISMS AND OCCURRENCE IN THE DIET OF FISH

Monthly shifts in the diets of the various species of fish are considered against shifts in relative abundance of the major food organisms in Fig. 6.

The following conclusions may be drawn:

1. The numerical peaks in the environment and in the stomachs of the fish broadly correspond in the case of Cladocera, Ostracoda, chironomid larva and pupae, Trichoptera larvae, and Mollusca. There was no correlation in the case of Copepoda, which are of little importance as fish food.

2. Numbers of Ephemeroptera and Odonata nymphs in the environment appear on the basis of the bottom counts to be relatively constant. Fish consumption increased in June and September in the case of the former, and June in the case of the latter. This suggests that the sampling technique was not adequate for catching these active organisms (Keast, 1965).

3. There was a lag in the fish response to the July build-up in Cladocera numbers. Fish consumption did not increase markedly until August, then remained high in September, i.e. after the environmental counts fell. The

fish were also slow to respond to increased numbers of Trichoptera larvae (August), feeding peaking in September, by which time the numbers in the environment started to fall.

4. The utilization of Cladocera as food by several fish species in August and September correspond with the period of "super-abundance" of these organisms. Amphipoda only figured prominently in the diets of fish in September, their peak month in terms of numbers. The three species of fish taking them were presumably not competing for them in the narrow sense, the resource being superabundant.

5. Chironomid larvae, the most important food in the bay (note number of fish species that consumed them) made up a constant volume of the diet of several species throughout the summer. Environmental counts (Fig. 6) were distinctly lower in July than in other months. Table 1 shows, however, that large-bodied larvae (i.e. with a diameter exceeding 1 mm) then predominated, making up 55% of the population, compared to only 5% to 26% during the other months. This is why, in terms of per cent volumes (see Figs 3-5), the amounts of chironomids consumed is no lower than during the other months.

6. Small Ephemeroptera nymphs (i.e. of less than 1 mm) reached their highest proportions in May and August (Table 1). The fish, however, consumed the largest numbers of these organisms in July and September (Fig. 6). Thus they were only utilized in numbers after they had grown to a certain minimum size. This is reflected in the per cent volumes consumed by fish of different sizes (Figs 3-5). Thus, only the 45-70 mm length group of *Ambloplites rupestris* took significant volumes of Ephemeroptera nymphs in May whereas this item made up important segments of the diets of four species in June, when they were larger. Again, there was a more widespread consumption in September when larger-bodied larvae predominated, four species of fish then taking them compared to one in August.

7. Trichoptera larvae were predominantly large-bodied in June, July and September, small-bodied in May and August. Fish took the greatest numbers in June and September, again when the hatches had reached a minimum size. Trichoptera larvae, however, are not important food organisms to the fishes of Birch Bay.

8. In the case of Mollusca small bodied individuals predominated in August, 50% of Gastropoda and 90% of Lamellibranchiata being under 1 mm in diameter. Numbers in the fish stomachs at this time peaked. The only fish species that consumed molluscs regularly, however, was *Lepomis macrochirus*.

NUMERICAL AND BIOMASS DIFFERENCES BETWEEN FISH SPECIES

Relative numerical abundance and biomass of fish species, and their size classes, are basic in assessing the "impact" that each is making on the food resources of the environment. Data of this kind for the various species and their size classes in Birch Bay are given in Keast (1965). Here data are contrasted for two typical months only, June and August (Table 2).

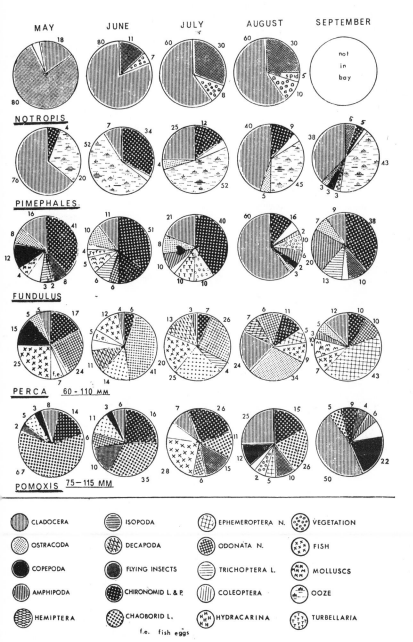

Fig. 3. Fishes of Lake Opinicon—monthly changes in diet expressed as per cent volumes: *Notropis heterodon*, *Pimephales notatus* (Cyprinidae), *Fundulus diaphanus* (Cyprinodontidae), *Perca flavescens* (Percidae), and *Pomoxis nigromaculatus* (Centrarchidae). Based on 250-350 fish of each species.

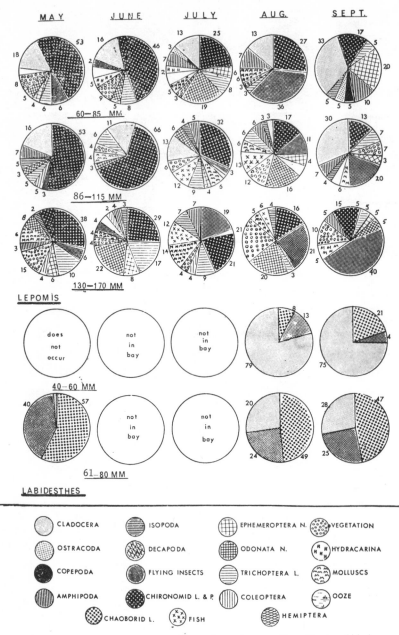

Fig. 4. Foods of fishes of Lake Opinicon: *Lepomis macrochirus* (Centrarchidae) and *Labidesthes sicculus* (Atherinidae). Based on 580 and 250 fish, respectively.

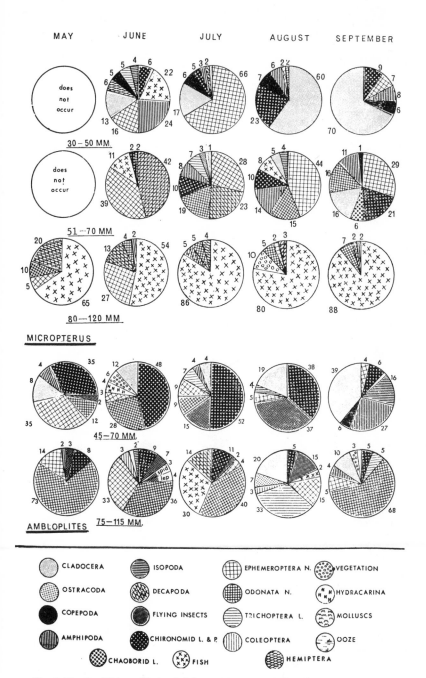

Fig. 5. Foods of fishes of Lake Opinicon: *Micropterus salmoides* and *Ambloplites rupestris* (Centrarchidae). Based on 400 and 380 fish, respectively.

TABLE 2.   Relative abundance and biomass, fishes of Birch Bay, June and August contrasted. Averaged at numbers per 1250 m² of seine netting. Biomass calculated on basis of mean weights of 10 fish in the middle of the size range

Species and size category		Weight (g)	June		August	
			No.	Biomass	No.	Biomass
Lepomis	60- 85 mm	4·3-8·0 (6·3)	34	211	86	522
macrochirus	85-115 mm	8·8-15·7 (11·7)	129	1548	111	1332
	115-135 mm	25-34 (30)	39	1170	32	960
	135-170 mm	69-99 (82)	64	6400	27	2025
	Total		266	9329	256	4838
Ambloplites	45-70 mm	2·9-6·0 (4·8)	6·0	28·8	32	154
rupestris	75-115 mm	10·8-18·3 (13·3)	8·4	118	7·2	96
	Total		14·8	147	39·2	250
Micropterus	30- 50 mm	0·8-1·4 (1·1)	19·2	21	62	68
salmoides	50- 70 mm	1·8-3·1 (2·5)	6·8	17	36	90
	80-120 mm	9·5-16·0 (12.7)	6·8	86	9·2	117
	Total		32·2	124	107	275
Pomoxis nigromaculatus	75-115 mm	6·8-13·9 (9·0)	7·2	65	9·8	86
Fundulus diaphanus*	40- 60 mm	1·2-2·4 (1·9)	79	166	39	74
Perca flavescens	60-110 mm	6·5-12·1 (9·7)	12	116	18	175
Pimephales notatus*	50- 75 mm	1·6-3·5 (2·8)	23	57	41	103
Notropis heterodon*	40- 70 mm	4·3-2·2 (1·9)	24	32	10	13
Labidesthes sicculus*	60- 80 mm	1·5-2·1 (1·8)	nil	nil	58	104

* Figures for *Fundulus*, *Pimephales*, *Notropis*, and *Labidesthes* in August do not include first year-class, which escape net.

The sunfish, *Lepomis macrochirus*, proved to be by far the most abundant species, with a count per 900 m² of 266 and a biomass of 9329 g in June, and a count of 256 (4836 g) in August. This greatly exceeded that of the other species—compare with *Fundulus diaphanus*, in which the counts were 79 and 39 for the two months, and biomasses 166 and 74 g; *Ambloplites rupestris* 15 and 39, 147 and 250 g; and small *Micropterus salmoides*, 32 and 107, 124 and 275 g. Taking the bay fauna as a whole the biomass of *Lepomis macrochirus* is about 20 times that of the other species: thus it is the most versatile and opportunistic feeder that is the most abundant. The specialized feeders, *Pimephales notatus*, *Notropis heterodon*, and *Labidesthes sicculus*, have only relatively small biomasses. The bay is only a marginal habitat for *Ambloplites rupestris*, *Pomoxis nigromaculatus*, and *Perca flavescens*; hence the figures here are not a true indication of their abundance in the lake as a whole.

A parallel analysis on a stream fauna (Keast, 1966) also showed a marked difference in the numbers and biomasses of the cohabiting species, counts differing by 15-30 times, and biomasses by 20 times. Again the more numerous species were versatile feeders: the cyprinid *Semotilus atromaculatus*,

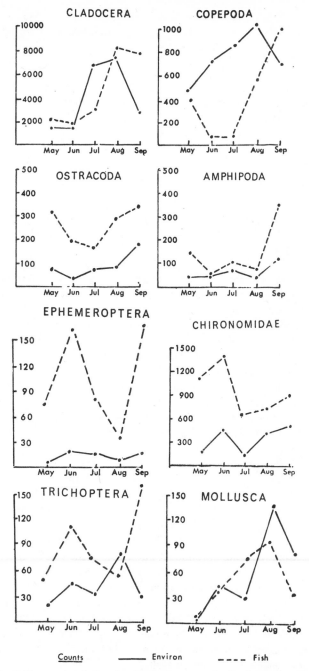

Fig. 6. Changes in relative abundance of seven major food organisms in the environment and in the stomachs of fishes, Birch Bay, Lake Opinicon.

stickleback *Eucalia inconstans* (Gasterosteidae), and sucker *Catostomus commersonni* (Catostomidae). The least common species were the organic detritus feeder *Pimephales promelas* (Cyprinidae) and the piscivore *Esox vermiculatus americanus* (Esocidae).

TROPHIC INTERRELATIONSHIPS OF THE DIFFERENT AGE AND SIZE CLASSES OF SPECIES

Whilst it has long been known that, during their growth, piscivorous fishes like *Micropterus salmoides* and *Esox lucius* go through periods of plankton feeding, then insect eating before becoming large enough to assume a fish-eating role (e.g. Hunt and Carbine, 1950), little comparative work has been done on dietary changes with age in fish. The defining of these changes and the duration of each of the phases are obviously important aspects of the bioenergetics of fish populations, especially since (see below) year classes may occupy different "ecological niches" from each other.

Examples of different kinds of dietary change with increasing body size are given in Figs 4, 5 and 7. The first three are from the Birch Bay assemblage but data on *Semotilus atromaculatus* (Cyprinidae) is drawn from the Jones Creek, Ontario, stream fauna.

1. *Lepomis macrochirus.* Fish of 60-85 mm total length, representing fish in their second year (i.e. ones hatched the previous June) differ in their diet from fish of 130-170 mm (age 3-4 years) only in changes in the relative volumes of different items taken, the elimination of Cladocera from the diet, and the occurrence of fish fry in the larger fish.

2. *Labidesthes sicculus.* Here, young of the year hatched in July, differ from adults (the species probably only lives 1-2 years) only in the greater proportion of planktonic Cladocera and less *Chaoborus*, larvae and adult Diptera consumed.

3. *Micropterus salmoides.* As this species goes through major dietary changes in the first few months of life (see above), first year fish were arbitrarily divided into larger and smaller individuals to bring out any differences in feeding with size. The young hatch in mid-June. The two groups shown (Fig. 5) represent earlier and later hatches, and rapid and slower growing individuals of the same hatch. A difference in the food organisms eaten is apparent. Cladocera were prominent in the diet of the smaller individuals as late as September. Larger individuals utilized these but little, eating mostly small Odonata and Ephemeroptera nymphs, and crayfish. Fish eating began in the first autumn. By late the following May, at body length of 80 mm, the species was largely piscivorous. Initially, small-bodied fishes like *Labidesthes sicculus, Pimephales notatus, Notemigonus crysoleucas,* and the smaller *Micropterus salmoides*, were eaten. Fish 2-5 years old consume largely sunfish, *Lepomis macrochirus*, and crayfish.

The above indicates that the first *Micropterus salmoides* to hatch and the more vigorously growing individuals, acquire the advantage of being able to utilize larger-bodied prey items, acquiring increased metabolic efficiency

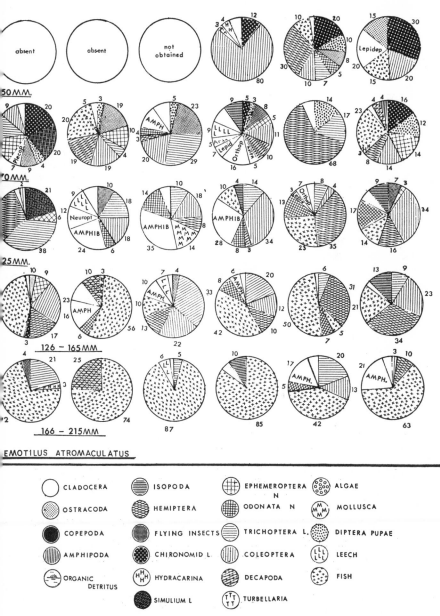

MAY   JUNE   JULY   AUGUST   SEPTEMBER   OCTOBER

EMOTILUS ATROMACULATUS

	CLADOCERA		ISOPODA		EPHEMEROPTERA N		ALGAE
	OSTRACODA		HEMIPTERA		ODONATA N		MOLLUSCA
	COPEPODA		FLYING INSECTS		TRICHOPTERA L.		DIPTERA PUPAE
	AMPHIPODA		CHIRONOMID L.		COLEOPTERA		LEECH
	ORGANIC DETRITUS		HYDRACARINA		DECAPODA		FISH
			SIMULIUM L		TURBELLARIA		

Fig. 7. Food of *Semotilus atromaculatus* (Cyprinidae), Upper Jones Creek, Ontario. The species is interesting because of the protracted nature of the dietary transition with age, the adult diet not being reached until about the 4th year. Study based on 830 fish. The symbol 'v' shown as 15% and 23% in the October series of the two smallest size classes stands for plant fragments.

thereby. There is thus selection for rapid growth. Because they concentrate on different items, the larger and smaller individuals are not in competition for food.

A rapid transition to fish-feeding is characteristic of all the more specialized piscivores. In *Esox lucius* the Cladocera-eating phase extends to a body length of 25 mm, the insect-eating phase from 25-50 mm, after which the species is piscivorous (Hunt and Carbine, 1950). Young yellow pickerel, *Stizostedion vitreum* (Percideae), raised in hatcheries, became piscivorous at 30-35 mm eating their fellows (Cheshire and Steele, 1962).

4. *Ambloplites rupestris*. Clearcut shifts in diet characterize the first three years. Fish in their second summer (length 45-70 mm, age about 11-15 months) consume mostly chironomid larvae, Amphipoda, Isopoda, Cladocera, and small flying insects; those in their third summer (length 75-115 mm, age 23-28 months) mostly Ephemeroptera and Odonata nymphs, Trichoptera larvae, some small fish and Decapoda (Fig. 5). From year III onwards the diet is very largely Decapoda.

5. *Semotilus atromaculatus*. The situation here contrasts with that in *Ambloplites rupestris* in that dietary changes with growth are gradual, and protracted. Whilst new items appear from time to time, successive year classes differ mainly in the different proportions of the same basic items consumed. It takes 4 to 5 years to become a true piscivore (Fig. 7).

The above examples emphasize the complexities involved in trying to assess the impact of a fish species on the resources of its environment. Whilst some fishes, particularly the small-bodied ones, utilize only a single food level or "niche", larger-bodied species may change their diets so much as they grow that they could be said to be simultaneously occupying two or more "food niches". In other words, in some cases, age classes may differ from each other ecologically as much as do cohabiting species. One important effect of this is to reduce intraspecific competition; but greater metabolic efficiency stemming from the capacity to utilize ever larger particle sizes is undoubtedly important. Dietary changes with growth may, as noted, be largely completed within the first year, or take several years to complete. The transition may involve a series of fairly basic shifts or, mainly, changes in the proportions of a basic series of organisms.

BODY AND MOUTH MORPHOLOGY RELATIVE TO FOOD AND FEEDING

The relationship of body morphology to ecological specialization and its significance in delimiting the ecological niche, and in restricting interspecific competition has been the subject of a detailed review by Keast and Webb (1966).

The 17 Lake Opinicon fish species range in body length, in adulthood, from 65 mm in the minnow *Notropis heterodon* to over 600 mm in the piscivore (*Esox lucius*), and they show a wide range of different body and mouth morphologies (Figs 1 and 2). A selection of the more important body types and their functional significance areas follows.

Notemigonus crysoleucas and *Notropis heterodon*. These minnows have compressed fusiform bodies, long and slender caudal peduncles, deeply-forked caudal fins, short-based high and mobile dorsal fins and subfalcate pectorals. Aquarium observations show this body form to be correlated with speed and manoevrability, the individuals of the school twisting and turning with facility and rapidity as they snap up individual plankters. The dorso-terminal mouth favours mid-water and surface feeding.

Labidesthes sicculus. The almost straight dorsal body line, small dorsal fin, and the dorso-terminal, beaklike mouth of this species are clearly linked with its habit of swimming immediately beneath the surface film of the water, from which position it snaps up flying insects as they alight on, or hover immediately over, the water. *Labidesthes* may leap into the air in securing prey. In compensation for the small dorsal fin the anal "keel" is longbased and markedly enlarged. Additional stability for the elongate tube-shaped body is gained from the enlarged and high-placed pectoral fins, which beat with a long sweeping motion.

Umbra limi. The mud minnow is an inhabitant of shallow, weedy ponds (swampy margins in Lake Opinicon). The foreshortened tubular body (definition in Keast and Webb, 1966), and rounded cudal fin, well adapted for pushing through vegetation, nevertheless considerably curtail manoevrability and speed. *Umbra* either rests on the bottom or "hangs" in mid-water, maintaining its position by gentle undulating movement of the permanently extended pectoral fins. When it moves forward it does so slowly and with a minimum of turbulence. Resting prey is taken from the weeds or bottom, the fish manoevring to within 1-2 cm and securing it by a quick strike. The terminal mouth is relatively wide (8 % of the standard length compared to 5 % in *Notemigonus crysoleucas* and 6 % in *Notropis heterodon*), scoop-like, and relatively strong (Fig. 2), adaptations that fit the diet of somewhat bulky and thick-skinned invertebrates (Keast, 1966).

Lepomis macrochirus and *Pomoxis nigromaculatus*. The sunfish and crappie have high, narrow, gibbose bodies, a form that is very stable since the large lateral area limits rolling. Pitching rotations are controlled by the large pectoral and pelvic fins. The gibbose body type is duplicated in various sub-orders of fishes and is characteristic of still water conditions. *Lepomis macrochirus* does not cruise in search of prey but hangs in loose aggregations in mid-water, rushing food organisms as they come into view. It is a very generalized feeder, obtaining food at all levels (Fig. 4). Much of the diet of *Pomoxis nigromaculatus*, by contrast, is obtained by means of the gill raker screen which has 25-29 rays on the first arch compared to 10-12 rays in the other centrarchids (Keast, 1968a).

Ictalurus nebulosus. Unlike the other species the bullhead is a chemo-sensory, not a visual, feeder (note barbels and reduced eyes): structural adaptations that enable it to feed at night (it is purely nocturnal in Lake Opinicon) and under muddy water conditions. The head is dorso-ventrally compressed, the mouth broad, and the hind part of the body laterally compressed: fitting it for cruising over the bottom at a shallow angle with the barbels in contact with the substrate.

Esox lucius and *Micropterus salmoides*. These piscivores are large-sized, have very large mouths, formidable dentition of enlarged and backwardly-directed teeth, and strong jaws. The body is adapted for prolonged cruising. The pike is obviously the better specialized not only in body shape but in the posterior position of the stiffened dorsal and anal fins which can thus supplement the thrust of the caudal fin in quick striking movements.

It is apparent that the above structural specializations adapt their owners for specific ecological roles, giving them marked advantages over species that do not have them. Body morphology, accordingly, enables predictions to be made about the way of life and food ecology of fish species. A moderate level of flexibility is, however, retained by Great Lakes fishes, presumably desirable in a cold temperate environment. Structural specializations, clearcut as they are, do not approach the extremes found in coral reef fishes and in those of the African rift lakes—note, for example, the bizarre tooth types of certain Lake Malawi fishes (Fryer, 1959).

DISTRIBUTIONAL PATTERNS WITHIN A LAKE: SIGNIFICANCE OF HABITAT AND FEEDING ZONE DIFFERENCES BETWEEN SPECIES

Fish species are commonly specialized in their habitat requirements. The Great Lakes fauna, for example, is composed of still-water and riffle, shallow water and pelagic, lake and stream fishes (Hubbs and Lagler, 1958). Differences in bathymetric distribution occur even in small lakes, as can be demonstrated by setting gill nets at different depths (e.g. Hile and Juday, 1941).

Initial observations in Lake Opinicon indicated a degree of grouping of species relative to different bottom types and vegetation zones. To determine the extent of this and assess its significance studies were initiated using plot counts, selective netting of different areas, and observations by face mask and snorkel. The counting plots, 2×2 metres in area, and marked by a yellow

TABLE 3. Distribution of fish by day relative to bottom type, Lake Opinicon, Ontario. Based on 10 000 counts of 4 m² plots set in water of 1-2 m depth, May-September, 1962-1965, expressed as numbers of fish of each species per 1000 m²

	Sandy bottom	Rocky bottom	Weedy bottom	Mixed bottom
Lepomis macrochirus	291·7	263·3	389·0	314·2
L. gibbosus	35·2	18·1	50·7	22·4
Ambloplites rupestris	5·8	20·7	38·2*	8·1*
Micropterus salmoides	6·8	3·0	15·8*	17·7*
Pomoxis nigromaculatus	0·1	0·2	0·4	0·3
Perca flavescens	0·3	1·8	0·6	0·8
Percina caprodes	—	23·3†	—	—
Fundulus diaphanus	208·4	—	91·8	95·0
Notropis heterodon	20·0	—	50·8	13·3
Pimephales notatus	133·3	—	74·3	41·0
Notemigonus crysoleucas	—	0·3	2·0	1·0
Labidesthes sicculus	5·0	0·5	2·3	3·5
Ictalurus nebulosus	—	—	2·3	2·0

* Partly year-class 1.
† Areas of small stones and pebbles only.

stone in each corner were laid out in water 0·8 to 2 metres deep along about a mile of foreshore in areas of five different bottom type: sandy, rocky, pebbly, weedy, and "mixed" (i.e. areas of accumulated leaves, sticks and weeds over sand or mud). Where possible 3-4 plots were laid out in close proximity to facilitate observations.

Counts were made from canoes, the method being to approach an area slowly, wait a few minutes, then make three counts of the fish in each plot at one minute intervals. The work was carried out from May to September in the years 1962-65, 10 000 counts being made. The alternative method of determining habitat preferences, by selectively netting areas of different bottom type, could be carried out only where the bottom was relatively homogeneous. The results duplicate those obtained by the plot counts. The netting method, of course, has the advantage in that it can be used by night as well as day.

Only the results of the plot counts are given here (Table 3), those for the various months being grouped. They are expressed as numbers of fish per 1000 square metres. The various species showed habitat preferences as follows:

Lepomis macrochirus and *L. gibbosus*—equally distributed over all bottom types;

Ambloplites rupestris—larger fish concentrated in weedy and rocky areas, smaller fish in weedy and mixed bottom but not rocky areas, i.e. where there was good cover;

Micropterus salmoides—larger fish free ranging, first year fish largely confined to areas of cover in the shallows;

Percina caprodes—totally confined to a few areas of pebbly bottom;

Fundulus diaphanus—mainly in sandy shallows where small areas of waterlogged leaves, sticks and weed provided a mosaic affect, some tendency to occur in weedy bays in late summer;

Notropis heterodon—mainly in weedy areas where the water was 0·5-2·0 m deep, sometimes over sand; and

Pimephales notatus—mainly concentrated over areas of hard-packed fine grain sand, a sand-mud bottom, or "mixed bottom" areas.

Selective netting and scuba observations subsequently revealed further data on habitat preferences as follows:

Umbra limi—strictly confined to areas of reed and marsh along the lake foreshore;

Ictalurus nebulosus—(nocturnal feeder) main concentrations in areas of soft oozy mud and sunken logs where the water is 2-5 m deep, dispersing more widely to feed at night;

Pomoxis nigromaculatus—(largely a nocturnal feeder) concentrates by day along shady areas of shoreline where ribbon grass, *Valisinaria americana*, extends through the water column, dispersing to feed at night;

Perca flavescens—concentrates mainly in the open water (2-5 m deep) in early summer and in weedy shallows in late summer though thinly dispersed along the foreshore at all times;

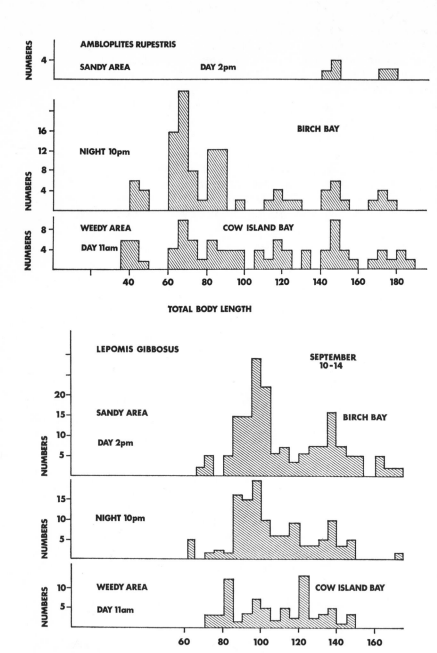

Fig. 8. Diurnal differences in the distribution of *Ambloplites rupestris* and *Lepomis gibbosus*, mid-September, 1968. The former species occurs in good numbers in the weedy bay (Cow Island Bay) by day, some remaining there at night. It is virtually absent from sandy and relatively open Birch Bay by day which becomes a feeding ground after dark. *Lepomis gibbosus* does now show any such diurnal movement pattern. Numbers of individuals are shown on the vertical axis body lengths on the horizontal.

Notemigonos crysoleucas—some preference for deeper weedy areas but schools widely ranging;

Labidesthes sicculus—close inshore in May, August, and September (especially at night), pelagic in mid-summer.

The above summary brings out that whilst some species may be found over areas of a particular bottom type throughout the summer others change their place of concentration seasonally. Again, especially in the case of nocturnal feeders, a particular zone or place may be utilized as a resting place by day, another for feeding at night. This is clearly brought out in the case of *Ambloplites rupestris* (Fig. 8). Comparative collections were made by day in weedy Cow Island Bay, and by day and night in sandy Birch Bay (September 10-15). It will be seen that whilst there is a good concentration (involving all

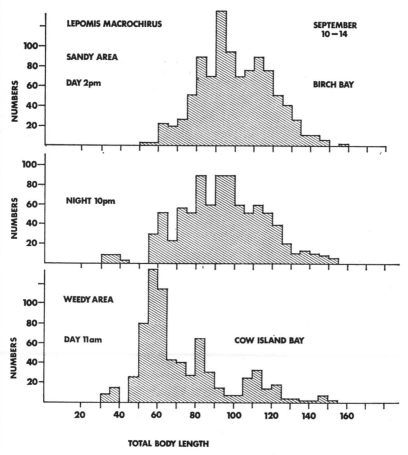

Fig. 9. Different distributions of large and small *Lepomis macrochirus* according to habitat. The first year fish concentrate in weedy areas like Cow Island Bay where there is abundant cover. The larger fish are not so restricted and are widely distributed, occurring in large numbers in open situations including Birch Bay. Mid-September 1968.

size classes) in the weedy bay by day (and to a reduced extent by night) the more exposed sandy bay is entered by numbers only after dark. The movement is a feeding one and involves both big and small individuals. Comparative data for *Lepomis gibbosus* is also shown on the graph. This species does not show any such movement pattern.

The small and large individuals of several species of Lake Opinicon fishes differ in their habitat preferences. Thus first year *Lepomis macrochirus* are only found close to, or within, weedy cover whilst adults occur at random through all the shallow water areas and form loose aggregations in clear water areas well away from the shore. These distributional differences were apparent in day collections made in Cow Island and Birch Bays in September (Fig. 9). The population structure, is quite different in the two bays. Fish less than 65 mm in total length then accounted for over 50% of the population in the weedy bay by day, but only 2% in the sandy bay.

Several conclusions may be drawn from the above habitat studies. It is obvious that different habitat preferences and distributional differences must serve to materially reduce interspecific and, to some extent, intraspecific interactions and competition. Two species taking chironomid larvae, for example,

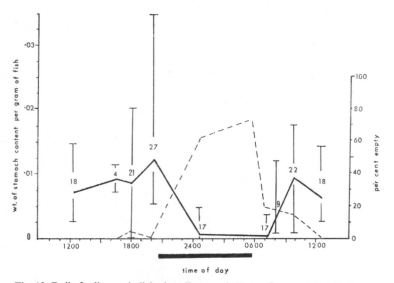

Fig. 10. Daily feeding periodicity in yellow perch (*Perca flavescens*, Percidae). Solid line, mean weights of stomach contents per g of fish; broken line, percentage of stomachs empty. Vertical lines above means, highest weights of stomach contents per g of fish; below means (where none of the stomachs was empty), lowest. Numerals in parenthesis are numbers of fish in samples (total 148). Solid bar shows hours of darkness. Data obtained over four 24-hour periods June 1, 2, 15, and 16, at water temperatures of 18-22° C. Body lengths of fish 90-130 mm, and body weights 7·8-26·5 g. The late afternoon and morning feeding periods in this species are, it might be noted, reflected in its diurnal movement patterns (Haslar and Bardach, 1949).

but living in essentially different areas are not in competition for them. The fact that fish species utilize a water body as a mosaic of different living areas greatly complicates the task of trying to construct meaningful food chains.

DIFFERENT FEEDING PERIODICITIES OF SPECIES AND THEIR ECOLOGICAL SIGNIFICANCE

Fish species may differ widely in their daily patterns of feeding activity (e.g. compare results of Spencer, 1929; Spoor and Schloemer, 1938; Darnell and Meierotto, 1962), but these patterns have been defined for only a few species.

In a series of monthly studies the feeding periodicities of several of the common Lake Opinicon fishes have been defined, the major objective being to determine to what extent different feeding periodicities influence interspecific contacts and determine the kinds of food eaten. The method used is to net series of 20-30 individuals of the commonest size class of each species

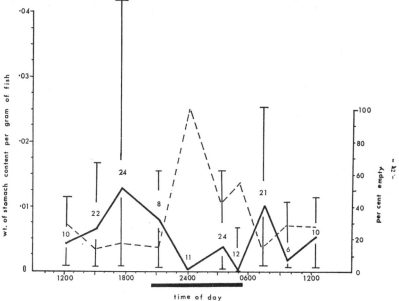

Fig. 11. Daily feeding periodicity in pumpkinseed (*Lepomis gibbosus*, Centrarchidae). Details as in Fig. 10. Number of fish, 148. Body lengths of fish, 100-140 mm, and weights 18·0-62·0 g.

every 2-3 hours over a series of 24-hour periods and to define changes in the mean weights of stomach contents. These are expressed as "weight of stomach contents per g of body weight" to equate fish of different sizes. They are plotted graphically along with the curve of "individual with the fullest stomach" in each series, and "percentage of fish with empty stomachs" in each series.

Figs 10 and 11 detail the results for *Perca flavescens* and *Lepomis gibbosus* for June, that is during the period of accelerated early summer feeding. Water temperatures at the time were 18-22°C and hours of darkness 9-9¼

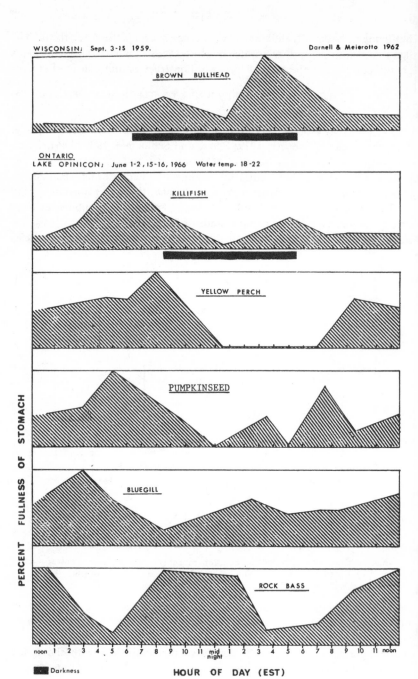

WISCONSIN; Sept. 3-15 1959.   Darnell & Meierotto 1962

BROWN   BULLHEAD

ONTARIO
LAKE OPINICON; June 1-2, 15-16, 1966   Water temp. 18-22

KILLIFISH

YELLOW PERCH

PUMPKINSEED

BLUEGILL

ROCK BASS

PERCENT FULLNESS OF STOMACH

noon 1 2 3 4 5 6 7 8 9 10 11 mid 1 2 3 4 5 6 7 8 9 10 11 noon
                                    night

Darkness

HOUR OF DAY (EST)

Fig. 12. Daily feeding periodicities of five cohabiting species compared. Data on
Ictalurus nebulosus is from Darnell and Meierotto (1962).

hours. In the former the mean weights of stomach contents proved to be greatest at dusk (8.00 p.m.) and mid-morning (10.00 a.m.) and in *Lepomis gibbosus* at about 5.30 p.m., 3.00 a.m., and 7.30 a.m. There are thus two periods of maximum stomach fullness in the former, and two major and one minor in the latter. The feeding periods are those periods when the stomachs are filling rapidly, that is the steeply upwards sloping lines prior to the peak. They are of about 2-3 hours duration. In both species the curves for the individuals with fullest stomachs in each series mirror the mean values. The curve for percentage of fish with empty stomachs is the reciprocal of these thus confirming the general correctness of the conclusions.

June feeding curves for three other Lake Opinicon species are given in Keast and Welsh (1968). All five species are compared with each other and with that of the nocturnal feeding *Ictalurus nebulosus* also common in the lake, in Fig. 12. Data on the latter are based on the studies by Darnell and Meierotto (1962) in Minnesota. In each graph the hours of darkness are indicated by the black line.

The feeding of the six species differs markedly. *Perca flavescens*, is a late afternoon and morning feeder. *Fundulus diaphanus* has a major feeding period in the afternoon and a lesser one before dawn. The two *Lepomis* have afternoon and morning peaks and a lesser one at night. *Ambloplites rupestris* is semi-nocturnal. One feeding period extends from the late afternoon through the first half of the night and there is another period of feeding in the morning. *Ictalurus nebulosus* in Lake Opinicon, as in Minnesota, is entirely nocturnal in its feeding. Possibly, in part, the differences in feeding times of these various fish species are linked to times of availability of their preferred food organisms. In other cases the different species are feeding on a common food (e.g. chironomid larvae) but obtaining them at different times of the day.

A comparison of the food invertebrates eaten by individual fish species at their different feeding peaks showed these to differ little in the case of purely diurnal feeders on the one hand, and purely nocturnal feeders on the other. In the two species of *Lepomis*, however, in which there are both day and night feeding periods, there was a marked difference. The major foods of *L. gibbosus* were: *afternoon*—chironomid larvae, pelecypods, and Anisoptera nymphs; *4.00 a.m.*—chironomid pupae, amphipods, Trichoptera larvae; *morning*—chironomid larvae, pelecypods, Zygoptera nymphs, Trichoptera larvae, and isopods. In *L. macrochirus* Cladocera were prominent in the stomachs by day but not at night. There is a partial correlation between the invertebrates consumed at different times and the activity cycles of these. Thus chironomid pupae and amphipods move into the water column at night presumably then becoming more accessible to the sunfish. Cladocera, it will be noted, were taken by these species by day, when they are concentrated low down, not at night when they are thinly dispersed through the water column.

FOOD UPTAKE RATES AND THE DAILY RATION

The ultimate objective of comparative food studies is not only to know what items each species is withdrawing from the environment but in what amounts. This requires study of the daily ration of each species and size class at different times of the year.

Daily feeding period studies have suggested a new method of calculating the daily ration (Keast and Welsh, 1968) by determining the differences between the mean peaks of stomach fullness and the succeeding troughs (greatest emptiness) and adding these. This gave an amount for the 7·8-26·5 g *Perca flavescens* in Fig. 10 of 0·020 g per g body weight (0·011 plus 0·009), or 2% of the body weight per day. Figures for the other fish studied were: *Lepomis gibbosus* (of 18·0-62·0 g body weight) and *L. macrochirus* (11·1-58·0 g), both about 2·5% of the body weight; *Ambloplites rupestris* (29·0-90·0 g), about 4%; and *Fundulus diaphanus* (2·5-5·4 g), about 1·5%. The figure for *Lepomis macrochirus* compares with the "maximum of 3% of the body weight" obtained by Gerking (1954) in aquarium fish housed at room temperature and fed mealworms in excess.

In the case of the first four species above, the samples were drawn from Year II and Year III age groups that were still rapidly growing: hence their food uptake requirements could be expected to be higher than those of *Fundulus diaphanus*, which were fully grown adults. The relatively lower daily ration figure for this species possibly also reflects the imminence of spawning (which takes place in late June), this being known to subdue feeding in some fishes, e.g. *Perca flavescens* (Pearse, 1919).

The above daily ration figures are presumably minimal ones since some digestion must take place whilst the fish are still feeding and some individuals continue to feed when the mean volumes are lessening. The method has limitations in the case of semi-continuous feeders like *Lepomis macrochirus* which do not show clearcut peaks. More precise calculations by this technique are obviously dependent on better data on digestion rates of the various species at different temperatures. These, however, are only available for *Lepomis macrochirus*. In this species Seaburg and Moyle (1964) record that, at 18-23°C, in fish under near-natural conditions, 50% of the stomach volume was digested in 5 hr, and 75% in 12 hr. Windell (1966) calculated figures for force-fed bluegills at 21°C as follows: 50% digestion in 6 hr, and 100% digestion in 18 hr. A voluntary meal, however, was 85% digested in 10 hr. A 75-85% emptying of the stomach in 10-12 hours accords well with the Lake Opinicon feeding periodicity data, most species having two feeding periods per 24 hours (Keast and Welsh, 1968).

These initial Lake Opinicon daily ration studies are inadequate to show whether or not species differ in daily ration at these temperatures (18-22°C). Species differences in daily ration will presumably prove to be much less than those between large and small individuals of species, however, judging from the differences in the latter described in the literature. Thus, hatchery brook trout, *Salvelinus fontinalis*, held at 16°C and fed a ration designed to give maximum food conversion and growth, consumed 6% of the body weight when 4-6 inches long, and 2% when 9-10 inches long (third (1952) edition of the New York State Hatchery Feeding Chart). Again, species may differ considerably in the amount of food consumed seasonally, irrespective of temperature. Seaburg and Moyle (1964), in the course of summer studies of the feeding of *Lepomis macrochirus*, *L. gibbosus* and *Pomoxis nigromaculatus* in some Minnesota lakes found that the daily ration fell off rapidly from high rates in May and June to about one-third that level in later summer and autumn.

SEASONAL DIFFERENCES IN FOOD UPTAKE RATES, PARTICULARLY BETWEEN SUMMER AND WINTER

At the latitude of the Great Lakes a high food uptake rate is characteristic of the warmer months and a low one of winter. Growth is purely a summer, or spring-summer-autumn phenomenon: such feeding as occurs in winter is generally recognized as having a purely maintenance function. There is ample experimental evidence that fish consume much less food at low than high temperature (Komarova, 1939; Hathaway, 1927; Markus, 1932; Pentelow, 1939), and that digestion rates are notably decreased then (Markus, 1932; Molnar and Tolg, 1952; Molnar, 1968). The winter feeding of fishes in the field has, however, received little attention. Field work by the writer in eastern Ontario has shown that cohabiting species may differ considerably in the amounts of food consumed at low temperatures and that fish in general are limited to a much smaller range of food types in winter than summer (Keast, 1966). Small-bodied fishes like *Fundulus diaphanus* and *Etheostoma exile* (Percidae) show clear evidence of emaciation towards the end of winter, i.e. they have a low "condition factor" or weight/length ratio: on the other hand *Umbra limi* remains robust (Keast, unpublished). Accelerated feeding by all species starts with the warming of the water in early spring. The resumption of feeding may, however, be delayed by spawning, *vide* in *Perca flavescens* (Pearse, 1919; Keast, 1968b). In *Semotilus atromaculatus* and others feeding is resumed earlier in the smaller prebreeding individuals than in adults (Keast, 1966).

Changing food uptake rates during the summer. A declining food uptake rate as the summer progresses, from a late spring early summer peak, has now been described in several freshwater fishes, e.g. *Esox lucius* in Lake Windermere, England (Johnson, 1966), and *Lepomis macrochirus*, *L. gibbosus*, and *Pomoxis nigromaculatus* in Minnesota (Seaburg and Moyle, 1964). In these latter species, as noted, the daily ration subsequently fell to about one-third of the May one but the precise time, and extent of the drop-off, varied with the lake. A high level of feeding as the waters warm is to be expected in animals that have been largely starved during the winter, just as is a tapering off as their physiological needs are satisfied. The phenomenon is probably fairly general. Incomplete studies on several Lake Opinicon species using the "daily feeding periodicity" approach are confirming the marked mid-summer drop-off in feeding recorded by Seaburg and Moyle (1964).

The termination of feeding in autumn and its commencement in spring. In the Centrarchidae in Lake Opinicon there is a progressive drop-off in feeding in late fall until, by late November, stomach content weights become uniformly low and show no diurnal fluctuation (unpublished). This is consistent with what would be expected from the slow digestion rates at such temperatures, for example Molnar (1968) found that, at 10°C, food remained in the stomach for over two days in several species.

The temperature at which feeding finally ceases in the Lake Opinicon fish probably varies somewhat with the species but this has not yet been studied. Small *Micropterus salmoides*, introduced into aquaria in November ceased feeding whenever the temperature was artificially dropped to below 12°C but resumed immediately it was raised to over 13°C: they responded this way repeatedly (Keast, unpublished). Adaptation downwards to declining temperatures presumably occurs eventually in all Great Lakes fishes. Seven species studied by the writer in indoor aquaria in December-January were all induced to take food at 5°C after careful adaptation over periods of up to 6 weeks. The species are those listed in Table 4, plus *Notemigonus crysoleucas* (Cyprinidae), *Lepomis gibbosus* and small *Micropterus salmoides*.

TABLE 4. Conversion efficiency at different temperatures, five common eastern Ontario freshwater fishes. Growth over an 8 week period at constant temperatures and on a 12 hour daylength. 10 fish to each temperature series of each species. Food figures are given as wet weight and expressed as g per g body weight. Apparent length increases of under 3% are within the range of measuring error

	Ictalurus nebulosus (bullhead) young		Umbra limi (mud minnow) adult		Notropis cornutus (common shiner) part-grown		Semotilus atromaculatus (creek chubb) part-grown		Fundulus diaphanus (killifish) adult	
Length ranges at beginning of experiment (mm)	60-110		60-100		60-100		100-130		70-95	
Weight ranges at beginning of experiment (g)	2·5-9·5		4·0-11·0		5·5-8·5		9·0-19·0		2·4-7·0	
Temperature (°C)	Food (g)	Length incr. %	Food (g)	Length incr. %	Food (g)	Length incr. %	Food (g)	Length incr. %	Food (g)	Length incr. %
5	1·7	(2·0)	1·6	6·0	1·4	(2·0)	1·6	(1·0)	1·8	(2·0)
10	2·0	(2·0)	2·7	6·0	3·3	(2·5)	2·2	(2·0)	3·1	(2·5)
15	3·4	(2·5)	2·7	6·0	8·0	15·0	4·5	6·2	4·1	(2·5)
20	5·0	25·0	4·1	8·0	6·5	10·0	7·0	19·0	10·1	11·0
25	9·4	25·0	9·1	4·0	9·0	12·5	6·5	11·0	13·2	14·0
30	9·4	10·0	9·4	(3·0)	19·0	15·0	7·6	3·5	21·5	17·2

Data on the resumption of feeding at the beginning of spring was obtained for two creek faunas in April-May, 1966 (full details in Keast, 1968b). In addition to differences in species composition, the fish used in the Little Cataraqui Creek study were large-bodied, and those in the Jones Creek study small-bodied. Netting was carried out on Little Cataraqui Creek immediately following the thaw, 20-30 individuals of the five commonest species being taken at two-day intervals over the 3-week period that the water temperature took to rise from 6°C to 11°C. The objective was to determine the precise temperature at which the different species commenced to feed. In the other study, carried out on the fauna of Upper Jones Creek, near Mallorytown, Ontario, good series of the 9 species of fish were netted at 8°C (April 28-30) and 15°C (May 13-15), to determine, amongst other things how, the mean weights of stomach contents compared at the two temperatures.

Only two of the Little Cataraqui Creek species were found to contain food at 6·0°C (April 11), when the study started: *Ictalurus nebulosus* (all individuals) and *Pomoxis nigromaculatus* (60% of individuals). The other species, *Lepomis gibbosus*, *Ambloplites rupestris*, and smaller numbers of *Micropterus salmoides*, *M. dolomieu*, *Morone americana*, and *Amia calva*, contained no food: the stomachs of the centrarchids, moreover, were shrunken, mucous-filled, and contracted far forward in the body cavity. Food was present in the stomachs of 50% of individuals of *Ambloplites rupestris* 8 days later, at a temperature of 8·5°C. Two days later, at 9·5°C, all the fish contained food. Food was present in the stomachs of 20% of the *Lepomis gibbosus* at 7·0°C, and 100% at 8·5°C. *Perca flavescens* did not commence feeding until April 30 after it had spawned. This species is, however, a winter feeder (Moffett and Hunt, 1945).

That fish species commence feeding at lower temperatures in spring than those at which they ceased feeding in the autumn was recognized by Pearse (1919). Thus in Wisconsin *Pomoxis nigromaculatus* ceases feeding at 10°C, cannot be caught by hook and line until mid-February, but begins feeding thereafter with the "first rise" in temperature. The temperature at which feeding commences in the spring may possibly vary with the year—note the early observation of Knauthe (1907) that carp normally begin feeding in spring at 6-8°C but that, after a hard winter, they will start at 3-4°C.

In the Jones Creek study the mean weights of stomach contents in most fish species averaged 3-4-fold higher at 15°C than 8°C. This conforms to the level of difference found for such temperature differences in experimental work, e.g. Hathaway (1927). Small members of species in Jones Creek contained more food, per g body weight, than large ones at both temperatures. This agrees with findings respecting small and large fish at higher temperatures (Pearse, 1924; Hathaway, 1927; Arnoldi and Fortunatova, 1937; Pearse and Achtenberg, 1920; and Thompson, 1941).

The feeding of fishes in winter. Comparative data on feeding under the ice (temperature about 4°C) were developed for the 6 small-bodied fishes making up the Fish Lake, Ontario, fauna, by making large collections at two-week intervals between January and March, 1966 and 1967.

The study, reviewed in Keast (1968b) provides clear data that cohabiting species within a fauna may differ considerably in degree of feeding at low temperatures. Thus, in *Umbra limi* 48-65% (depending on size class) of the 264 fish examined contained food, compared to 30-37% in *Notemigonus crysoleucas* (186 fish), 24-43% in *Ictalurus nebulosus* (200 fish), 26% in *Eucalia inconstans* (180 fish), 22% in *Etheostoma exile* (145 fish), and only 9% in *Fundulus diaphanus* (190 fish).

Within species a higher percentage of small-bodied individuals contained food than did larger ones, and they also tended to contain more food. Figures for mean weight of alimentary contents per g body weight ranged from 0·005-0·02 g for the small fish, and from 0·002-0·005 g for the larger ones. This equals 0·5-2·0% of the body weight in the case of the former, and 0·2-0·5% in the latter.

A real appreciation of the striking differences in food uptake of the various fish species between winter and summer is gained if digestion rates at these low temperatures be taken into account. Thus Molnar (1968) found that, in several species of European fish studied by him, it took about 5 days to clear the stomach at 5°C: this contrasts with 12-18 hour figure for *Lepomis macrochirus* at 20°C. This suggests that in the Fish Lake species, except for *Umbra limi* that is cold adapted, the amount of food eaten in mid-winter may only be 1/20th that of summer.

The low food uptake, of course, is matched by much lower metabolic needs. This being so it is possible that interspecific competition for food is no greater than in summer, despite the greater food overlaps between species in winter.

Optimal temperatures for feeding and growth, experimental comparisons between species

The above indicates that, within assemblages of fish, species and the size classes of species are not utilizing the food resources of the environment equally and maximally at all times but that this varies with the season and environmental temperatures. The matter can best be further investigated experimentally by comparing feeding and growth responses in a series of controlled temperature experiments.

Five common species belonging to four families (Table 4) were netted in late fall at environmental water temperatures of 10-12°C, divided into samples of 10, and placed in tanks measuring 87 × 60 × 60 cm for acclimation to temperatures of 5°C, 10°C, 15°C, 20°C, 25°C, and 30°C, respectively. The daylength was standardized at 12 hours.

Commercial "brown-nosed redworms" (Oligochaeta) were fed in excess. Acclimation downwards was at the rate of one degree every three days and upwards at the rate of one degree daily. The final temperatures were reached at the end of 3 weeks, after which the fish were given a week to settle down. Each individual was then measured and weighed and the experiment started. Weighed quantities of oligochaetes, in excess of requirements were provided daily, the uneaten portion being removed and weighed each morning prior to the new supply being provided. The experiments were continued for 8 weeks, the fish being weighed and measured every two weeks and at the end of the experiment.

Table 4 shows the wet weight of food eaten, and the total percent increase in body length of all the fish in each tank over the 8-week period. The species are seen to differ considerably in the temperatures at which they achieved the best growth and conversion efficiencies. In *Umbra limi* (adults) these were 10-20°C, *Semotilus atromaculatus* and *Notropis cornutus* (half grown), 15-25°C, *Fundulus diaphanus* (adults), 20-25°C, and small *Ictalurus nebulosus*, 20-30°C. *Notropis cornutus* and *Fundulus diaphanus* achieved good growth levels at 30°C but in doing so used higher amounts of food than in attaining an equivalent amount of growth at 20°C and 25°C. *Umbra limi* alone achieved some length increase at 5°C and 10°C (the 2-3% length increases suggested in the other species are within the range of measuring error). *Umbra* responded poorly at 30°C however, confirming that it is a cold adapted species.

These differences in optimum temperatures for feeding and growth obviously have important ecological implications. They not only help explain distributional differences (i.e. tendency to occur in colder or warmer bodies of water) but, as the field data suggested, must serve to reduce competition for food. Only at intermediate temperatures (e.g. 20°C) do all the species apparently feed equally.

The explanation for these physiological differences apparently lies in the origins of the Great Lakes fish fauna. This fauna was derived in post-glacial times (8000-10 000 years before present) from two main sources: the Mississippi-Missouri basin and the glacial front lakes, and/or Alaska-Yukon refuge, respectively. One or two of the latter, e.g. *Esox lucius* are presumably late colonizers from the Paleoarctic region by way of the Bering Strait land bridges. The fauna is thus a blend of "warm water" and "cold water" fishes. The two elements are partly separable on the temperatures at which they ovulate, the former tending to breed in mid-summer, and the latter in early spring or fall.

ENERGY REQUIREMENTS FOR GROWTH AT DIFFERENT TEMPERATURES, EXPERIMENTAL WORK ON THE BULLHEAD, *Ictalurus nebulosus*

In recent decades the study of productivity and energy transformations has become one of the exciting frontiers of fisheries biology (e.g. Winberg, 1956; Mann, 1965). In any aquatic system it is important to know the relative amounts of energy each component fish species and size class are withdrawing. Amongst the needs are comparative data on small-bodied species with a rapid population turnover (e.g. *Fundulus diaphanus*) and larger, slower maturing ones like *Ambloplites rupestris* which do not breed until the second or third year (note growth data of Beckman, 1949, in Centrarchidae). There are, however, formidable difficulties in the way of answering such questions by field work alone.

One possible way of getting data on the energy various species are withdrawing from the natural system would seem to be to try to duplicate natural growth rates in the laboratory under conditions where measured amounts of food of known calorific value are fed and where calories lost in excretion (and respiration) can be exactly measured. A series of such experiments has been started by the writer and will be reported upon in detail elsewhere. The results of one of these only will briefly be reviewed here to illustrate the approach being used.

Ictalurus nebulosus is one of the commonest and most successful fish in eastern Ontario and has a high growth rate. A series of fish in their second year (total length 98-175 mm) were netted in November and acclimated to six different temperatures as outlined in the previous section. The experiment was run for 8 weeks, the same procedures being applied. Digestive efficiencies at temperatures of 10°C, 20°C, and 30°C, were investigated in an independent series of experiments in which the calorific value of faeces resulting from the consumption of weighed amounts of food of previously determined calorific value was determined. Standard calorimetric procedures were used in the work, a Phillipson Microbomb Calorimeter being used.

The results of the study are summarized in Table 5. The food uptake (wet weight) of the first and second four-week periods, weight of total food consumed, total kcals assimilated, and the percentage length and per cent weight increases over the 8-week period, at the six temperatures, are given. It will be seen that only nominal amounts of food were eaten at 5°C and 10°C, and

TABLE 5. Growth at different temperatures relative to kcals assimilated. *Ictalurus nebulosus* (Ictaluridae), length 98-175 mm at beginning of experiment. Duration of experiment, 8 weeks, at constant temperatures and on a 12 hour daylength

	Temperature (°C)					
	5	10	15	20	25	30
No. of fish	10	11	13	10	13	10
Lengths of fish at beginning of experiment (mm)	112-170 (130)	106-156 (126)	98-149 (125·5)	114-145 (127)	140-175 (153)	111-147 (127)
Weights at beginning of experiment (g)	13·9-50·5 (23·8)	16·1-45·6 (25·5)	9·4-32·8 (20·2)	15·4-30·4 (20·7)	32·9-74·4 (47·3)	13·6-29·9 (21·0)
Food (wet weight) consumed in first 4 weeks of experiment (g/g body wt.)	0·105	0·305	1·327	2·429	3·059	4·851
Food consumed in second 4 weeks of experiment (g/g body wt.)	0·112	0·287	0·994	1·589	2·552	2·394
Total wet weight of food consumed (8 weeks) (g/g body wt.)	0·217	0·592	2·366	4·018	5·611	7·245
Total kcals assimilated (8 weeks)	0·122	0·358	1·682	2·688	3·557	4·963
Lengths of fish at end of experiment (mm)	117-177 (133·5)	108-165 (129)	89-154 (130)	127-157 (143)	152-218 (175)	119-155 (143)
Weights of fish at end of experiment (mm)	14·4-50·4 (24·3)	16·3-47·5 (25·8)	12·2-47·1 (29·3)	28·1-51·4 (38·6)	55·2-153·4 (88·9)	22·6-47·5 (39·2)
% increase in length	nil	nil	nil	12·5	14·0	13·0
% Increase in weight	nil	nil	45	86	88	87

there was no growth. At 15°C the food uptake quadrupled and there was a 45% increase in body weight, but no increase in length. At 20°C and 25°C calorific intakes amounting to about twice that at 15°C produced mean length increases of 12·5% and 14%, and weight increases of 86% and 88%, respectively. At 30°C it required a rather greater energy intake to achieve equivalent increases, indicating a higher cost of body maintenance at this temperature.

The greater amounts of food consumed in the first 4-week period than in the second at temperatures from 20°C to 30°C are typical of experiments of this type: they are probably partly due to initial hunger and partly to the stimulating effect of increased temperature. As noted, in natural fish populations a rapid rise in early summer is followed by a decline in mid-summer.

Work on natural growth rates of *Ictalurus nebulosus* in the field in eastern Ontario is incomplete. These figures, however, compare favourably with growth rates in Fish Lake, where the species is common, over a corresponding period of summer. Because of basic differences in conditions, nature of food, and availability of food, the energy outlay required to produce given levels of growth in the laboratory cannot be expected to exactly duplicate that in nature. Such figures, nevertheless, provide a basis for calculations, and for comparisons with other size classes and species housed under equivalent conditions.

ECOLOGICAL INTERRELATIONSHIPS OF COHABITING FISH SPECIES, AN INTEGRATION

The present study shows that, in any fauna or assemblage of fishes the component species differ in their body size and morphological characteristics, that is they are structurally modified for different ways of life. They show a moderately high degree of ecological exclusion in foods eaten, differ in being specialized as compared to generalized feeders, and in the levels in the water at which the food is obtained. Some do not show marked seasonal changes in diet but others do. Species differ in their habitat preferences and requirements. They vary greatly in numerical abundance and biomass in individual areas, and in the lake as a whole. Each species has its characteristic feeding times, and these differ between species. There are apparently minor differences in the periods of year when the different species feed and, as has been shown experimentally, they may differ markedly in optimum temperatures for feeding and growth. Some species are always found only in certain areas but others differ in distribution diurnally and seasonally; thus *Labidesthes sicculus* is concentrated inshore in early and late summer, and pelagic in between (see also Hubbs, 1921). To sum up, the species utilize or manipulate the environment in a wide variety of different ways.

Compounding the complexity of the system are the year classes of species. These may have near-similar diets, or differ from each other as much as different species do from each other. The transition from juvenile to adult fish may take only a few months, or several years. Age classes sometimes occupy different habitats. Even adjacent bays, especially if they differ slightly as living areas, may have quite different age-class abundances. Obviously, a dramatic change in the interrelationships of fish and their food invertebrates, and between the species and age-classes of fish, follows the annual recruitment of new, small-bodied individuals into the population. The ecological demands of these young fish are quite different from those of adults.

It is apparent that the interrelationships of cohabiting species within a fish fauna are highly complex and sophisticated. An idealized "food web", or "food chain" cannot apply equally to all areas of a particular body of water, or to all seasons. Feeding and bioenergetic interrelationships must be worked out independently for different parts or areas, taking such factors as relative biomasses into account, before interrelationships can be properly expressed for the water-mass as a whole.

With adequate detailed work one will be able to determine, or "contour", the factors of basic importance to the different fish species. The significance of biomass differences between them can be assessed, and the energy requirements for maximum growth of a range of species, or different ecological types of fish, determined. Ultimately, with the accumulation of sufficient data of this type, it will be possible to generalize and predict the pattern of interrelationships that will follow the bringing together of various combinations of species.

ACKNOWLEDGEMENTS

The research described in this paper has been carried out under grants supplied by the National Research Council of Canada, with supplementary support from the Ontario Research Foundation and Ontario Hunters and Anglers Association. Figs 1, 2, 10 and 11 are republished, with permission, from *J. Fish. Res. Bd Can.*, and Figs 3, 4, 5 and 6 from *Proc. Gt Lakes Res. Inst.*, Ann Arbor.

REFERENCES

ALLANSON, B. R., and KERRICH, J. E. 1961. A statistical method for estimating the number of animals found in field samples drawn from polluted rivers. *Verh. int. Verein. theor. angew. Limnol.*, 14, 491-94.

ARNOLDI, L. V., and FORTUNATOVA, K. R. 1937. A contribution to the experimental study of the nutrition of Black Sea Fishes. *Dokl. Akad. Nauk. SSSR*, 15, 513-16.

BECKMAN, W. C. 1949. The rate of growth and sex ratio for seven Michigan fishes. *Trans. Am. Fish. Soc.*, 76, 63-67.

CHESHIRE, W. F., and STEELE, K. Z. 1962. Rearing pickerel in bass ponds. *Prog. Rep. Ont. Dep. Lands Forests*, Dec. 1962, 3 pp.

DARNELL, R. M., and MEIEROTTO, R. R. 1962. Determination of feeding chronology in fishes. *Trans. Am. Fish. Soc.*, 91, 313-20.

FRYER, G. 1959. Some aspects of evolution in Lake Nyasa. *Evolution*, 13, 440-51.

GERKING, S. D. 1954. The food turnover of a bluegill population. *Ecology*, 34, 490-98.

HARTLEY, P. H. T. 1948. Food and feeding relationships in a community of fresh water fishes. *J. Anim. Ecol.*, 17, 1-14.

HASLER, A. D., and Bardach, J. E. 1949. Daily migrations of perch in Lake Mendota, Wisconsin. *J. Wildl. Mgmt*, 13, 40-51.

HATHAWAY, E. S. 1927. The relation of temperature to the quantity of food consumed by fishes. *Ecology*, 8, 428-34.

HILE, R., and JUDAY, C. 1941. Bathymetric distribution of fish in lakes of the northeastern highlands, Wisconsin. *Trans. Wis. Acad. Sci. Arts Lett.*, 33, 147-87.

HUBBS, C. L. 1921. An ecological study of the life history of the fresh water atherine fish, *Labidesthes sicculus. Ecology*, 1, 262-76.

Hubbs, C. L., and Lagler, K. F. 1958. Fishes of the Great Lakes region. *Bull. Cranbrook Inst. Sci.* (26), 213 pp.

Hunt, B. P., and Carbine, W. F. 1950. Food of young pike, *Esox lucius*, and associated fishes in Peterson's Ditches, Houghton L., Michigan. *Trans. Am. Fish. Soc.*, **80**, 67-83.

Johnson, L. 1966. Experimental determination of food consumption of pike, *Esox lucius*, for growth and maintenance. *J. Fish. Res. Bd Can.*, **23**, 1495-1505.

Keast, A. 1965. Resource subdivisions amongst cohabiting fish species in a bay, Lake Opinicon, Ontario. *Proc. 8th Conf. Gt Lakes Res., Univ. Michigan*, 106-32.

Keast, A 1966. Trophic interrelationships in the fish fauna of a small stream. *Proc. 9th Conf. Gt Lakes Res., Univ. Michigan*, 51-79.

Keast, A. 1968a. Feeding biology of the black crappie, *Pomoxis nigromaculatus*. *J. Fish. Res. Bd Can.*, **25**, 285-97.

Keast, A. 1968b. Feeding of some Great Lakes fishes at low temperature. *J. Fish. Res. Bd Can.*, **25**, 1199-1218.

Keast, A. and Webb, D. 1966. Mouth and body form relative to feeding ecology in the fish fauna of a small lake, Lake Opinicon, Ontario. *J. Fish. Res. Bd Can.*, **23**, 1845-74.

Keast. A, and Welsh, L. 1968. Daily feeding periodicities, food uptake rates, and dietary changes with hour of day, in some lake fishes. *J. Fish. Res. Bd Can.*, **25**, 1133-44.

Knauthe, K. 1907. *Das Susswasser*. Neudamm, Germany, 663 pp.

Komarova, I. V. 1939. The feeding of the long rough dab (*Hippoglossoides platessoides*) in the Barents Sea in relation to its food supply. *Trudy vses. nauchno-issled. Inst. morsk rÿb. Khoz. Okeanogr.*, **4**, 297-320.

Larkin, P. A. 1956. Interspecific competition and population control in freshwater fish. *J. Fish. Res. Bd Can.* **8**, 164-77.

Mann, K. H. 1965. Energy transformations by a population of fish in the River Thames. *J. Anim. Ecol.*, **34**, 253-75.

Markus, H. C. 1932. The extent to which temperature changes influence food consumption in large-mouth bass (*Huro floridana*). *Trans. Am. Fish. Soc.*, **62**, 202-210.

Moffett, J. W., and Hunt, B. P. 1945. Winter feeding habits of bluegills, *Lepomis macrochirus* Raffinesque, and yellow perch, *Perca flavescens* Mitchell in Cedar Lake, Washtenaw County, Michigan. *Trans. Am. Fish. Soc.*, **73**, 231-42.

Molnar, G. Tamassy, E., and Tolg, I. 1967. The gastric digestion of living, predatory fish. In *The biological basis of freshwater fish production*. Ed. by S. D. Gerking. Oxford, Blackwell Scient. Publs, 135-149.

Molnar, G., and Tolg, I. 1962. Experiments concerning gastric digestion of pike perch (*Lucioperca lucioperca* L.) in relation to water temperature. *Acta biol. hung.*, **13**, 231-39.

Pearse, A. S. 1919. Habits of the black crappie in inland lakes of Wisconsin. *Doc. U.S. Bur. Fish.* No. 867, 5-16.

Pearse, A. S. 1924, Amount of food eaten by four species of fresh-water fishes. *Ecology*, **4**, 254-58.

Pearse, A. S., and Achtenberg, H. 1920. Habits of yellow perch in Wisconsin lakes. *Bull. Bur. Fish., Wash.* **36**, 295-366.

Pentelow, F. T. K. 1939. The relation between growth and food consumption in the brown trout (*Salmo trutta*). *J. exp. Biol.*, **15**, 446-73.

Seaburg, K. G., and Moyle, J. B. 1964. Feeding habits, digestive rates, and growth of some Minnesota warm water fishes. *Trans. Am. Fish. Soc.*, **93**, 269-85.

Spencer, W. P. 1929. Day and night periodicity in the activity of four species of freshwater fishes. *Anat. Rec.*, **44**, 197.

Spoor, W. A., and Schloemer, C. L. 1938. Diurnal activity of the common sucker, *Catostomus commersonnii* (Lacepede), and the rock bass, *Ambloplites rupestris* (Rafinesque) in Muskellunge Lake. *Trans. Am. Fish. Soc.*, **68**, 211-20.

Thompson, D. H. 1941. Fish production of inland streams and lakes. In *A Symposium on hydrobiology*. Ed. Madison. Univ. Wisconsin Press, 206-17.

Winberg, G. G. 1956. Rate of metabolism and food requirements of fish. In *Intensivnost obmena i pischevye petrebrosti rÿb. Trudÿ belorussk. gos. Univ. Minsk.* Fish. Res. Bd Can., Trans. Ser. No. 194.

Windell, J. T. 1966. Rate of digestion in the bluegill sunfish. *Invest. Indiana Lakes Streams*, **7**, 185-214.

Part Six

THEORETICAL PROBLEMS

Introduction

JOHN STEELE

The study of natural food chains, or more correctly, food webs, involves the examination of systems with a very large number of simultaneously varying components. In an experimental situation this complexity can be reduced by keeping most of the variables constant and studying the changes in only one or two parameters. Typically the relations between growth, metabolism and temperature are studied in this way under laboratory conditions. The main difficulties which arise in applying these results to natural systems concern the natural activity of the organism in its normal environment. This problem of activity is merely one aspect of our lack of estimates of behavioural effects as distinct from the relative abundance of physiological or metabolic information.

At the same time, studies of natural food webs can define some of the limits possible to such systems. For example, interrelations are observed between the amplitude of cycles in the physical environment, the stability of populations and the species diversity (Dunbar, this volume; MacArthur, 1955; Margalef, 1961). These types of interrelations suggest that limits can be set to the range of variables, or even that, without detailed knowledge of the internal structure of the system observations of the external features of the system can have predictive value (Brocksen et al., this volume).

Often there is no exact division between experimental models, linguistic descriptions and mathematical formulations since they can all be parts of the exposition of a theoretical problem. In any of these systems the generality achieved in describing natural systems depends on the empirical content on which the hypotheses are based and on the extent to which the conclusions are empirically falsifiable. It has always been one of the major dangers of mathematical models that it is possible to increase the apparent generality by decreasing the empirical content of the assumptions without producing deductions that are testable over a correspondingly general range of environments.

For these reasons theoretical models have tended to be set up for those aspects of the marine environment which are best known. In the field of primary production the work of Riley and others relating physical, chemical, plant and herbivore facets which I have reviewed (Steele, 1959) showed that some success is possible in relating the productivity of phytoplankton to the physical environment. Where such models tend to break down is in the construction of postulates describing the effects of grazing by the herbivorous zooplankton, One limitation on the further development of such models lay in the inability to describe quantitatively the growth and reproduction rates of zooplankton as functions of the natural levels of food supply. Usually it was necessary to assume fixed rates of filtration and very simple relations between the rate of production and biomass. The problems that can arise from such assumptions are discussed in the work of Greze (this volume) where ratios of production/biomass derived from one region give unexpected results when applied to other regions with very different physical environments. Several papers in this volume and elsewhere illustrate the dangers of the excessively simple nature of the hypotheses used. The importance of patchiness (Parsons et al., this volume), the effects of vertical migration of zooplankton through a thermocline on their own growth and reproduction (McLaren, 1963), the consequences of diurnal migrations together with variable feeding rates on prey density (McAllister, this volume), all illustrate the kinds of complexity that need to be incorporated into future models.

Perhaps the most successful models in the marine field have been those linking fish stocks with their human predators (Ricker, 1954; Beverton and Holt, 1957). Their success depends not only on the fact that there are extensive data on the fish populations and on their metabolism but in this special case some information on predator behaviour is available. In fact this type of model may be a useful basis for generalization to other levels (Paloheimo and Dickie, this volume). Yet in the area of commercial fisheries, it is evident (Gulland, this volume) that our lack of knowledge of the interaction of the fish and *their* food may inhibit our development of theoretical models of exploited fisheries.

Thus the theoretical problems are merely one aspect of the general problem posed by this symposium—what extra information do we have and do we need linking primary production and commercial yields? It is possible that no significant degree of generality can be achieved and that a separate picture is needed for each community so that a search for general rules comparable to that undertaken in the physical sciences is basically inapplicable. It is unlikely that this can be tested except by trial and error since such a hypothesis can only be falsified by finding successful models.

Apart from this pessimistic approach there are a variety of possibilities. As already mentioned there is the topological approach of MacArthur and Margalef, as distinct from the more usual models based on sets of differential equations. So far, no completely theoretical picture of natural food chains has yet been attempted but certain basic rules have been suggested. Slobodkin (1962) proposed that the ratio of energy removed from a population by its predators to the energy taken in by the population was nearly constant at 10%. This kind of concept, if acceptable, has the enormous advantage of reducing greatly the number of rates we need to measure, and increasing the usefulness of measurements of comparatively easy static factors like biomass. Many of the papers in this volume are relevant to the assessment of this type of hypothesis in the marine environment.

Another, rather different type of generalization (Hairston *et al.*, 1960) attempts to typify the response of very broad groups with hypotheses such as "Herbivores are seldom food-limited [and] appear most often to be predator limited"; and "The predators and parasites, in controlling the populations of herbivores must thereby limit their own resources, and as a group they must be food limited." This paper and the ensuing controversy (Murdoch, 1966; Ehrlich and Birch, 1967; Slobodkin *et al.*, 1967) were concerned with problems of energy exchange and questions of hypothetical tests of the hypotheses. At the same time Holling (1959, 1965) and Tinbergen (1960) were developing theoretical approaches to very similar problems but based on a much more behavioural or functional analysis of detailed situations illustrating the interactions of herbivores and carnivores. Thus the evidence for generalizations may be expressed in energetic units but will depend on detailed investigations of the common features of behaviour. These arguments have been concerned solely with terrestrial environments, and the experimental work with insects, small mammals and birds. The marine aspects have been largely ignored, mainly because underwater events, although they can permit, and almost encourage, specious generalizations about energy flow (Steele, 1965), are much less amenable to detailed hypotheses concerning predator or prey behaviour. It is likely that the papers in this volume give some of the evidence needed. Certainly, any theoretical formulation of general rules about food chains will have to include aquatic ecosystems.

Thus in summary there is the expected cycle where the earlier apparently elegant and simple models describing events mainly within a trophic level have been superseded by a search among a welter of new facts for common patterns covering many levels. So far this search has not produced a new synthesis but two features may be apparent; although the units used are likely to be energetic the formulae may characterize common behaviour patterns rather than similar rates of energy flow. As a consequence the hypotheses may condense around functional rather than systematic groups.

REFERENCES

BEVERTON, R. J. H., and HOLT, S. J. 1957. On the dynamics of exploited fish populations. *Fishery Invest., Lond., Ser.* 2, **19**, 533 pp.
EHRLICH, P. R., and BIRCH, L. C. 1967. The "balance of nature" and "population control". *Am. Nat.*, **101**, 97-107.

HAIRSTON, N. G., SMITH, F. E., and SLOBODKIN, L. B. 1960. Community structure, population control, and competition. *Am. Nat.*, **94**, 421-25.

HOLLING, C. S. 1959. Some characteristics of simple types of predation and parasitism. *Can. Ent.*, **91**, 385-98.

HOLLING, C. S. 1965. The functional response of predators to prey density and its role in mimicry and population regulation. *Mem. ent. Soc. Can.*, (45), 5-60.

MACARTHUR, R. 1955. Fluctuations of animal populations and a measure of community stability. *Ecology*, **36**, 533-36.

MCLAREN, I. A. 1963. Effects of temperature on growth of zooplankton and the adaptive value of vertical migration. *J. Fish. Res. Bd Can.*, **20**, 685-727.

MARGALEF, R. 1961. Communication of structure in planktonic populations. *Limnol. Oceanogr.*, **6**, 124-28.

MURDOCH, W. W. 1966. Community structure, population control, and competition—a critique. *Am. Nat.*, **100**, 219-26.

RICKER, W. E. 1954. Stock and recruitment. *J. Fish. Res. Bd Can.*, **11**, 559-623.

SLOBODKIN, L. B. 1962. *Growth and regulation of animal populations.* New York, Holt Rinehart-Winston, 184 pp.

SLOBODKIN, L. B., SMITH, F. E., and HAIRSTON, N. G. 1967. Regulation in terrestrial ecosystems, and the implied balance of nature. *Am. Nat.*, **101**, 109-124.

STEELE, J. H. 1959. The quantitative ecology of marine phytoplankton. *Biol. Rev.*, **34**, 129-58.

STEELE, J. H. 1965. Some problems in the study of marine resources. *Spec. Publs. int. Commn NW Atlant. Fish.*, (6), 463-76.

TINBERGEN, L. 1960. The natural control of insects in pine woods. I. Factors influencing the intensity of predation by song birds. *Arch. néerl. Zool.*, **13**, 265-343.

Zooplankton rations, phytoplankton mortality and the estimation of marine production

C. D. McALLISTER
Fisheries Research Board of Canada
Pacific Oceanographic Group
Nanaimo, B.C., Canada

ABSTRACT. New data relating zooplankton rations to phytoplankton concentration are presented and older data reviewed. Expressions describing the relations are here modified to define the rate of grazing mortality on growing populations of phytoplankton in terms of the concentrations of phytoplankton and zooplankton present. It is shown that the relation of ration and plant mortality per unit grazer to plant concentration can differ markedly between nocturnally and continuously grazing copepods and that failure to take this into account may result in serious errors in estimates of production from conventional field data. The significance of these factors for production in the sea was examined in 10-day experiments in which phytoplankton were grown in a light–dark cycle and exposed to either continuous or nocturnal grazing by marine copepods. The consequences of the various experimentally determined relationships were also explored using a theoretical model which generated sequences of plant and animal stock ensuing from selected initial values as a result of growth and the relationships of zooplankton ration and plant mortality to concentrations of organisms in the two trophic levels. The models also tested the effects of random variations in the plant growth rate and in zooplankton stock.

INTRODUCTION

Mathematical models of plankton production (e.g. Cushing, 1959) appear to assume implicitly that grazing by planktonic herbivores operates continuously whereas some statements about the interpretation of field measurements of primary production (e.g. Steeman-Nielsen, 1963) imply a belief that grazing on the phytoplankton must be largely nocturnal. McAllister (unpublished) investigated the consequences of these two conflicting assumptions for the estimation of secondary production from field data obtained at Ocean Station P in the north-east Pacific. Measurements of primary production and standing stocks of phytoplankton were used to estimate the instantaneous

419

rate of growth by the plants in the absence of grazing. The plant growth rate, with observed changes in monthly mean phytoplankton stocks, was used, with an estimate of the rate of nocturnal respiration by the plants, to estimate the rate of grazing mortality exerted on the phytoplankton, which in turn enabled the calculation of the rations obtained by the zooplankton. The rations, corrected for assimilation and respiration by the observed zooplankton stocks, permitted estimates of secondary production, first assuming continuous grazing, in which the estimated mortality was partitioned between day and night and then assuming nocturnal grazing, under which the plants

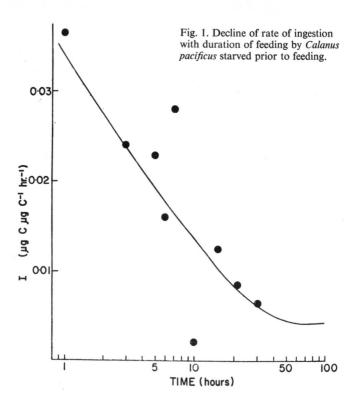

Fig. 1. Decline of rate of ingestion with duration of feeding by *Calanus pacificus* starved prior to feeding.

grew unimpeded during the daylight hours. In addition, the observation (Fig. 1 and Mullin, 1963) that the rate of ingestion by copepods starved for 12 to 24 hours prior to feeding may be initially very high but declines during the course of a night's feeding was taken into account, since copepods feeding nocturnally at a constant rate will ingest a smaller fraction of the plant material which would otherwise be lost to plant respiration than those feeding with an initially high but declining rate whose mean is equal to the constant rate. Secondary production was estimated assuming two different rates of decline in addition to the assumption of the constant nocturnal rate of ingestion, and to continuous grazing.

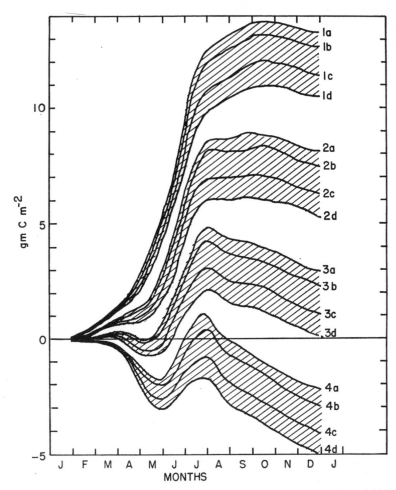

Fig. 2. Cumulative totals of monthly secondary production estimated from field data from Ocean Station P, on the assumption of different grazing schemes, and zooplankton respiratory rates. Curves 1, 2, 3, and 4 show the production estimated on the assumption of zooplankton respiratory rates of 4, 6, 8 and 10% of the body weight, per day, respectively. Curves marked *a* and *b* were estimated assuming nocturnal grazing, each with a different rate of decline of the feeding rate (see text). Curves *c* assume nocturnal grazing at a constant rate equal to the mean of the variable rate. Curve *d* assumes continuous grazing (from McAllister, unpublished).

Results obtained under the four different grazing schemes for each of four different assumed rates of zooplankton respiration are reproduced in Fig. 2 and indicate that for a given apparent rate of primary production and a given total rate of grazing mortality, markedly different estimates of secondary production may be obtained, depending on the way in which the grazing mortality is applied. The relative effects of assuming different grazing schemes were found to vary markedly with relatively small changes in the value of the rate of zooplankton respiration (Fig. 2) and also with changes or errors in the estimates of the plant growth constant and other parameters. The following is a summary of two other approaches to the problem of assessing the implications of interactions between zooplankton feeding and the growth of the phytoplankton. A brief account of attempts to culture the two trophic levels together is given, and some examples of the application of a computer simulation model to some other recent experimental results are presented.

METHODS

Feeding experiments were carried out using *Calanus pacificus* isolated from net tows made in the vicinity of Georgia Strait, British Columbia, and cultures of diatoms isolated from the same waters and reared in filtered seawater enriched as described by Jitts *et al.* (1964) although to a lower level of nutrients.

Experiments were carried out with durations ranging from 2 hours up to 11 days. The longer experiments were done in a light–dark cycle, with phytoplankton cultures grazed either continuously or nocturnally by the copepods. In the latter case the stock of copepods was added to the phytoplankton at laboratory sunset and removed at sunrise to be kept in filtered seawater until the next grazing period.

The rate of grazing mortality on the plants was estimated using the following expression:

$$P_1 = P_0\, e^{(k-g)t}$$

where P_1 is the concentration of phytoplankton observed at the end of the grazing period, and P_0 is that at the beginning; t is the time interval in hours; g is the instantaneous rate of grazing mortality and k is the instantaneous rate of increase of the phytoplankton. The plant growth constant was evaluated from ungrazed control cultures held in the same conditions as the grazed cultures. In the longer experiments the controls were withdrawn from the grazed cultures twice daily, and were returned after use.

The rate of ingestion per unit concentration of copepods was estimated as the product of the rate of grazing mortality exerted per unit concentration of copepod and the mean stock of phytoplankton during the period of grazing, calculated following Cushing (1959).

Most of the observations of the phytoplankton were of the volume of plants present, obtained using a Model B Coulter Counter. A few earlier experiments employed microscope determinations of the numbers and dimensions of cells present, in order to estimate the plant concentration as volume. Rates of ingestion, as volume, were converted to equivalent carbon using either direct determination of the carbon content in a Coleman Carbon-Hydrogen Analyser, or a relation proposed by Mullin *et al.* (1966). The carbon content of the copepods was estimated using relationships found between metasome length, dry weight and carbon content. Results were finally expressed as the rate of ingestion of phytoplankton carbon per unit concentration of zooplankton carbon per hour.

The simulation model is described in the section following the results of the experimental work.

RESULTS

Figs 1 and 3 show examples of the decline in the rate of ingestion of phytoplankton with the duration of the feeding period for copepods starved for 12 to 24 hours before the start of the experiment. These and other data suggest that the rate of ingestion in the twelfth hour of feeding at a given concentration of food reaches a value of about $\frac{1}{3}$ of that in the first hour, on

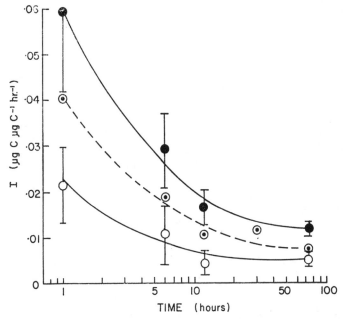

Fig. 3. Other examples of decline in rate of ingestion of phytoplankton by copepods with duration of feeding.

the average. However, the equilibrium rate of feeding is not approached until 30 to 50 hours after the onset of feeding and has a value averaging about $\frac{1}{3}$ of the mean during the first 12 hours. This suggests that in a constant concentration of phytoplankton nocturnally feeding copepods would ingest more per day than those feeding continuously at the equilibrium rate.

The observation is pertinent to the design of feeding experiments. Much of the data on feeding by copepods has been obtained in experiments lasting about 24 hours, and the exact feeding prehistory is not always made clear. The results of such measurements would be expected to overestimate the rate of feeding by continuous grazers and to underestimate that of nocturnal grazers, providing that the animals had been starved before the experiment.

In taking animals from nature for field experiments it would appear important that the duration of the experiments and the time of sampling be matched to the *in situ* feeding behaviour of the animals.

It was mentioned above that McAllister (unpublished) has calculated that the initially high rate of ingestion by nocturnal copepods would permit a greater use of the fraction of plant material which would otherwise be consumed by plant respiration than would a constant rate equal to the mean of the variable rate. In addition, if cell numbers or plant volumes increase during the night as was observed in all the present experiments, the animals would maximize the food intake per unit capture by feeding most rapidly early in the night, since the ratio of carbon per cell or per unit volume must decrease during the night if volume or cells numbers increase, and also as a consequence of respiration by the plants. It would appear that the assumption of simple exponentiality in the grazing relations between phytoplankton and zooplankton may be a source of error, both in estimating from field data, and in evaluating experimental results.

The rate of ingestion was also found to vary with the concentration of phytoplankton in a manner similar to that found by Rigler (1961), and McMahon and Rigler (1963) working with *Daphnia magna*, Reeve (1963) in a study of feeding by *Artemia*, Adams and Steele (1966) in field experiments with *Calanus* and by Parsons *et al.* (1967) with several species of marine herbivorous zooplankton.

Means and standard deviations of a large number of observations by the writer of rates of ingestion in several ranges of phytoplankton concentration are shown for nocturnal and continuous grazing in Fig. 4. Data from the individual experiments suggest that feeding does not begin until the phytoplankton have attained a critical concentration of phytoplankton, here termed P'. Above P' the rate of ingestion increases with increasing food up to a maximum value above which the rate of ingestion remains constant with further increases in food concentration.

The rates of ingestion were corrected for the effects of nocturnal volume increase in decreasing the ratio of carbon to volume. However, the effect of the variable rate of grazing in altering the partition of plant stock between plant respiration and grazing was not taken into account in estimating ingestion, as carbon, by nocturnal feeders.

It has been shown that such ingestion curves are fitted by a modification (Parsons *et al.*, 1967) of an expression due to Ivlev (1945),

$$I = I_M (1 - e^{-\delta(P - P')})$$

where I is the rate of ingestion per unit concentration of grazer at phytoplankton concentration, P (the mean concentration during the period of grazing); I_M is the maximum rate of ingestion attainable by the zooplankters; δ is a constant defining the rate of change of ingestion with food concentration; and P' is the concentration of phytoplankton at which feeding begins. The standard deviations about the means in Fig. 4 are large and it is obvious that there must be considerable uncertainty attached to any Ivlev constants

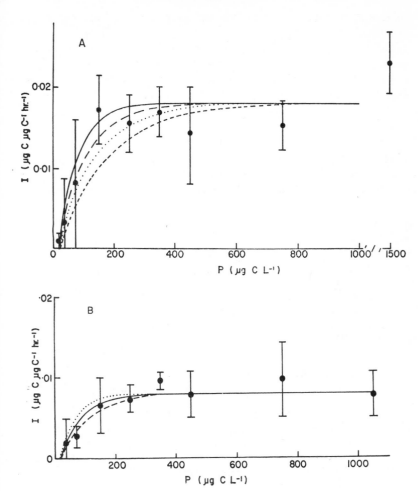

Fig. 4a. Ingestion rate by nocturnally feeding copepods and phytoplankton concentration. Circles with bars are means and standard deviations. Lines were fitted assuming I_M, = 0·0180, P' = 15. Solid line, δ = 0·0153; long dashes, δ = 0·0109; dots δ = 0·0081; short dashes, δ = 0·0063.
Fig. 4b. Ingestion rate by continuously feeding copepods, and phytoplankton concentration. Circles with bars are means and standard deviations. Lines were fitted assuming I_M = 0·0080, P' = 15. Solid line, δ = 0·0140, dotted line, δ = 0·0184; dashed line, δ = 0·0098.

assigned to the aggregate data. The curves shown in the figure are those obtained by estimating different values of the constant, δ, which in effect specifies the concentration at which the maximum rate of ingestion is attained.

The usefulness of such curves for assessing feeding conditions for given species of zooplankton from observations of phytoplankton at particular times or places has been examined by Parsons et al. (1967). It can also be pointed out that such curves specify the relation of growth and growth efficiencies to food concentrations for given values of assimilation and respiration by the zooplankton.

The ingestion curves may also be used to relate the instantaneous rate of grazing mortality exerted on the plant stock per unit concentration of grazer to the concentration of phytoplankton, simply by dividing the hourly rate of ingestion by the plant concentration,

$$m = \frac{I}{P} = \frac{I_M}{P}\left(1 - e^{-\delta(P - P')}\right)$$

where m is an estimate of the instantaneous rate of grazing mortality exerted per unit concentration of grazers and the other symbols are as before. Curves of m for the nocturnal ingestion curves shown in Fig. 4 are depicted in Fig. 5.

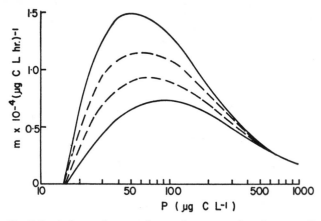

Fig. 5. Rate of mortality exerted per unit concentration of nocturnally feeding zooplankton vs. concentration of phytoplankton. I_M and P' are the same, 0·0180 and 15, respectively for each curve. The values of δ for the curves in descending order are 0·0153, 0·0109, 0·0081 and 0·0063.

The value of m increases to a maximum at food concentrations somewhat less than that at which I_M is attained, and declines with further increases in phytoplankton concentration. Changes in the value of I_M result in proportional changes in the maximum value of m, whereas changes in δ (Fig. 4) alter both the magnitude and the position of the maximum in m. Moderate changes in the value of P' do not appear very significant in plots of ingestion against food concentration but can be seen to have a very important effect on the variation of mortality with plant concentration (Fig. 6).

Thus, while some combinations of values of I_M, δ, and P' would appear to specify much better rations and therefore growth, at particular concentrations of phytoplankton and would thus appear to be more valuable to the grazers than other sets of constants, the apparently better set might also specify greater mortality on the plants with the result that excellent rations obtained in one period might seriously impair subsequent feeding conditions.

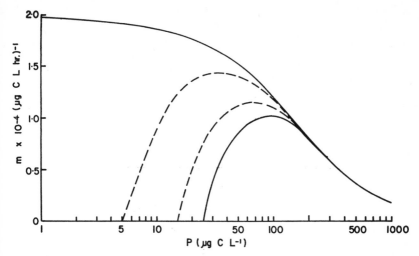

Fig. 6. Nocturnal grazers. Mortality exerted per unit concentration of grazer against phytoplankton concentration for different values of P'. I_M and δ are the same, 0·0180 and 0·0109, respectively, for each curve. Values of P' for the curves in descending order are 0, 5, 15, and 25.

In order to compare the implications of different grazing schemes, or different ingestion curves, under different combinations of standing stocks of the two trophic levels, the grazing relationships must be considered in connection with such parameters as rates of photosynthesis, respiration by the phytoplankton and zooplankton, and the assimilation of the ration by the zooplankton over periods of time significant with respect to seasonal events in nature. An example of such an attempt, employing field data, was cited above, although it did not incorporate the ingestion curves. The effects of grazing schemes can also be compared in cultures of phytoplankton grown in a light – dark cycle and subject to grazing. With such an approach, the various factors and their interactions are *implicit* in the experimental conditions. An example follows.

Two cultures of the diatom, *Thalassiosira rotula*, initially at a concentration of 175 μg C/l, were grown under conditions giving a mean plant growth constant of about 0·02/hr, equivalent to a doubling time of about 36 hours. One culture was nocturnally grazed and the other continuously, by a concentration of 4 copepods per litre, initially equivalent to 160 μg C/l.

Fig. 7 shows that the diatom culture under nocturnal grazing attained much higher stock levels than that continuously fed upon, apparently a consequence of the fact that the plants grew unimpeded by grazing during the daylight hours. It can be seen that the daylight increase in the continuously grazed culture was markedly depressed. The growth rate of the nocturnal feeders was estimated to average about 16% per day whereas that of the continuous feeders was about 7%. Although the rate of growth of the continuous feeders was less, the stock apparently increased sufficiently to raise the total grazing mortality above the plant growth rate, and the plant stock declined for the latter half of the experiment. In spite of the higher growth rate of the nocturnally grazing copepods, the unimpeded growth by the plants during the daylight appeared to allow the plant stock to maintain an approximately logarithmic rate of increase for the duration of the experiment.

However, the differences may not have all been due to the different grazing schemes. The dry weight and carbon content of the zooplankton can be measured directly at the end of such experiments, whereas that at the beginning must be estimated indirectly and is subject to wide error. Thus, it is possible that the disparity in the development of the two cultures (Fig. 7) was partly

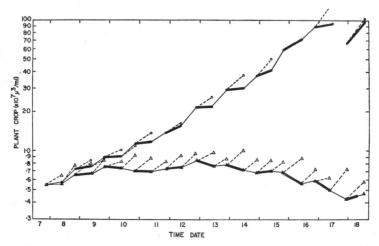

Fig. 7. Comparison of effects of continuous (lower curve) and nocturnal grazing (upper curve) on initially identical phytoplankton cultures, by initially identical zooplankton stocks. Heavy portions of curves denote laboratory night, and the fine lines, daylight. The short dashed line segments indicate the growth of the control cultures.

due to differences in the initial biomass of the zooplankton and that the disparity in zooplankton growth rates was greater than estimated, on the assumption that the initial stock of nocturnally grazing copepods was actually less than that in the continuously grazed culture. While the results of other such experiments tend to confirm the general conclusion that nocturnal grazing results in higher production over ten-day periods, the results are not all clear cut, and the *exact* consequences of the two grazing schemes for given initial stocks would be difficult to determine.

The effects of interactions of grazing schemes with other parameters has also been examined using a simulation model programmed for an IBM 1130 computer.

The model is shown schematically in Fig. 8 in which the numbers refer to equations listed in the appendix. The model permits the calculation of the sequences of stocks of phytoplankton and zooplankton ensuing from selected initial stocks as a result of the interactions among zooplankton feeding (and therefore phytoplankton mortality) specified by the modified Ivlev relation, daylight growth and nocturnal respiration by the phytoplankton, and assimilation and respiration by the zooplankton.

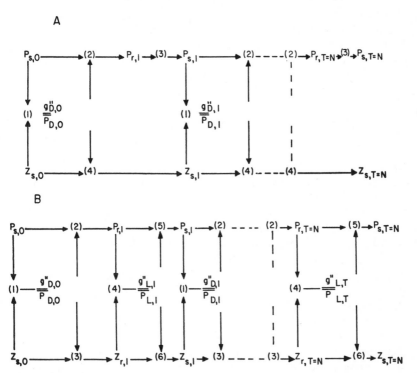

Fig. 8. Schematic depiction of operation of plankton simulation model. Numbers refer to equations listed in Appendix I, where the symbols are also described. A. Nocturnal grazing. B. Continuous grazing.

Briefly the model operates as follows. At the beginning of each sequence of calculations, the following input is specified:

1. The constants, I_M, δ, and P' which specify the Ivlev relation to be used in the estimation of the grazing mortality on the phytoplankton and the rations obtained by the zooplankton.

2. The initial sunset stocks of zooplankton and phytoplankton, as microgrammes of carbon per litre (μg C/l).

3. The instantaneous rates of daylight growth and nocturnal respiration by the phytoplankton.

4. The rate of assimilation of the ration by the zooplankton and their rate of respiration expressed as the fraction of the sunset stock required per day for maintenance.

5. The length of day and night in hours.

6. The duration of the sequence of calculations, in days, up to a maximum of 50.

7. The grazing scheme, either continuous or nocturnal.

8. The value of a factor, which may be used to model mortality or recruitment on the zooplankton.

In addition, two other options may be specified. The plant growth constant may either be assigned a constant value as implied above, or may be varied randomly, or as a function of time, with or without random fluctuations about the values specified by the function. The other option permits similar variable factors to be applied to the sunset stock of zooplankton.

The initial sunset stocks of phytoplankton and zooplankton are used, with the Ivlev constants, to estimate the rate of grazing mortality, and with the nocturnal rate of plant respiration, the nocturnal decline in the stock of plants, and the ration obtained by the zooplankton. The latter is used, in the case of nocturnal grazing, to estimate the stock of zooplankton developed by the next sunset as a result of feeding and respiration. The plant growth constant, taken to denote the rate of increase of particulate plant carbon during the daylight hours regardless of its physiological interpretation, is applied to the calculated stock of phytoplankton at the first sunrise to estimate that at the subsequent sunset.

Under continuous grazing the nocturnal ration is used to estimate the sunrise stock of zooplankton, and the sunrise stocks of both trophic levels are used to estimate the daylight rate of grazing mortality, which, combined with the plant growth constant permits the estimation of the daylight ration obtained by the zooplankton and the sunset stocks of the two trophic levels. The calculations are repeated for each day of the run.

The ration curves presented above are based on the *mean* stock of phytoplankton during the period of grazing whereas the model supplies the sunset and sunrise stocks of phytoplankton, i.e. those at the beginning of each grazing period. Therefore, successive approximations of the mean stock of plants and the rate of grazing mortality are made for each period of grazing. The rate of grazing mortality is assigned a value of zero if the approximated mean stock of phytoplankton in a grazing period is less than P', in order to prevent the incorporation of negative grazing rates. However, in the version of the model presented here the zooplankton are not prevented from grazing the phytoplankton down to stocks below the value of P' when the zooplankton stock is very high, at critical concentrations of phytoplankton (Fig. 9e). However, usually only a few such values are obtained in a given sequence and only in certain conditions. This fault in the model has since been corrected.

Unless otherwise specified, the following parameters are used in the model calculations presented here: plant growth constant = 0·0216, plant respiratory constant = 0·010, zooplankton assimilation = 0·80, zooplankton respiration = 0·04, day = night = 12 hours. Comparisons of sequences of stocks of the two trophic levels ensuing from an initial zooplankton stock of 100 μg C/l for different initial stocks of phytoplankton, under three different sets of Ivlev constants for nocturnal grazing and two under continuous grazing are shown in Fig. 9. The initial stock of zooplankton is equivalent to about one stage V *Calanus* per litre.

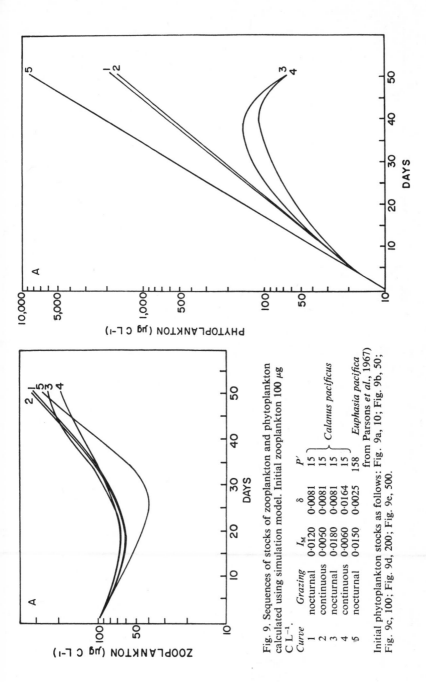

Fig. 9. Sequences of stocks of zooplankton and phytoplankton calculated using simulation model. Initial zooplankton 100 µg C L^{-1}.

Curve	Grazing	I_M	δ	P'	
1	nocturnal	0·0120	0·0081	15	
2	continuous	0·0050	0·0081	15	*Calanus pacificus*
3	nocturnal	0·0180	0·0081	15	
4	continuous	0·0060	0·0164	15	
5	nocturnal	0·0150	0·0025	158	*Euphasia pacifica*

from Parsons *et al.*, 1967)

Initial phytoplankton stocks as follows: Fig. 9a, 10; Fig. 9b, 50; Fig. 9c, 100; Fig. 9d, 200; Fig. 9e, 500.

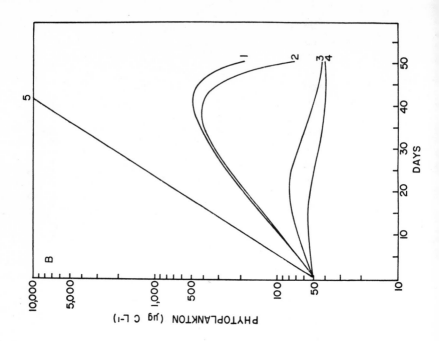

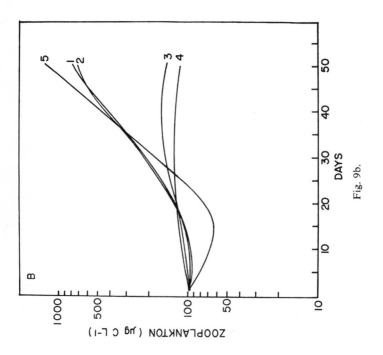

Fig. 9b.

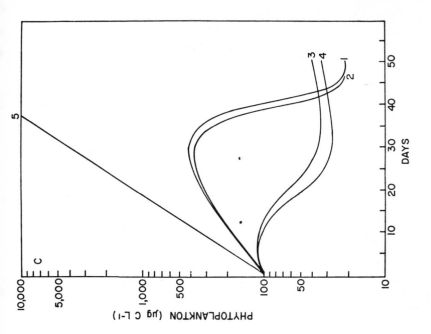

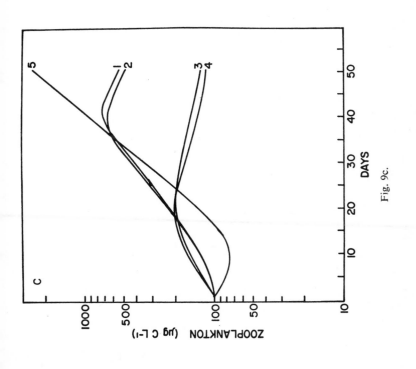

Fig. 9c.

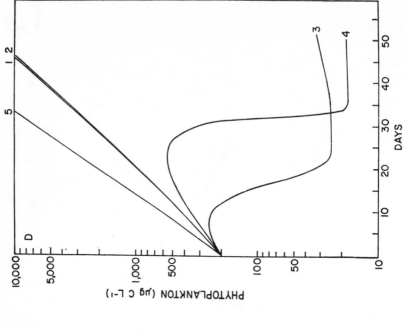

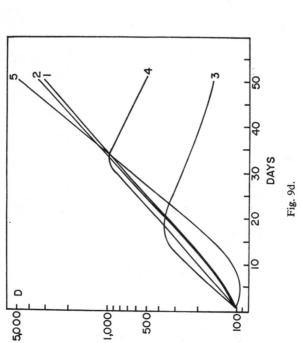

Fig. 9d.

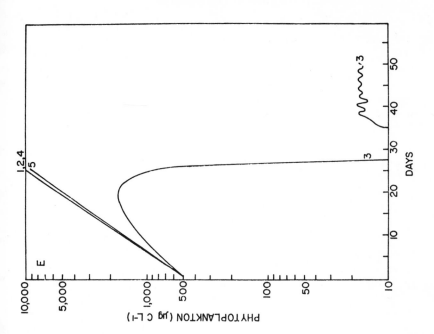

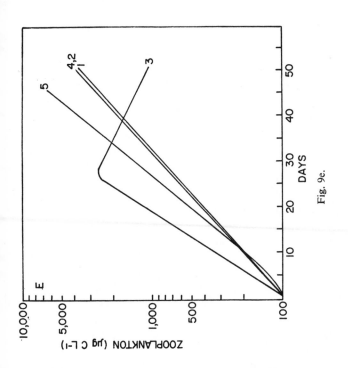

Fig. 9e.

Curves 1 and 2 in each figure are for nocturnal and continuous grazing, respectively, with ingestion curves assigned the same values for δ and P' but with I_M such that the same daily ration would be obtained by both types of grazers under a constant concentration of phytoplankton, assuming a 12 hour nocturnal grazing period. This is, in effect, the same partition of a total grazing mortality used in comparing grazing schemes in the analysis of field data cited above. Contrary to what might have been expected from the analysis of field data (Fig. 2) the differences in the effects of the two grazing schemes are minor with production slightly higher under continuous grazing. The latter partly results from the fact that the computation begins with the sunset stocks, and under nocturnal grazing, with its higher value of I_M, the plants are reduced more than under continuous grazing, and this in effect sets the pattern of differences between production under the two schemes. If the initial sunset stocks of phytoplankton are adjusted so that the same stocks of phytoplankton are achieved at the first sunrise under both schemes, nocturnal grazing does result in more production than the continuous scheme. However, the difference is small. The present model does not incorporate the effects of the variable rate of nocturnal grazing. If this was done, the partition of the plant production between plant respiration and grazing would be changed, and initial secondary production under nocturnal grazing would be increased considerably at some stock levels. However, the net effect over a period of time could be to increase the stock of zooplankton more rapidly with the possible result that (see below) plant production would eventually be retarded by increased grazing, thus decreasing final production.

It should be noted that with initial phytoplankton concentrations of less than 50 μg C/l, the zooplankton do not succeed in controlling the increase in the plant stock in the 50-day period, partly due to the decline in the stocks of zooplankton in the early part of the sequence. The decline accelerates the increase in the phytoplankton and by the time the zooplankton have started to grow rapidly the gap between the two trophic levels is too great to be closed by grazing alone. This is also a consequence of the plateau on the ration curve which specifies a declining rate of mortality per unit concentration of grazer as plant stocks increase. With ratios of zooplankton to phytoplankton below a critical value the plant stock can increase more rapidly than the total grazing mortality exerted by zooplankton stock as a whole. Of course, in nature, nutrient limitation could ultimately slow the growth rate of the phytoplankton, permitting the zooplankters to control the stock of plants.

However, with initial phytoplankton stocks of 50 to 100 μg C/l, the Ivlev constants used in curves 1 and 2 (Fig. 9) are such that the animals are able to exert sufficient grazing mortality on the plants from the beginning to succeed in controlling the plant growth. With initial plant stocks of 200 μg C/l and greater the zooplankton ration is at or near the maximum, but the rate of mortality per unit grazer has started to decline and in the absence of nutrient limitation of the plant growth rate the animals are again unable to control the increase of the phytoplankton. Thus, with the parameters used to estimate sequences 1 and 2 in Fig. 9, the animals are able to exert effective control on the development of the phytoplankton stocks only with initial stocks of between about 50 and 100 μg C/l, but as a consequence limit their own production and there is little change in final zooplankton production with changes in initial plant stock in this range.

Curves 3 and 4 in Fig. 9 present another comparison of nocturnal and continuous grazing in which the effects of specifying the observed mean Ivlev constants for the two grazing schemes are examined, with I_M, δ, and P' being 0·0180, 0·0081 and 15, respectively, for nocturnal grazing, and 0·0060, 0·0164, and 15, respectively, for continuous grazing. Use of these constants gave stock sequences markedly different from those depicted by curves 1 and 2 in the same figures. Also, the differences between sequences 3 and 4 are much greater than those between 1 and 2, and differences between curves 2 and 4 change markedly with initial plant stocks greater than 100 μg C/l. With initial plant stocks up to this value higher stocks of both trophic levels are attained under nocturnal grazing, but not as much as would be expected considering the ration curves alone. Due to the higher value of I_M under nocturnal grazing and the greater value of δ under continuous grazing, and hence higher values of m, secondary production is much less than with the previous two sets of Ivlev constants, and the changes in plant stock with time are not pronounced.

With an initial plant stock of 200 μg C/l (Fig. 9d), the stock of copepods feeding nocturnally begins to decrease near day 16, but the continuous grazers continue to increase and attain a higher maximum stock. The higher value of I_M assigned to the nocturnal grazers gives almost maximum mortality on the plant stock very early in the sequence, resulting in the decrease of the plant stock and early limitation of the growth of the zooplankton. With continuous grazing the sharp decrease in the plant stock is delayed and higher secondary production can occur. Although the mortality exerted per unit concentration of grazers decreases as the plant stock increases, the ratio of the two trophic levels is in a range such that as the total stock of zooplankton increases, the total grazing mortality exerted increases sufficiently to overcome the increase of the plants. With an initial plant stock of 500 μg C/l (Fig. 9e) secondary production is much greater under nocturnal grazing in the first part of the sequence but is exceeded by that of the continuous grazers in the latter part.

Curves 5 in Fig. 9 were computed using Ivlev constants obtained by Parsons et al. (1967) for nocturnally feeding Euphausia pacifica, with a very low value for δ and a high value for P'. As a result the phytoplankton are little affected by grazing until they reach very high levels, by which time the euphausiids have decreased to the point that they are unable to control the phytoplankton bloom even though growing very rapidly.

It will be noted that although in some cases the phytoplankton decrease very rapidly, they are not grazed to extinction (Fig. 9d and 9e) but approach a low level set by the value of P' and the stock of zooplankton present, and thereafter slowly increase. Due to the defect in the model mentioned earlier the plant stock is grazed to a level much less than the value of P' in curve 3,

Fig. 9e. However, once the plants are less than P' grazing ceases and the plants rapidly return to concentrations near P' and enter a period of short term oscillations.

It is apparent that the ratios of production under the two grazing schemes can vary markedly with time and with the starting stocks. However, using the final values of stocks in one sequence as input for the next sequence, and repeating, it appears that slowly oscillating steady states are approached. Fig. 10 shows such a sequence in which the ratios of the stocks under the two grazing schemes approach an approximately steady state, with stocks higher under nocturnal grazing.

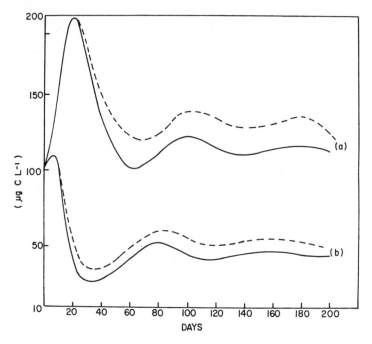

Fig. 10. Modelled comparison of nocturnal (dashed lines) and continuous (solid lines) grazing for 200 days. Curves (a) denote zooplankton and (b) phytoplankton.

Grazing	I_M	δ	P'
nocturnal	0·0180	0·0081	15
continuous	0·0060	0·0164	15

The results obtained with the model are thus not in complete agreement with those obtained in the treatment of field data (Fig. 2). The difference results from the fact that the sequence of stocks in the field study were specified by the observations, and the production for a given set of data is determined solely by interpretation of the grazing. In the model, the stocks are free to vary as dictated by the changing interactions of the stocks and the other parameters. The difference is a significant one. Assumptions which appear beneficial when estimating secondary production from given data may

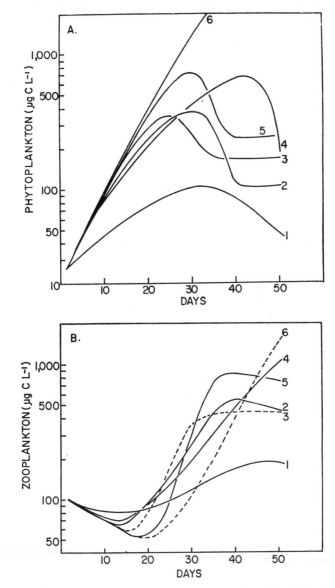

Fig. 11. Modelled stocks of phytoplankton (A) and zooplankton (B) calculated using Ivlev constants derived from Parsons *et al.* (1967) and Parsons and LeBrasseur (personal communication) and one from the present work.

Curves	Grazing	I_M	δ	P'	Source	
1	nocturnal	0·0180	0·0081	15	C. D. McAllister	*Calanus pacificus*
2	continuous	0·0083	0·0107	81	Parsons & LeBrasseur	*Calanus pacificus*
3	continuous	0·0500	0·0013	124	Parsons & LeBrasseur	*C. plumchrus* III, IV
4	continuous	0·0067	0·0110	57	Parsons & LeBrasseur	*C. plumchrus* V
5	continuous	0·0204	0·0025	190	Parsons & LeBrasseur	*Pseudocalanus, Oithonas*
6	nocturnal	0·0188	0·0025	158	Parsons *et al.*	*Euphasia pacifica*

have the opposite consequences when predicting the stocks and production developing from a set of *initial* conditions.

With respect to the development of populations from specified initial values, the differences between the effects of nocturnal and continuous grazing may be due much more to the Ivlev constants assigned to the two grazing schemes than the effects of the grazing schemes *per se*. The extremely marked differences between sequences 2 and 3 in Fig. 9 are due to the differences in I_M, and it appears that lower values of I_M may be of more ultimate advantage to a grazer, contrary to what one might expect from casual inspection of

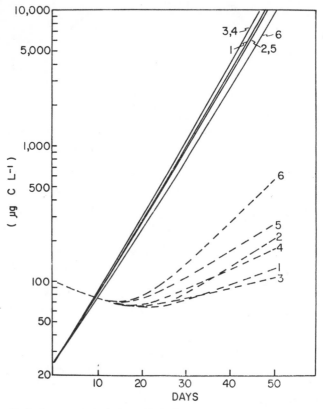

Fig. 12. Sequences of zooplankton (dashed curves) and phytoplankton (solid curves) modelled using different sets of Ivlev constants, all derived from a ration curve by Conover (1966), assuming continuous grazing.

Curve	I_M	δ	P	Carbon conversion factor	Method
1	0·0035	0·0039	0	from graph	from parabola
2	0·0046	0·0029	0	from text	from parabola
3	0·0029	0·0074	20·0	from graph	from mean points
4	0·0039	0·0056	39·6	from text	from mean points
5	0·0044	0·0072	20	from graph	from upper limit
6	0·0059	0·0054	39·6	from text	from upper limit

ration curves. The marked differences between curves 1 and 4 are due to the change in the value of δ. While the higher value of δ may give more rations at lower plant stocks, the effect on the development of populations over a period of time may be to reduce the production in both trophic levels.

Shown in Fig. 11 are sequences of stocks estimated using ration curves observed by Parsons et al. (1967) and Parsons and LeBrasseur (personal communication), compared with one sequence obtained using one of the present sets of Ivlev constants. Although all the sequences seem plausible it should be noted that each ingestion curve has markedly different consequences for the development of stocks of both trophic levels. With the generally higher values of P' found by Parsons et al. (1967) the phytoplankton after heavy grazing, attain plateaus at rather high stock levels, in contrast to the results obtained by the author. The pronounced differences between curves 5 and 6 are due mainly to a small change in the value of I_M.

Stock sequences, estimated using different sets of Ivlev constants fitted to ingestion curves drawn through data of Conover (1966, his Fig. 2A), are shown in Fig. 12. Constants were fitted to the first limb on Conover's parabola, to a curve fitted by eye to his mean points, and to a curve drawn through the upper limits shown to his data, using both the carbon conversion factor implicit in the Conover figure, and that given in his text, since the latter differed from the former. Fig. 12 shows that except for the sequence estimated using the upper limits of Conover's (1966) data and the higher of his two carbon factors, zooplankton growth was very slow, even at high concentrations of phytoplankton. Similar sequences, with slow zooplankton growth and little control of the plant stock, were obtained using Ivlev constants fitted to the first limb of two parabolic ingestion curves shown by Mullin (1963). The problem of controlling the plant increase and of zooplankton growth would have been worsened had the parabolic relation been used in the model, since rations would have actually decreased at moderate to high concentrations of phytoplankton.

The above results are not necessarily a criticism of the data by Mullin (1963) and Conover (1963). The phytoplankton concentrations given by the authors appear to be those at the beginning of feeding rather than the mean during the grazing period. Converting the concentrations to the mean would have had the effect of shifting I_M to the left and increasing the value of the constant, δ, and therefore increasing the mortality exerted on the phytoplankton per unit concentration of grazer. Transforming the published data to equivalent carbons was another source of possible error. Nevertheless, the computation serves to indicate the use of the model for testing given sets of data, as did the use of ration curves due to Parsons et al. (1967), and Parsons and LeBrasseur (personal communication) (Fig. 11).

Differences in I_M and δ were suggested to be more important in producing differences between sequences than the difference between nocturnal and continuous grazing per se. The use of four different values of P', with I_M and δ equal to 0·0180 and 0·0081 respectively, showed that the sequences of stocks are also extremely sensitive to the choice of value of P' (Fig. 13).

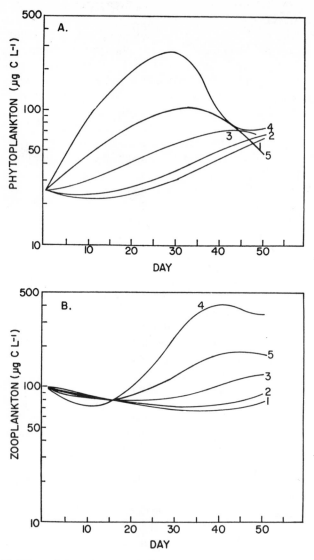

Fig. 13. Effects of changes in P' on modelled sequences of phytoplankton (A) and zooplankton (B). Nocturnal grazing, $I_M = 0.0180$, $\delta = 0.0081$. Curves 1, 2, 3, 4, and 5 calculated with $P = 0, 1, 5, 50$ and 15 respectively.

The same values of P' tested under continuous grazing with the same values of I_M and δ used in curve 4 of Fig. 9 produced an even wider spread amongst the stock sequences, as did applying random variations to the plant growth constant. With random variations applied to the zooplankton of Fig. 13, the maximum spread was not affected but the duration of the maximum spread was increased to about 20 days (Fig. 14). It may be noted that sharp

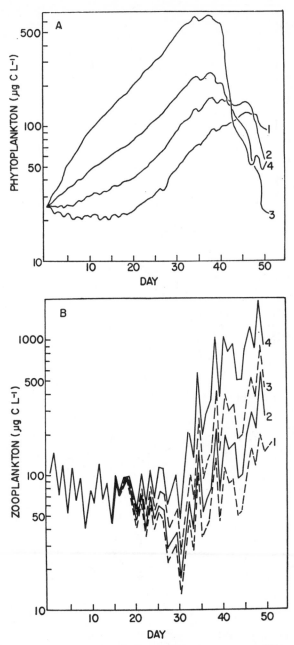

Fig. 14. Effects of changes in P' on modelled sequences of phytoplankton (A) and zooplankton (B), with random variation applied. Nocturnal grazing, $I_M = 0.0180$, $\delta = 0.0081$. Curves 1, 2, 3, and 4 estimated with $P' = 0$, 5, 15 and 50 respectively.

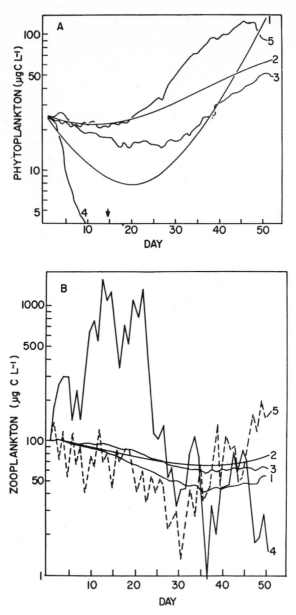

Fig. 15. Effects of variability applied to plant growth rate, or zooplankton, with nocturnal grazing, $I_M = 0.0180$, $\delta = 0.0081$, $P' = 0$. A. Phytoplankton, arrow indicates date of extinction of phytoplankton. B. Zooplankton.

Curve	Plant growth rate	Zooplankton
1	linear increase 0·0144 to 0·288	smooth
2	constant 0·0216	smooth
3	random about 0·0216	smooth, but randomly varied rations
4	random about 0·0216	random series 1
5	constant 0·0216	random series 2

fluctuations in zooplankton did not result in correspondingly large variations in phytoplankton.

Fig. 15 depicts stocks computed with P' equal to zero and with different schemes of variation applied to the plant growth constant and the stock of zooplankton. The application of one set of random variations to the zooplankton resulted in the phytoplankton being grazed to extinction (curves 4, Fig. 15). A minor change in the order of the factors in the random series prevented extinction and resulted in much less marked fluctuations in the stock of zooplankton. This and other data not presented here indicate that changes in perturbation applied to the zooplankton may have profound effects. Different schemes of variation applied to the plant growth constant and the imposition of random fluctuations on the ration obtained by the grazers (curves 1, 2, and 3, Fig. 15) made appreciable differences to the development of the phytoplankton but had much less effect on the sequences of zooplankton stock. However, with $P' = 50$, marked differences between zooplankton sequences 1, 2 and 3 did occur, and of course the phytoplankton

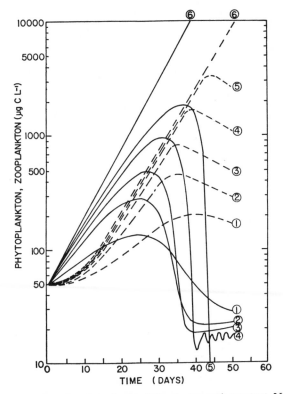

Fig. 16. Effects of varying the estimate of the plant growth constant. Nocturnal grazing: $I_M = 0.0180$; $\delta = 0.0031$; $P' = 15$; plant respiration 0.005. Solid lines, phytoplankton. Dashed lines, zooplankton. Plant growth constant for curves 1, 2, 3, 4, 5 and 6 are 0.0144, 0.0175, 0.0190, 0.200, 0.0205 and 0.0216 respectively.

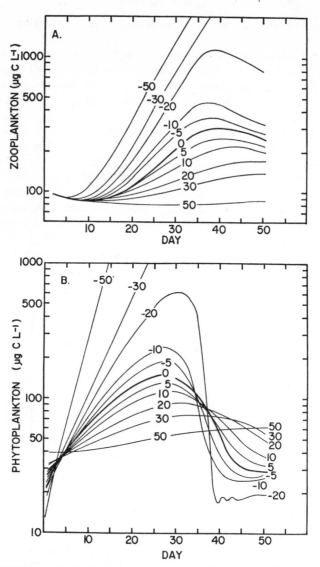

Fig. 17. Effects of errors in plant growth constant resulting from errors in estimate of phytoplankton stock. The numbers indicate the percentage error above or below the true initial plant stock, assumed in calculating the sequences. Nocturnal grazing, $I_M = 0.0180$, $\delta = 0.0081$, $P' = 15$. True plant growth constant $= 0.0216$. A. Phytoplankton. B. Zooplankton.

could not be grazed to extinction. It is apparent that the choice of P' can profoundly affect the sequences of stocks in a manner not readily apparent from inspection of ration curves, but that the effects of changes or errors in the estimate of P' can vary strongly with the accompanying parameters.

The original intent of the model was to examine the implications of ration curves for the development of stocks and to facilitate comparisons of the effects of the two different grazing schemes. However, it may be used in other ways. For instance, Fig. 16 shows the effects of varying k, the plant growth constant, within a total spread of $\pm 15\%$ about a value of 0·0175. As k increases, the sequences of stock become more sensitive to small variations in the plant growth constant. Thus, increasing k from 0·0200 to 0·0205 has a very large effect, equal to that obtained by the change from a value of 0·014 to 0·0175. The $\pm 15\%$ range about the value of 0·0175 produces a range in the estimate of secondary production of about an order of magnitude. This is rather disturbing, since it is unlikely that the plant growth constant can be estimated in nature within even this degree of uncertainty.

One source of error in estimating the plant growth constant from field data on primary production would be error in the estimate of the standing stock of phytoplankton. In addition to giving false estimates of the plant growth constant, such errors will give errors in the estimate of the grazing mortality and the ration obtained by the zooplankton, when ingestion curves are applied as in the present work. Fig. 17 shows sequences of stocks of the two trophic levels obtained using k values estimated on the assumption of various errors in the estimate of the initial plant stock. Thus, a -20% error in initial plant stock, combined with the resulting error in k, overestimates the stocks by a maximum of about 4-fold, whereas overestimating the initial plant stock by 20% results in underestimating the stocks by a maximum of about 30%. The figure indicates that the error in the predicted stocks from this source is non-linearly related to the error in the estimate of the initial plant stock, and that the magnitude and direction of the error may vary with the duration of the run. The magnitude and direction of the error also were found to change markedly with the grazing scheme, with the Ivlev constants assigned, and the other parameters, and would have to be evaluated separately for each set of circumstances.

The output of the model can also be treated as if it was field data and used to estimate production by various techniques such as that employed by McAllister (unpublished) when making the estimates presented in Fig. 2. For example the first and last 25 days of a computer run were averaged, and the given plant growth constant applied, with the rate of respiration by the plants, to estimate the rate of grazing mortality from the difference in the averages for the two periods, and thence the total secondary production for the 50-day period. Secondary production was also estimated from each "monthly" average, calculating the grazing mortality on the assumption that a steady state obtained for each 25-day period. Thirdly, weekly averages of the stocks, the plant growth rate and the grazing mortality actually tabulated in the computer output were taken and used to estimate secondary production. The results of these calculations are compared (Table 1) with the true secondary production for a sequence with random variations applied to the zooplankton and for a run with no perturbations applied to the zooplankton. It is evident that the use of averaging of time series data can lead to serious errors in attempts to estimate secondary production from field data, and that the estimates of secondary production depicted in Fig. 2, taken from McAllister (unpublished), may be in serious doubt as a result of the averaging used, although his discussion of the importance of grazing schemes is not necessarily affected.

TABLE 1. Secondary production estimated from data in computer outputs

Method	Random zooplankton	Smooth zooplankton
Changes between 25-day means	792 μg C/l	201 μg C/l
25-day steady states	561	198
Weekly means, observed grazing rates	570	109
True production	475	90

When random variations are applied to the stock of zooplankton in the model the observed change in stock does not necessarily represent the true production, even in the absence of predation since zooplankters actually produced by grazing on the phytoplankton are arbitrarily removed, and others produced elsewhere, are added to the stock. Hence, to obtain the true production resulting from consumption of the "local" phytoplankton, the daily ration taken by the animals was estimated from the data in the computer output and corrected for assimilation and respiratory losses by the observed daily stocks of the zooplankton.

The range and standard deviation of the random factor applied to the zooplankton in the model were selected to approximately match daily variability in catches of zooplankton observed at Ocean Station P. The manner in which the variability is imposed in the model implies a coherent stock of phytoplankton in the upper layers of the ocean, beneath which a variable stock of nocturnally migrating zooplankters were being advected, relative to the upper layer. Zooplankton migrating up to graze on the plant stock on one night would move down at dawn and be carried away to be replaced by a different body of grazers on the next night. Such an explanation does not appear too far fetched except that a plant stock of the required coherence seems rather unlikely, and the above reasoning applies only to the case of nocturnal grazers. In a recent modification to the model, patchiness can be applied to the phytoplankton as well as the zooplankton, mitigating the phytoplankton coherence requirement and facilitating the interpretation of the modelling of continuous grazing with random variations applied to the zooplankton.

The above deals with the possible effects of averaging time series data. Modelling may also be of use in assessing the effects of spatial averaging. Assume two "patches" of water with differing initial stocks of phytoplankton and/or zooplankton. We may average the initial stocks and compute the production ensuing from the mean, or we may estimate the production in each patch and average the end result. Four different initial distributions, each giving an average stock of 50 μg C/l of both the zooplankton and phytoplankton, are listed in Table 2, and the estimates of secondary production by the two methods for the different distributions are shown in Table 3.

The results suggest that the averaging of spatial distributions can lead to serious error. The assumption that the two patches retain their identity with no exchange between them, is a serious oversimplification, but case 4 and other examples not presented here indicate that the errors can become serious within a week or ten days. The correct modelling of patchiness and the assessment of the penalties of averaging data cannot be attained until we

TABLE 2. Postulated distributions of zooplankton and phytoplankton between two patches

Case	Trophic level	Patch 1 (μg C/l)	Patch 2 (μg C/l)
1	zooplankton	25	75
	phytoplankton	50	50
2	zooplankton	50	50
	phytoplankton	25	75
3	zooplankton	25	75
	phytoplankton	25	75
4	zooplankton	25	75
	phytoplankton	75	25

TABLE 3. Secondary production ensuing from stocks listed in Table 2

Day	From mean of initial stocks μg C/l	Case	From means of final production 1	2	3 μg C/l	4
10	22		17	22	27	-1
30	466		380	422	251	305
50	380		1981	476	1635	1065

know much more of the nature of patchiness in the sea. It is obvious that if a patch really consists of a convergent area, in which the phytoplankton are advected in to the zooplankton accumulated in the convergence, the simple form of modelling, and method of assessment of the effects of averaging used above will require considerable revision. Nevertheless the examples serve to indicate that indiscriminate averaging may produce large errors, and that the consequences of averaging should be examined in each case before accepting the results.

GENERAL DISCUSSION

The examples given of three approaches to the assessment of concepts and data on zooplankton feeding relations are fragmentary and incomplete. The examination of field data cited (McAllister, unpublished) required many approximations and assumptions in the absence of certain knowledge, and for convenience in manual calculations. Application of the computer simulation model suggested that the estimates of secondary production themselves (Fig. 2) may be in serious doubt due to the method of averaging. Data were not available for partitioning the zooplankton between the two grazing schemes, and the relative error due to incorrect partitioning was found to vary markedly with zooplankton respiration (Fig. 2), much studied, but in any given situation known only within limits such that it is impossible to assess the real as opposed to the possible errors due to the grazing scheme assumed. The paucity and variability of data on phytoplankton respiration are an equally severe impediment to such studies. Assessment of the effects of possible errors in the estimate of plant stock on the calculation of the phytoplankton growth constant (Fig. 17) using the computer model demonstrated another serious source of error. Nevertheless, the main conclusion (McAllister, unpublished) that in estimating secondary and higher production from time series measurements of plankton stocks and primary production it is necessary to take into account the partition of the grazers between the two schemes appears warranted as does the more general conclusion that our habit of disregarding the phase relations between different processes, *within days*, is a serious source of error.

The greater estimates of secondary production obtained by applying a *given* rate of mortality nocturnally in the field study, tempted the conclusion that nocturnal grazing was of more benefit to grazers capable of performing the required diurnal vertical migrations than continuous grazing. Results of the 10-day grazing experiments and casual inspection of the variation of the rate of feeding with time both seemed to confirm this. However, use of the simulation model (curves 1 and 2, Fig. 9) showed that this is not necessarily so when the stocks are free to vary as a result of the various interactions, instead of being *specified* as in the case of field data, and that the conclusion may be reversed in some circumstances, if different ingestion curves are specified for the two schemes. It would appear that the only certain benefit of nocturnal grazing is to the biological oceanographer hoping for high estimates of secondary production from field data. Things are not so simple when advantage to the organism is considered.

The study of field data did not take into account either the ingestion curves or the observation that nocturnally and continuously feeding copepods may not exert the equivalent total daily mortality on the plants. The estimation of the grazing rates from the field data itself has the advantage that the rates so estimated are consistent with the observations of primary production and the observed changes in the standing stock of phytoplankton. If laboratory observations on zooplankton feeding are to be applied to field data it needs to be demonstrated that the implied mortality on the phytoplankton is consistent with the changes in the stock and its growth rate. Thus, the observed changes in the Station P stocks studied by McAllister (unpublished) could not be accounted for by the application of the ingestion curves presented here. The converse, the application of the mortality on the phytoplankton exerted per unit concentration of zooplankton (e.g., Fig. 5) to estimate the stock of zooplankton from the rate of total grazing mortality calculated from the field data, badly overestimated the zooplankton stocks (McAllister, unpublished). Nevertheless this would appear to be a useful independent method of estimating zooplankton stocks and warrants further investigation.

When the observed mean ingestion curves for the two grazing schemes were tested using the simulation model it was found that differences in the sequences of stocks were due to the Ivlev parameters rather than to the choice of grazing scheme *per se*, that the ratios of production under the two schemes varied markedly with the duration of the sequence and the ratio of the two stocks, that the sequences of stocks calculated were very sensitive to small changes or errors in the Ivlev constants (e.g. Fig. 15), and that such errors were very sensitive to the accompanying parameters and daily variability in zooplankton stock. Since the errors would be difficult to assess for any given situation by inspection of the component parameters, it would appear that some form of computer modelling is necessary.

The simulation model presented here contains many deficiencies. One of the most obvious is that nutrient limitation of plant growth is not taken into account and as a result impossibly high stocks can be predicted, in both trophic levels. In dealing only with concentrations of carbon, such factors as recruitment and reproduction are not adequately dealt with. No direct provision was made for changes in the ingestion curves with the growth of the animals. However, this could be incorporated indirectly by calculating shorter sequences and changing the Ivlev constants (and such factors as rates of respiration) between runs. While mortality on the zooplankton can be modelled through the application of arbitrary factors, the feedback between the resulting growth of the implied predators and the two lower trophic levels is not taken into account.

Because of these deficiencies and others which could be listed, and the fact that so many of the parameters with which we deal are so poorly known or subject to serious error, it might be felt that there is little point in attempting to develop such models. I would support an opposite view, that knowing our measurements and assumptions to be subject to wide error from a wide variety of sources, we would be negligent in *not* seeking by whatever means possible to assess their implications for what we claim we are trying to do. Because we deal with complex systems, the consequences of errors, conflicting assumptions, alternate choices, and decisions as to whether or not, or how to average data can rarely be divined by conventional inspection, whereas computer modelling would seem to be a rapid and useful way of doing so.

Some other aspects of food chain data which might benefit from examination in the context of trophic relationships using computer models may be cited. The present model, and others (Riley, 1963), assume plant growth to be exponential, whereas a number of studies (e.g. Lorenzen, 1963) demonstrating diurnal variation in photosynthetic potential cast doubt on the assumption. McAllister (unpublished) has speculated that if prey size selectivity by herbivorous zooplankton enables them to select larger cells from within species of phytoplankton, then it follows that cells about to divide suffer preferential mortality, and our present techniques for estimating grazing by the zooplankton and mortality on the plants will require drastic revision. Furthermore, different cycles and degrees of synchrony in cell division and volume increase, combined with the fact that increase in nutritive content is limited to the daylight hours, may also have implications for trophic relations. McLaren (1963) concluded that diurnal vertical migration by zooplankton conferred increased fecundity and growth when carried out in the presence of appropriate temperature gradients, whereas Petipa (1966) concluded that such migrations were metabolically very expensive. Odum *et al.* (1963) appear to have concluded that the rate of phytoplankton respiration may decline through the night. All these observations could have important implications for the study of trophic relations which would be difficult to assess or compare without some form of modelling and would be inflexible and exceedingly tedious by manual computation. Until they have been assessed in relation to other processes operating, we are not justified in excluding them in our attempts to estimate and predict production or study the factors controlling it. The desirability of long term culture experiments, whether aimed at life history studies (e.g. Mullin and Brooke, 1967) or productivity oriented seems self-evident and requires little discussion.

The foregoing is not intended to portray computer modelling as an alternative to experiment and observation but as a useful, and perhaps essential adjunct of which much more use might profitably be made.

ACKNOWLEDGEMENTS

It is a pleasure to express my gratitude to J. A. Thomson and the staff of the computing centre at the Fisheries Research Board of Canada Biological Station, Nanaimo, B.C., for the development of the IBM 1130 computer program which enabled the use of the simulation model presented here. However, any errors in conception and formulation must be attributed to me.

REFERENCES

ADAMS, J. A., and STEELE, J. H. 1966. Shipboard experiments on the feeding of *Calanus finmarchicus* (Gunnerus). In *Some contemporary studies in marine science*, edited by H. Barnes. London, George Allen and Unwin, 19-35.
CONOVER, R. J. 1966. Factors affecting the assimilation of organic matter by zooplankton and the question of superfluous feeding. *Limnol. Oceanogr.*, 346-54.
CUSHING, D. H. 1959. On the nature of production in the sea. *Fishery Invest., Lond.*, Ser. 2, **22** (6), 40 pp.
IVLEV, V. S. 1945. The biological productivity of waters. *Usp. sovrem. Biol.*, **19** (1), 88-120.
JITTS, H. R., and others. 1964. The cell division rates of some marine phytoplankters as a function of light and temperature. *J. Fish. Res. Bd Can.*, **21**, 139-57.
LORENZEN, C. J. 1963. Diurnal variation in the photosynthetic activity of natural phytoplankton populations. *Limnol. Oceanogr.*, **8**, 56-62.
MCALLISTER, C. D. (Unpublished). Manuscript submitted to *J. Fish. Res. Bd Can.*
MCLAREN, I. A. 1963. Effects of temperature on the growth of zooplankton, and the adaptive value of vertical migration. *J. Fish. Res. Bd Can.*, **20**, 685-727.
MCMAHON, J. W., and RIGLER, F. H. 1963. Mechanisms regulating the feeding of *Daphnia magna* Straus. *Can. J. Zool.*, **41**, 321-32.
MULLIN, M. M. 1963. Some factors affecting the feeding of marine copepods of the genus *Calanus*. *Limnol. Oceanogr.*, **8**, 239-50.
MULLIN, M. M., SLOAN, P. R., and EPPLEY, R. W. 1966. Relationship between carbon content, cell volume, and area in phytoplankton. *Limnol. Oceanogr.*, **11**, 307-11.
MULLIN, M. M., and BROOKE, E. R. 1967. Laboratory culture, growth rate, and feeding behaviour of a marine planktonic copepod. *Limnol. Oceanogr.*, **12**, 657-66.

ODUM, H. T., BEYERS, R. J., and ARMSTRONG, N. E. Consequences of small storage capacity in nanoplankton pertinent to measurement of primary production in tropical waters. *J. mar. Res.*, **21**, 191-98.

PARSONS, T. R., LEBRASSEUR, R. J., and FULTON, J. D. 1967. Some observations on the dependence of zooplankton grazing on cell size and concentration of phytoplankton blooms. *J. oceanogr. Soc. Japan*, **23**(1), 10-17.

PETIPA, T. S. 1966. On the energy balance of *Calanus helgolandicus* (Claus) in the Black Sea. In *Physiology of marine animals*. Moscow. Acad. Sci. Publ. House. Transl. by M. A. Paranjape, Dep. Oceanogr, Univ. Wash., 60-81.

REEVE, M. R. 1963. The filter feeding of *Artemia*. I. In pure cultures of plant cells. *J. exp. Biol.*, **40**, 195-201.

RIGLER, F. H. 1961. The relation between concentration of food and the feeding rate of *Daphnia magna* Straus. *Can. J. Zool.*, **39**, 857-68.

RILEY, G. A. 1963. On the theory of food chain relations in the sea. In *The Sea*. 2. Edited by M. N. Hill. N.Y., Interscience Publ., 438-63.

STEEMAN-NIELSEN, E. 1963. Productivity, definition and measurement. In *The Sea*. 2. Edited by M. N. Hill. N.Y., Interscience Publ., 129-64.

APPENDIX: EQUATIONS

NOCTURNAL GRAZING

The subscripts s, D and r refer to sunset, darkness and sunrise respectively. The number following a comma immediately after one of the above subscripts indicates the day, T, in the sequence, and N is the total number of days in the run. The numbers of the equations below are those indicated in Fig. 8.

Equation 1

The rate of ingestion at a particular concentration of phytoplankton, divided by that concentration, is taken as an estimate of the instantaneous rate of grazing mortality exerted on the plants per unit grazer. Hence the total rate of grazing mortality may be taken as the product of the rate of ingestion, as estimated by the expression given above, and the ratio of the stock of zooplankton to that of phytoplankton. However, the values of phytoplankton stock given in the model are those at the onset of a given grazing period, whereas the expression used here to relate ingestion to plant concentration requires the *mean* concentration during the period of grazing.

The given value for the stock of phytoplankton is therefore used to approximate a value for the grazing mortality which is then used to give a first approximation to the mean stock of plants. This value is then used to give a second approximation to the grazing mortality. Another approximation of the mean is made and used to calculate a final estimate of the rate of grazing mortality.

The expression used for estimating the mean was taken from Cushing (1959).

$$(a)\quad g_{D,T} = \frac{Z_{s,T}}{P_{s,T}} I \left(1 - \exp\{-\delta[P_{s,T} - P']\}\right)$$

$$(b)\quad \bar{P}_{D,T} = P_{s,T} \frac{(1 - \exp\{-[n + g_{D,T}]t_D\})}{(n + g'_{D,T})t_D}$$

$$(c)\quad g'_{D,T} = \frac{Z_{s,T}}{\bar{P}_{D,T}} I_M \left(1 - \exp\{-\delta[\bar{P}_{D,T} - P']\}\right)$$

$$(d)\quad \bar{\bar{P}}_{D,T} = P_{s,T} \frac{(1 - \exp\{-[n + g'_{D,T}]\, t_D\})}{(n + g'_{D,T})\, t_D}$$

$$(e)\quad g''_{D,T} = \frac{Z_{s,T}}{\bar{\bar{P}}_{D,T}} I_M \left(1 - \exp\{-\delta[\bar{\bar{P}}_{D,T} - P']\}\right)$$

where $g_{D,T}$, and $g''_{D,T}$ are succcessive approximations to the instantaneous rate of grazing mortality on the phytoplankton during the night beginning on day T; $Z_{s,T}$ and $P_{s,T}$ are the stocks of zooplankton and phytoplankton, respectively, at sunset on day T; $\bar{\bar{P}}_{D,T}$ and $P_{D,T}$ are successive approximations to the mean stock of plants during the night beginning on day T; I_M is the maximum possible rate of ingestion attainable by the zooplankton; δ is a constant; P' is the concentration of phytoplankton below which the animals cease to feed; n is the instantaneous rate of nocturnal respiration by the phytoplankton and t_D is the duration of the night in hours.

When the concentration of phytoplankton, $P_{D,T}$ is less than P', negative values of the grazing mortality are obtained. Hence, g'' is arbitrarily assigned a value of zero when the mean stock of plants is less than the value of P'. This prevents the inclusion of negative rates of grazing mortality but the present version of the model does not prevent sunset or sunrise estimates of the plant stock from falling below P' as a result of grazing by very high stocks of zooplankton. However, usually only one such value is attained in a given run, and this error has since been rectified.

Equation 2

The sunrise stock of plants remaining after grazing and plant respiration during the night was taken as,

$$P_{r,T} = P_{s,T-1} \exp\{-(n + g''_{D,T-1})\, t_D\}$$

where $P_{r,T}$ is the concentration of phytoplankton at sunrise on day T, and $P_{s,T-1}$ is that at the previous sunset; n is the instantaneous rate of nocturnal plant respiration; $g''_{D,T-1}$ is the instantaneous rate of grazing mortality approximated from the plant and animal stocks at sunset on the previous day, $T-1$; and t_D is the duration of the night in hours. The values of n and t_D are assigned at the start of each run.

Equation 3

The sunset stock of phytoplankton is estimated as,

$$P_{s,T} = P_{r,T} \exp k t_L$$

where $P_{s,T}$ and $P_{r,T}$ are as above; t_L is the duration of the daylight period in hours; and k is the instantaneous rate of net growth by the phytoplankton during the daylight. The value of t_L is assigned at the start of each run. Values of k, either constant throughout the run, or randomly varying, or as a function of time with or without random fluctuations are also given at the start of each run.

Equation 4

The stock of zooplankton achieved at sunset on day T, as a result of the ration, $g''\bar{P}_D$, obtained during the previous night and respiration during the 24 hours is approximated as,

$$Z_{s,T} = (Z_{s,T-1}\,(1-c) + t_D\,Ag''_{D,T-1}\,\bar{\bar{P}}_{D,T-1})\,F_T$$

where $Z_{s,T}$ is the concentration of zooplankton at sunset on day T; $Z_{s,T-1}$ is that at the previous sunset; c is the rate of respiration by the zooplankton, expressed as a fraction of the sunset stock on day $T-1$, per day; A is the fraction of the ration obtained which can be used; and $g''_{D,T-1}$ and $\bar{\bar{P}}_{D,T-1}$ are the grazing mortality and the mean stock of plants during the night, respectively. The product of the latter two items gives an estimate of the ration obtained by the animals. The factor, F_T, may be assigned any constant value, or varied during a run. Factors greater than 1·0 may be taken to crudely model recruitment. Those less than 1·0 can be interpreted as mortality on the zooplankton, or if a random series is used it may be considered that patchiness is crudely modelled. The increments or decrements are arbitrarily applied at sunset and their exact interpretation is not simple as noted in the general discussion of some of the results observed with the model. The growth expression itself is a rather crude one, implying as it does that zooplankton growth increments do not begin to respire until the sunset following their aquisition, and thus may overestimate the growth rate.

CONTINUOUS GRAZING

The continuous model differs from the nocturnally grazed situation only in that the values of the stock of zooplankton are estimated for each sunrise as well as for each sunset, and in that the effects of daylight grazing must be taken into account in the estimation of the sunset stocks of both the zooplankton and the phytoplankton. There are slight modifications to the system of equations, but the list of input parameters and the basic calculations are the same.

Equations (1) as in equations (1) of the model containing only nocturnal grazing.

Equation (2) as in equation (2) of the model containing only nocturnal grazing.

Equation (3) Estimation of sunrise stock of zooplankton:

$$Z_{r,T} = Z_{s,T-1} (1 - \frac{t_D}{24}c) + Ag''_{D,T-1}\bar{\bar{P}}_{D,T-1}t_D$$

where the symbols are as in the previous section.

Equations 4

Successive approximation of the grazing mortality and mean stock of phytoplankton during the daylight hours.

(a) $g_{L,T} = \dfrac{Z_{r,T}}{P_{r,T}} I_m (1 - \exp\{-\delta[P_{r,T} - P']\})$

(b) $\bar{P}_{L,T} = P_{r,T} \dfrac{\exp\{(k - g_{L,T})t_L\} - 1}{(k - g_{L,T}) \, t_L}$

(c) $g'_{L,T} = \dfrac{Z_{r,T}}{P_{L,T}} I_m (1 - \exp\{-\delta[P_{r,T} - p']\})$

(d) $\bar{P}_{L,T} = P_{r,T} \dfrac{\exp\{(k - g'_{L,T})t_L\} - 1}{(k - g'_{L,T})t_L}$

(e) $g''_{L,T} = \dfrac{Z_{r,T}}{P_{L,T}} I_m (1 - \exp\{-\delta[P_{r,T} - P']\})$

where the subscript, L, refers to daylight and the other symbols are as before. Again, if the calculation yields a negative estimate of the final approximation to the grazing mortality it is arbitrarily assigned a value of zero, although in the version presented here, application of g can in some circumstances graze the plants to a value less than P'.

Equation 5

Estimation of the sunset stock of phytoplankton

$$P_{s,T} = P_{r,T} \exp\{(k - g''_{L,T}) \, t_L\}$$

where the symbols are as before.

Equation 6

Estimation of the sunset stock of zooplankton

$$Z_{s,T} = (Z_{r,T} [1 - \frac{t_L}{24}c] + t_L Ag''_{L,T}\bar{\bar{P}}_{L,T})F_T$$

where the symbols are as before.

The program has now been modified to list in the output the cumulative totals of nominal phytoplankton production (that which is observed in present field techniques in which the measurements are made in bottles in the absence of grazing by the zooplankton); the true net plant production (taking into account the effects of daylight grazing in reducing daylight production in continuously grazed populations as well as the effects of nocturnal respiration of the plants, corrected for the effects of nocturnal grazing on the stock of plants). In addition, the cumulative totals of the rations obtained by the zooplankton are listed. These features are useful in calculating the efficiencies with which zooplankton are produced, and also in estimating the true secondary production in situations where F_T may be giving increments or decrements to the animal stock which are not production in the usual sense.

The biomass and production of different trophic levels in the pelagic communities of south seas

V. N. Greze
Institute of the Biology of the Southern Seas
Academy of Sciences of the Ukraine SSR
Sevastopol, USSR

ABSTRACT. The ratio of the biomass of phytoplankton, herbivorous and carnivorous zooplankton in the Black and Mediterranean Seas and in the tropical region of the Atlantic is considered. The values of primary production and secondary production of herbivores and carnivores determined by the rate of growth and the age composition of populations are compared. The comparison shows the intensity of utilization of energy at the successive trophic levels in different geographical latitudes, at different depths and under different ecological conditions.

Trophodynamics is one of the most important problems of modern oceanography, and many different approaches are needed for its study, such as feeding habits of marine animals, the quantitative physiological regularities of food utilisation, food webs and many other questions. But there is a more general point of view on this problem which summarizes the other approaches. I have in mind the creation and transformation of organic matter in the sea as a whole. We have to study the predator-prey relation, the utilization of food and so on but must not forget that the results are valuable in the end as new data for solving the main problem of the biological productivity of the ocean.

Modern ecology and marine ecology especially can consider the diversity of feeding relations in the community as a comparatively simple generalized food chain which includes only 3-5 links. It is important for the development of our presentation on the dynamics of organic matter and hence on the productivity of the sea to measure the ratios between these generalized elements of the food chain in different localities and seasons. Only after these data are obtained can the problem of the estimation of biological production

TABLE 1. The calculation of the biomass and production on various trophic levels in the neritic zone of the Black Sea off Sevastopol (summer)

Trophic levels and their components	Dry matter (%)	Cal/mg dry weight	P/B coeffi-cient	Biomass			Daily production		
				Wet weight mg/m³	Dry weight mg/m³	Cal.	Wet weight mg/m³	Dry weight mg/m³	Cal.
1	2	3	4	5	6	7	8	9	10
I Phytoplankton	8	2·4	0·80	450·00	36·00	66·40	360·00	28·80	53·10
II Phytophages				74·77	8·82	36·09	10·45	1·37	5·58
Cladocera (*Penilia, Evadne*)	15	4·1	0·19	12·04	1·81	7·42	2·31	0·35	1·44
Paracalanus parvus	12	4·2	0·09	18·85	1·41	5·92	1·63	0·19	0·79
Pseudocalanus elongatus	13	4·6	0·14	17·67	2·30	10·58	2·48	0·32	1·47
Centropages ponticus	13	4·6	0·09	1·0	0·13	0·59	0·09	0·01	0·05
Acartia clausi	12	4·1	0·12	12·33	1·46	6·00	1·48	0·18	0·75
Oithona similis	14	3·8	0·08	0·50	0·07	0·26	0·08	0·01	0·04
O. minuta	14	3·8	0·11	3·59	0·50	1·90	0·85	0·12	0·45
Oikopleura dioica	10	3·0	0·30	1·48	0·15	0·45	0·44	0·04	0·12
Mollusca larvae	15	3·0	0·15	3·83	0·57	1·71	0·57	0·09	0·27
Polychaeta larvae	12	3·0	0·15	3·48	0·42	1·26	0·52	0·06	0·18
III Carnivores				35·68	4·07	16·55	6·69	0·67	2·83
Acartia clausi	12	4·1	0·12	1·98	0·24	0·99	0·19	0·02	0·08
Oithona similis	14	3·8	0·08	2·09	0·29	1·10	0·11	0·01	0·04
O. minuta	14	3·8	0·11	8·65	1·21	4·59	0·83	0·12	0·45
Sagitta setosa	8	4·0	0·20	22·26	1·78	7·12	4·45	0·35	1·40
Pisces larvae	15	5·0	0·30	3·70	0·55	2·75	1·11	0·17	0·85
IV Detritus feeders *Noctiluca miliaris*	4	4·1	1·20	213·20	8·52	44·92	255·63	10·21	41·85

of all levels in the ocean be studied and solved correctly. For this purpose, investigations were made of the biological structure in the pelagic communities in the Black Sea, Mediterranean and equatorial Atlantic.

The Black Sea is especially convenient for this kind of exploration since its zooplankton includes only a few species. Most of them were studied well and their biology and ecology are known. Furthermore, prolonged investigations in 1960-67 on the dynamics of the populations throughout the year were carried out and made it possible to calculate the production rate in various animal populations. Dividing them according to their biology into categories corresponding to successive trophic levels, we succeeded in presenting a general outline of the trophic structure, production and rate of transformation of organic matter in the pelagic community.

The plankton materials from the Mediterranean and the Atlantic were treated in the same way. In these regions, the composition of plankton is more complex, the biology and feeding habits of many species is not clarified, but for many of them we have the required information (Lebour, 1922, 1923; Conover, 1960; Geinrikh, 1958; Beklemishev, 1954; Marshall and Orr, 1961, 1966; Vinogradov, 1962; Anraku and Omori, 1963; Ponomareva, 1959; Tchindonova, 1959; Wickstead, 1962). Thus it is possible here also to present a trophodynamic picture of the pelagic community with a good approximation.

For calculation of the production, the P/B-coefficients (Production/Biomass) obtained in our investigations in the Black Sea were used here. This approximation can lead to errors in calculations for the other seas, but as the investigation had shown correspondingly narrow limits of changes in these coefficients the possible error cannot be very considerable.

An example of these calculations is shown in Table 1. In its first column all the main species of zooplankton in the neritic zone of the Black Sea are arranged, according to their feeding habits, into three trophic groups. Some of them—*Acartia clausi, Oithona similis, O. minuta*—consume a mixed food and therefore are present both in the II and III trophic levels. The biomass of these populations and their production is attributed to the herbivorous or carnivorous group according to the proportion of vegetable or animal food in their diet.

According to Petipa (1959) the food composition of *Acartia clausi* changes during the year and thus its contribution, and the contribution of other omnivorous species, to the biomass and production of carnivores and herbivores changes in various seasons.

To the detritophagous group in the Black Sea we have attributed only *Noctiluca miliaris* though, after Mironov (1954), phytoplankton or zooplankton sometimes plays a considerable role in its nutrition. On the other hand detritus is consumed to a certain extent by the carnivorous and herbivorous animals so such simplification in calculation cannot misrepresent considerably the real interrelations in the trophic structure of zooplankton.

The average figures of the biomass of different populations (Table 1) obtained during the seasonal observations on four stations of the 10-mile

TABLE 2. The biomass and production on various trophic levels (in mg/m³ dry weight and in cal/m³) in the upper layer of different seas

Region	Sevastopol Bay		The Black Sea off Sevastopol		Atlantic		Ionian Sea	
Season	Summer (VI-IX)		Summer (VI-IX)		Spring (III-IV)		Summer (VIII-IX)	
Layer, m	0-14		0-40		0-50		0-50	
	mg	cal	mg	cal	mg	cal	mg	cal
Biomass								
I Phytoplankton	2100·0	404·0	450·0	66·4	19·0	3·7	6·8	1·3
II Herbivores	191·9	102·5	74·8	36·1	36·6	20·3	48·7	15·3
III Carnivores	28·6	13·9	35·7	16·5	17·4	7·5	9·7	3·7
IV Detritus Feeders	43·3	7·1	213·2	44·9	9·2	4·9	2·2	1·3
Production								
I Phytoplankton	1321·0	254·4	360·0	53·1	40·6	7·8	11·6	2·2
II Herbivores	28·9	15·6	10·4	5·6	5·5	3·0	8·8	2·5
III Carnivores	2·6	0·8	6·7	2·8	1·6	0·7	1·1	0·4
IV Detritus feeders	51·8	8·5	255·6	41·8	0·8	0·4	0·2	0·1

TABLE 3. The seasonal changes of the biomass and production on various trophic levels (cal/m³)

Trophic levels	Sevastopol Bay (0-14 m)				Black Sea off Sevastopol (0-40 m)				Atlantic (0-50 m)	
	Spring	Summer	Autumn	Winter	Spring	Summer	Autumn	Winter	Spring (III-IV)	Autumn (IX)
Biomass										
I Phytoplankton	400·0	404·0	168·8	15·4	32·1	66·4	14·7	7·1	3·7	4·5
II Herbivores	24·0	102·5	11·6	11·5	8·3	36·1	9·4	8·0	20·3	23·0
III Carnivores	3·9	13·9	9·5	0·5	29·7	16·5	12·3	8·5	7·5	14·4
IV Detritus feeders	8·6	7·1	—	3·7		44·9			4·9	2·7
Production										
I Phytoplankton	148·0	254·4	92·9	25·5	5·4	53·1	1·8	0·6	7·8	9·4
II Herbivores	3·6	15·6	1·2	0·7	0·5	5·6	1·1	1·3	3·0	3·4
III Carnivores	0·2	0·8	0·5	0·1		2·8			0·7	1·3
IV Detritus feeders	11·2	8·5	—	1·7	38·8	41·8	7·4	6·7	0·4	0·3

section off Sevastopol were used for further calculation. The biomass in mg/m³ wet weight was evaluated in mg/m³ dry weight and in cal./m³ after the data of Vinogradova (1959, 1961), Vinogradova *et al.* (1962) and Kisewetter (1954). The coefficients assumed in these calculations are given in columns 2 and 3 of Table 1. The determination of the daily production of the zoo-plankton populations was made according to the P/B-coefficients given by Greze and Baldina (1964), Greze (1968), Greze *et al.* (1968), and Zaika (1969), and presented in column 4, Table 1. Primary production and phytoplankton biomass were calculated from data presented by Kondratjeva 1967).

In the examination of the trophic structure in the Mediterranean and equatorial Atlantic the results of our investigations on zooplankton (Greze 1963, 1963a) and the data by Kondratjeva (1967), Finenko (1965) and others were used. In the zooplankton of these regions a large group of species was attributed to omnivores and then an assumption was made that $\frac{1}{3}$ of their total biomass can be considered as attributed to the herbivorous group, $\frac{1}{3}$ to the carnivorous and $\frac{1}{3}$ to the detritophagous group.

The results of all these calculations are presented in Table 2, where the trophic structure of the plankton community in the upper layer of different seas is compared. The calorific values of the biomass are the most comparable. They demonstrate the variation of phytoplankton content from 1·3 cal/m³ in the Ionian Sea to 404 cal/m³ in Sevastopol Bay. The biomass of the second trophic level changes from 15·3 to 102·5 cal/m³ respectively and that of the third trophic level from 3·7 cal/m³ in the Ionian Sea to 16·5 cal/m³ in the Black Sea off Sevastopol.

These figures show first that the Ionian Sea is the poorest one in the region we deal with. The Black Sea off Sevastopol and Sevastopol Bay especially is the most productive as was to be expected owing to the neritic position of these places. Secondly the biomass and production on the lower trophic levels fluctuate more than on the upper ones. The difference between the highest and lowest values of the biomass observed in various regions are up to 30 times in phytoplankton, 6-7 times in herbivores and only 4 times in carni-vorous zooplankton. Thus we can suggest a great stability on the upper trophic levels in plankton communities as compared to lower ones, so that drastic changes in the number of phytoplankton arising at the "blooms" are buffered in the following links of the food chain where the reproduction rate is lower.

In Table 3 the annual variations in biomass and production are shown. The comparison of these figures demonstrates that in Sevastopol Bay the annual oscillation is most pronounced. The highest values of the biomass of phytoplankton in summer exceed the lowest winter figures by about 25 times, the herbivorous ones by 9-10 times. The production values change corres-pondingly in scale. In the Black Sea off Sevastopol these oscillations are moderate and in the more stable conditions in the tropical Atlantic they cannot be definitely observed if the figures in spring and autumn are compared.

In the Atlantic and the Ionian Sea we can determine the changes in biomass

TABLE 4. The biomass and daily production (in mg/m³ wet weight and in cal./m³) on various trophic levels and in different depths in the Tropical Atlantic and Ionian Sea

Region		Atlantic							Ionian Sea				
Layer (m)		0-50		50-200		200-500		0-50		50-200		200-500	
		mg	cal	mg	cal	mg	cal	mg	cal	mg	cal	mg	cal
Biomass													
I Phytoplankton		19·0	3·7	11·9	2·3	6·0	1·5	6·8	1·3	2·9	0·6	2·0	0·4
II Herbivores		36·6	20·3	6·0	3·4	1·0	0·6	48·8	15·3	10·7	6·5	2·2	1·4
III Carnivores		17·4	7·5	3·9	1·5	0·3	0·2	9·7	3·8	10·7	3·0	2·3	0·7
IV Detritus feeders		9·2	4·9	1·2	0·8	0·2	0·1	2·2	1·3	1·7	1·0	0·6	0·4
Production													
I Phytoplankton		40·6	7·8	23·8	4·6	6·6	1·6	11·6	2·2	4·3	0·8	1·80	0·36
II Herbivores		5·5	3·0	1·0	0·5	0·1	0·05	8·8	2·5	1·6	0·9	0·20	0·13
III Carnivores		1·6	0·7	0·4	0·2	0·015	0·010	1·1	0·4	1·1	0·4	0·14	0·04
IV Detritus feeders		0·8	0·4	0·2	0·1	0·010	0·005	0·2	0·1	0·2	0·1	0·03	0·02

TABLE 5. The fraction of the biomass and daily production (in calories) of different trophic levels (in per cent of primary level). The caloric values of the I level assumed 100%

Region	Season	Layer (m)	Biomass				Production			
			I	II	III	IV	I	II	III	IV
Sevastopol Bay	spring	0-14	100·0	6·0	1·0	2·1	100·0	2·4	0·1	7·5
	summer		100·0	25·3	3·4	1·8	100·0	6·1	0·3	3·5
	autumn		100·0	6·9	5·6	—	100·0	1·3	0·5	—
	winter		100·0	74·6	3·2	24·0	100·0	0·3	0·4	6·6
Black Sea off Sevastopol	summer	0-40	100·0	54·1	24·8	67·5	100·0	10·5	5·2	78·4
The Equatorial Atlantic	autumn	0-50	100·0	510·0	319·0	60·0	100·0	36·2	14·8	3·2
	spring	0-50	100·0	548·0	202·5	134·5	100·0	38·5	9·0	5·1
	spring	50-200	100·0	148·0	65·1	34·7	100·0	10·8	4·3	2·2
	spring	200-500	100·0	40·0	13·3	6·7	100·0	3·1	0·6	0·3
The Ionian Sea	summer	0-50	100·0	1175·0	292·0	100·0	100·0	113·3	18·2	4·5
		50-200	100·0	1081·0	500·0	166·5	100·0	112·5	50·0	12·5
		200-500	100·0	350·0	175·0	100·0	100·0	36·0	11·1	5·5

of different trophic groups at various depths (Table 4). The production figures given in Table 4 for the lowest layer, 200-500 m, must be considered as an approximation because the rate of reproduction of organisms inhabiting this depth is not clear. In our calculation we assumed that the reproduction of phytoplankton in this layer is a half of reproduction in the upper layer owing to the approximately twofold decrease of the temperature with depth. Even if this inaccuracy is taken into consideration, Table 4 demonstrates the relative increase of biomass and production of the carnivorous organisms as compared to the phytoplankton and herbivores. This becomes more obvious when we take into consideration that a number of the phytoplankton organisms in the deep layers cannot function as primary producers but consume organic matter and, strictly speaking, should be placed in the second trophic level.

Since the relation of biomass and production at different trophic levels is the most important task of our study, we have summarized the data in Table 5. The biomass and production at all trophic levels, expressed as calorific values, is given as a percentage of the biomass or production of phytoplankton. The examination of these relations shows large changes. First of all a predominance of phytoplankton can be marked in neritic regions, especially during the spring and autumn. In winter the zooplankton biomass increases but in Sevastopol Bay, even in this season, the biomass of zooplankton did not exceed 100% of the phytoplankton biomass. In the open Atlantic and the Ionian Sea our data show an opposite relation. Zooplankton contributes the most part of the biomass.

It is remarkable that the biomass of herbivores and in many cases even carnivores exceeds considerably the biomass of phytoplankton. In the upper layers of the tropical Atlantic, the biomass of herbivores makes up to 500% and carnivores up to 200-300% of the phytoplankton biomass. In the Ionian Sea the corresponding figures rise up to 1100 and 300-500%.

Another point to be noted is that in the upper layer of the Atlantic and the Ionian Sea the ratio of the upper trophic levels to the biomass of phytoplankton is higher than in deep layers. For example in the tropical Atlantic the biomass of herbivores makes up to 500% of the phytoplankton biomass in the 0-50 m, 150% in 50-200 m and 40% in 200-500 m layer. The changes in the other trophic groups are in the same order. This fact allows us to suggest that in the lower layers, the reproduction rate of the autotrophic or heterotrophic phytoplankton organisms decreases much more than the rate of reproduction of the herbivores and other consumers. So the unit of phytoplankton biomass can maintain a smaller population of zooplankton than above.

The reality of this relation in different cases will be considered from an energetic point of view and in this respect it is more interesting to compare the possible results of the productive activity of these biomasses in various trophic levels. It is known that the production on a given trophic level can never be equal to the production on the previous one. On the right side of Table 5 the relation of production on these levels is given and hence the utilization of energy in successive levels of the food chain in marine plankton can be described.

In the Black Sea and Sevastopol Bay the production on the second trophic level makes up no more than 10 % of the primary production. The production of carnivores in some cases increases up to 50 and even 100 % of the production of the previous level. But in the last case a considerable production of detritophagous animals occurs in the plankton community and hence can be used by the carnivores. If one takes into consideration this source of their food, the production of carnivores never exceeds ⅓ of the production of their prey.

It was shown recently by Petipa (1967) and Pavlova (1967) that *Calanus helgolandicus*, *Acartia clausi* and *Penilia avirostris* in the Black Sea use for growth about 20-50 % of the consumed food, therefore the relations between the amounts of production on various trophic levels obtained in our calculations show a good agreement with the energetic possibilities of the plankton community.

In the tropical Atlantic the corresponding relation demonstrates more intensive utilization of the energy accumulated on a given trophic level by the organisms of the following one. In the upper layer, 0-50 m, the production of herbivores amounts to 35-40 % of the primary production and production of carnivores 25-35 % of the production created by herbivores and detritus feeders.

In the Ionian Sea the calculations gave an unexpected result when the secondary production of the herbivorous zooplankton exceeded the primary production. The reason for such incorrectness may lie in the inadequacy of the data for phytoplankton and primary production in the region studied. The figures for its biomass in Tables 2 and 4 are obtained on three stations only (Kondratjeva, 1967). Also, it is not unlikely that microplankton in this region is more abundant than can be registered by the method used (Bernard, 1961). Nevertheless this tentative data demonstrate a very high utilization of the primary production in the Mediterranean.

CONCLUSIONS

The energy created by the organisms of the plankton community on various trophic levels is utilized intensively.

This intensity increases from neritic waters, where biological effectiveness on the second trophic level in the Black Sea is equal to 2-3 % and on the third level about 12 %, to the open ocean waters where in the tropical Atlantic the corresponding figures reach 30-35 and 40-45 %.

In oligotrophic tropical and subtropical seas the secondary production achieves its highest possible values in the plankton communities from an energetic point of view.

REFERENCES

ANRAKU, M., and OMORI, M. 1963. Preliminary survey of the relationship between the feeding habit and the structure of the mouth-parts of marine copepods. *Limnol. Oceanogr.*, **8**, 116-26.

BEKLEMISHEV, K. V. 1954. Feeding of some mass species of the plankton copepods in far-eastern seas (in Russian). *Zool. Zh.*, **33** (6), 1210-29.

BERNARD, F. 1961. Problèmes de fertilité elémentaire en Méditerrenée de O à 3000 mètres de profondeur (Campagne de la 'Calypso', juillet 1955). *Annls. Inst. océanogr., Monaco*, **39**, (5), 61-160.

CONOVER, R. J. 1960. The feeding behaviour and respiration of some marine planktonic crustacea. *Biol. Bull. mar. biol. Lab. Woods Hole*, **119**, 399-415.

FINENKO, Z. Z. 1965. Primary production in the Black and Azov Seas and in tropical parts of the Atlantic. Dissertation.

GEINRIKH, A. K. 1958. On the feeding in marine copepods in trophic region (in Russian). *Dokl. Akad. Nauk SSSR*, **119**, (5).

GREZE, V. N. 1968. The rate of production in populations of heterotrophic marine organisms. *2nd Int. oceanogr. Congr., Moscow* (in press).

GREZE, V. N., and BALDINA, E. P. 1964. Dynamics of populations and annual production of *Acartia clausi* Giesbr. and *Centropages kröyeri* Giesbr. in neritic zone of the Black Sea (in Russian). *Trudy sevastopol. biol. Sta.*, **17**, 249-61.

GREZE, V. N., BALDINA, E. P., and BILJOVA, O. K. 1968. Production of plankton copepods in neritic zone of the Black Sea in 1960-66 (in Russian). *Oceanology*, **8** (6), 1066-1070.

KISEWETTER, I. V. 1954. On the fodder value of plankton in the seas of Okhotsk and of Japan (in Russian). *Izv. tikhookean, nauchno-issled. Inst. ryb. Khoz. okeanogr.*, **39**.

KONDRATJEVA, T. M. 1967. The production and daily changes of phytoplankton in south seas. Dissertation.

LEBOUR, M. V. 1922. The food of plankton organisms. *J. mar. biol. Ass. U.K.*, **12**, 644-77.

LEBOUR, M. 1923. The food of plankton organisms. 2. *J. mar. biol. Ass. U.K.*, **13**, 70-92.

LINDEMANN, R. L. 1943. The trophic-dynamic aspect of ecology. *Ecology*, **23**, 399-418.

MARSHALL, S. M., and ORR, A. P. 1961. Food and feeding in copepods. *I.C.E.S., C.M. 1961, Zooplankton Symposium*, Pap. no. N.23.

MARSHALL, S. M., and ORR, A. P. 1966. Respiration and feeding in some small copepods. *J. mar. biol. Ass. U.K.*, **46**, 513-30.

MIRONOV, G. N. 1954. Feeding of plankton carnivores. I. Feeding of *Noctiluca miliaris* (in Russian). *Trudy sevastopol. biol. Sta.*, **8**.

PAVLOVA, E. V. 1967. Food consumption and transformation of energy by the cladocera population in the Black Sea (in Russian). In *Structure and dynamics of water communities and populations*. Kiev, Naukova dumka.

PETIPA, T. S. 1959. Feeding of *Acartia clausi* Giesbr. and *A. latisetosa* Kritcz. in the Black Sea (in Russian). *Trudy sevastopol. biol. Sta.*, **12**, 130-52.

PETIPA, T. S. 1967. On the effectiveness of utilisation of energy in pelagic ecosystem in the Black Sea (in Russian). In *Structure and dynamics of water communities and populations*. Kiev, Naukova dumka.

PONOMAREVA, L. A. 1959. The euphausids of the seas of Okhotsk and Bering (in Russian). *Trudy Inst. Okeanol.*, **30**, 115-47.

TCHINDONOVA, Y. G. 1959. Feeding of some groups of deep-sea macroplankton in North-West Pacific (in Russian). *Trudy Inst. Okeanol.*, **30**, 166-89.

VINOGRADOV, M. E. 1962. Feeding of the deep-sea zooplankton. *Rapp. P.-v. Réun. Cons. perm. Explor. Mer*, **153**, 114-20.

VINOGRADOVA, Z. A. 1959. The biochemical composition of zooplankton in north-west part of the Black Sea (in Ukrainian). *Nauk. zap. Odess. biol. Sta.*, **1**.

VINOGRADOVA, Z. A. 1961. Peculiarities of the biochemical composition and caloricity of zooplankton in north-west part of the Black Sea in 1955-1959 (in Ukrainian). *Nauk. zap. Odess. biol. Sta.*, **3**.

VINOGRADOVA, Z. A., and others. 1962. The biochemical composition and caloricity of phyto- and zooplankton in the Black Sea (in Ukrainian). *Nauk. zap. Odess. biol. Sta.*, **4**.

WICKSTEAD, J. H. 1962. Food and feeding in pelagic copepods. *Proc. zool. Soc. Lond.*, **139**, 545-55.

ZAIKA, V. E. 1969. Production of *Oikopleura dioica* and *Sagitta setosa* in neritic zone of the Black Sea. In *Biological production in the south seas.* Kiev. Nauk, dumka.

Analysis of trophic processes on the basis of density-dependent functions*

R. W. BROCKSEN, G. E. DAVIS and C. E. WARREN
Department of Fisheries and Wildlife
Oregon State University
Corvallis, Oregon, U.S.A.

ABSTRACT. Ecologists, whether working at the population or community level, recognize that there are density-dependent factors operating which cause fluctuations in numbers. One such density-dependent factor that contributes to this fluctuation in numbers is the nutritional base of the population, that is the food supply. One can reasonably conclude that the density and distribution of an organism's food resource in its immediate surroundings determine the amount of food the animal can obtain in a brief period of time. The energy obtained in this food will ultimately be reflected in the birth, death, consumption, growth and production of the organism. Density of a food resource is an outcome of its rate of production or introduction into a system, and the rates at which it is consumed, decomposed and leaves the system. Thus, changes in any of these should be reflected in changes in the density or amount of the food resource. The mean density or amount of this resource over a period of time only sufficiently long for measurable changes to occur in an organism of interest would appear to be one useful measure of the resource availability causing such changes.

When relationships can be found that show growth rate or reproduction rate of a species of interest to be functions of the density of its nutrient resource, and that the density of nutrient resource is some function of the density of the species of interest, these relationships can be incorporated into simple models permitting prediction of the biomass of the species of interest that would maximize its production or reproduction. Density measurements are being made continuously by ecologists. The application of such measurements in the analysis of trophic processes, however, has usually been limited to their use in estimating production. Examples suggesting the possibility of application of density-dependent relationships in the analysis of interactive processes within and between trophic levels will be discussed.

* A contribution from the Pacific Cooperative Water Pollution Laboratories, Oregon State University. Technical Paper 2515, Oregon Agricultural Experiment Station.

THE STUDY OF TROPHIC RELATIONS

Lindeman's (1942) magnificent contribution has, in very general terms had two effects on the study of trophic relations, both important but both not good. His emphasis on the ecosystem with energy as a common denominator has encouraged a broad view and a search for principles in trophic studies. However, his paper has encouraged others to publish and discuss trophic level efficiencies in the absence of knowledge not only of the food consumption and production but even of the food habits of important species.

We believe that during the past 25 years the study of trophic relations has somehow left the mainstream of ecological and evolutionary thought. Perhaps the emphasis on ecosystems has drawn attention away from the trophic relations which are so important in the evolution of species. We all accept that a species evolves only as a part of an ecosystem in which it comes to occupy a niche. But lumping of species and emphasis on only the dynamic aspects of trophic relations blurs the problem of what an individual faces and the meaning of this to the individual and its population in terms of natural selection. Just as it is hard to conceptualize how species compete for a niche rather than for an existing resource, it is hard to conceptualize how species compete for the production rate of a food resource rather than for existing biomasses of food. Moreover, as valuable as the concepts of niche and production may be, we must try to visualize the immediate problems individuals of coexisting species must face in satisfying their needs.

Ivlev (1945) emphasized the need to focus attention on one "product of interest" at a time rather than to diffuse effort over many species. Ivlev's equation accounting for the production of one species in terms of its utilization of the production of the principal food species makes use of "ecotrophic", or cropping efficiencies, and growth efficiencies. The great task of making trophic analyses demanded even by Ivlev's approach has perhaps caused some to dismiss it, but this approach makes clear some of the real problems yet to be faced in the study of energy and material transfer in natural communities. The approach provides the ecologist with a basis for determining how best to apportion his efforts. But when one considers that most carnivores utilize at least several species of prey organisms, we are left with the problem of determining several prey production rates, even with Ivlev's less encompassing approach. There is no evidence in the literature of the past 25 years that we can often hope to explain the production of a predator of interest by making production rate determinations on even its more important prey species. Moreover, it is not at all clear that the production rates of several prey species provide the most useful measures of their actual availability to a particular predator, for prey production usually passes into many trophic pathways. If we are to be able to relate changes in the rates of food consumption, growth, and production of an animal to changes in its food resources, it appears to us that approaches alternative to the determination of the production rates of prey species are needed.

One such alternative view is that prey density and distribution rather than production rate immediately determine the amount of food a predator can obtain and the energetic cost of its pursuit. It is this alternative view which we intend to develop in this paper. Density of a food resource is an outcome of its rate of production or introduction into a system and the rates at which it is consumed, decomposed, and leaves the system. Thus, changes in entry, production, consumption, decomposition, and exit should be reflected in changes in the density of the food resource. The mean density of a food resource over a period of time only sufficiently long for measurable growth to occur in the consumer of interest would appear to be one useful measure of the resource availability permitting such growth. Density measurements can be routinely made whereas production measurements usually cannot. Moreover, the summing of biomasses of food organisms of different trophic types does not present the conceptual and analytical difficulties presented by summing the production rates of food organisms of different trophic types.

We intend to present evidence in this paper not only that the growth rates of consumers in nature are often simple functions of the densities or biomasses of their food organisms but also that the densities of the food organisms are often simple functions of the densities of the consumers. When such relations can be defined, the production of a predator can be modelled as a function of its biomass, the biomasses of its competitors, and the density of prey organisms without resorting to any assumptions as to the mathematical nature of the relations involved. The use of such a model to estimate the biomass of a consumer that would maximize its production should insure that important interactions of that predator, its principal competitors, and their food organisms are taken into account.

Since publication of Lindeman's paper, those who have studied trophic relations have largely ignored if not belittled the importance of density or biomass except as it has been needed in the estimation of production. It is indeed curious that they should have done so, particularly during this period of time, for publication of papers by Nicholson (1933) and Smith (1935) led to great emphasis in ecological thought on the importance of density-dependent phenomena in population control. Earlier, Howard and Fiske (1911) viewed population regulation much as did Nicholson. And we know that the essay by Malthus (1798) on human population led Verhulst (1838) to formulate the logistic equation and was important in Darwin's (1859) conceptualization of natural selection. Students of trophic relations have limited their thinking in failing to include density-dependence in their concepts of production processes.

Students of population dynamics have usually tried to explain changes in birth and death rate in terms of changes in the numbers of the organisms of interest and have rarely related them at the same time to determinations of environmental conditions. This is evident in nearly all population models and in the design of both laboratory and field studies. Still, implicit if not explicit in the presentation of their findings is the idea that the nutritive base of the population is one of the more important density-dependent factors. Where it is, consumption will influence food density, and food density and distribution will influence consumption. The quantity of energy obtained in food will ultimately be reflected in birth, death, growth and production of the species of interest.

The trophic dynamic or production point of view so stimulated by Lindeman has vitalized an important area of ecological thought and has shown its relevance to man's problem of food production. In suggesting an alternative point of view, or, more properly, an additional point of view that is necessary, we are in no way demeaning the importance of dynamic production processes; they underlie all trophic phenomena we observe and measure. Rather, we are suggesting that we cannot afford to limit our conceptualization of the ways in which these processes function. It could be suggested that the present general orientation of trophic studies is an example of the evolution of a "Ruling Theory" as explained by Chamberlin (1897, p. 841): "a premature explanation passes first into a tentative theory, then into an adopted theory, and lastly into a ruling theory". Chamberlin's antidote was "The Method of Multiple Working Hypotheses", and he believed that, in addition to counter-acting a poison, his antidote could lead to the explanation of phenomena too complex to be explained by a single hypothesis. But we are not really here concerned just with hypotheses, we are concerned with points of view. No one questions the importance of the dynamic aspects of production or the value of this point of view. Likewise, few question the importance of density-dependent relations in ecological systems, but density-dependent relations have fallen into disuse in trophic studies. Both points of view are needed in trophic studies if they are to contribute to the mainstream of ecological and evolutionary thought and to the solution of man's problems on earth.

FOOD CONSUMPTION, GROWTH, AND PRODUCTION OF JUVENILE
SOCKEYE SALMON AND THE DENSITY OF ZOOPLANKTON

In any ecosystem in which the limiting resource of a species is food, an increase in the biomass of that species will lead to a reduction in the amount of food each individual can obtain; and this will lead to a reduction in individual growth rate (Fig. 1). The production of any age class of an animal population in any period of time is usually defined as the product of its mean growth rate and its mean biomass. Thus, with growth rate declining with increasing biomass, production will increase from a low level at some low biomass to a maximum at some intermediate biomass; and, then, with further increases in biomass, production will decline toward zero as growth approaches zero (Fig. 1).

Over the past decade in studies of stream communities in which trout (*Salmo clarki*) and sculpins (*Cottus perplexus*) were our carnivores of interest, we have been fairly successful in elucidating the biological basis of this superficially simple relationship between biomass and production (Warren *et al.*, 1964; Davis and Warren, 1965; Brocksen *et al.*, 1968). For any food-limited species, the hypothetical relationship described is logically sound and is accepted by most biologists. It has been demonstrated in the pond culture of

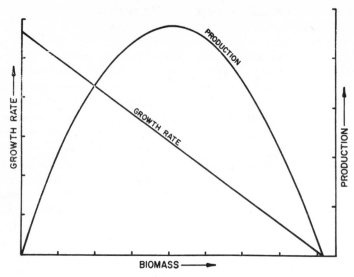

Fig. 1. Theoretical relationships of growth rate and production to biomass.
Production derived as product of growth rate and biomass.

fish (Walter, 1934; Wolny, 1962), in trout streams (Backiel and LeCren, 1968),
and Ivlev (1947) has provided a mathematical formulation of this relationship. However, the production of fish in large natural ecosystems has not
often been estimated, and certainly not at different biomasses so as to define
the relationships between biomass and production. Thus, there is a paucity
of information to provide a basis for discussing this fundamental production
relationship in most marine ecosystems. We have chosen to discuss the
trophic relations of large lake systems in which juvenile sockeye salmon
(*Oncorhynchus nerka*) are the carnivore of interest, since there is some
pertinent information available on these systems, and because principles
operating in plankton food chains in lakes should also operate in marine-
plankton food chains, a major topic of this symposium. But appropriate data
even on food chains leading to sockeye production are sparse, and it has been
necessary for us to make various conversions of data which we would rather
were not necessary; we have done this solely to illustrate the principles which
we believe to be operating.

Johnson (1961) studied juvenile sockeye salmon and their zooplankton
food resources in a series of seven basins in the Babine and Nilkitkwa lake
systems of British Columbia. From the mean individual weights which he
gives for the fish from the different basins in mid-October, we have estimated
their mean growth rates for the mid-June to mid-October period. These we
have plotted against estimates of the mean biomasses of sockeye in different
parts of the lake system during this period (Fig. 2). Over the range of sockeye
biomasses which existed, growth rates decline with increasing biomass.
Maximum production probably would have occurred at a biomass near the
maximum existing ones (Fig. 2). Extrapolation of the growth line is necessary

to suggest the probable form of the sockeye production-biomass curve, which resembles the theoretical one shown in Fig. 1.

Decline in the growth rate of a consumer with increases in its biomass may result from some controlling influence the consumer exerts on its food resource; but whatever the biological basis of the decline, it should be elucidated. The first step in this elucidation is perhaps to determine whether or not the growth rate of the consumer is controlled by the availability of its food resource in any definable manner. We can later turn to the other question: whether or not the consumer controls to an important extent the availability of its food resource.

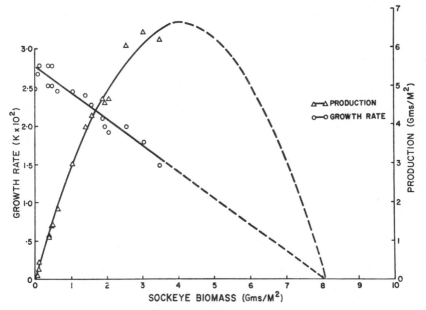

Fig. 2. Relationships of growth rate and production to biomass of sockeye salmon in the Babine-Nilkitkwa Basins during a four-month period (June-September) in 1956 and 1957. Production computed as product of growth rate and biomass. Dashed portion of production curve based upon extrapolated values of growth rate. (Data of Johnson, 1961.)

One can reasonably conclude that the density and distribution of an animal's food organisms in its immediate surroundings determine the amount of food the animal can obtain in a short period of time and the energy cost of obtaining this food. Ivlev (1961) used zooplankton biomass as a measure of food availability in his examination of the growth of the bleak (*Alburnus alburnus*); and we have shown the growth of sculpins to be controlled by the biomass of their food organisms in the benthos of laboratory streams, and the growth of trout to be controlled by the biomass of drifting food organisms in these streams (Brocksen *et al.*, 1968).

The mean growth rate of the juvenile sockeye salmon studied by Johnson (1961) appears to be a simple function of the mean biomass of zooplankton during the mid-June to mid-October period in which most of the annual growth occurs (Fig. 3). The prey biomass or density at which growth rate approaches zero, which might be estimated by an extrapolation of the growth curve, is of considerable biological interest. Such a zooplankton density perhaps approximates the density at which the sockeye could not be successful if this density were primarily due to low zooplankton production or intensive interspecific competition. Were this low density primarily due to intensive intraspecific competition, sockeye biomass would presumably decline to a level permitting the existence of zooplankton densities more favourable to sockeye growth. As one would expect, though there are lakes having low zooplankton densities in which the growth of sockeye is very low, we have found no instances of sockeye growth closely approaching zero during the growth period of the year.

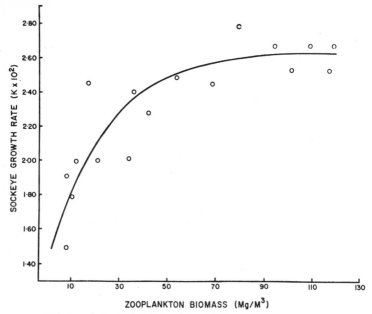

Fig. 3. Relationship between mean instantaneous growth rate of sockeye salmon and mean biomass of zooplankton in the Babine-Nilkitkwa basins for the period June to September during 1956 and 1957. (Data of Johnson, 1961.)

We can now turn to the question of whether or not the sockeye in the Babine-Nilkitkwa lake system exert a controlling influence on the availability of their zooplankton food resource; and we find that indeed they do. Mean zooplankton biomasses during the mid-June to mid-October period appear to be a simple function of sockeye biomasses in the different parts of this lake system and exhibit a regular decline with increasing sockeye biomass, with two possible exceptions (Fig. 4). Of course, zooplankton production as well as fates other than consumption by sockeye must determine the zooplankton level of this relationship. The stickleback (*Gasterosteus*) is considered to be an important competitor of sockeye for the zooplankton food resources in some lakes. The relationship between zooplankton density and sockeye density in the Babine-Nilkitkwa system appears to be adequate without taking possible competition by sticklebacks into account. But were consumption by sticklebacks to be an important fate of zooplankton production, a definable relationship could require that this be done. Zooplankton density might be found to be some simple function of the combined biomass of the two competitors, as we found to be the case for the benthic food resource of sculpins and their stonefly (*Acroneuria*) competitor (Brocksen *et al.*, 1968).

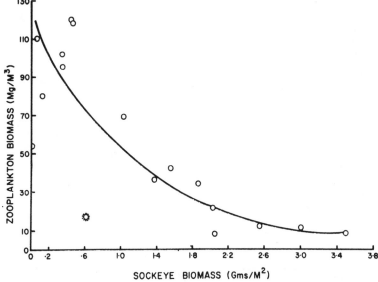

Fig. 4. Relationship between mean zooplankton biomass and mean sockeye salmon biomass in the Babine-Nilkitkwa basins for the period June to September during 1956 and 1957. (Data of Johnson, 1961.)

For the Babine-Nilkitkwa lake system, then, we have at least sketchy information on the relationships between sockeye growth rate and zooplankton density and between zooplankton density and sockeye biomass which explain the decline in sockeye growth rate with increasing biomass. These relationships define the predator-prey interactions which determine the form and position of the sockeye production-biomass curve for this system. In a later section we will describe how such relationships can be incorporated into a model appropriate for predicting the biomass of a consumer that would maximize its production.

It appears obvious to us that relations between food density and consumer biomass and between consumer growth rate and biomass should be different in ecosystems having different capacities for food production so long as food is the limiting resource, even if the consumer is the same species. For the sockeye in particular, we should expect that lakes having high capacities for zooplankton production would maintain high zooplankton densities, which could sustain higher sockeye growth rates at higher sockeye biomasses than would be possible in lakes having low capacities for zooplankton production. Information even less satisfactory than that for the Babine-Nilkitkwa lake system is available for Owikeno Lake (Ruggles, 1965) in British Columbia and for Lake Dalnee in Kamchatka (Krogius, 1961; Krogius and Krokhin, 1948). Here again, conversions of data are necessary to illustrate the expected relationships, conversions which are tenuous but which could not in themselves lead to these relationships.

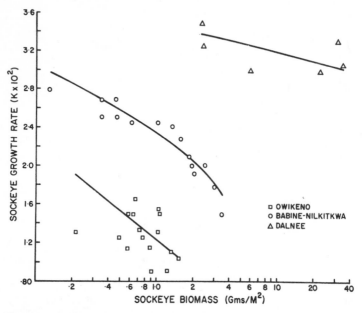

Fig. 5. Relationships between mean instantaneous growth rate and mean sockeye salmon biomass for Owikeno Lake, Babine-Nilkitkwa basins, and Lake Dalnee. Salmon growth rates in Owikeno Lake and Babine-Nilkitkwa system based upon a graphically-estimated mid-June mean weight of 0·2 grams and upon measured mid-October mean weight. Mean biomass based upon estimates of mid-August numbers and mean weight (Ruggles, 1965; Johnson, 1961). Growth rates in Lake Dalnee based upon mid-June mean weight of 0·3 grams and mid-October mean weights computed by multiplying mean smolt weight by 0·75. Mean sockeye biomass in Lake Dalnee based upon mid-August mean weight and mid-August geometric mean of numbers of fry entering lake and numbers of migrating smolts (Krogius and Krokhin, 1948).

Even at very low biomasses, sockeye growth rates in Owikeno Lake are very low, much lower than the growth rates of sockeye at higher biomasses in the Babine-Nilkitkwa system (Fig. 5). And in Lake Dalnee, sockeye growth rates at very high biomasses still are very much higher than those in the other lakes. In the Babine-Nilkitkwa system, growth rates clearly decline with increasing biomass. This decline is not as well defined in the two other lakes. Relationships established between sockeye production and biomass for the different lakes provide a comparison of the relative capacities of the

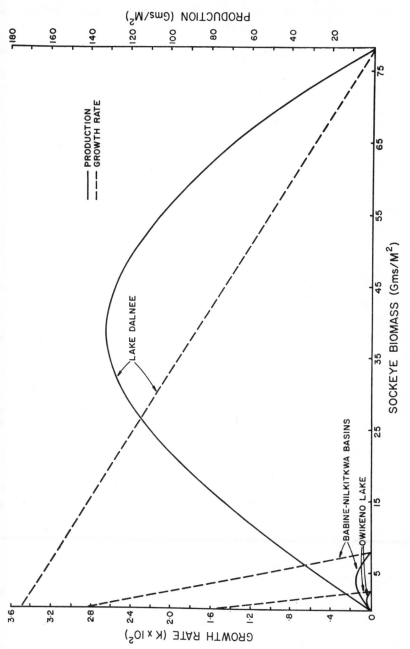

Fig. 6. Comparison of growth rate, biomass and production relationships for sockeye salmon in Owikeno, Babine-Nilkitkwa, and Dalnee lakes. Computed from data shown in Fig. 5.

lakes to support sockeye biomass (Fig. 6). The impact of increasing sockeye biomass on sockeye growth rate in Lake Dalnee is very markedly less than in the other two lakes. This suggests that the zooplankton food resources in Lake Dalnee are much less sensitive to utilization by the sockeye than are those resources in the Babine-Nilkitkwa system and in Owikeno.

The mean growth rates of sockeye during the mid-June to mid-October period in these three lake systems when plotted against mean zooplankton densities lie remarkably near the same line (Fig. 7). Lake Dalnee apparently

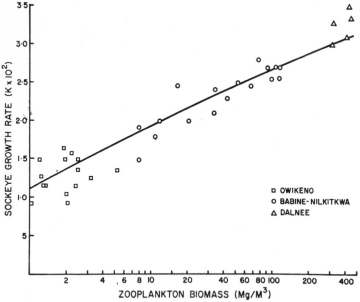

Fig. 7. Relationship between mean instantaneous growth rate of sockeye salmon and mean biomass of zooplankton for Owikeno, Babine-Nilkitkwa, and Dalnee lakes (data from Ruggles, 1965; Johnson, 1961; Krogius and Krokhin, 1948, respectively).

is able to maintain very high zooplankton biomasses, and here sockeye growth rates are highest; Owikeno Lake appears only to be able to support low zooplankton densities, and, in consequence the growth rates of its sockeye are low. The Babine-Nilkitkwa system is intermediate in this relationship. It is of considerable biological interest that the growth rates which juvenile sockeye salmon can maintain at different densities of zooplankton are apparently more characteristic of this salmon than of the particular lake ecosystem. Consideration of the costs of food capture and utilization in the growth process might lead one to the same conclusion: whatever are the production levels and fates of prey that lead to particular prey densities, these densities determine the growth rate of the predator.

One can reasonably conclude that ecosystems having the highest basic capacity to produce food should maintain the highest biomasses of a consumer, so long as food is the limiting resource. Higher rates of food produc-

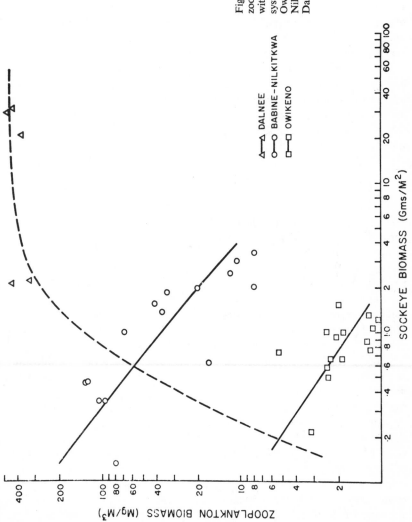

Fig. 8. Relationships between biomass of zooplankton and sockeye salmon biomass within systems (solid lines) and between systems (dashed line) based upon data from Owikeno Lake (Ruggles, 1965). Babine-Nilkitkwa basins (Johnson, 1961) and Lake Dalnee (Krogius and Krokhin, 1948).

tion should lead to higher food densities and these to higher consumer growth rates and biomasses. Annual differences in the capacity of a particular ecosystem to produce food should also lead to annual differences in the consumer biomasses that ecosystems can support. The relationship between food density and consumer biomass can be expected to be negatively correlated only within systems of constant basic productive capacity, even if the consumer exerts some control over the availability of its food. Thus, we must suppose that the various parts of the Babine-Nilkitkwa system have the same basic productive capacity, since zooplankton density was negatively correlated with sockeye biomass (Fig. 8). At a much lower level of zooplankton density, a similar relationship appears to exist in Owikeno Lake. At a much higher level of zooplankton density in the very productive Lake Dalnee, sockeye biomasses over a very wide range appear to exert little or no control over zooplankton density (Fig. 8). Perhaps it is space in Lake Dalnee that limits sockeye production (Fig. 5). This can only occur in a system having a very high capacity for food production, a system in which large changes in biomass of the consumer would have relatively little influence on either the biomass or production of food, as we will explain in the next section. This may be the reason for the apparent absence of a controlling influence on food density by sockeye in Lake Dalnee. Control of food density by the consumer would appear to be inevitable in the Babine-Nilkitkwa and Owikeno systems, which apparently have a much lower capacity for food production.

ZOOPLANKTON REPRODUCTION, GROWTH, AND PRODUCTION AS FUNCTIONS OF PHYTOPLANKTON DENSITY

The production process in the zooplankton, although presenting no problems of conceptualization, most certainly presents serious problems of measurement. Fish live longer, and the growth rates and biomasses of individual age classes can often be estimated, making it possible to calculate production rates. The short lives, the high birth and death rates, and the simultaneous existence of several generations of zooplankton species may make measurement of their rates of production impossible, at least by ordinary procedures of determining products of growth rates and biomasses of individual age classes. A large part of the production of any age class must be represented by sex product and offspring; and estimates of reproduction rates may sometimes be useful approximations of minimal or relative production rates. But the tissue elaboration of all age classes as they grow to reproductive size after release from the parents must also be included in any estimate of actual production. Furthermore, we often speak of the production of the zooplankton without regard for the species that are present. In so far as all of the species are herbivores, this presents no conceptual problem. Nevertheless, without knowledge of the competitive and other trophic interactions of these species, we might erroneously attribute to the zooplankton as a whole production relationships we have reason to believe are valid for individual species. In our following attempt to provide a conceptual framework for considering the relationships between the reproduction, growth, and production of zooplankton and the density of phytoplankton, we are well aware of all of these difficulties and dangers. Although we can probably never obtain all of the measurements that might be desirable, even a tentative way of viewing these trophic interactions may have value in guiding our attempts to understand trophic processes in marine ecosystems.

Edmondson (1962) has shown that the birth rate of the rotifer (*Keratella cochlearis*) in four lakes in England increased with increases in the density or biomass of the planktonic algae upon which it fed. At 10 to 15°C, increases in birth rate with increasing phytoplankton density beyond 200 μg/l were not great (Fig. 9). At temperatures over 15°C, increases in birth rate were more

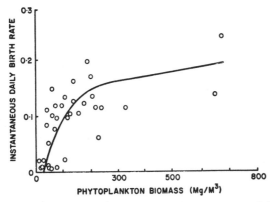

Fig. 9. Relationship between instantaneous daily birth rates of the rotifer *Keratella cochlearis* and biomass of phytoplankton for four English lakes. (Redrawn from Edmondson, 1961.)

nearly linear up to phytoplankton concentrations of about 400 μg/l. Increases in food availability to some level, as measured by phytoplankton biomass, should be expected to lead to increases in zooplankton birth rate. And to some point, increases in birth rate must mean more individuals are growing to reproductive size and reproducing, leading to more production, whatever the values may be. With crowding, growth rates and perhaps size at reproductive maturity could be reduced before birth rates were reduced, leading to a reduction in production. On the other hand, birth rates could be reduced in a compensatory manner.

From the idea that birth rate of zooplankton species is a function of the density of the phytoplankton upon which they feed, it follows that food consumption and perhaps growth rates also are functions of phytoplankton density; and there are those who hold this view. Parsons, LeBrasseur, and Fulton (1967) found that the food consumption rates of *Calanus* and *Euphausia* were a function of the density of their phytoplankton food organisms (Fig. 10), this function being best described by a modification of Ivlev's (1945, 1961) equation relating ration to food density. The animals, soon after removal from the sea, were held at different food densities in aquaria.

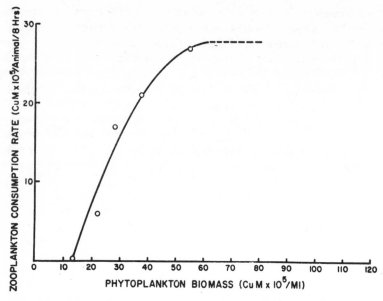

Fig. 10. Relationship between consumption rates of *calanus* and *Euphasia* and biomass of phytoplankton. (Redrawn from Parsons, LeBrasseur and Fulton, 1967.)

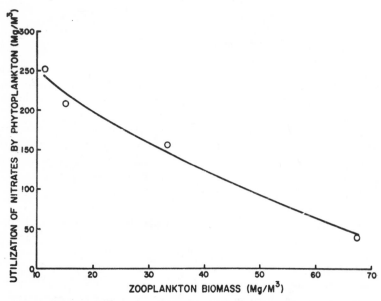

Fig. 11. Relationship between utilization of nitrates by phytoplankton and biomass of zooplankton in the Southern Barents Sea from 1953 through 1956. (Redrawn from Zelikman and Kamshilov, 1960.)

Many believe, and it may certainly reasonably be argued, that consumption by the zooplankton is often the principal fate of phytoplankton production (Cushing, 1958; Cushing and Vucetic, 1963; Conover, 1964). When it is, increases in food consumption by zooplankton will lead to decreases in phytoplankton density. Given, then, ecosystems having the same basic capacity to produce phytoplankton, the ones supporting the highest zooplankton biomasses will have the lowest phytoplankton densities. The Southern Barents Sea is considered to be very stable from one year to the next in the characteristics determining its phytoplankton production capacity. Here, Zelikman and Kamshilov (1960) found a high negative correlation between an index to phytoplankton biomass and zooplankton biomass on four consecutive years (Fig. 11). This is the same kind of dependence of food organism biomass on consumer biomass that we described for sockeye salmon (Fig. 4).

If, however, we consider a series of ecosystems having different basic capacities to produce phytoplankton, those systems having the highest phytoplankton production capacity should be able to sustain higher phytoplankton densities at higher zooplankton densities; and a positive correlation should exist between phytoplankton biomass and zooplankton biomass. Lake

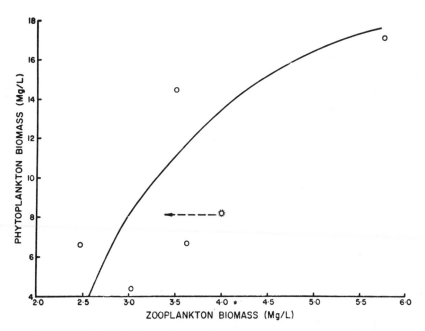

Fig. 12. Relationship between phytoplankton biomass and zooplankton biomass in Lake Dalnee for the years 1937 to 1942. The dotted circle represents a year when a plankton net of finer mesh was used and the arrow indicated a probable position of this point in relation to the others. (Data from Krogius and Krokhin, 1948.)

Dalnee is believed to vary greatly from year to year in the biogenic substances leading to phytoplankton production; and Krogius and Krokhin (1948) found that on successive years the highest phytoplankton biomasses existed along with the highest zooplankton biomasses (Fig. 12).

We must, then, in considering density-dependent relations of phytoplankton and zooplankton, distinguish between two sets of conditions, one leading to negative correlations and the other leading to positive correlations. For a given capacity for phytoplankton production, increases in zooplankton biomass should lead to decreases in phytoplankton biomass, so long as consumption by zooplankton is a principal fate of phytoplankton production. But, when different basic capacities for phytoplankton production exist, the maintenance of higher phytoplankton densities will permit the existence of higher zooplankton biomasses. We mean here also to distinguish clearly between the basic capacity of a system to produce phytoplankton and the actual rate of phytoplankton production going on during any part of a productive season. There is every reason to believe that the production rate of phytoplankton is some function of its biomass; and that phytoplankton production is different at different points along the curve shown in Fig. 11, even though the productive capacity of the system is probably the same. That such a relationship can exist, when phytoplankton production rates vary, is evidence of the establishment of equilibria between phytoplankton production rates and zooplankton consumption rates at the various zooplankton biomasses.

In Fig. 13, by means of a series of theoretical curves, we have attempted to illustrate the two kinds of relationships which we believe could exist between

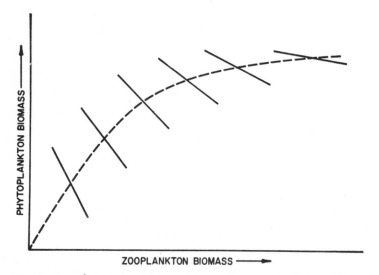

Fig. 13. Theoretical relationships between phytoplankton biomass and biomass of zooplankton within particular systems (solid lines) and between systems (dashed line).

phytoplankton biomass and zooplankton biomass, relationships much like those between zooplankton and sockeye salmon biomasses which we illustrated in Fig. 8. Each curve drawn with a solid line (Fig. 13) represents the decline in phytoplankton biomass occasioned by increasing zooplankton biomass in an ecosystem having a constant basic capacity to produce phytoplankton or in systems having the same capacity. The different curves represent ecosystems having different productive capacities and are thus placed higher on the phytoplankton biomass scale as the ranges of zooplankton biomasses sustained in the system are higher. The broken line drawn upward through these curves represents the increase in the zooplankton biomass that can be supported at higher phytoplankton biomasses made possible by increased capacity to produce phytoplankton. We have drawn the solid lines with successively decreasing negative slopes as they represent ecosystems of successively greater productive capacity. We believe that in systems of very high productive capacity, consumption by zooplankton will become a relatively less important fate of phytoplankton production. Thus, the inability of the zooplankton to benefit proportionally from phytoplankton increases at high levels, and even possibly deleterious effects on the zooplankton, would tend to decrease the controlling effect of the zooplankton and lead to the disappearance of density-dependent relationships.

We have, in the lakes we used as examples in our discussion of sockeye salmon production—Owikeno, Babine-Nilkitkwa and Dalnee—three ecosystems of greatly different basic productive capacity. This is readily apparent from the great differences between the biomasses of zooplankton the three lakes support. In the previous section of this paper, we discussed the relationships between the biomasses, growth rates, and production rates of juvenile sockeye salmon in the three lakes (Figs 2, 5, 6). This discussion had to be based on rather scanty information, and any similar discussion for the zooplankton in these lakes must be based on even more scanty information. Fully recognizing the dangers, we believe there is real value in thinking about possible relationships between zooplankton biomasses, growth rates, and production rates in systems of greatly different productive capacities.

To start with, we have little more information than zooplankton biomasses for the three systems, these being very high in Dalnee, much lower in Babine-Nilkitkwa, and very much lower in Owikeno (Figs 8, 14). We have attempted to estimate the relative differences in magnitude of zooplankton production in the three lake systems (Fig. 14) by estimating very crudely for different years the zooplankton consumed by the juvenile sockeye salmon which migrated from the lakes. This, of course, does not take into account zooplankton production passing into other trophic pathways, and Fig. 14 must be taken only as a crude representation of possibilities. Zooplankton production must, of course, be zero at zero biomass in all three lakes; and, theoretically, at some relatively high biomass in each lake, zooplankton reproduction and growth would be reduced to zero and production would again be zero. The positions and slopes of the growth rate lines were determined by dividing production estimates by the associated biomasses.

Estimated by these procedures, growth inevitably will decline with increasing biomass. Nevertheless, not only are there reasons for believing growth rates of zooplankton organisms should decline with increasing biomass and competition, but also there is some evidence to support this interpretation (Figs 9-12). There are still other weaknesses in our representation: dealing with all age classes and species simultaneously; and considering only growth rates rather than somehow considering reproduction as well.

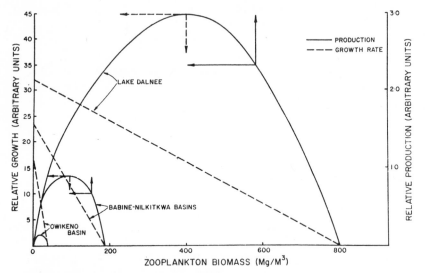

Fig. 14. Theoretical relationships of zooplankton growth rate and production of zooplankton biomass for Dalnee, Babine-Nilkitkwa, and Owikeno lakes. Vectors indicate the changes in zooplankton production that could be expected to result from changes in zooplankton biomass in the different lake systems.

But, even with all of these difficulties, we believe Fig. 14 can serve to call attention to certain possibilities of importance in explaining the dynamics of production in marine-plankton food chains. The vectors drawn on the production curves illustrate the most important of these possibilities. In a system of very high basic productive capacity, such as Lake Dalnee, a very large change in zooplankton biomass is necessary to bring about much decrease (dashed vector), or increase (solid vector) in zooplankton production rate. This perhaps explains the lack of control of zooplankton density exhibited by large biomasses of sockeye in Lake Dalnee (Fig. 8). In systems having low basic productive capacity, such as Owikeno Lake and the Babine-Nilkitkwa system, changes in zooplankton biomass much smaller than those in Lake Dalnee can have relatively more profound effects on zooplankton production rates (dashed and solid vectors). Particularly in the Babine-Nilkitkwa system, sockeye salmon have been shown to have a strong controlling influence on zooplankton density at sockeye biomasses much lower than those existing in Lake Dalnee.

NUTRIENTS, BIOMASS, AND PRODUCTION OF PHYTOPLANKTON

Perhaps in most plankton communities, the upper limits of zooplankton production are set by the levels of phytoplankton production possible with existing plant nutrient resources. Over much of the year in temperate and arctic regions, light and temperature may control phytoplankton production, but usually during some season one or more nutrients become limiting.

Photosynthetic plants generally represent the only reasonably well defined trophic level. Unlike for animals, then, methods of measuring phytoplankton production are not subject to criticism because they do not distinguish between species having different nutritional requirements. And measurements of the photosynethetic rate of the phytoplankton component of the community, when properly made, are usually accepted as being indicative of the general level of phytoplankton production. Various kinds of bottle experiments provide perhaps our best estimates of phytoplankton production, but always there remains the problem of how well the bottle microcosm models the ecosystem. This is of particular concern when we seek to determine the concentrations at which certain nutrients become limiting under different conditions.

Because of the nature of the photosynthetic process and the relative convenience of measuring its rate in the water medium, in measuring phytoplankton production we do not usually concern ourselves with the biomass or mean growth rate of the plants present, as we must in estimating the production of animals. It is, of course, fortunate that this is not necessary; for it often would be more difficult to measure phytoplankton biomass and mean growth rate than to measure its rate of photosynthesis or production. But, as we will attempt to explain, it may make less clear our concept of the interactions of phytoplankton populations with their environments that determine the possible levels of production.

First, we should perhaps clarify what we mean by the growth rate of a phytoplankton population. There should be no difficulty with the idea of production, either gross or net, as measured by photosynthetic and respiration rates, or with the idea of phytoplankton biomass or density. We must, however, distinguish between the total elaboration of new tissue by an existing plant biomass over a given period of time and the elaboration of new tissue per unit of that biomass during the same period of time. It is this latter that we shall refer to as the mean growth rate of a phytoplankton population, and it can perhaps most simply be measured and expressed as the ratio of mean production rate to mean biomass.

For an ecosystem having a given resource of plant nutrients, increases in phytoplankton biomass can be expected to decrease the concentration (density) of the limiting nutrient and its availability to the growing, reproducing plant cells. We should expect, then, the mean growth rate of the phytoplankton population to decline, with phytoplankton density increasing over some range. In kind, this decline in growth rate with declining density of nutrients is really not different from the decline in growth rate of sockeye salmon or zooplankton populations with decline in the density of their food resource, as we have discussed (Figs 3, 8, 9, 10). The outcome, then, of these two density-dependent relationships—the decline in nutrient concentration with increasing phytoplankton biomass, and the decline in mean phytoplankton growth rate with limiting nutrient concentration decline—should, in theory at least, be no different from that of animal populations (Figs 1, 2). That is, with phytoplankton biomass increasing from zero to some relatively high level, phytoplankton production should first increase from zero to some maximum, and then it should decline toward zero with further increases in biomass. This is for an ecosystem having a given resource of plant nutrients.

However, among ecosystems having different levels of plant nutrients, we must suppose that, other conditions being similar, those ecosystems having the highest levels of plant nutrients will sustain the highest biomasses of phytoplankton. For a given system, then, the relationship between plant nutrients and phytoplankton biomass can be expected to be negative; whereas between systems this relationship can be expected to be positive. Here again, as regards these relationships between populations and their nutrient resources, we may find that the same principles operate in both plant and animal populations (Figs 8, 13). If this is so, the relative effects on the production of phytoplankton occasioned by zooplankton grazing would be much less in ecosystems having the higher levels of plant nutrients, much like the relationship we might expect between zooplankton production and its utilization by sockeye salmon (Fig. 14). This would appear to be the basis for the great differences in the productivity of Owikeno Lake, the Babine-Nilkitkwa Lakes, and Lake Dalnee for sockeye salmon (Fig. 6) and zooplankton (Fig. 14).

A PRODUCTION MODEL BASED ON DYNAMIC-BIOMASS EQUILIBRIA

For a system in which the growth rate of a consumer can be defined as some function of the biomass of its population (Figs 1, 2), we can obviously develop a model of production as a function of biomass, production being the product of growth and biomass. From such a model, we can predict for a system, so long as the nature of the system remains unchanged, the biomass of a consumer that would maximize its production. We cannot, however, from such a model, predict the kinds of changes in the consumer or in its ecosystem which will alter the function that its production is of its biomass. This, of course, is because the apparently very simple relationship between biomass, growth rate, and production is in reality biologically very complex. The identification of the kinds of changes in the consumer, its food organisms, and its competitors that would alter this function could be very important; and a production model for a consumer which would incorporate not only consumer biomass functions but also functions of the biomasses of its competitors and food organisms would be much more valuable.

It was as a result of our studies of fish and insect production in simple laboratory stream communities (Brocksen *et al.*, 1968) that we first became aware that the growth rates of consumers are sometimes simple functions of the density of their food organisms and that the density of food organisms is sometimes a simple function of consumer biomass. It now appears that, for a given physiological state of the consumer, its growth rate is often determined by the density and distribution of its food organisms (Figs 3, 7, 10). Thus, within the consumer, bioenergetic equilibria become established between the rate of food consumption possible at different food densities, the cost of obtaining and utilizing this food, and the resulting growth. Furthermore, it now appears that the density or biomass of food organisms is often a rather simple function of the biomass of the consumer and any of its important competitors (Figs 4, 8). Here, dynamic-biomass equilibria between food organisms and consumer must exist over moderate periods of time, or relationships would not be apparent.

The feedback mechanisms leading to these dynamic equilibria are intriguing to contemplate, for they lead to growth rate and production being functions of consumer biomass. Some of these feedback mechanisms are more readily apparent than others. For example, if decreases in food density from low production or high consumption of food lead the growth rate of the consumer to approach zero (Fig. 3), then consumer biomass will tend, because of weight loss or mortality, to decrease. But the decrease in consumer biomass will permit the biomass or density of the food organisms to increase (Fig. 4). Thus, new equilibria between consumer biomass and food organism density, and between food organism density and consumer growth rate will tend to become established. In mechanisms such as these lie the biological bases of the density-dependent functions we have been discussing.

The production of a consumer population is dependent on the growth rates and biomasses of its age groups. In computing production, age groups should be handled separately, for their growth rates will differ. Moreover, the growth rates may be different functions of food density, and this must be taken into account in developing a model of production. It will be simplest to develop and explain a production model for a population having a single age group. If, however, the necessary density-dependent functions can be defined for each age group in other populations, we can suppose a useful production model for such populations can be developed.

The production of a consumer having a single age group can be represented by the equation:

$$P_{cons} = G_{cons} \times B_{cons} \qquad [1]$$

where

P_{cons} = production of consumer of interest,
G_{cons} = growth rate of consumer of interest, and
B_{cons} = biomass of consumer of interest.

As we have shown for the sculpin and the trout in laboratory stream communities, the relationship between the growth rate of a predator and the density of its prey may be fairly simple (Brocksen *et al.*, 1968). Data for sockeye salmon suggest similar relationships may sometimes exist for plankton-feeding fish (Figs 3, 7). We may find for other species as well that the relationship between the growth rate of the consumer and the density of the food can be represented by the equation:

$$G_{cons} = f_1(B_{food}) \qquad [2]$$

where

$$B_{food} = \text{biomass of food organisms.}$$

The function f_1 can be expected to change if behavioural changes of the consumer alter the efficiencies with which it crops its food at one or more food organism densities, or if physiological or bioenergetic changes in the consumer alter the efficiencies with which it utilizes for growth the food consumed at one or more food densities. For example, in different seasons of the year or in different ecosystems, the growth rate of a particular consumer species may be a different function of its food organism biomass. This function should not change with changes in food organism production or in either intraspecific or interspecific competition, if the effect of the competition is *only* to change the density of the food.

By substituting $f_1(B_{food})$ for G_{cons} in equation [1], we can express the production of the consumer as a function of food density and consumer biomass as follows:

$$P_{cons} = f_1(B_{food}) \times B_{cons}. \qquad [3]$$

This equation should define the production of the consumer unless behavioural or physiological changes in the consumer alter the relationship between consumer growth rate and food organism density, as discussed for equation [2].

When the consumer of interest has no important competitors, the biomass of the food organisms may be some fairly simple function of the biomass of the consumer (Figs 4, 8). It may then be possible to represent the biomass of the food organisms as follows:

$$B_{food} = f_2(B_{cons}). \qquad [4a]$$

The function f_2 can be expected to change if there are changes at one or more consumer biomasses in the cropping efficiencies of the consumer, in the production of the food, or in the proportions of food production passing into different trophic pathways. Such changes might occur in different seasons of the year or in different ecosystems for the same consumer species.

When the consumer of interest has important competitors, the biomass of the food can perhaps be represented by some function of the biomasses of the consumer and its competitors. The biomass of the prey organisms appears to be a simple function of the combined biomasses of sculpins and stonefly naiads when these two predators are present in laboratory streams; and, when trout are also present, the density of their prey organisms drifting in the current may be a function of trout biomass and the combined biomasses of sculpins and stoneflies (Brocksen *et al.*, 1968). Sticklebacks are believed to compete for zooplankton with juvenile sockeye salmon, and zooplankton density may be found to be a function of the combined biomasses of these fish species when both are present in important numbers. Perhaps food biomass when competitors are present can be represented:

$$B_{food} = f_{12}(B_{cons}, B_{comp}) \qquad [4b]$$

where

B_{comp} = biomass of competitors of consumer of interest.

Function f_{12}, like function f_2 in equation [4a], may change if at one or more consumer or competitor biomasses there are changes in the cropping efficiencies of either the consumer of interest or its competitors, in the production of food, or in the proportions of food production passing into different trophic pathways.

The production of a consumer having no important competitors can now be represented as functions of its biomass by substituting $f_2(B_{cons})$ from equation [4a] for B_{food} in equation (3) giving:

$$P_{cons} = f_1[f_2(B_{cons})] \times B_{cons}. \qquad [5a]$$

This equation should define the relationship between the production of a consumer and its biomass in different seasons of the year or in different ecosystems unless f_1 of f_2 change as explained for equations [2] and [4a].

Where a consumer has important competitors, we can represent its production as functions of its biomass and the biomasses of its competitors by substituting $f_{12}(B_{cons}, B_{comp})$ from equation [4b] for B_{food} in equation [3] giving:

$$P_{cons} = f_1[f_{12}(B_{cons}, B_{comp})] \times B_{cons}. \qquad [5b]$$

Equation [5b] should define the relationship between the production of a consumer and its biomass, and the biomasses of its competitors in different seasons or in different ecosystems unless f_1 or f_{12} change as explained for equations [2] and [4b].

By bringing functions together, we can write equation [5a] as:

$$P_{cons} = f_3(B_{cons}) \times B_{cons} \qquad [6a]$$

or as

$$P_{cons} = f_4(B_{cons}). \qquad [7a]$$

Similarly, we can write equation [5b] as:

$$P_{cons} = f_{13}(B_{cons}, B_{comp}) \times B_{cons}. \qquad [6b]$$

or as

$$P_{cons} = f_{14}(B_{cons}, B_{comp}). \qquad [7b]$$

Since our earliest laboratory stream experiments (Davis and Warren, 1965), we have been aware that the production of sculpins is a function of their biomass and the biomass of the competing stonefly naiads and aware that seasonal and other differences in prey availability and in the behaviour and physiology of the fish changed this function. Establishing the existence of relationships between prey density and the growth rate of the sculpins, and between sculpin and stonefly biomass and prey density encouraged us to formulate production equations involving these relationships (Brocksen et al., 1968). The separation of the relationships involved has facilitated the identification of the kinds of changes in a predator, its competitors, its prey, or its ecosystem that result in changes in the function that its production is of its biomass. Further laboratory stream experiments have shown the density of herbivorous insects to be simple functions of algal density, and algal density to be a function of light level and current velocity (Brocksen, unpublished data). We have in this paper presented evidence that relationships like these exist in some natural ecosystems.

The ideas we have presented here are a departure from the viewpoint that has dominated studies of trophic relations since Lindeman (1942) published his paper on the trophic dynamic aspect of ecology. Those who have been interested in trophic relations have stressed the need to have information on production of the species involved and have rarely attempted to use biomass information except in estimating production. But during these 25 years, the production rates of only a few fish populations and very few invertebrate populations have been satisfactorily measured in nature. We are aware of no study in which the production of even one fish population has been related to the production of its prey organisms. With sufficient effort, it is quite possible to measure the production of almost any fish population; and this is extremely valuable and interesting information to have if it can be related to the conditions making that production possible. Likewise with sufficient effort, the production of some invertebrate populations can probably be determined. But most carnivorous fish utilize several species of prey organisms, and we cannot usually make measurements of the production rates of all of the prey populations. It will be helpful if we can find other parameters than prey production for relating the production of particular predator species to the availability of their food resources.

We believe that when relationships between food density and consumer biomass and growth can be defined, there is a clear alternative to dependence on food production information for relating consumer production to food availability. Whereas it is usually not feasible to determine the production rates of several food organism populations, the literature is full of information on the densities of such populations. Not many have attempted to relate both the biomass and the growth rates of consumers to the densities of their food organisms; but, as we have discussed, there is evidence that rather simple relations may sometimes exist. When such relations can be defined, the production of a consumer can be modelled as functions of its biomass, the biomasses of its competitors, and the density of the food organisms without resorting to any assumptions as to the mathematical nature of the relations involved. Modern computers now make it possible to define and combine more complex functions than those used in classical population models. The simplicity of classical models has made them appealing to students of population dynamics, but the assumptions upon which they are based severely limit their application. For any system in which the density relations we have been discussing can be defined with reasonable precision, the appropriateness of a production model combining these relations appears to us to be self-evident and inarguable. The use of such a model to estimate the biomass of a predator that would maximize its production should insure that important interactions of that predator, its principal competitors, and their prey organisms are taken into account.

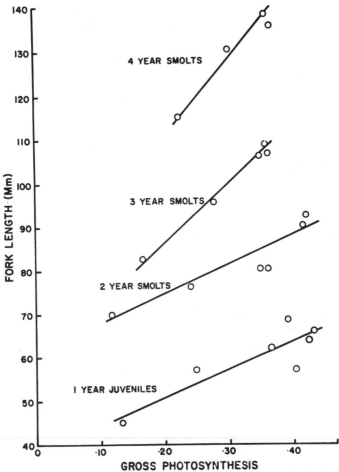

Fig. 15. The relationship between rates of gross photosynthesis and the growth of four age groups of sockeye salmon. (Redrawn from Nelson, 1957.)

It might at first appear that a production model developed for a consumer in one system would not be suitable for that species in other systems; and such might be the case. A relatively few measurements of consumer biomass and growth rate and food density for these other systems, if made over a range of food densities, could give some indication of how appropriate the established model might be for these systems. Depending on the importance of correctly knowing the production relations of the consumer in the other systems, further studies could be instituted to confirm the appropriateness of the original functions or to lead to new functions. Similarly, whether or not a production model developed for a consumer for a particular season of the year could be correctly used for other seasons of the year could be evaluated.

TROPHIC PERSPECTIVES

The upper limits of production of animals in aquatic ecosystems are often supposed to be set by the biogenic substances present in these systems. Except perhaps in pond fish culture, close relationships between the quantities of these substances present and the levels of fish production maintained have not usually been apparent. Still, there is some evidence that even in rather large systems close relationships of this kind exist.

Nelson (1958) has reported that fertilizing an Alaskan lake annually with inorganic nitrate and phosphate led to increased gross production of phytoplankton. And growth rates of the four age groups of juvenile sockeye present in the lake were closely and positively correlated with the algal production (Fig. 15). We must suppose, then, that the production of the zooplankton upon which these sockeye fed was limited by the production of phytoplankton. Furthermore, we should expect that appropriate measurements would have made apparent density-dependent relationships between the successive links of this food chain.

There is little direct evidence of the existence of density-dependent relationships between successive links of entire food chains, not so much, we suspect, because they do not exist but because they have not been sought in appropriate ways. Nonetheless, Izhevskii (1961) does present evidence that the growth rate of bream (*Abramis brama*) in the Caspian Sea is a function of the biomass of their benthic food organisms (Fig. 16), and that the biomass of these organisms is a function of the runoff of the Volga (bringing nutrients into the Sea).

When relationshps between plant nutrient levels and the growth rate that particular consumers maintain at different consumer biomasses can be defined, it should be possible to develop production models for these consumers that would incorporate nutrient functions. These functions would permit prediction of the effects of nutrient change on the function that consumer production is of biomass. The existence of a relationship between the production of a consumer and the level of plant nutrients would most certainly suggest that intermediate links in the food chain are appropriately taken into account, even though they are not explicitly involved in the model.

But, of course, when density-dependent functions can be defined between successive intermediate links in a food chain, greater understanding of the trophic processes leading to consumer production can be achieved. And when such functions can be defined, there exists the possibility of developing a production model for a consumer of interest that will explicitly incorporate not only all of the links of the food chain but its competitive interactions with

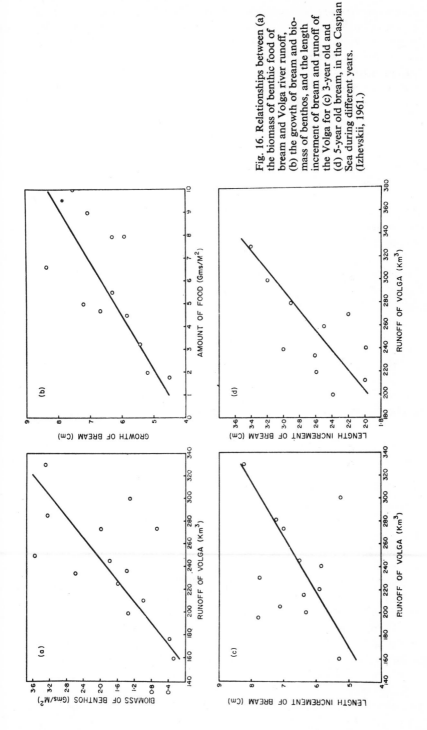

Fig. 16. Relationships between (a) the biomass of benthic food of bream and Volga river runoff, (b) the growth of bream and biomass of benthos, and the length increment of bream and runoff of the Volga for (c) 3-year old and (d) 5-year old bream, in the Caspian Sea during different years. (Izhevskii, 1961.)

other species as well. Initially, the greatest value of the development of such models will perhaps lie in any meaningful order it might bring to our thinking about trophic processes. Ultimately, such models might prove to be useful tools in the management of aquatic resources.

Even before we have the information necessary to develop production models incorporating intermediate links in food chains, information of considerable value in resource management may be expected to become available. Definition of the function that production is of the biomass of a consumer, even without much understanding of the trophic processes leading to this function, would permit the maximization of the production of the consumer by manipulation of the level of exploitation. From such a model, however, good prediction of the level of exploitation that would maximize consumer production would depend on the constancy of the ecosystem. Changes in the ecosystem would change the biomass-production function. And, increasing the production of the consumer of interest beyond this maximum would require manipulation of the trophic processes existing in the ecosystem.

Before the expenditure of extensive effort in the manipulation of trophic processes would be warranted, considerable information on the competitive interactions of the consumer of interest and on intermediate links in the food chain would be needed. Control of competing species might be found to be economically justifiable, and, were utilization of competing species to be found possible, there could be added economic incentives beyond that of maximizing the production of the consumer of primary interest.

Ultimately, there always remains the possibility of increasing the primary productive base of an ecosystem through addition of plant nutrients. This, of course, is being done in aquiculture in fresh waters, and extension of this practice to larger bodies of fresh water is to be expected. Application in marine aquiculture is similarly promising. More remote, perhaps, is the possibility of enriching larger marine bodies. Still, if it is conceivable that the removal of fish year after year from the North Sea could deplete critical nutrients, is it inconceivable that domestic sewage from coastal cities could be introduced in ways better to counteract this depletion? Victor Hugo reminds us that the nutrients in the wastes from Paris, even in 1862, were not monetarily inconsequential. In *Les Miserables*, he begins the book entitled *The Intestine of a Leviathan*: "Paris throws five millions a year into the sea."

Lindeman (1942) contributed much breadth to trophic perspectives. Those of us who have followed him have contributed all too little substance to the broad picture he sketched. As he emphasized, our studies and discussions of trophic relations must conceptually encompass ecosystems, for only in this context can we come to understand the persistence and production of populations. But a holism that does not give due attention to parts is intellectually sterile. Studies of trophic relations must advance knowledge of the feeding, bioenergetics and growth of individual organisms, must increase our knowledge of the trophic pathways leading to products of interest, and must enlarge our understanding of the competitive interactions that are so important in the production of particular species and the structure of communities, before these studies can contribute much to man's attempts to manage the resources of the sea.

REFERENCES

BACKIEL, T., and LeCREN, E. D. 1968. Some density relationships for fish population parameters. In *Biological basis for freshwater fish production*, edited by Shelby D. Gerking. Oxford, Blackwell, 261-93.

BROCKSEN, R. W., DAVIS, G. E., and WARREN, C. E. 1968. Competition, food consumption, and production of sculpins and trout in laboratory stream communities. *J. Wildl. Mgmt*, **32**, 51-75.

CHAMBERLAIN, T. C. 1897. The method of multiple working hypotheses. *J. Geol.*, **5**, 837-48.

CONOVER, R. J. 1964. Food relations and nutrition of zooplankton. *Occ. Publs. Symp. exper. mar. Ecol.*, **2**, 81-91.

CUSHING, D. H. 1958. The effect of grazing in reducing the primary production: A review. *Rapp. P.-v. Réun. Cons. perm. int. Explor. Mer*, **144**, 149-54.

CUSHING, D. H., and VUCETIC, T. 1963. Studies on a *Calanus* patch. III. The quantity of food eaten by *Calanus finmarchicus*. *J. mar. biol. Ass. U.K.*, **43**, 349-71.

DARWIN, C. R. 1859. *The origin of species by means of natural selection or the preservation of favored races in the struggle of life*. London, Murray (numerous editions).

DAVIS, G. E., and WARREN, C. E. 1965. Trophic relations of a sculpin in laboratory stream communities. *J. Wildl. Mgmt*, **29**, 846-71.

EDMONDSON, W. T. 1962. Food supply and reproduction of zooplankton in relation to phytoplankton population. *Rapp. P.-v. Réun. Cons. perm. int. Explor. Mer*, (*Symp. Zooplankton, Pap. no. 22*), **153**, 137-41.

HOWARD, L. O., and FISKE, W. F. 1911. The importation into the United States of the parasites of the Gypsy Moth and the Browntail moth. *Ent. Bull. U.S. Dep. Agric.*, **91**, 312 pp.

IVLEV, V. S. 1945. The biological productivity of waters. (in Russian). *Usp. sovrem. Biol.*, **19**, 98-120.

IVLEV, V. S. 1947. Effect of density of planting on the growth of carp. (In Russian). *Byull-mask. Obshch. Ispyt. Prir.*, (*Biol.*), **52** (1), 29-38.

IVLEV, V. S. On the utilization of food by plankton-eating fishes (In Russian). *Trudy sevastopol. biol. Sta.* **14**, 188-201. (*Fish. Res. Bd. Can., Trans. Ser.*, (447). Prelim trans. by W. E. Ricker.

IZHEVSKII, G. K. 1961. *Oceanological principles as related to the fishery productivity of the seas*. Jerusalem, Israel program for scientific translations, 186 pp.

JOHNSON, W. E. 1961. Aspects of the ecology of a pelagic, zooplankton-eating fish. *Verh. int. Verein. theor. angew. Limnol.*, **14**, 727-31.

KROGIUS, F. V., and KROKHIN, E. M. 1948. On the production of young sockeye salmon (*Oncorhynchus nerka* Walb) (in Russian). *Izv. tikhookean. nauchno-issled. Inst. ryb. Khoz. Okeanogr.*, **28**, 3-27. (*Fish. Res. Bd Can., Trans. Ser.*, (109). Trans. by R. E. Foerster).

KROGIUS, F. V. 1961. On the relation between rate of growth and population density in sockeye salmon (in Russian). *Trudy Soveshch. ikhtiol. Kom.*, **13**, 132-46. (*Fish. Res. Bd Can., Trans. Ser.*, (411). Trans. by R. E. Foerster).

LINDEMAN, R. L. 1942. The trophic-dynamic aspect of ecology. *Ecology*, **23**, 399-418.

MALTHUS, T. B. 1798. *First essay on population*. London, Macmillan & Co., Ltd. (reprinted 1926), 396 pp.

NELSON, P. R. 1958. Relationship between rate of photosynthesis and growth of juvenile red salmon. *Science, N.Y.*, **128**, 205-206.

NICHOLSON, A. J. 1933. The balance of nature in animal populations. *J. Anim. Ecol.*, **2**, 132-78.

PARSONS, T. R., LeBRASSEUR, R. J., and FULTON, J. D. 1967. Some observations on the dependence of zooplankton grazing on the cell size and concentration of phytoplankton blooms. *J. oceanogr. Soc. Japan*, **23**, 10-17.

RUGGLES, C. P. 1965. Juvenile sockeye studies in Owikeno Lake, British Columbia. *Can. Fish Cult.*, **36**, 3-21.

SMITH, H. S. 1935. The role of biotic factors in the determination of population densities·
J. econ. Ent., **28**, 873-98.

VERHULST, P. F. 1838. Notice sur la loi que la population suit dans son accroissement.
Corresp. math. phys., **10**, 113-21.

WALTER, E. 1934. Grundlagen für algemeinen Fisherieliechen. Produktions Lehre.
Handb. Binnenfisch. Mitteleur. **4** (5), 480-662.

WARREN, C. E., WALES, J. H., DAVIS, G. E., and DOUDOROFF, P. 1964. Trout production
in an experimental stream enriched with sucrose. *J. Wildl. Mngmt*, **28**, 617-60.

WOLNY, P. 1962. The influence of increasing the density of stocked fish population on the
growth and survival of carp fry (in Polish, English and Russian summaries). *Roczn.
Naukro In* **81-B**, (2), 171-88.

ZELIKMAN, E. A., and KAMSHILOV, M. M. 1960. Long-term dynamics of the plankton
biomass in the Southern Barents Sea and the factors determining it. *Trudÿ murmansk.
biol. Inst.*, **2**, 68-113. (Israel Program for Scientific translations, IPST Cat No. 947).

Production and food supply

J. E. PALOHEIMO
University of Toronto
Ramsay Wright Zoological Laboratories
Toronto, Ontario

L. M. DICKIE
Fisheries Research Board of Canada
Marine Ecology Laboratory
Dartmouth, N.S.

ABSTRACT. It is clear that production in any biological population ultimately depends on the energy derived from its food. However, equilibrium fish populaiions are often treated in simpler fashion as a biomass or group of individuals having characteristic rates of turnover. These rates are described in terms of the population parameters, recruitment, growth and mortality. The biomass is established at various equilibrium levels when subjected to different amounts of fishing or predation. The effects of changes in the food supply have been explicit considered in very few theoretical or experimental situations. Where they have been, it has been concluded as a first approximation that the yield at given fishing mortality is proportional to food supply. An appreciation of the possible generality of this conclusion, for predator-prey systems, and of those food-chain situations in which there may be significant deviations, can only be obtained from a study of the effects of food supply on the physiological and behavioural mechanisms determining the population parameters.

This paper will give a simplified description of population biomass, productivity and yield in terms of empirically derived physiological and behavioural parameters. The parameters express relations between population abundance, structure and feeding. In the simplest case the population parameters other than the "fishing" mortality are constant and independent of the food supply. Situations are then elaborated where growth, recruitment or grazing ability are variable but dependent on the structure of the population and the amount of food available. The results will be examined for the case of a single food source for the prey. Models examined incorporate concepts of "feedback" between environmental circumstances and population structure. They also permit the study of the probability of the population reaching an "equilibrium" state.

Preliminary examination of the more realistic models gives evidence for deviations from expectations based on the simple theoretical models: deviations of the sort which are encountered in both experiments and nature. Conclusions from the study lead to inferences about the significance of different "types of population dynamics" for relations between yield to the predator and food of the prey.

499

Production in a fish population depends on the food supply, and longterm yield to a predator is dependent on production. However, at a given time, the predator acts on the standing stock or biomass. Therefore, besides the specification of population parameters, construction of models of production and yield in relation to basic food supply evidently requires a knowledge of conversion rates and efficiencies linking the energy transfer processes to the organic material of the populations.

Ecology is not yet in the position of those physical and technological systems where sufficient information is available to permit complete modelling of the system. Furthermore, in view of the complexity of kinds of animals involved and the number of different exchanges which take place, it is doubt-ful if detailed modelling of the biological system is economically practicable or possible. In such a case approximate models appear to have at least two important functions. They provide useful estimates of the magnitude of the change in production and yield in terms of certain parameters of population change. More importantly, however, they can also demonstrate the relative importance of certain parameters and structural features, permitting a re-grouping and reframing of concepts which can form the basis for a higher degree of generality in understanding factors underlying population regula-tion.

Past attempts to construct generalized models of population production systems may be classified on the basis of the length of food-chain and the degree of structuring which was considered. Thus, the numerical model first described by Baranov (1918) was distinguished by a high degree of size-structuring of the population, but considered only the interaction of predator and prey, per unit of prey. It thus side-stepped a measure of absolute biomass and implied a food supply in proportion to the requirements of the animals. Later developments by Ricker (1940, 1944) and Beverton and Holt (1957) retained the highly desirable structural detail and worked towards release of some of the restrictions on biomass changes by considering possible responses of the production processes of growth and recruitment to relative biomass changes. But food supply effects were still implicitly in terms of requirements by the animals. A later model by Baranov (1926), and models by Lotka and Volterra (Volterra, 1928), Nicholson and Bailey (1935), and more recently by Schaefer (1954, 1967), introduced the concept of a limited food supply, to the extent that this is implied by an asymptotically realized maximum biomass in a particular universe, but lacked the structural detail of the earlier analytic models, making the fitting and testing of parameters a very difficult task. Explicit consideration of the steps in the chain from food through production to yield in terms of the structure of the system have received treatment com-paratively recently in the work of Odum and Smalley (1959), Slobodkin (1959, 1960), Silliman and Gutsell (1958) and Silliman (1968).

In this paper we attempt a natural elaboration of this latter development, beginning with a numerical-analytical model where structure is defined in terms of size, and developing it towards a statement in energetic terms in relation to the physiological and behavioural specializations of the use of energy for production processes. This is accomplished by the use of the empirically derived parameters of energy uptake and conversion which formed the growth curve described by Paloheimo and Dickie (1965). The model is used to calculate equilibrium yield, biomass and production in relation to food supply in various hypothetical situations. The results indicate the possible relative importance of various physiological properties of organisms for production and of various conversion efficiencies, and may have value in specifying relations which can be considered in experimental or observational tests on real systems.

A SIMPLE POPULATION MODEL

We begin by constructing a simple analytical population model much like that originally outlined by Baranov (1918), and since elaborated and studied by Ricker (1940, 1944, 1958), Beverton and Holt (1957) and others. To keep the formulations as simple as possible we shall initially assume that the parameters of population change: growth, recruitment, natural mortality and predatory or fishing mortality are constant and operate throughout the life span of the animals. Such obvious over-simplifications will be relaxed as appropriate later on. If we adopt the notation:

$r\,\Delta t$ births in numbers per time interval Δt,

M instantaneous natural mortality rate,

F instantaneous rate of predation or fishing mortality rate,

$\Delta N(t)$ number of individuals of age t to $t+\Delta t$ where Δt is an infinitesimally short interval,

then from the definition of instantaneous rates

$$\Delta N(t) = r\,\Delta t e^{-(F+M)t}.$$

The total number of animals present at any time, N, is then the integral over all ages.

$$N = \int_0^\infty re^{-(F+M)t}dt$$

$$= \frac{r}{F+M}.$$

From this it appears that as long as the rates of mortality and recruitment (= birth rates) are constant, the population size in numbers is independent of time (i.e. constant).

We may also calculate the biomass, assuming growth rate per individual is constant:

G (exponential) rate of growth,

w_0 initial size of the newborn,

$w(t)$ weight of an individual at time t.

From the definitions

$$\frac{dw(t)}{w(t)dt} = G$$

whence

$$w(t) = w_0 e^{Gt}$$

but biomass

$$B = \int_{\substack{\text{over} \\ \text{all} \\ \text{ages}}} \left(\begin{array}{c}\text{weight per individual} \\ \text{of age } t \text{ to } t+\Delta t\end{array}\right) \times \left(\begin{array}{c}\text{no. of individuals of} \\ \text{corresponding age}\end{array}\right)$$

$$= \int_0^\infty w_0 e^{Gt} r e^{-(F+M)t} dt$$

$$= \frac{rw_0}{F+M-G}.$$

The yield to a predator, Y, is the fraction $F/F+M$ of the total deaths or simply

$$Y = F \times B = \frac{Frw_0}{F+M-G},$$

and the loss to "natural" mortality (i.e. non-predatory mortality) is similarly calculated, as the fraction $M/F+M$ of the total mortality.

Other important quantities which we will wish to discuss later may be calculated by observing that

$$\text{Production} = \text{reproduction} + \text{growth}$$

and rate of replacement of the equilibrium stock per unit time per unit of stock = Production/Biomass.

To summarize, for the simple population model we have the following six quantities defined in terms of the parameters of population change:

Total number $= N = \dfrac{r}{F+M}$ [1]

Biomass $= B = \dfrac{rw_0}{F+M-G}$ [2]

Yield to the predator in time Δt (not necessarily small) $= \dfrac{Frw_0}{F+M-G} \Delta t$ [3]

"Non-predatory" deaths in time Δt $= \dfrac{Mw_0 r}{F+M-G} \Delta t$ [4]

$$\text{Production} \qquad = \text{Prod.} = \left(rw_0 + \frac{Grw_0}{F+M-G}\right) \qquad [5]$$

$$= \frac{(F+M)rw_0}{F+M-G}$$

Rate of replacement of biomass in time Δt

$$= \text{Repl.} = \frac{\text{Prod.}}{B} = F+M. \qquad [6]$$

Writing the equations in this form makes clear some of the relationships between biomass, production and yield which are sometimes confused in discussion. Thus, for example, we may define the turnover rate as the inverse of the mean life span. But

$$\text{mean life span} = \int_{\substack{\text{over} \\ \text{all} \\ \text{ages}}} (t \times \text{No. } t\text{-year-olds})/N = 1/(F+M)$$

whence the turnover rate $= F+M$. That is, it is the same as the replacement rate (equation [6]).

From equation [2] we may then write for the biomass

$$B = \frac{rw_0}{(\text{turnover rate}) - G} \qquad [2']$$

from which it is seen that biomass is inversely related to the turnover rate. Moreover the production

Prod. $= (\text{turnover rate}) \times \text{biomass} = (\text{rate of replacement}) \times \text{biomass.}$ [5']

This shows in relation to [2'] that an increased turnover rate or rate of replacement does not necessarily imply an increased production unless biomass is constant, or conversely that biomass is not a reliable index of production unless the turnover rate is approximately the same.

It is somewhat perplexing to find in equation [6] that the rate of replacement of the biomass is independent of the rates of growth and reproduction, the two mechanisms which are responsible for it. The reason for this in this simple formulation is obviously a result of the fact that we have assumed constant rates, in which growth and recruitment affect production and biomass in precisely the same way hence cancel out. In the more general case where parameters are no longer constant, the dependence on growth and recruitment will be affected by population structure, hence the terms describing them do not cancel out in such obvious fashion. In a balanced stock, however, it is still clear that rate of replacement will be the same as the total averaged mortality rate as shown by equation [6].

The formulation also indicates some of the limitations of such a simple model. For example, to reach a finite biomass in equation [2], it is apparent that $F+M>G$, otherwise B would increase without limit. But in the situation $F+M>G$, the biomass at the moment of birth is at the maximum for the cohort. This leads to the conclusion that yield increases with F, becoming greatest at $F = \infty$, that is, when animals are all cropped the moment they are born. Such technical absurdities are, of course, a direct result of the use of constant parameters in the model. In this simple form the model is chiefly useful for purposes of definition.

A SIMPLE PRODUCTION MODEL

Until recently, researches in population dynamics of fishes have been largely concerned with age or size specific elaborations of the simple system described above, for the purpose of defining various parameters in a manner which would permit their measurement in nature. Given sufficient observations it might then be possible to establish the nature of relations among the values taken by various parameters, under particular circumstances (e.g. Beverton and Holt, 1959; Holt, 1962), hence build up a predictive system of probable relationships. Unfortunately, variability in nature and vagaries of sampling procedures together give rise to such high statistical variances for almost all the parameters that most such attempted developments have seemed more a display of brilliant intuitions and courageous speculations, than a framework of scientific observation. It is in the face of such results that attempts have been made to reformulate the system to incorporate recognizable reference points, in particular an enlargement of the system to take account of the fact that, ultimately, the population production processes depend upon the food supply. With this enlargement to consideration of one full step in a food chain, it becomes possible to use the simple population model as a basis for describing a simple production system and defining particular quantities in energetic terms. In what follows we show that even at a rather simple level, the results from such a model suggest a number of relationships which are potentially useful in the study of natural processes. The simple system also provides a basis for constructing and studying more realistic model systems, in which the descriptive equations become a good deal more complicated. We outline our conclusions from the study of one such model.

In what follows we will initially assume that food supply is constant, and that all the food supplied is ingested. We will also consider that the efficiencies of use of the food for growth, K_1, and reproduction, K_2, are constant and equal and that the rate of food intake by individuals is proportional to their body weight. In this situation the total food intake of the population, I, is proportional to the biomass.

From equation [5], the production due to growth in a balanced stock is given by $Grw_0/F + M - G$. Hence the food intake per unit time for growth

$$I_G \Delta t = \frac{1}{K_1}\left(\frac{Grw_0}{F+M-G}\right)\Delta t.$$

Similarly the food required for reproduction

$$I_r \, \Delta t = \frac{1}{K_2} (rw_0)\Delta t.$$

Where $K_1 = K_2$, the food required to sustain fixed rates of G and r

is
$$I\Delta t = \frac{1}{K}\left(\frac{(F+M)rw_0}{F+M-G}\right)\Delta t$$

where we define $K = K_1 = K_2$.

The population production processes in the balanced stock may respond to a change in food supply through a change in growth or recruitment. If the response were by a growth change, the rate of growth may be expressed explicitly in terms of food intake as

$$G = \frac{(F+M)\,(KI-rw_0)}{KI}. \tag{7}$$

The comparable expression for reproduction is

$$r = \frac{(F+M-G)K}{(F+M)w_0}\, I. \tag{8}$$

It now appears, however, that whether expression [7] or [8] is used, that is, whether the change in the production process results from a change in growth or reproduction or both, we arrive by simple algebraic manipulation at a single set of definitions for equilibrium biomass, yield and production in terms of food intake. The results are as follows:

$$\text{Biomass} \quad = B = \frac{KI}{F+M} \tag{9}$$

$$\text{Yield} \quad = Y = \frac{F}{F+M}\, KI \tag{10}$$

$$\text{Production} \quad = \text{Prod.} = KI \tag{11}$$

That is, at least as long as the food conversion efficiencies are the same, it makes no difference to the equilibrium biomass, yield and production whether the population utilizes the growth or reproductive pathways.

Given the definitions arrived at from the simple production model, we are in a position to calculate what Slobodkin (1960) has called the ecologically significant energy relations.

Ecological efficiency
$$= E_e = \frac{\text{Yield}}{\text{Food intake}} = \frac{Y}{I} = \frac{KF}{F+M} \tag{12}$$

Caloric cost of maintaining the population per unit biomass (calories)
$$= C = \frac{\text{Food intake}}{\text{Biomass}} = \frac{I}{B} = \frac{F+M}{K} \tag{13}$$

$$= \frac{F}{K} + \frac{M}{K}$$

where evidently F/K is the total energy cost of producing the equilibrium yield and M/K is the energy cost of maintaining the residual stock (after predation).

$$\text{Population efficiency} = E_p = \frac{\text{Yield}}{(\text{Cost of producing yield}) \times \text{Biomass}} = \frac{Y}{\frac{F}{K} \times B} = K. \qquad [14]$$

With this simplified food-limited production model, we are justified in considering that we have approached the nature of a real population more closely than in the numerical population model, where maintenance of the steady state depended on arbitrarily defined values of the parameters. It is still grossly over-simplified because of the assumption of constant rates. Nevertheless, at this stage several noteworthy conclusions emerge. Their generality is checked later when some of the simplifying assumptions are relaxed.

ECOLOGICAL EFFICIENCY

Equation [12] shows that ecological efficiency must increase with fishing mortality rate. But this increase will be at a decreasing rate, asymptotically approaching the value of K, which we have defined as the average gross conversion (growth + reproductive) efficiency. Slobodkin (1959) distinguished between ecological efficiency as defined here, and food-chain efficiency which is similar to equation [12] except that instead of food intake, the food supply is used in the denominator. In his experimental situations there was little difference between the two efficiencies at low fishing mortalities. At intermediate and high mortalities, the mortality rate seemed to impair the rate of reproduction by the *Daphnia* to the point where the stock was no longer able to consume all the food supplied. Thus at the higher mortalities food-chain efficiency decreased markedly as food-supply increased. Furthermore the maximal food-chain efficiency for a particular feeding level occurred at lower and lower fishing mortalities as the food supply level was increased.

On the other hand, Slobodkin found that the experimentally determined ecological efficiency was independent of feeding level, as would be expected from equation [12], although it increased faster at higher fishing mortalities than would be expected from equation [12] with constant parameters. (Note that Slobodkin's definition of F is not the same as used here. From his data it appears that his $F = 0.9$ is equivalent to approximately $F = 0.5$ to 0.6 in our terminology.) This would be the result if natural mortality in the *D. pulex* population increased with age, since the decreased life span at high F would mean that as F increased the average M decreased. Smith (1963) found that M was dependent on age in *D. magna* populations and survival data on *D. galeata* by Hall (1964) confirmed "the near-rectangular shape (of the survivorship curve) reported for *Daphnia*" by earlier authors. According to equation [12] an increase in ecological efficiency with increase in F would also result if the change in population structure brought about a progressive increase in K. In fact, both M and K may have been affected in the real population. The effects of a changing K are studied further in the more complex models considered below.

RATE OF REPLACEMENT AND MORTALITY

Equation [13] defines the cost of replacing the population explicitly in terms of mortality rates and conversion efficiencies. It suggests that one can consider costs of replacing the yield and the residual stock separately, a concept which Slobodkin (1959) applied to the *Daphnia* population experiments. His results when compared with our formulation imply some relationships of potentially great practical interest.

Slobodkin calculated a replacement *rate* for the residual stock (i.e. average biomass remaining after predation) from measures of biomass, birth rates (calculated), and conversion efficiencies. He found that the result was numerically equal (*circa*) to the percentage weight loss of starved *Daphnia*, as well as to the inverse of the calculated cost of maintaining the residual population. From equation [6], Repl. = Prod./Biomass = $F + M$. Hence, by analogy with equation [13] Slobodkin's "definition" that all energy not going to replace yield is used to replace the residual stock, is equivalent in our formulation to identifying replacement rate of the balanced residual stock with M, the average natural mortality rate. The foregoing equations may be used as an aid to understanding the possible relations among the experimentally derived replacement rate, weight loss and maintenance cost.

To understand the relation between replacement rate or mortality rate and percentage weight loss by starved *Daphnia*, we may recall the basic energy equation for a balanced population in the form:

$$I = T + \text{Prod.} \tag{15}$$

where we define T as the total metabolic expenditure of the population. In a balanced feeding and producing population we may then define metabolism in terms of calories as

$$T = I - KI, \tag{15'}$$

whence metabolic rate

$$\frac{T}{B} = (1 - K)\left(\frac{F + M}{K}\right) \tag{16}$$

or in the absence of predation

$$\frac{T}{B} = \frac{1 - K}{K}(M) = (1 - K)\left(\frac{M}{K}\right). \tag{16'}$$

That is, where the population system is in balance with its food supply, the metabolic rate is proportional to the mortality rate and may be measured as the energy equivalent of the mortality rate $F + M/K$ which according to equation [13] is the energy cost of replacing the stock multiplied by the complement of the gross food conversion efficiency.

Given the situation described by equations [15] and [15'], if the external food supply were suddenly removed so that $I = 0$, then the animals in the population would be dependent on their "endogenous" or stored food supply.

By equating I to zero in [15] or [15'] we get $T = -$Prod. and hence by omitting the minus sign can write

$$\frac{T}{B} = \frac{\text{Prod}}{B} = (F+M).$$ [17]

That is by purely algebraic manipulation we find that the metabolic rate of starved animals appears to be equivalent to the rate of replacement for the stock. This seems to explain the coincidence between replacement rate and weight loss of starved *Daphnia* found by Slobodkin.

On the other hand Richman (1958) reported that when *Daphnia* were starved over a six-day period there was no apparent change in their rate of oxygen consumption which remained the same as it had been during feeding. That is when the food supply was cut off, T remained unchanged. Richman did find that the respiratory quotient of unfed animals was slightly lower than for fed animals, indicating the shift from metabolism of carbohydrate to metabolism of stored fat or protein, but with appropriate corrections for body weight equivalents of O_2 consumption Richman's experiments suggest that for the starved *Daphnia* $T = (1-K) I$, where I is the feeding level prior to starvation, that is

$$\frac{T}{B} = \frac{1-K}{K} (F+M).$$ [17']

Although equation [17] explains Slobodkin's experimental results other experiments suggest that the metabolic rate of starved animals should be given by [17'], i.e. by $(1-K)(F+M)/K$ rather than by [17]. The two values agree only when $K = 0.5$. Reference to the experimental results again suggest some important considerations for production studies.

Richman (1958) fed *Daphnia* at various rates varying between 25 000 to 100 000 cells/ml/day. The volume of water swept clear per *Daphnia* per day was strongly dependent on body size, but virtually independent of food concentration, averaging about 4.0 ml for an intermediate sized animal. Under these experimental conditions, he found that while the ingested energy increased in proportion to food concentration in the water so too did the rate of egestion. The percentage of the *consumed* energy which showed up as production therefore decreased with increased feeding rate from a maximum of 17% to a minimum of 10% averaged over the 40-day life-history. By contrast the percentage of the *assimilated* energy which showed up as production increased from 55% at the lowest feeding level where respiration was a large fraction of the total energy exchange, to over 70% at the highest feeding level. The energy going to adult growth was virtually constant, despite differences in feeding level; the bulk of assimilated energy directed to production went into reproduction which increased with feeding level.

The Slobodkin experimental populations were given predetermined fixed food rations at 4-day intervals, the basic ration being $6 \cdot 2 \times 10^6$ cells/50 ml of culture, and the maximum ration 5 times this amount. If this ration was consumed at near the rates found by Richman, it would appear that in the cultures subjected to lower rates of predation, where the residual stock was numbered in hundreds, average food amounts available per day, even for the maximum ration, were in the lower part of the range used by Richman. At the highest predation rates, residual stocks were much smaller and, as Slobodkin indicates, were unable to consume all the food. In these cases available food concentrations may have been very high. In the main, however, it appears that the Richman conversion efficiencies, probably characterized Slobodkin's populations.

Two inferences may be drawn from these observations in relation to our simple production formulation. The first is that in a filter-feeder of the *Daphnia*-type, where equation [17] holds true, the appropriate index of efficiency, K, for measuring the relation of respiration and production to food supply, must be some function of the assimilated rather than the ingested food. If the population respiration and conversion was determined by the food ingested rather than food assimilated, it is difficult to understand how Slobodkin's measured replacement rates could have been so close to Richman's starvation weight losses. In his paper for this symposium, Suschenya independently comes to the same conclusion in his fitting of a growth curve to Richman's *Daphnia* data.

A corollary of this first inference is that production processes in animals like *Daphnia*, having such a restricted or inflexible metabolic response to opportunities for feeding, would have to be treated rather differently from animals where the metabolic responses are not so rigid. Among fishes, for example, the work of Brown (1946, 1957), has shown remarkable adaptations of the metabolic level to starvation food levels, and Paloheimo and Dickie (1966a) found that metabolic levels at a given temperature, changed over at least a four- to fivefold range in response to changes in feeding opportunities above the maintenance level. Such observations suggest that in appropriate experiments with fish it is possible to define a "standard" metabolic level for a maintenance feeding rate, I_s (Fry, 1957; Beamish and Dickie, 1967) where Prod. = 0. At this maintenance feeding rate standard total metabolism

$$T_s = I_s,$$

whence for the short-term experimentally determined metabolic rate

$$\frac{T_s}{B} = \frac{I_s}{KI(F+M)} = \frac{F+M}{K} \times \frac{I_s}{I}. \qquad [18]$$

This may be compared with equation [16] describing the balanced energy system for a population feeding in excess of maintenance. If a measure of I_s/I is available, equation [18] differs from equation [16] by the unknown factor $(1 - K)$. Both equations are however evidently subject to experimental determination in a metabolically adaptable animal. It thus appears that appropriate experiments relating metabolic rate to feeding level may provide important information on conversion or production efficiencies and replacement costs. Correct interpretation of the metabolic results would depend on a knowledge of the degree of flexibility in metabolic response in the subject animals.

It remains in this section to compare the replacement rate and rate of weight loss by starved *Daphnia* with the inverse of the maintenance cost, the third of the three quantities which Slobodkin found to be the same (see p. 507). In this case we have replacement rate, which from its equivalence to $(F + M)$ has the dimension 1/time, equated to the inverse of maintenance cost which from equation [13] we write as $K/F + M$. Since K is by definition dimensionless, the ratio has the dimension of "time". It is difficult to explain how two quantities with the dimensions of time and 1/time could be the same except by coincidence resulting from a particular choice of time period. Alternatively the coincidence may arise through difficulties associated with experimental measurement of K because of the ingestion-assimilation situation discussed above. There appears to be no further significance which can be readily attributed to this observation of coincidence.

POPULATION EFFICIENCY

The definition of population-efficiency derived in equation [14] is the same as that used by Slobodkin (1959, 1960) although the manner of derivation in terms of constant population parameters shows that, in the simple situation we have assumed, population efficiency reduces to the logical equivalent of the gross conversion or production efficiency. We can thus verify, at least in the situation where growth and reproductive efficiency are approximately constant throughout the life of the animals, Slobodkin's conclusion that population efficiency is constant and independent of the rate of predation. Paloheimo and Dickie (1966b) suggested that in fishes the assumption of constancy of the growth efficiency may not be justified for prolonged feeding on one type of food although Kerr and Martin (this volume) have suggested that in real populations this effect may be offset by a change in the trophic level at which the animals feed. Complex situations, in which the average growth efficiency of the population will change with population structure, are studied in the more elaborate models developed below. At this simple level it is easy to further verify Slobodkin's (1960) conclusion that population efficiency is always greater than ecological efficiency, although from our results it appears that ecological efficiency increases with fishing or rate of predation and asymptotically approaches population efficiency at high rates of predation.

A SIMPLE EXTENSION—THE GRAZING COEFFICIENT

We have been discussing models which, on the basis of the numerical definitions given in equations [1] to [6], are potentially highly structured with respect to age or size. However, by assuming constant parameters and a rate of food intake proportional to body size, this structure is of no particular significance to our conclusions. That is, with the simplifying conditions adopted above each unit weight of animal is of equivalent value to the population as a "production machine", regardless of whether the units come in large or small, old or young packages. Under such conditions it is not especially surprising to find relatively simple expressions for biomass, yield and production in terms of the energy of food intake. In this section we will explore further the extent to which these population measures and the ecologically significant energy relations derived from them remain independent of population structure as some of the over-simplifying assumptions are relaxed.

It has long been known that metabolism per unit weight of animal decreases with body size, and while the food intake must supply both metabolism and growth, the evidence for decreasing rates of growth with increasing body size clearly indicates that unless there are marked changes in growth efficiency, the food requirement is also likely to decrease with increasing size. This situation was studied by Paloheimo and Dickie (1966a, b) who concluded that in general the relationship between metabolism, body size and rations could be expressed as

$$T = \alpha w^\mu = R(1 - e^{-a - bR}) \qquad [19]$$

where T is now the total metabolism of an individual fish, α and μ are the fitted parameters of the metabolism-body-weight relationship and R is the individual food intake or ration. The expression $(1 - e^{-a - bR})$ describes a rate of change of growth efficiency with ration R. In the models used up to this point $b = 0$, hence this expression is equivalent to $(1 - K)$, where K is the (constant) growth efficiency of the individual fish. Given a particular value of μ, equation [19] indicates the size of ration required to maintain a body weight, w, at a particular level of metabolism, α. That is, given constant feeding conditions the ration intake is proportional to w^μ, rather than w as was assumed in the previous discussion.

If food intake is not simply proportional to body weight, growth rate is itself not such a simple function of body size, hence the relationships between biomass, yield, production and food intake are dependent on size-compositions. These relationships can be calculated as follows:

From equation [19] the rate of growth

$$\frac{dw}{dt} = \alpha K w^\mu$$

where we have redefined α slightly as $\alpha/(1 - e^{-a})$. It has been repeatedly shown that a value of $\mu = 0 \cdot 8$ is a generally acceptable first approximation for fishes, whence the growth curve

$$w(t) = (\alpha K t / 5 + W_0)^5.$$

But biomass

$$B = \int w(t)\, N(t)\, dt$$

where $N(t)$ is the number of fish of age t. By putting $w_0 = 0$ as a first approximation we arrive by integration at the expression

$$B = \frac{(\alpha K)^5}{(5)} \frac{5!\, r}{(F+M)^6} \qquad [20]$$

From equation [3], the yield

$$Y = F \times B \qquad [21]$$

and from equation [5'] or equations [9] and [12]

$$\text{Prod.} = (F+M) \times B. \qquad [22]$$

Since in this approximation we have put $w_0 = 0$, these expressions ignore the (small) contribution of the recruitment to the biomass and production at a given instant. This may be included if exact equations are desired, but the quantity will generally be small compared with total biomass, the complete integration formulae are complex, and the exact form unnecessary for present purposes.

Given the relation of biomass to the size-composition, we are in a position to express the relationship of biomass to food intake. For this, two situations arise from consideration of equation [19] and the conditions under which it was derived. The first situation is trivial and simple. That is, if the population is administered a given amount of food, and consumes all of it, expressions [20] to [22] become equal to the right hand sides of equations [9] to [11]. That is, in this laboratory-type situation, of controlled food supply, the division of the production, yield and biomass among the size elements of a stock is more or less complicated, depending on the relationship of the food intake and growth efficiencies to body weight. The overall population biomass production or yield are however, functions of the imposed conditions of constant food supply. The ecological energy relationships derived from the structurally more complex populations are similarly unaffected, a conclusion which is verified by the Slobodkin experiments which were adequately described by equations [9] to [14] despite the fact that Richman (1958) notes that at the imposed constant feeding levels used by himself and Slobodkin food intake per unit weight of *Daphnia* was significantly lower for the larger sized animals.

The situation in which the total amount of food allowed to a population per unit time is controlled by the experimenter is simple and easily understood as an experimental device. However, as has been pointed out by Paloheimo and Dickie (1965, 1966a), it does not seem likely to describe most situations in nature, and did not satisfactorily describe the results of a number of fish feeding experiments. They found that when the fish were presented with a constant *opportunity* for feeding, it appeared that μ of equation [19] was constant and equal to values which have been obtained from respiration experiments. Hence, with a given "availability" of food, fish of different sizes in the experiments ate different total amounts, and were feeding in such a way as to maintain a value of α which characterized that feeding opportunity. A change in opportunity for feeding led to a change in the average value of α. If this situation were generally true in nature, we would be led to the conclusion that the grazing efficiency of an individual fish changes with its size, and is in fact proportional to w^{μ}.

It follows that the grazing power of the population may itself change with the size composition of the stock. We therefore define a grazing power per unit biomass as a relative grazing coefficient of the population. This relative grazing coefficient is given by

$$\int w(t)^{\mu} \, N(t)dt/\text{Biomass}. \qquad [23]$$

From our conclusion that $\mu = 0.8$, and again putting $w_0 = 0$ and $b = 0$, this becomes approximately equal to $(F+M)/\alpha K$. If we consider that at any instant the fish population exerts a grazing mortality on its food population, the fraction of the available food which is consumed is given by $1 - e^{-g(F+M)}$, where we have absorbed the constant $1/\alpha K$ in the proportionality factor g. In other words, if we define a maximum ration I_M which the fish population can take from a given supply without decreasing the long-term food production, then the food intake of the population

$$I = I_M(1 - e^{-g(F+M)}). \qquad [24]$$

In equation [24] we are implying that the grazing efficiency of a particular sized fish is constant (defined by α and K) relative to its body weight. The grazing coefficient $g(F+M)$ therefore refers to the average grazing efficiency of the feeding population or stock. This population grazing coefficient will change as a reflection of a change in the population structure, and will therefore lead to a change in the fraction of the available food supply which is taken in by the stock.

The production from the stock is now determined by

$$\text{Prod.} = \int \frac{dw}{dt} \, N(t)dt + rw_0 = KI_M(1 - e^{-g(F+M)}) \qquad [25]$$

while biomass

$$B = \frac{\text{Prod.}}{F+M} = \frac{KI_M(1 - e^{-g(F+M)})}{F+M}, \qquad [26]$$

and yield

$$Y = F \times \text{Biomass} = \frac{F}{F+M} \, (\text{Prod.}). \qquad [27]$$

From these equations it is apparent that the response of the population production to a change in, for example, predation becomes complex. If we assume that changes in our population do not significantly affect the production of their food, then an increase in predation will release part of the supply which was being cropped. It becomes available for distribution among the residual population, but production will depend importantly on how this is divided among the processes of growth and reproduction. If the primary response is an increase in growth rate, higher predation will not alter the long-term population structure or grazing efficiency, hence will not lead to significant long-term yield changes. It is possible that some species of fish which show high growth plasticity may be of this type. If the primary response was, however, a recruitment response there would be a major effect on population structure; grazing efficiency, production and ultimately the yield would be enhanced by a factor $(1 - e^{-g(F+M)})/(1 - e^{-g(M)})$, which is greater than 1·0. This may be the situation which holds for the metabolically less flexible zooplankton organisms which are so successfully exploited as food organisms by complex and abundant higher trophic levels. Some of these relations are illustrated in the more complex model developed below.

The introduction of physiological and behavioural feeding specializations which are reflected in the grazing coefficients begins to give an appreciation of the significance of structure as a determinant of biomass, production and yield from a given food supply level. As a final step in this first extension, it is worth pointing out that the ecological energy relationships described for the simple model by equations [12], [13] and [14], remain unaffected. That is, the complexities of interactions of sizes with production, yield and food supply, do not by themselves give any greater yield per unit of food consumed, do not affect the cost of replacement, and do not affect the population efficiency which is still a constant and equivalent to the average gross production efficiency. From consideration of grazing effects suggested by equation [19], however, it is clear that the situation of food intake, *I—not* equal to the maximum available food I_M—is not necessarily the trivial situation that it was for Slobodkin's experiments and his original definitions. In particular the "food-chain efficiency" (or yield/food supply) and overall production efficiency (Prod./I_M) become of considerable interest in relation to the size-composition of the stock. They may change with predation rate provided recruitment is responsive to changes in feeding opportunities.

A MORE COMPLEX MODEL—VARIATIONS IN PRODUCTION EFFICIENCY

In the foregoing discussion we have already noted the importance of differentiating between growth and reproduction as parts of the production mechanism. The importance of the fecundity as a measure of the reproductive capacity is poorly understood, but it is at least clear that in all cases of reproduction other than binary fission, the growth and reproductive processes are partly or wholly separated in time, reproduction being generally excluded from production among the smaller and younger stages. In the simpler models we have introduced size-age structuring of the population in relation to feeding. Evidently this "specialization" must be extended to the reproduction and growth if the model is to represent an approach to the understanding of a real ecosystem. In addition, it is conceded in general that the efficiencies of various metabolic processes may change with age. This is obvious for most warm-blooded animals which cease major growth early in life. Paloheimo and Dickie (1966b) found evidence for changing efficiencies of somatic growth among fishes, and the same may be true among some invertebrates. While we know relatively little about the scale of such changes, it is important to consider their potential effects on overall production if only as a guide to the importance of further investigation in nature. These basic deficiencies of the simple model are taken into consideration in the further development described below.

CONSTRUCTION OF THE MODEL

For study of population production in energy units it is essential as a minimum to describe the process of individual growth explicitly in terms of the food supply. An appropriate model for somatic growth was derived from experimental data by Paloheimo and Dickie (1965), and while its generality remains to be tested, it provides us with a suitable starting point. Paloheimo and Dickie postulated that the growth of individual fish could be adequately described by the following set of equations:

$$T = \alpha w^{\mu} = R(1 - e^{-a - bR}) \tag{19}$$

$$\frac{\Delta w}{\Delta t} = Re^{-a - bR}. \tag{28}$$

That is, according to equation [19], with constant temperature and feeding opportunities on one type of food, a fish of size w consumes a ration R, sufficient to maintain it at a constant level of metabolism, α. Given this ration intake per unit time, fish growth follows equation [28], that is, is proportional to the amount of the ration but with the changing efficiency $(e^{-a - bR})$. At the end of the time interval, Δt, equation [19] describes the ration which would be taken by the fish at its new weight w.

In their discussion of food-growth curves Paloheimo and Dickie (1965) also tested the validity of approximating the individual ration intake by the simpler expression

$$R = \alpha w^{\mu}$$

and found that the resulting growth curves were hardly distinguishable from the exact equation. In setting up a population model for this study we therefore replace equation [28] by the differential equation

$$\frac{dw}{dt} = \alpha w^{\mu} \exp(-a - b\alpha w^{\mu}) \tag{29}$$

$$= \alpha K w^{\mu} \exp(-b\alpha w^{\mu}).$$

There is no information known to us which shows whether or not the production of gametes follows the same function, relative to food intake, as does somatic growth. To avoid unnecessary complications and still to distinguish realistically between the somatic and reproductive production processes, we have therefore divided the life history into two phases. That is, we have assumed that for a given time at first maturity, TM, the growth of immatures, where $t < TM$, follows equation [29]. Adult growth occurs where $t \geqslant TM$ and follows the function

$$\frac{dw}{dt} = p\alpha K w^\mu \exp(-b\alpha w^\mu).$$
[30]

That is, given the ration αw^μ the growth increment for immatures is described by equation [29]. At maturity, the fraction, p, of the growth energy goes into an increase in body weight, equation [30], while the fraction $(1-p)$ becomes gametes. In the models described below we have assigned a value to $p = 0.5$. This value is similar to values found for perch by LeCren (1951) and slightly higher than found by Toetz (1967) for blue gill sunfish. Preliminary examination of the model showed that even major changes in p do not affect the generality of conclusions discussed below.

For given values of α, K, b, TM and p the above equations specify the growth of individuals in the population. The population numbers for given recruitment are

$$N(t) = re^{-Mt} \qquad \text{for } t < TF$$

and $\qquad = re^{-Mt - F(t - TF)} \qquad \text{for } t > TF$

where TF is the age of first capture. If r, the recruitment, is specified we may calculate population production parameters as

$$\text{Prod.} = \int_0^\infty \alpha K w \exp(-b\alpha w^\mu)$$
[31]

$$B = \int_0^\infty w(t)N(t)dt$$
[32]

$$Y = \int_{TF}^\infty Fw(t)N(t)dt.$$
[33]

Note that we are actually calculating these integrals, specifying the age at maturity. That is, if we write $G = \alpha K w^\mu \exp(-b\alpha w^\mu)N(t)$, then in equation [31] we calculate

$$\text{Prod.} = \left\{ \int_0^{TM} Gdt + p \int_{TM}^\infty Gdt \right\} + (1-p) \int_{TM}^\infty Gdt$$

where the first term describes the growth production and the second egg production.

There remains the problem of specifying a value for reproduction, r, in the population. Very little in detail is known about this situation in nature but we do know that in order to maintain a balanced population, that part of the production which goes to reproduction must be sufficient to produce the recruitment to it; that is it must at least be equal to rw_0 in equations [2] to [5]. In terms of our equations this means that for the balanced stock

$$w_0 \leqslant \int_{TM}^{\infty} Kw^\mu \exp(-b\alpha w)^\mu \left\{ \begin{array}{l} e^{-Mt}, \ t \leqslant TF \\[2mm] e^{-Mt - F(t-TF)}, \ t > TF \end{array} \right\} = w_1. \qquad [34]$$

Where we have the situation $w_0 < w_1$, each member of the adult stock produces on the average more than one recruit in its life-time. The population would then increase *ad infinitum* were there no limitation on its food supply. Before this situation was reached in nature there would be a tendency towards restoration of population balance, perhaps initially by a reduction in the feeding opportunities which in our terminology is equivalent to a reduction in α, which would have the effect of putting the time at maturity at a later age. In the models we have generally therefore specified TM and also calculated a value for WM, the weight at first maturity for different growth curves. The results are discussed briefly below. For present purposes it is important only to note that by fixing the level of α we are fixing the level of food available to both growth and reproduction. Hence we have in our model a method of establishing a relation between a balanced stock and a fixed food supply.

In calculating recruitment to our populations we assume that there is a fixed amount of food available to the population. If α is fixed and the inequality of equation [34] is valid, then in the presence of predation the recruitment must expand to make use of the food released by the formerly competing adults. Food acquired by the population is given by

$$\text{Food} = r \int \alpha \, w(t)^\mu \left\{ \begin{array}{l} e^{-Mt} \\[1mm] e^{-Mt - F(t-TF)} \end{array} \right\} dt. \qquad [35]$$

Therefore if the inequality of equation [34] exists, r continues to increase until total food consumed becomes equal to the assigned value, or to the value specified by equation [24]. The recruitment therefore reaches an upper limit. In fact, at this point, assuming that there is no change in α (or in b) we still have $rw_0 < rw_1$ and the population appears to be increasing. By fixing r so that food is a given constant we are therefore assuming that from any total reproduction only the fraction w_0/w_1 actually survive, resulting in an r which is specified by equation [35]. This value would be an equilibrium value for the balanced stock at fixed α. It should be noted that in a fully dynamic model, the rate at which a population would tend to re-establish a balance after a change in the rate of fishing or predation, F, would be determined by the ratio w_0/w_1, which might therefore be compared with the intrinsic rate of increase of the population.

For very high values of F we eventually reach the situation in the models where the adult population becomes small enough so that

$$w_0 > w_1.$$

At this point the population can no longer persist. Except for this, we do not deal further with the stock-recruitment problem.

In natural populations it seems to be rarely the case that a population consumes all the food available. We therefore consider the two situations described in the last section. In the first we consider that the food intake of the population is constant. As pointed out above, this situation is only trivially different from that in which all food is consumed. In the second situation we consider that the grazing power of an individual varies with w^{μ}, hence express the food intake of the population by equation [24].

Finally, it may be noted that we consider the natural mortality rate M to be constant, and that the predation or fishing mortality rate is constant after a given age at first capture specified by TF. The model may, of course, be used to investigate other situations. The results described below, however, use mortality functions identical with those employed by other fisheries workers.

RESULTS

Given the functional relationships among the physiological and ecological parameters discussed above it is possible to calculate relative values for production, biomass and yield for a wide series of hypothetical populations. At this stage of study we do not have values of metabolic levels, weights, food amounts or of the sizes of animals at birth, which would permit us to estimate the energy relations in absolute terms; although it should soon be possible to do this for selected organisms. Meantime the results even in rather arbitrary relative units suggest relations of sufficient interest to be considered here and to justify further study in real units.

In Fig. 1 we show an example of a fish growth curve which has been scaled so that with an arbitrarily defined value of $\alpha = 2\cdot5$, $b = 0\cdot11$ and a weight at first maturity $= 5\cdot0$ the animal involved would show a rather typical fish growth pattern through sizes ranging up to 24 units of weight in time of 20 units. In this curve the weight at first maturity of $5\cdot0$ corresponds with an age (time) of approximately $4\cdot9$, or about $\frac{1}{4}$ of the maximum "life-span" of the animals. The side panels show, for comparison with the results of other production models, the effects on yield of different amounts of predatory or fishing mortality at different ages at first capture with natural mortality held constant at $0\cdot2$ throughout. In panel B, which shows the relative change of yield with increasing age at first capture, as exploitation begins at earlier and earlier ages, the yield continues to increase, rather rapidly as TF is reduced from higher ages to an age of about $2\cdot0$. Below this the yield curves flatten out in keeping with the flatter early part of the hypothetical growth curve. The trends are similar for all three fishing mortality rates, yield being higher at higher mortalities. Two sets of lines are shown on the panel. The solid lines are for models in which the population food intake was held constant. The dashed lines represent the situation where the grazing ability of the population is determined by equation [24]. As would be expected there is almost no appreciable effect of grazing power changes when the age of animals is kept large by a high age at first capture. At low ages at first capture

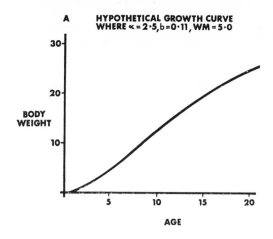

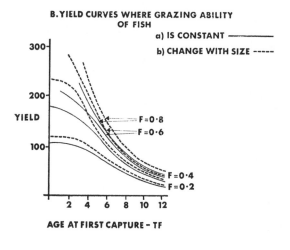

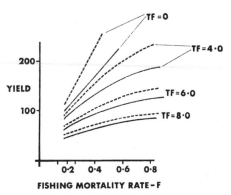

Fig. 1. Growth curve and predator (= fishing) yield for "standard" parameter values and with food supply constant. For further explanations see pp. 518-20.

such effects become significant, and are especially large at high F where the size-composition of the stock would be most markedly altered in favour of the small animals.

Panel C of Fig. 1 shows the conventional type of yield-fishing mortality plot. With low ages at first capture yield rises rapidly as fishing mortality increases. It should be noted, however, that while at the higher fishing mortality rates ($F = 0.6$ and $F = 0.8$), the yield curve continues to rise steeply, when TF was reduced below about 2.0 the populations failed to maintain themselves.

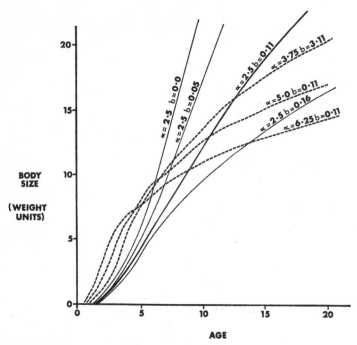

Fig. 2. Growth curves derived from equations (19) and (29) with food supply constant. The heavy central line is the reference curve used in Fig. 1 for calculating yields. Lighter solid lines are for constant α, but change in the growth efficiency exponent b. Dashed lines are for increasing α at constant b.

Production and biomass curves for these populations have also been calculated (not shown). Where the food intake is held constant the total production by the stock is more or less uniform at ages of first capture younger than the age at first maturity. As TF increases the production decreases much more slowly than the yield up to about 8 or 9, then tends to level off at a value about 30 to 50% lower than its initial value. Total production is at higher levels for successively higher levels of fishing mortality, showing an increase of about 80% from $F = 0.2$ to $F = 0.6$ at $TF = 0$, while yield increased over 150%. Where the grazing power changes with size, this tends to magnify the effects of fishing. As might be expected, trends in biomass are intermediate between the production and yield trends.

Fig. 2 shows a series of growth curves which result from assigning various constant values to the physiological-ecological parameters α and b in equations [19] and [28] or [29]. The central curve, drawn in a heavy solid line, is the same as used in Fig. 1 and provides a convenient reference point for further discussions. If b, the slope of the growth efficiency curves (cf. K-curves of Paloheimo and Dickie, 1965) is held constant and we consider that the metabolic level α increases, the result is an abrupt increase in growth production at early ages, but an earlier inflexion point leading to a smaller final size, hence lower total production among the older age groups. Also shown in Fig. 2 are a series of growth curves which result when the level of α is fixed at the reference value of 2·5 and b varies from 0 to 0·16. When the efficiency of utilization of food for production processes does not change with food intake ($b = 0$), and the individual ration is proportional to w^u, the resulting growth is nearly exponential. As can be seen in Fig. 2 the early growth is largely dominated by the value of α, but changes in the value of b have a marked effect on growth at later ages.

In Fig. 3 we present a series of diagrams which summarize the effects of different growth curves on the yield, biomass and production, choosing an intermediate value of $TF = 6·0$ and $F = 0·6$ on Fig. 1 as reference point. These points from Fig. 1 are marked with an open circle on Fig. 3a. Each of the sub-lettered graphs is divided into two panels. The left-hand panel is for the situation where food intake is held constant. The right-hand panel shows the changes resulting from a change in grazing ability with animal size. The curves in Fig. 3 thus show effects on yield, biomass and production of altering values of α, b and WM separately, while the other two parameters are held constant at the reference value.

As might be expected from Fig. 2, Fig. 3a shows that as the level of metabolism, α, rises, the yield from the population decreases, at first rather rapidly. Comparison with Fig. 3b indicates that this results in a similar decrease in the biomass, which falls off steadily as α increases. These effects are undoubtedly a result of the fact that the production is progressively concentrated among younger and younger members of the population. The total production of the stock also decreases although not nearly as much as the yield. Between $\alpha = 2·5$ and 3·75, the effect is especially marked where grazing varies with size, again indicating the effect of the shift to smaller sizes. The yield decrease reflects partly the fact that we have restricted the fishing to age 6·0 and over. If TF is lowered the yield still falls with increasing α but not nearly as rapidly as at $TF = 6·0$.

Fig. 3 also shows the effects of changes in b. As it decreases towards its lower limiting value $b = 0$ all three population attributes—production, biomass and yield—increase, the effect of changes in b being somewhat less than for changes in α over the range of values tested.

The figures indicate that as the weight or age at first maturity increases the yield, biomass and production all show a slight increase, apparently a

3A. EFFECTS ON YIELD

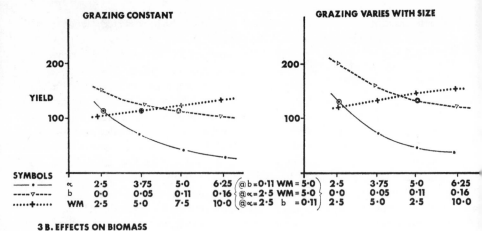

SYMBOLS		2·5	3·75	5·0	6·25	(@b=0·11 WM=5·0)	2·5	3·75	5·0	6·25
—•—	$\propto$	2·5	3·75	5·0	6·25	@$\propto$=2·5 WM=5·0	2·5	3·75	5·0	6·25
---▽---	b	0·0	0·05	0·11	0·16	@$\propto$=2·5 b=0·11	0·0	0·05	0·11	0·16
····+····	WM	2·5	5·0	7·5	10·0		2·5	5·0	2·5	10·0

3B. EFFECTS ON BIOMASS

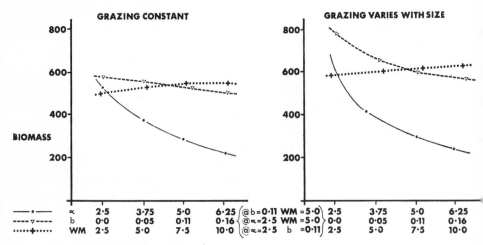

		2·5	3·75	5·0	6·25	@b=0·11 WM=5·0	2·5	3·75	5·0	6·25
—•—	$\propto$	2·5	3·75	5·0	6·25	@$\propto$=2·5 WM=5·0	2·5	3·75	5·0	6·25
---▽---	b	0·0	0·05	0·11	0·16	@$\propto$=2·5 b=0·11	0·0	0·05	0·11	0·16
····+····	WM	2·5	5·0	7·5	10·0		2·5	5·0	7·5	10·0

3C. EFFECTS ON PRODUCTION

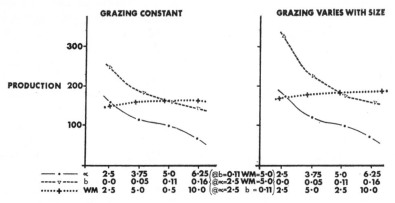

		2·5	3·75	5·0	6·25	@b=0·11 WM=5·0	2·5	3·75	5·0	6·25
—•—	$\propto$	2·5	3·75	5·0	6·25	@$\propto$=2·5 WM=5·0	2·5	3·75	5·0	6·25
---▽---	b	0·0	0·05	0·11	0·16	@$\propto$=2·5 b=0·11	0·0	0·05	0·11	0·16
····+····	WM	2·5	5·0	0·5	10·0		2·5	5·0	2·5	10·0

Fig. 3. Effect of changing the parameters of the growth curves (see Fig. 2) on Yield (Fig. 3a), Biomass (Fig. 3b), and Production (Fig. 3c), holding the food supply and the other population parameters constant: age at first capture, $TF = 6·0$, $F = 0·6$, $M = 0·2$. Circled values are from Fig. 1.

reflection of the fact that at $TF = 6$ all depend primarily on somatic growth which in this model is increased if the onset of maturity is delayed.

DISCUSSION

The results of this consideration of production processes through the food chain from food supply, to grazer, to top carnivore suggest a set of production-yield relationships which are quite different from those which have been derived from numerical models in which recruitment is held more or less constant. The results are hardly surprising, depending as they do both on the assumption that recruits are able to fully utilize the food supply, and on the postulate that grazing itself may change with the size-composition of the stock, being more efficient among the smaller animals. The experimental population evidence supporting this is still very meagre. Slobodkin's *Daphnia* populations and Silliman's guppy populations were both approximately proportional to the food supply at lower levels of exploitation. In both animals the intrinsic rate of increase appears to be low with the production of live young at an advanced state of development. Therefore, at high rates of exploitation, both biomass and yield in these populations declined severely. Our models showed similar phenomena including failure of the populations to maintain themselves under extreme exploitation. It is not without interest, in addition, that Silliman's data (1968) also show a significantly increased yield efficiency for decreases in the average sizes of fish in his populations.

The indication in our review of the simpler food-growth models, of differences in the metabolic response of *Daphnia* and certain fishes to feeding opportunities suggests however, that animal groups at various levels of the food chain may show fundamental differences in production and population regulation mechanisms. The Richman and Slobodkin results clearly illustrate that in the situations they studied, there was a relative metabolic inflexibility of the adult *Daphnia*, with correspondingly low response of adult growth rate to change in feeding opportunity. A large fraction of the increased food consumed at high densities was simply egested although this did not prevent a major relative change in reproductive rate, with the result that at least at low rates of predation, population size was proportional to food supply. A similar predictability of production potential, more or less independent of food supply above a minimum value, and thus measurable by temperature alone, was suggested by McLaren (1963, 1965) for an important zooplankton component, the copepod *Pseudocalanus*, and for *Sagitta*, indicating that *Daphnia* may not be an isolated instance. On the other hand, Deevey (1964) found evidence for at least limited influence of food abundance on adult growth in three out of five species of copepods studied off Bermuda, and Conover (1959, 1960) found some evidence for metabolic adaptation to varying environmental conditions and for influence of feeding. Conover and Corner (1968) found seasonal changes in metabolism within an approximate two- to threefold range which are probably at least partially independent of temperature in the cold stable environments in which the zooplankton lived.

Calanus-type animals appeared to have the greater metabolic plasticity which the authors termed "partially independent" of environment as opposed to a "totally dependent" carnivorous type of zooplankton. It thus appears that although the zooplankton cannot be regarded as very simple production "machines" their response to changing conditions is certainly specialized in the direction of reproduction rather than somatic growth, and that the specialization is reflected in a low metabolic response to feeding. This would appear to contrast with the situation in fishes, where reproductive products are generally a lower proportion of the total production process, their adult metabolism shows a flexibility in response to feeding conditions which appears to reflect their growth response. Such observations suggest that the measure of metabolic response to feeding may offer information on the productive capacities of the animals, hence of their significance to food-chain systems. Such possibility of characterizing the "type" of population dynamics of an animal group provides strong motives for pursuing the energetics approach to the study of vital statistics of populations.

From the point of view of the practical fisheries manager differences among various animal groups, together with results of yield calculations given in the more complex models, suggest some need for caution in maintaining high ages at first capture, where the objective of the fishery is clearly that of obtaining the maximum overall yield. There seems to be little doubt that food released by the removal of the larger, relatively unproductive adults is utilized by some other component of the population. However, community food webs are complex and there is no *a priori* reason to expect that the young of the exploited species will always benefit from the increase in available food which results when predation removes some of the older animals. This was illustrated experimentally by Paine (1966) who showed increased overall community production after introduction of a top predator, but considerable shifts in the representation of various species in the newly established community. The situation is further complicated by the fact that a change in the grazing population will almost certainly affect production of its food population as well. We have not considered such complications here but with greater knowledge of the physiological characteristics of animals at different trophic levels it would evidently be possible to treat longer food-chains in reasonably realistic fashion. The results appear likely to have a bearing on methods appropriate to management objectives.

In connection with the simple population models we have discussed the changes in production, biomass and yield in relation to the ecological and population efficiencies. For the more complex model the first of these, defined as yield per food intake, is derived very simply from yield curves like those shown as solid lines in Fig. 1*b* and 1*c*. In this version of the model we have held the food intake constant so that the ecological efficiency curves are simply the yield curves multiplied by a constant. Reference to Fig. 1*c* shows that ecological efficiency increases with fishing mortality, rather steeply for $TF = 0$, less so as age at first capture is raised.

Population efficiency is more complex. Calculations from this model indicate that it decreases as α increases, and increases as b increases. In the simpler model we found that population efficiency was the same as the average conversion efficiency, which is given by the ratio of the production to food consumed. In this version of the model we find that population efficiency follows the same trend as the average conversion efficiency if we hold b constant and increase α. However, it displays an opposite trend for α constant and increasing b. The reason for this opposite trend seems to be that with increasing b, the biomass decreases more rapidly at $F = 0$ than it does when there is predation on the stock. The calculations also showed that population efficiency tended to be rather irregular with changes in TF and F. It was more or less independent of level of predation where the total metabolic level is very high, but in general when $b = 0$ we did not verify Slobodkin's suggestion that population efficiency is independent of the level of predation. It appears then that his concept of the "prudent predator" which takes only those forage animals about to die naturally is of significance in relation to types of "predation" represented by disease and parasitism, or for species with rectangular survivorship curves. It does not appear to apply to spedators such as man if the exploited populations have a constant mortality rate over a series of age groups and follow growth curves of the type reprerented by our reference curve in Fig. 1. Of possibly greater significance in understanding this type of predation is the production per unit biomass. From the foregoing this appears to increase with a decrease in α or b, as well as with decreasing age at first capture and with increasing rates of fishing, up to the limit of the power of the reproduction to maintain population balance.

The stock-recruitment problem clearly remains as a problem to be further explored by whatever model and observational tools we can find. The models developed to this stage suggest, however, that physiological and behavioural parameters, of the sort specified here in terms of α and b, are likely to be of central importance in the development of predictive theory in ecology. The fact that such parameters are open to experimental study, implies that the energetics approach which uses them holds promise for a rapid increase in our understanding of population regulation mechanisms as well as effects of exploitation of food-chains.

REFERENCES

BARANOV, F. I. 1918. On the question of the biological basis of fisheries. *Izv. nauchno issled. ikhtiol. Inst.*, **1**, 81-128.
BARANOV, F. I. 1926. On the question of the dynamics of the fishing industry. *Byull. rybn. Khoz.*, 1925 (8), 7-11.
BEAMISH, F. W. H., and DICKIE, L. M. 1967. Metabolism and biological production in fish. In *The biological basis of freshwater fish production*. Edited by S. D. Gerking, Oxford, Blackwell scient. Publ., 215-42.
BEVERTON, R. J. H., and HOLT, S. J. 1957. On the dynamics of exploited fish populations. *Fishery Invest., Lond.*, ser. 2, **19**, 533 pp.

BEVERTON, R. J. H., and HOLT, S. J. 1959. A review of the life-spans and mortality rates of fish in nature, and their relation to growth and other physiological characters. In *The lifespan of animals. CIBA Fdn Coll. Ageing*, 5. Edited by G. E. W. Wolstenholme and M. O'Connor. London, J. and A. Churchill Ltd., 142-80.

BROWN, M. E. 1946. The growth of brown trout (*Salmo trutta* Linn.). II. The growth of two-year-old trout at a constant temperature of 11·5°C. *J. exp. Biol.*, 22, 130-44.

BROWN, M. E. 1957. Experimental studies on growth. In *The physiology of fishes*, 1, *Metabolism*. Edited by M. E. Brown. New York, Academic Press, 361-400.

CONOVER, R. J. 1959. Regional and seasonal variation in the respiratory rate of marine copepods. *Limnol. Oceanogr.*, 4, 259-68.

CONOVER, R. J. 1960. The feeding behavior and respiration of some marine planktonic crustacea. *Biol. Bull. mar. biol. Lab.*, *Woods Hole*, 119, 399-415.

CONOVER, R. J., and CORNER, E. D. S. 1968. Respiration and nitrogen excretion by some marine zooplankton in relation to their life cycles. *J. mar. biol. Ass. U.K.*, 48, 49-75.

DEEVEY, G. B. 1964. Annual variations in length of copepods in the Sargasso Sea off Bermuda. *J. mar. biol. Ass. U.K.*, 44, 589-600.

FRY, F. E. J. 1957. The aquatic respiration of fish. In *The physiology of fishes*, 1. *Metabolism*. Edited by M. E. Brown. New York, Academic Press, 1-63.

HALL, D. J. 1964. An experimental approach to the dynamics of a natural population of *Daphnia galeata Mendotae*. *Ecology*, 45, 94-112.

HOLT, S. J. 1962. The application of comparative population studies to fisheries biology—an exploration. In *The exploitation of natural animal populations*. Edited by E. D. LeCren and M. W. Holdgate. New York. John Wiley & Sons, 51-71.

KERR, S. R., and MARTIN, N. V. 1970. Trophic-dynamics of lake trout production systems. This volume, pp. 365-376.

LeCREN, E. D. 1951. The length-weight relationship and seasonal cycle in gonad weight and condition in perch (*Perca fluviatilis*). *J. Anim. Ecol.*, 20, 201-219.

McLAREN, I. A. 1963. Effects of temperature on growth of zooplankton, and the adaptive value of vertical migration. *J. Fish. Res. Bd Can.*, 20, 685-727.

McLAREN, I. A. 1965. Some relationships between temperature and egg size, body size, development rate, and fecundity, of the copepod *Pseudocalanus*. *Limnol. Oceanogr.*, 10, 528-38.

NICHOLSON, A. J., and BAILEY, V. A. 1935. The balance of animal populations. I. *Proc. zool. Soc. Lond.*, 551-98.

ODUM, E. P., and SMALLEY, A. E. 1959. Comparison of population energy flow of a herbivorous and a deposit-feeding invertebrate in a salt marsh ecosystem. *Proc. natn. Acad. Sci. U.S.*, 45, 617-22.

PALOHEIMO, J. E., and DICKIE, L. M. 1965. Food and growth of fishes. I. A growth curve derived from experimental data. *J. Fish. Res. Bd Can.*, 22, 521-42.

PALOHEIMO, J. E., and DICKIE, L. M. 1966a. Food and growth of fishes. II. Effects of food and temperature on the relation between metabolism and body weight. *J. Fish. Res. Bd Can.*, 23, 869-907.

PALOHEIMO, J. E., and DICKIE, L. M. 1966b. Food and growth of fishes. III. Relations among food, body size, and growth efficiency. *J. Fish. Res. Bd Can.*, 23, 1209-48.

RICHMAN, SUMNER. 1958. The transformation of energy by *Daphnia pulex*. *Ecol. Monogr.*, 28, 273-91.

RICKER, W. E. 1940. Relation of "catch per unit of effort" to abundance and rate of exploitation. *J. Fish. Res. Bd Can.*, 5, 43-70.

RICKER, W. E. 1944. Further notes on fishing mortality and effort. *Copeia*, 1944 (1), 23-44.

RICKER, W. E. 1958. Handbook of computations for biological statistics of fish populations. *Bull. Fish. Res. Bd Can.*, (119), 300 pp.

SCHAEFER, M. B. 1954. Some aspects of the dynamics of populations important to the management of the commercial marine fisheries. *Bull. inter-Am. trop. Tuna Commn.*, 1, 27-56.

SCHAEFER, M. B. 1967. Fishery dynamics and present status of the yellowfin tuna population of the eastern Pacific ocean. *Bull. inter-Am. trop. Tuna Commn.*, 12, 89-136.

SILLIMAN, R. P. 1968. Interaction of food level and exploitation in experimental fish populations. *Fishery Bull. Fish Wildl. Serv. U.S.*, 66, 425-39.

SILLIMAN, R. P., and GUTSELL, J. S. 1958. Experimental exploitation of fish populations. *Fishery Bull. Fish Wildl. Serv. U.S.*, 58 (133), 215-52.

SLOBODKIN, L. B. 1959. Energetics in *Daphnia pulex* populations. *Ecology*, 40, 232-43.

SLOBODKIN, L. B. 1960 Ecological energy relationships at the population level. *Am. Nat.*, **94**, 213-36.

SMITH, F. E. 1963. Population dynamics in *Daphnia magna* and a new model for population growth. *Ecology*, **44**, 651-63.

TOETZ, DALE,W. 1967. The importance of gamete losses in measurement of fresh water fish production. *Ecology*, **48**, 1017-20.

VOLTERRA, V. 1928. Variations and fluctuations of the number of individuals in animal species living together. *J. Cons. perm. Explor. Mer*, **3**, 1-51.

Marine ecosystem development in polar regions

M. J. DUNBAR
Marine Sciences Centre
McGill University
Montreal, Quebec, Canada

ABSTRACT. The thesis is put forward that selection favouring adjustment to the annual oscillation of marine environments in high latitudes operates in two directions which may be to a certain extent in conflict. These directions are: (*a*) adaptation to the fact of high oscillation in the availability of nutrients and plant food, and (*b*) adaptation towards greater ecosystem stability. The first favours slow growth, large body size, small number of species and high fecundity; the second favours faster growth, small body size, lesser fecundity, and larger number of species.

The first type of development is straightforward and is readily seen to have occurred; the second type is to be expected on general theoretical grounds and evidence is offered that it is in fact proceeding. This second development, towards greater stability, requires diversification in food chains and the sharing of the available resources among incipient species and diverging populations.

Ecosystems evolve, in the same sense as other living units evolve. Food-chains, being one property of ecosystems, become modified as the ecosystems themselves change. In the tropical and subtropical seas of the world, the ecosystem, or ecosystems, are well established and may be taken as having come to equilibrium a long time ago. That is to say, they have reached saturation in terms of diversification; no more new species or immigrant species, or very few such new species, can find a viable niche in the system except by displacement of species already there, which will be difficult. In temperate, and especially in polar, environments, the probability of ecosystem saturation is much less, owing to the comparatively recent upheaval caused by the Pleistocene changes of an uncertain date in the past, of the order of from two to four million years, and also by climatic and general environmental oscillations since that time; therefore continued diversification of

528

temperate and polar systems is to be expected, with concomitant complication and sophistication of the food-chains.

This theory, that high latitude systems are younger and therefore less diverse than tropical systems, was first put forward in 19th century terms by Alfred Russel Wallace (1878), and it has been revived quite recently by several authors (Fischer, 1960; Dunbar, 1960, 1968).

It has been widely held, and in fact it has often been considered to be axiomatic, that the decisive difference between polar and tropical environments, the factor responsible for the differences in their living communities, is the low temperature level characteristic of the high latitudes. This belief I have recently challenged (Dunbar, 1968), particularly where aquatic environments are concerned, but applied to terrestrial regions as well. It has become increasingly clear, since the first suggestions of Krehl and Soetbeer (1899) and Krogh (1916), and more so since the experimental studies of Thorson (1936), Spärck (1936), Fox (1936), Fox and Wingfield (1937), Dehnel (1955), and others, that living organisms, poikilotherms just as well as homotherms, can and do adapt to low temperatures very readily, especially in terms of metabolism and locomotor activity, and also, if the situation demands it, in terms of growth rate. That is, when polar, temperate and tropical populations of the same or closely related species are compared, the Q_{10} law is flagrantly and happily disobeyed. Examples of this are now very numerous, and have been summarized in general reviews (Bullock, 1955; Fry, 1958; Dunbar, 1957).

If low temperature is not a serious problem in adapting to Pleistocene conditions in high latitudes (and I take it we are agreed that we live in the Pleistocene period today), then the explanation for the very great differences between polar and tropical ecosystems, manifested in species diversity, growth rates, body sizes and so on, must be sought elsewhere. We cannot point to low temperature as being effective at the ultimate level at all, and only to a limited degree at the proximate level. In considering these problems, involving as they do physiological responses to external factors and the evolution of ecosystems, it is extremely important to keep the proximate and ultimate levels of causation and of explanation strictly separate; to confuse them leads to logical anarchy.

Besides the generally lower temperature regimes of the high latitudes, the Pliocene-Pleistocene change brought other important consequences, such as:

1. The presence of sea ice.
2. An Arctic Ocean regime highly stable in the vertical column and therefore very unproductive.
3. A very unstable, and therefore productive, Antarctic Ocean;
4. A highly oscillating basic food supply (inorganic nutrients for plants and plant food for herbivores) the oscillation being annual in period. This is particularly marked in the Arctic, owing to the great extent of sea ice and the intense vertical density stratification, which keep the nutrient concentrations low in the euphotic zone and causes them to be rapidly used up in the spring. This applies also to fjords and ice-covered bays

in Antarctica, but the whole pattern of oceanic circulation is quite different at the two poles, resulting in far more freely circulating water in the south.

For the purposes of this paper, it is this last of the legacies of the Pleistocene event, the seasonally oscillating environment, that is most significant, especially when considered together with the generally low productivity in the Arctic. For the herbivorous animals depending on the phytoplankton for their livelihood, the short plant production season, which may be only three to five weeks long, and the very low food supply during the rest of the year, make demands which encourage the following developments or adjustments:

1. Growth rates from egg to maturity which fit the periods of phytoplankton abundance. For most planktonic herbivores this means a considerable extension of the life cycle when compared with the same species in temperate waters, so that spawning is delayed until the time of the next phytoplankton bloom. It is probable, as pointed out by Conover (1964) that microzooplankton and organic matter in the water form an important secondary food source for these herbivores at potentially all times of year; nevertheless they breed not more than once a year in the Arctic.

2. Large body size, which is associated logarithmically with fecundity (high egg number) in the female. This assures an adequate breeding stock despite the mortality during the year. If the egg number is not large, then the eggs themselves must be large and yolky, so that survival rates are increased. In either case the energy reserves must be high.

3. A small number of species; if the number of species is too large the survival of all species will be in hazard, assuming that the total food supply is severely limited, owing to the natural mortality during the long period during which spawning does not occur.

These adjustments are in fact found in the ecosystems of the high latitudes; slow growth rates, large body size, and few species. And food chains tend, in consequence, to be short and simple: diatom, herbivorous zooplankter, fish or whale or seal. One of the standard specific examples is from the Antarctic: *Fragilariopsis antarctica* → *Euphausia superba* → blue whale.

Such simple systems, however, tend to be unstable, especially where the biomass is low, as in the Arctic, and it is reasonable to wonder whether they are in fact anywhere near saturation. They are subject to serious population oscillations, again particularly if there is low total energy or biomass, since both lags and over-productions in specific populations occur readily, leading to local extinction and breakdown of the community. The standard examples of this are from terrestrial environments, which are easier to observe, such as that in which the lemming is involved; but there is evidence from our work in the far north (Nansen Sound complex) that the same thing is demonstrable in marine systems.

Greater ecological stability is achieved by increasing the number of species (species diversity), increasing the biomass, reducing specific fecundity, increasing growth rates, and the spreading of the feeding load over a much larger portion of the year; this implies that herbivores must develop a degree of omnivorous habit, using detritus, microzooplankton and dissolved organic material. It is to be noted, in passing, that these objectives are not the same as those already evolved in response to the oscillating environment; in fact the two evolutionary processes, if they exist together, are to a great extent in competition with each other.

Evidence suggesting that selective evolution toward greater ecological stability in polar and cool temperate regions is in fact going on around us has been marshalled elsewhere (Dunbar, 1968). In the particular context of food chains, with which we are concerned here, the interest lies in the possibility that differentiation of morphological pattern and food habit, within species and probably sympatrically, is an active process in these regions.

The number of species in tropical environments, both terrestrial and marine is at first bewildering in its lavishness, and many a stranger to tropical biology has been appalled by the tasks faced by his tropical colleagues. Closer examination of tropical faunas, however, reveals that the species themselves are well disciplined taxonomically, that is to say they do not present serious problems of decision to the systematic expert other than by their sheer number. In temperate faunas, and especially it seems in subpolar areas, intraspecific variation, mutants and "varieties", plague the systematist at almost every turn. That some such intraspecific variation exists also in warmer waters is admitted, as for instance the example of certain Mediterranean copepods described by Battaglia (1958), but these are much less impressive than the taxonomic jungle of the higher latitudes. I am excluding from consideration here, as irrelevant to the argument, local variant populations on a small geographic scale, such as changes in intertidal populations corresponding to degree of exposure to the air or to wave action, and also obvious allopatric geographic races such as distinguish Hudson Bay from the Beaufort Sea.

Examples which are relevant and which may be offered are the following:

1. The genus *Gammarus*. The former "*Gammarus locusta*" has been shown to consist of several species, forming a complex in which overlap of geographic range is often very large indeed. *G. locusta*, in the modern sense, is confined to the shores of north-west Europe, but including Iceland; *G. oceanicus* is temperate and subarctic on both sides of the Atlantic; *G. setosus* extends further north into Arctic water, and *G. wilkitzki* is more strictly Arctic still. There are also other species.

2. The genus *Calanus*. The splitting up of the former *Calanus finmarchicus* into a number of species has much in common with the *Gammarus* case, but here we are dealing with planktonic populations, which perhaps increases the theoretical ecological interest. Both cases suggest very strongly a recent evolutionary divergence with the effect of increasing the ecological diversity. In the broad regions of overlap between *C. helgolandicus* and *C. finmarchicus* (s.s.), and between *C. finmarchicus* and *C. glacialis*, one wonders how the Gausean principle of exclusion could apply, unless significantly different ecological, perhaps feeding, demands have also been evolved. In addition

to these three, in the North Atlantic and the Arctic areas, there is *C. hyperboreus*, Subarctic and Arctic in distribution, and very similar morphologically to the others.

3. The genus *Parathemisto*. There are at least two separable forms, or "morphs", of *P. gaudichaudi*, formerly considered to be separate species, known as "*bispinosa*" and "*compressa*"; both are known from the northern and southern hemispheres, and there is a third form, *thomsoni*, known in the south only. They differ in the shape and setation of the third and fifth peraeopods and in the development of dorsal spines.

4. The genus *Thysanoessa*. *T. inermis*, *T. raschii*, and to a lesser extent *T. longicaudata*, form a close trio of euphausids occupying much the same regions and water masses. *T. longicaudata* has elongated second thoracic legs, and there is a variant of *inermis*, known as "*neglecta*", which has similar modified second legs.

5. The genus *Hyperia*. There is a variant, apparently a somewhat rare variant, of *Hyperia galba*, known as "*H. spinigera*", known from Svalbard, northern Norway, Labrador, Hudson Strait, the west coast of Ireland and the south coast of England, and from the Antarctic.

6. The genus *Aglantha*. The Trachymedusan *Algantha digitale*, very abundant in northern waters, exists in several colour variants which have so far not been explained.

7. Chaetognatha. *Sagitta elegans* and *Eukrohnia hamata* in the Arctic Ocean possess well developed "irridescent spots", so-called, on the fins. These are well supplied nervously, and may well be similar to the ciliary tufts described by Horridge and Boulton (1967) in *Spadella cephaloptera*. In the Arctic Ocean itself, a very high proportion (85-97%) of the individuals of *E. hamata* possess these organs, and in the northern Barents Sea some 20% of individuals of both *E. hamata* and *S. elegans* possess them; in the southern Barents Sea, however, they are absent from both species (Newbury, personal communication). These characters, associated as they probably are with the perception of vibrations in the water, may be associated with the lesser intensity of light which prevails in the Arctic Ocean. In general, however, they may also have other significances, and in particular, the types of prey organisms eaten.

8. The genus *Eucalanus*. Johnson (1939) drew attention to the systematic problems of the group "commonly lumped more or less indiscriminately by some authors under the name *Eucalanus elongatus* without reference to varieties established by Giesbrecht (1892)". Several forms, now raised to specific rank, are found in the North Pacific, recorded under the specific or subspecific names *bungii*, *elongatus*, *californicus* and *inermis*. This situation is not unlike that of the former *Calanus finmarchicus*.

To these examples should be added others of a somewhat different sort, namely examples of anomalies of geographic distribution which suggest the evolution of mutant stocks which differ physiologically but not morphologically from the main body of the specific population; "cryptomorphs" which might well differ also in food habits. Such examples are to be found in the euphausid *Meganyctiphanes norvegica*, an Atlantic species which is found unexpectedly in north-east Greenland waters; the north-east Greenland population of *Gammarus oceanicus* and of *Thysanoessa longicaudata*; and *Parathemisto gaudichaudi* in northern Foxe Basin.

In the face of these interesting variations and anomalies, the next step is to discover in what ways they differ from their cousins, or conspecific partners, and especially, in the present context, in their food habits. This unfortunately has not yet been done, nor is it likely to be done for some time, for the material and the problems are both large; it is one of the facets of the I.B.P. study we are beginning at McGill in the Gulf of St. Lawrence, and there is a good deal of work being done along these lines in several countries. Despite the lack of detailed information, one may be excused for indulging in a little pleasurable anticipation, and discussing what few facts are at hand. Differentiation in ecological demands and necessities, probably including food requirements and food-getting abilities, is to be expected, involving sophistication of the food-chain pattern within the system as a whole.

An excellent recent summary of what is known or suspected about the food habits of marine organisms, especially the plankton, has been provided by Raymont (1963), from whose work it emerges that "our knowledge of specific dietary requirements is . . . fragmentary". The pattern that appears so far is that most planktonic animals display considerable lability in food habits. Even the most clearly herbivorous forms, such as species of *Calanus* and *Euphausia*, appear to ingest animal food occasionally, and others, such as individual species of *Thysanoessa*, may be herbivorous, carnivorous, or omnivorous (Raymont, 1963; Nemoto, 1967). Dr Nemoto's paper is an example of the precise work that is needed, the study of food habits in planktonic groups in relation to morphological characters. The Hyperiid amphipods appear to be predominantly carnivorous (e.g. Dunbar, 1946, for *Parathemisto libellula*); nothing is known of any difference in food habits between the two (or three) morphs of *P. gaudichaudi*.

Omnivores are common in the zooplankton. Raymont (1963) writes that "the animals probably change their diet to a considerable extent depending upon the availability of food". In ecological theory, omnivorous habit can mitigate the restricting effects of the simple ecosystem, because it allows change of diet to respond either to the scarcity or the superabundance of any given food organism, thus acting as a dampening influence on specific population oscillation. This is interesting in itself, for it suggests that too intense attention to "trophic levels" in marine ecology may not be the most productive approach to the economy of the sea. It does not, however, reduce the weight of the argument present here, which is, in brief, that the evidence suggests strongly that there is a continuing present evolution toward greater diversification and subdivision of niches, including that part of the niche represented by food habit and position in the general food web.

REFERENCES

BATTAGLIA, B. 1958. Ecological differentiation and incipient intraspecific isolation in marine copepods. *Colloques, I.U.B.S.*, Ser. B, (24), 259-68.
BULLOCK, T. H. 1955. Compensation for temperature in metabolism and activity of poikilotherms. *Biol. Rev.*, **30**, 311-42.
CONOVER, R. J. 1964. Food relations and nutrition of zooplankton. *Occ. Publs. Proc. Symp. exp. mar. Ecol.*, (2), 81-89.
DEHNEL, P. A. 1955. Rates of growth of gastropods as a function of latitude. *Physiol. Zool.*, **28**, 115-44.
DUNBAR, M. J. 1946. On *Themisto libellula* in Baffin Island coastal waters. *J. Fish. Res. Bd Can.*, **6**, 419-34.
DUNBAR, M. J. 1957. The determinants of production in northern seas: A study of the biology of *Themisto libellula* Mandt. *Can. J. Zool.*, **35**, 797-819.
DUNBAR, M. J. 1960. The evolution of stability in marine environments; natural selection at the level of the ecosystem. *Amer. Nat.*, **94**, 129-36.
DUNBAR, M. J. 1968. *Ecological development in polar regions.* London, Prentice-Hall, 119 pp.
FISCHER, A. G. 1960. Latitudinal variations in organic diversity. *Evolution*, **14**, 64-81.
FOX, H. MUNRO. 1936. The activity and metabolism of poikilothermal animals in different latitudes. *Proc. zool. Soc. Lond.*, 945-55. Also *ibid.*, 1938, 501-505, and 1939, 141-56.
FOX, H. MUNRO, and WINGFIELD, C. A. 1937. The activity and metabolism of poikilothermal animals in different latitudes, II. *Proc. zool. Soc. Lond.*, 275-82.
FRY, F. E. J. 1958. Temperature compensation. *A. Rev. Physiol.*, **20**, 207-20.
HORRIDGE, G. A., and BOULTON, P. S. 1967. Prey detection by Chaetognatha via a vibration sense. *Proc. R. Soc. Ser. B.*, **168**, 413-19.
JOHNSON, MARTIN W. 1939. The study of species formation in certain *Eucalanus* copepods in the North Pacific. *Proc. 6th Pacif. Sci. Congr.*, **3**, 565-68.
KREHL, L., and SOETBEER, F. 1899. Untersuchungen über die Temperaturabhängigkeit von Lebensprozessen bei verschiedenen Wirbellosen. *Pflügers Arch. ges. Physiol.*, **77**, 611.
KROGH, A. 1916. *Respiratory exchange of animals and man.* London: Longmans, Green and Co., 173 pp.
NEMOTO, T. 1967. Feeding pattern of euphausids and differentiations in their body characters. *Inf. Bull. Planktol. Japan*, **13** (Suppl.), 151-71.
RAYMONT, J. E. G. 1963. *Plankton and productivity in the oceans.* New York, Pergamon Press., 660 pp.
SPÄRCK, R. 1936. On the relation between metabolism and temperature in some marine lamellibranchs and its ecological and zoogeographical importance. *Biol. Meddr*, **13**, 1-27.
THORSON. G. 1936. The larval development, growth, and metabolism of Arctic marine bottom invertebrates compared with those of other seas. *Meddr. Grnl.*, **100** (6), 1-55.
WALLACE, A. R. 1878. *Tropical nature and other essays.* London and New York, Macmillan, 356 pp.

Part Seven

SUMMARY OF THE SYMPOSIUM

L. B. Slobodkin

Summary

L. B. SLOBODKIN

Perhaps the clearest statement that can be made with reference to the meeting as a whole is that the simplified constant of 10% ecological efficiency cannot reasonably be accepted at face value in any situation where there is a practical difference or a practical importance to a fisheries' regulation decision. Something of the order of 10% is perhaps approximately correct, but, as Gulland has pointed out, rather small differences in the efficiency estimate for each trophic level become more important logarithmically with trophic level. Therefore, the assumption of a particular constant value is a delicate matter and may result in order of magnitude errors with rather serious consequences.

The fact that the available data did not really converge in any convincing way on the 10% figure throws extremely grave doubts on the validity of the use of this approximation by fisheries' regulation agencies or for estimations of the total food supply available from the ocean. It is, in the long run, a good thing to have this point clarified since it is more dangerous to have a false sense of assurance in dealing with critical problems than to be aware of areas of ignorance or areas of difficulty and attempt to take steps to solve the difficulties. In place of an inexplicable, and as the case turned out incorrect constant, we now have a family of curious biological phenomena implying that each fishery has to be studied biologically in its own right, in order to develop adequate predictive regulatory procedures.

Were we simply to conclude that further study is required this would seem a singularly sterile and unrewarding conclusion from a great deal of rather expensive and arduous labour. Fortunately, we can do somewhat better than that and the kinds of study that will be of significance can be indicated in a broad way.

First, we can clearly rule out on general grounds certain kinds of theories as being impossible to use. Specifically, any theory stated in terms of exclusively extensive variables without clear intensive correlates cannot be used to develop enforcable regulations for fisheries.

This concept is implicit in the papers of this symposium and has been demonstrated by the history of past fisheries' regulations, but has not previously, to my knowledge, been stated in as explicit a form as it is being stated now. It therefore deserves a certain amount of precise explication. Specifically, if a description is made of a natural situation, it can be made either in the form of an extensive or intensive variable and it is of great importance to make this distinction abundantly clear. To use a homely example, when I enter a bathtub, I am concerned with the intensive variable of temperature. That is, a large swimming pool full of ice cubes will have a greater total heat content than a small tub full of warm water. But my immediate concern is the contact zone between the water and my skin and I am not in any way concerned, nor am I really seriously influenced by the summation of temperatures over the entire body of water. That is, extensive variables are sums in some sense over a large temporal or geometric area. They typically are determined by some procedure which integrates over the point specific properties of the system.

In the process of evolution, one is concerned with gene frequency change for a population, which is an extensive variable, but this is the integral of a complex set of probabilities of survival for individual animals in individual situations that are actually encountered or may be actually encountered in the intensive variables from which the gene frequency distribution changes are derived.

Similarly, in the case of fishery, if a regulation is made of the form, "The total catch from such and such a fishery should not exceed x tons" then the individual fisherman (assuming that he wishes to follow the regulation) must be able to assess his own activities in terms of the activities of all the other fishermen in the fishery. To the degree that all the fishermen approach the set limit, he must in turn limit his fishing activity. Now, no one can reasonably expect an individual fisherman to behave in this way. That is, a regulatory statement of the form "The total catch from a particular fishery should be x tons," unless it also explicates the precise relationship that must exist between the individual fisherman and his daily fishing operation, is a regulation stated in extensive variable form, and cannot be enforced even with the best will in the world on the part of all parties concerned. However, a fishery regulation which says: "All the nets in this fishery shall have a mesh size of not smaller than so and so centimetres" defines some properties of the fisherman-fish interaction, wherever this interaction occurs. In that sense it represents a set of intensive variables which the fisherman can in practice follow and which in practice, or at least in principle, can be enforced by suitable inspection of fishing boats.

It may be the case that the regulatory agency is concerned with an extensive variable such as total catch, but its regulation, in order to be coercive on the individual fisherman in any realistic way must be couched in terms of intensive variables. In precisely the same sense, the concept of ecological efficiency, while it may receive various complex definitions, typically relates to the total energy income into a particular population or trophic level and the ratio of that to the total energy output from the same population or trophic level. To the degree that these are extensive variables, neither the fish nor the fishermen can be expected to modify his local behaviour in terms of them and what is clearly required both for purposes of comprehension and for purposes of regulation in actual fisheries is the translation of the concept of ecological efficiency or similar concepts into terms of intensive variables, that is into terms of what occurs when an actual predator meets an actual prey.

We have seen in this meeting that certain rather fascinating things occur. Among others we have seen that the notion of trophic levels is a blurred and confused one. That is, organisms nominally on trophic level n will find themselves consuming either at different stages or at different times, organisms on trophic levels n-1, n-2, and so on. The effect of this is two fold. On one level it obviously increases the food supply to the organisms in question. On a slightly more subtle level, short circuiting the food chain sequence by animals of high trophic levels acts to make the ultimate estimate of the permissible yield from any high trophic level less sensitive to particular differences in estimates of ecological efficiency. That is, just as the errors in estimating ecological efficiency introduce a multiplicative error into a food chain, this error is removed by suitable short-circuiting of the food chain.

One of the most fascinating things that has appeared in this meeting is the possibility that prey actually permit predators to take a certain portion of their bodies without permitting the predator to kill them outright. That is, the Tellinid clams in Loch Ewe and the Thiona in the Great Banks (referred to by Dr Tyler) both have strictly autotomizable anatomical parts which seem to suffice for the predators.

In addition, arguments have been presented with the general sense that predators are not simply blind machines grabbing the next food that comes along but that, as the philosopher Wittgenstein has been reported to have said to his weekend hostess, 'I don't care what I eat so long as it is always the same." That is, predators seem capable of shifting their food desires in such a way as to feed upon the most common prey species leaving the relatively rare prey species alone. This has the salutary effect of stabilizing the predator-prey system which in simple minded numerical or laboratory analyses is highly unstable.

The fact that at least some of the marine copepod populations may not be in a steady state at all, but may simply be washouts from some home area, requires an immediate revision of our ideas of energy-transport and steady-state conditions in the sea. That is, there should be no consideration given to the role of these animals as portions of a self-sustaining population.

This raises the interesting possibility that other populations of this sort exist in nature and that optimal fishing procedure both for man and other organisms may, in certain localities and for certain populations, involve taking entire sub-populations of particular species and in that sense liberating the fishery procedure from the normal requirements of a rational fishery which leaves behind enough material to maintain itself. It is at least suspected that the spiny lobster in Bermuda is in this state. This again, emphasizes the necessity of a detailed, biological understanding of those individual populations which are to be subjected to fisheries' pressure and our inability to rely on simplified models or on quick and easy routine guesses. In a sense we can now see the form that future research will take in this area. Classical, theoretical analyses of fisheries problems by Ivlev, Ricker, Riley, Winberg, Nikolsky and others can be seen in a kind of retrospect as appropriate way stations or starting points for the observational study that will provide operational procedures for the management of the seas. We can see the form that these operational theories will take, namely that of models built in terms of intensive variables and perhaps, more important, we can visualize the form that they cannot take, namely, they cannot be in the form of overall integrations over large areas and wholesale extrapolations from one taxonomic group to another. This however, is not simply a refutation or denial of significance of the work that has been done in this field, but a much more optimistic statement to the effect that our information about the seas and its organisms has finally become sufficient to test extant theories and to require the development of new theories and also, this can be taken as evidence that the precision of data from the seas can at least in principle exceed the precision of our theoretical ideas. Most of the arguments that have filled the literature for the last 20 or 30 years can now finally be relegated to the history of science. New phenomena have been described inviting new empirical data and the development of new kinds of theories, in addition to which it is clear that the quality of work that is done in the field has increased markedly in the last 10 years. It is no longer possible to consider the marine biologist and fisheries biologist to be rather unsophisticated seagoing naturalists but quite the reverse; the level of sophistication demonstrated by the papers presented in this symposium gives every reason to hope for major intellectual breakthroughs in the future.

Subject Index

Fulvic acids, 11

Grazing, plankton, 419–457
  coefficient, 510, 511, 513, 514
  effect of light on, 427, 428, 429
  efficiency, 128, 513, 514
  experiments, 328, 329, 331, 332, 333, 340,
    419, 420–422, 429, 448–451
  mortality, 421, 422, 426, 429, 430, 436, 437,
    447, 450, 451, 513
  relations, 424, 427
Growth, 97, 137, 138, 426, 428
  coefficient, 136
  curves, 136, 139, 501, 511, 519–522, 525
  effect of food, 111, 112, 173, 174, 183, 184
  effect of temperature, 74, 88, 172, 173,
    183, 186, 187, 406–409
  efficiency, 74, 83, 88, 93, 107, 137, 138,
    177, 181, 186, 187, 369, 370, 372, 374,
    375, 425, 469, 510, 511
  energy expenditure, 107, 108, 127, 128,
    132, 136
  energy requirements, 332, 337, 340, 407–
    409
  formula, 102, 103
  in bivalves, 261, 262, 288–293
  in chaetognaths, 170–173, 180, 183, 184,
    186, 187
  in copepods, 74–93, 427
  in crustacea, 127, 132, 134–138, 140
  in fish, 267, 270, 271, 272, 275, 277, 288,
    289, 303, 305–307, 309, 332, 337, 369,
    372–374, 403, 404, 406–409, 472–478,
    489, 492, 494, 495, 511–516
  in phytoplankton, 422, 427, 430, 436, 441,
    444, 447, 450, 487, 488
  in polychaetes, 97, 102, 103, 109–112
  in shrimps, 97, 102, 103, 109–112
  in zooplankton, 147, 148, 428, 436, 441,
    451, 452, 480, 481, 486
  metabolism, 127, 134–137, 140
  model, 500–505, 515–518
  polar populations, 530

Herring, 244, 298, 300, 303, 312, 313, 347 (see
    Digestion, Feeding, Excretion, etc.)
  larvae, 345–360
Heterotrophic nourishment,
  in plankton, 144
  levels, 127, 158, 162
Heterotrophy,
  in bacteria, 21, 31, 217
  in diatoms, 25
  in a sandy ecosystem, 20, 21, 25, 31
  nature of, 21
Homotherms, adaptation to low tempera-
    ture, 529
Humic acids, 8, 11
  extraction, 8
  fractionation, 8
  gel filtration, 8

Hydrogen content, in copepods, 115, 118–
    121, 125

Ingestion,
  constants, 425, 426, 429, 436–443
  curves, 423–427, 436, 441, 447, 450, 451
Intestines,
  flagellate population, 235, 236
  length in fish, 234, 235
Isopods, in fish diet, 382, 385, 386, 387, 391,
    392, 401
Ivlev's coefficient,
  aggregation, 337
  energy growth, 136 137, 139
Ivlev's electivity index, 337
Ivlev's equation, food relationships, 327,
    328, 338, 349
  modifications, 327, 424, 429, 481
  parameters, 451

K-line, production efficiency, 372, 373,
    374
Krogh correction curve, 136

Lipids, in copepods, 186
  extraction, 92
Lipid protein ratio, zooplankton, 84
Low-molecular compounds, 6, 17, 19
  dissolved in sea water, 6, 17

Macrobenthos, 63–65, 209–215
  biomass, 209–215
  competition for food, 63, 64, 65
  consumers, 63, 64
  distribution, 214, 215
  in an estuary, 63
Macroflora,
  input of organic matter, 55
  production, 55, 56, 57, 208, 223
Macromolecular compounds, in sea water,
    6, 9–13, 17, 18, 19, 57
  adsorption, 6–9, 12, 13, 16, 17, 29
  aggregation, 57
  desorption, 13, 16, 17
  extraction, 7, 11
  fractionation, 7–11
  hydrolysis, 13
  molecular weights, nature, 9, 10, 11
Marine humus, 217
Meiofauna, in a sand ecosystem, 21, 26, 208,
    287
  carbon requirements, 27, 29, 30
  competition for food, 63, 64, 65
  consumers, 63
  metabolism of, 58
  oxygen demand, 21, 60–63
Metabolism, 107–111, 127, 134–137, 369–
    371, 508, 511
  and growth, 127, 134–137, 140
  and food, 127, 137, 332

Systematic Index